Annals of Mathematics Studies

Number 97

RIEMANN SURFACES AND RELATED TOPICS: PROCEEDINGS OF THE 1978 STONY BROOK CONFERENCE

EDITED BY

IRWIN KRA
AND
BERNARD MASKIT

PRINCETON UNIVERSITY PRESS

AND

UNIVERSITY OF TOKYO PRESS

PRINCETON, NEW JERSEY

1981

Published in Japan exclusively by
University of Tokyo Press;
In other parts of the world by
Princeton University Press

Printed in the United States of America
by Princeton University Press, Princeton, New Jersey

Library of Congress Cataloging in Publication data will
be found on the last printed page of this book

CONTENTS

CONTENTS

vii

PREFACE

This volume contains papers and abstracts by participants of the
Conference on Riemann Surfaces and Related Topics, which was held at
the State University of New York at Stony Brook, June 5-9, 1978. This
was the fourth in a series of conferences on more or less the same subject
(Tulane 1965, Stony Brook 1969, Maryland 1973). We invited papers from
all the Conference participants, with acceptance for publication subject to
refereeing. All the manuscripts were indeed refereed by participants, and
not all were accepted.

As usual, thanks are due to the National Science Foundation for
financial support, the State University of New York at Stony Brook for its
hospitality, and Princeton University Press for providing a series where
these Proceedings could be published (volumes 66 and 79 contain the
Proceedings of the previous two Conferences). Most of all we thank the
participants in the Conference who wrote these papers and who refereed
them, who gave invited lectures and seminar talks, and who talked mathematics and created the atmosphere of excitement that made our publishing
effort worthwhile.

We were particularly pleased by the appearance (both at the Conference
and in these Proceedings) of many new (both young and old) faces. Mathe-
maticians from many diverse fields are now interested in Riemann Surfaces
and Kleinian groups. We are delighted that the old classical theory of
functions of one complex variable still shows so many signs of vitality.

I. Kra
B. Maskit

MAY 1979

ix

Riemann Surfaces and Related Topics

A GEOMETRIC PROPERTY OF BERS' EMBEDDING
OF THE TEICHMÜLLER SPACE

William Abikoff

In this short note we prove a geometric property of the Bers embedding
of the Teichmüller space. To fix the notation, let G be a finitely generated Fuchsian group of the first kind acting in the unit disc Δ. The Bers
embedding of $T(G) = T(\Delta/G)$ represents $T(G)$ in the space B of bounded quadratic differentials ϕ for G in the exterior E of Δ. In the
usual way we associate to each $\phi \epsilon B$, the normalized solution

$$f_\phi(z) = \frac{1}{z} + \sum_1^\infty b_n(\phi) z^{-n}$$

of the Schwarzian differential equation $\{f_\phi, z\} = \phi$. It is important to
note that f_ϕ, hence $b_n(\phi)$ is a holomorphic function on B. Set $G_\phi = f_\phi G f_\phi^{-1}$ and let $i : T(G) \to B$ be the Bers embedding. If $\phi \epsilon \overline{T} = \overline{i(T(G))}$
then f_ϕ is schlicht and G_ϕ is a b-group. Let $\Lambda(G_\phi)$ denote the limit
set of G_ϕ and $m(\phi)$ be the area of $\Lambda(G_\phi)$.

We prove the following

THEOREM. *If $\phi \epsilon \partial i(T(G))$ and $m(\phi) = 0$, then $\phi \epsilon \partial$ Ext $i(T(G))$.*

Proof. The tripartite classification of b-groups shows that if $\phi \epsilon \partial i(T(G))$
then G_ϕ is either totally degenerate or has accidental parabolic transformations. The two cases must be handled separately.

© 1980 Princeton University Press
Riemann Surfaces and Related Topics
Proceedings of the 1978 Stony Brook Conference
0-691-08264-2/80/000003-03 $00.50/1 (cloth)
0-691-08267-7/80/000003-03 $00.50/1 (paperback)
For copying information, see copyright page

If G_ϕ has accidental parabolic transformations, then there is some $\gamma_\phi \in G_\phi$ so that $\tau(\phi) = \mathrm{tr}^2\, \gamma_\phi = 4$, but $\tau(\phi)$ is a nonconstant holomorphic function on B. Thus near ϕ, τ takes on all values sufficiently close to 4. It follows that there are groups G_ψ arbitrarily close to G_ϕ which have elliptic elements of infinite order. Such groups are not Kleinian and $\psi \notin \partial\, i(T(G))$.

We proceed to the case where G_ϕ is totally degenerate. Set

$$A : B \to [-\infty, \pi]$$

$$\psi \mapsto \pi\left(1 - \sum_1^\infty n|b_n(\psi)|^2\right).$$

Since $b_n(\psi)$ is holomorphic on B, A is plurisuperharmonic. Further, Gronwall's Area Theorem says that if f_ψ is schlicht then $A(\psi)$ is the area of $C \setminus f_\psi(E)$. Assume $G_\phi \in \mathrm{Int}\ \overline{i(T(G))}$. Then any holomorphic map

$$h : \overline{\Delta} \to B$$
$$0 \mapsto \phi$$

satisfies

$$A(\phi) \geq (2\pi)^{-1} \int_{\partial\Delta} A(h(e^{i\theta}))\, d\theta \geq 0.$$

But we may choose a holomorphic disc $h(\Delta)$ with center ϕ and intersecting $i(T(G))$ along a nontrivial boundary arc and such that $h(\overline{\Delta}) \subset \overline{i(T(G))}$. It follows from the above inequality that $A(\phi) > 0$. But for totally degenerate groups, $A(\phi) = m(\phi)$ and we have assumed $m(\phi) = 0$. We have the desired contradiction.

SOME REMARKS

1) The theorem is a finite dimensional version of Gehring's theorem that the universal Teichmüller space is the interior of the Schwarzians of schlicht functions.

2) At this conference, Thurston announced a proof that $m(\phi) = 0$, for all boundary groups of the Teichmüller space. This result eliminates the need for our main hypothesis.

3) The second part of the proof may be repeated verbatim to prove the following statement. Given any holomorphic mapping of the punctured disc into Teichmüller space (or equivalently, a holomorphic family of finite Riemann surfaces over the unit disc), then the puncture (or central fiber) cannot be filled in by a totally degenerate group whose limit set has zero area.

UNIVERSITY OF ILLINOIS
AT URBANA-CHAMPAIGN

PLANE MODELS FOR RIEMANN SURFACES ADMITTING CERTAIN HALF-CANONICAL LINEAR SERIES, PART I[*]

Robert D. M. Accola[**]

1. *Introduction*

Let W_p be a Riemann surface of genus p. In considering vanishing properties of the theta-function at half periods of the Jacobian associated with W_p, one is led naturally, via Riemann's vanishing theorem, to half-canonical linear series on W_p, that is, to linear series whose doubles are canonical. A theorem of Castelnuovo assures us that a half-canonical g^r_{p-1} must be composite if $p < 3r$, and this leads directly to the existence of automorphism of period two on W_p [2, Part III]. In this paper we are concerned with surfaces where $p = 3r$ and W_p admits a simple g^r_{p-1} (which must necessarily be half-canonical). We show that such surfaces exist for all r and we investigate the consequences. By another theorem of Castelnuovo it follows that, except for $r = 5$, the existence of a simple g^r_{p-1} on W_{3r} insures the existence of a g^1_4 without fixed points. This in turn implies that such a Riemann surface has a plane model where the half-canonical g^r_{3r-1} is easily seen. From these models one easily calculates the dimension of such Riemann surfaces in Teichmüller space. The methods developed here also allow us to characterize, for such surfaces, when the divisors of the g^1_4 are the orbits of an automorphism group which is non-cyclic of order four.

[*]The author wishes to express his thanks to Dr. Joseph Harris for valuable discussions concerning the material of this paper.

[**]Research supported by the National Science Foundation

It turns out that these methods also apply to W_{3r+2}'s admitting two simple half-canonical g^r_{3r+1}'s and to W_{3r+3}'s admitting four simple half-canonical g^r_{3r+2}'s whose sum is bicanonical. We shall consider these cases in Part II of this paper.

2. *Notation, definitions, and preliminary results*

A compact Riemann surface of genus p will be denoted W_p. A linear series on W_p of dimension r and degree n will be denoted g^r_n. Such a series may have fixed points, may be simple or composite, and may be complete or incomplete. For $x \in W_p$, $g^r_n - x$ will denote the linear series of degree $n-1$ of divisors of g^r_n passing through x, not counting x. If x is not a fixed point of g^r_n, then $g^r_n - x = g^{r-1}_{n-1}$.

For the convenience of the reader we include the following definitions [6, p. 257]. A linear series g^{r*}_n will be defined to be *simple* if for a general choice of x, $g^r_n - x$ is without fixed points. In this situation it is known that for a general choice of x, $g^r_n - x$ will also be simple. A linear series g^{r*}_n will be defined to be *composite* if for any choice of x, $g^r_n - x$ has fixed points. In this latter situation W_p is a t-sheeted covering of a surface of genus q, W_q, and a divisor of non-fixed points of g^r_n is a union of the fibers of the map $\phi: W_p \to W_q$. In such a case W_q admits a $g^r_{(n-f)/t}$ where f is the degree of the divisor of fixed points of g^r_n, and for x not fixed for g^r_n, $g^r_n - x$ has $t-1$ additional fixed points, the other points in the fiber of ϕ containing x. If g^r_n is complete on W_p, then so is $g^r_{(n-f)/t}$ on W_q.

If g^r_n is a linear series, a second series g^s_m is said to impose t (linear) conditions on g^r_n if there is a linear series g^{r-t}_{n-m} so that

$$g^r_n = g^s_m + g^{r-t}_{n-m}.$$

This means that if D is any divisor of g^s_m of m distinct points, then there are t points of $D, x_1, x_2, \cdots, x_t$ so that

$$g^r_n - (x_1 + x_2 + \cdots + x_t) = g^{r-t}_{n-m} + D - (x_1 + \cdots + x_t)$$

---

*Without fixed points.

has $D - (x_1 + \cdots + x_t)$ among its fixed points. Also $x_1, x_2, \cdots, x_t$ impose independent conditions; that is, for each k there is a divisor in g_n^r containing all the x_i, $i = 1, 2, \cdots, k-1, k+1, \cdots, t$, but not containing x_k.

If g_m^1 imposes one condition on g_n^r, then $g_n^r = rg_m^1 + D_{n-rm}$ where D_{n-rm} is the divisor of fixed points of the composite g_n^r; for whenever a divisor of g_n^r contains a point x, it must contain all of the unique divisor of g_m^1 containing x.

We will use the classical fact that since a g_n^1 $(n \leq p)$ without fixed points imposes $n-1$ conditions on the canonical series, it imposes at most $n-1$ conditions on any special linear series. The extension of this is that a simple special g_m^s without fixed points will impose at most $m-s$ conditions on any other special linear series whose dimension is at least $m-s$.

If g_n^r is simple $(r \geq 2)$ and without fixed points, then W_p can be realized as a curve in P^r and the hyperplane sections cut out the divisors of g_n^r. In such cases we will say that g_n^r has a k-fold singularity if the curve in P^r does. In case g_n^2 is simple and without fixed points, W_p admits a plane model of degree n. If d represents the number of double points suitably counted, then

$$p = \frac{(n-1)(n-2)}{2} - d .$$

To compute the dimension R of all plane curves of degree n with s given ordinary singularities of multiplicities $k_1, k_2, \cdots, k_s$, we use the formula

$$R \geq \frac{n(n+3)}{2} - \sum_{j=1}^{s} \frac{k_j(k_j+1)}{2} .$$

Often, this formula is precise. A singularity of multiplicity k will be called a *k-fold point* of the curve or linear series.

A surface will be called *q-hyperelliptic* $(q \geq 0)$ if it is a two-sheeted cover of a surface of genus $q' \leq q$. Thus rational and elliptic surfaces are q-hyperelliptic for all q.

A divisor D_{p-1} of degree $p-1$ on W_p will be called *half-canonical* if $2D_{p-1}$ is canonical.

We now discuss the preliminary results upon which the later sections of this paper depend.

THEOREM 2.1 (Castelnuovo's Theorem [5] or [6, p. 295]). *Let* W_p *admit a simple* g_n^r. *Then*

$$p \leq \frac{n-r+\varepsilon}{r-1} \cdot \frac{n-1-\varepsilon}{2}$$

where $0 \leq \varepsilon \leq r-1$ *and* $n-r+\varepsilon \equiv 0 \pmod{r-1}$.

In the proof of Castelnuovo's theorem he shows that if g_n^r is a simple linear series then $2g_n^r = g_{2n}^R$ where $R \geq 3r-1$. In [4] he goes on to prove that if $2g_n^r = g_{2n}^{3r-1}$ $(n > 2r)$ then the model C of W_p given in $\mathbf{P}^r$ by g_n^r lies on a surface of degree $r-1$. It follows that if $r \neq 5$ this surface is a rational normal scroll and the rulings cut out on C a linear series g_m^1 which imposes two conditions on g_n^r. If $r = 5$ the surface could be the Veronese surface. In our case, $n = 3r-1$, he asserts that $m = 4$; however, no proof is given so we shall establish this fact later. Since equality in Castelnuovo's theorem implies that $2g_{3r-1}^r = g_{6r-2}^{3r-1}$ we can summarize this discussion as follows.

THEOREM 2.2. *Suppose we have equality in Castelnuovo's theorem,* $r \neq 5$, *and* $n > 2r$. *Then* W_p *admits a* g_m^1 *which imposes two linear conditions on* g_n^r.

The following three lemmas for half-canonical series are the main results upon which depend much of the later development. The first lemma is essentially proven in [3].

LEMMA 2.3. *Let* W_p $(p \geq 5)$ *admit a* g_3^1 *without fixed points. Let* g_{p-1}^r *be a half-canonical linear series with* $r \geq 1$. *Then* g_3^1 *imposes one condition on* g_{p-1}^r *and so* g_{p-1}^r *is composite.*

Proof. Let $x + y + z$ be a divisor in g^1_3 containing three distinct points and no fixed points of g^r_{p-1}. Now any divisor of g^r_{p-1} containing two of the points of $x + y + z$ must contain the third since g^r_{p-1} is special. If the lemma is false, we can find a divisor D_x of g^r_{p-1} containing x but not y nor z and we can find a D_y in g^r_{p-1} containing y but not x nor z. But $D_x + D_y$ is canonical and so must contain z. We have reached the desired contradiction. q.e.d.

LEMMA 2.4. *Let* g^1_m *be a linear series without fixed points. If* $m \le 2r+1$, *then* g^1_m *imposes at most* $[m/2]$ *conditions on any half-canonical linear series* g^r_{p-1}.

Proof. Let D_m be a divisor of g^1_m of m distinct points $x_1 + x_2 + \cdots + x_m$. We may suppose that $x_1, x_2, \cdots, x_t$ impose t independent conditions on g^r_{p-1} where t is maximum with this property. Then $t \le r$. We want to show that $t \le [m/2]$.

Suppose $t > [m/2]$. Then $m - t < m - [m/2] = [(m+1)/2]$, and so $m - t \le r$. Let D_0 be a divisor of g^r_{p-1} containing $x_{t+1} + \cdots + x_m$. For $k = 1, 2, \cdots, t$ let D_k be a divisor of g^r_{p-1} containing all the x_i's, $i = 1, 2, \cdots, t$ except x_k. Then the canonical divisor $D_0 + D_k$ contains all the points of D_m except possibly x_k. Therefore, $D_0 + D_k$ contains x_k and so D_0 contains all points of D_m. Thus D_m imposes at most $m - t$ conditions on g^r_{p-1}, and we have $t \le m - t$ or $t \le [m/2]$, a contradiction. q.e.d.

THEOREM 2.5. *Let* g^s_m *be a simple linear series without fixed points with* $m - s \le 2r$. *Then* g^s_m *imposes at most* $[(m-s+1)/2]$ *conditions on any half-canonical* g^r_{p-1}. *Thus any such half-canonical* g^r_{p-1} *must be simple.*

Proof. Let $s - 1$ distinct points $x_1, x_2, \cdots, x_{s-1}$ be chosen so that $g^s_m - (x_1 + x_2 + \cdots + x_{s-1}) = g^1_{m-s+1}$ is without fixed points. g^1_{m-s+1} imposes at most $[(m-s+1)/2]$ conditions on g^r_{p-1}. Since g^r_{p-1} is special, any divisor of g^r_{p-1} containing a divisor of g^1_{m-s+1} must also contain $x_1 + x_2 + \cdots + x_{s-1}$. g^r_{p-1} must be simple since it contains the simple g^s_m. q.e.d.

The following three lemmas will be useful later on.

LEMMA 2.6. *Suppose* W_p *is q-hyperelliptic. If* W_p *admits a simple half-canonical* g^r_{p-1}, *then* $r \leq q$.

Proof. We show that if $r \geq q+1$, then any half-canonical g^r_{p-1} must be composite and is lifted from a $g^r_{(p-1-f)/2}$ on $W_{q'}$, $q' \leq q$ where f is the number of fixed points of g^r_{p-1}. Let $n = p-1-f$. Let $\phi : W_p \to W_{q'}$ be the two-sheeted cover. It suffices to show that every divisor of g^r_n with n distinct points contains a complete fiber of ϕ. For then there is an algebraic relation between a pair of points in each divisor of g^r_n which by analytic continuation is seen to hold for other pairs of points in each divisor of g^r_n. We argue by contradiction.

Suppose D is a divisor of g^r_n with n distinct points in n distinct fibers of ϕ. If E is a divisor of $q'+1$ such points, then $\phi(E)$ determines a linear series of dimension at least one on $W_{q'}$ and so $\phi^{-1}\phi(E)$ determines a $g^1_{2q'+2}$ on W_p. Since $r \geq q+1 \geq q'+1$, we can assume that the points of E impose independent conditions on g^r_n (by the original choice of E). But $g^1_{2q'+2}$ imposes at most $(q'+1)$ conditions on g^r_n and so $\phi^{-1}\phi(E)$ must be in D, a contradiction. q.e.d.

LEMMA 2.7. *Suppose* W_{3r} *admits a simple half-canonical* g^r_{3r-1} *and a* g^1_4. *Suppose* g^1_4 *imposes at most two conditions on a* g^s_m *where* $s \leq r$ *and* $m < 4s$. *Then* g^s_m *is simple.*

Proof. Suppose g^s_m is composite. Then there is a cover of t sheets $W_{3r} \to W_q$ and a $g^s_{(m-f)/t}$ on W_q which lifts to the non-fixed points of g^s_m. Also g^1_4 is lifted from W_q and we see that $t = 2$ or 4. If $t = 4$ then $q = 0$ and g^1_4 imposes one condition on g^s_m. Consequently $g^s_m = sg^1_4 + D_f$ and $m = 4s+f$, a contradiction. Suppose $t = 2$. If

$g^s_{(m-f)/2}$ is special on W_q then by Clifford's theorem we have $(m-f)/2 \geq 2s$, a contradiction. If $g^s_{(m-f)/2}$ is not special then $(m-f)/2 - s \geq q$. But by Lemma 2.6 $q \geq r$ and by hypothesis $r \geq s$, so we arrive at the final contradiction $(m-f)/2 - s \geq s$. q.e.d.

LEMMA 2.8. *Suppose* g^1_4 *and* h^0_4 *are complete linear series and* g^1_4 *is without fixed points. Suppose that* $2g^1_4 = g^2_8$ *is also complete. Suppose finally that* $2h^0_4 \equiv 2g^1_4$. *Then there are disjoint integral divisors of degree two,* P *and* Q, *so that* $h^0_4 = P+Q$ *and* $|2P| = |2Q| = g^1_4$.

Proof. Let $h^0_4 = x_1 + x_2 + x_3 + x_4$ and $2(x_1 + x_2 + x_3 + x_4) = D_1 + D_2$, where $|D_1| = |D_2| = g^1_4$ and D_1 and D_2 are disjoint. Now the result follows by examining the various possibilities for D_1 and D_2. q.e.d.

We conclude this section with some results on W_{3r}'s admitting g^r_{3r-1}'s.

LEMMA 2.9. *Suppose* W_{3r} *admits a simple* g^r_{3r-1}. *Then* g^r_{3r-1} *is complete, without fixed points, and half-canonical.*

Proof. That g^r_{3r-1} is complete and without fixed points follows from Castelnuovo's inequality since the inequality is now an equality. Also $2g^r_{3r-1} = g^R_{2p-2}$ where $R \geq 3r-1 = p-1$. By the Riemann-Roch theorem $R = p-1$ and so $2g^r_{3r-1}$ is canonical. q.e.d.

LEMMA 2.10. *On* W_{3r} *a simple* g^r_{3r-1} *is unique.*

Proof. If h^r_{3r-1} is a second half-canonical series then g^r_{3r-1} imposes at most r conditions on h^r_{3r-1} by Theorem 2.5. Consequently each divisor of g^r_{3r-1} is contained in a divisor of h^r_{3r-1}. It follows that $g^r_{3r-1} = h^r_{3r-1}$. q.e.d.

LEMMA 2.11. *If* W_{3r} $(r \geq 2)$ *admits a unique complete* g^r_{3r-1} *then* g^r_{3r-1} *must be simple.*

Proof. Suppose g^r_{3r-1} is composite. The only way that this can happen is for W_{3r} to be a two-sheeted cover of a surface of genus q and on W_q

there is a complete $g^r_{(3r-1-f)/2}$ where f is the degree of the divisor of fixed points of g^r_{3r-1}. If $g^r_{(3r-1-f)/2}$ is special then by Clifford's theorem we have the contradiction $(3r-1-f)/2 - 2r \geq 0$. This series is not special and so $q = (r-1-f)/2$. By [2, p. 51] it follows that the q-hyperelliptic surface W_{3r} admits at least 4^q composite half-canonical g^r_{3r-1}'s. If q is zero the result also follows. q.e.d.

THEOREM 2.12. *W_{3r} admits a simple g^r_{3r-1} if and only if g^r_{3r-1} is unique. Consequently, W_{3r} admits a simple g^r_{3r-1} if and only if the theta-function for W_{3r} vanishes at precisely one half-period to order $r+1$.*

The second statement of the theorem is just a translation of the first statement into the language of vanishing properties of the theta-function via Riemann's vanishing theorem.

3. *The plane models*

LEMMA 3.1. *Suppose W_{3r} $(r \geq 4)$ admits a simple g^r_{3r-1} and a g^1_m which imposes two linear conditions on g^r_{3r-1}. Then $m = 4$.*

Proof. If $x \in W_{3r}$, then $g^r_{3r-1} - x$ has a singularity of multiplicity $m-1$, namely the other points in the divisor of g^1_m containing x; for if y is one of these $m-1$ points then $(g^r_{3r-1}-x)-y$ has the remaining $m-2$ points as fixed points. Now choose a divisor $D_{r-2} = x_1 + x_2 + \cdots + x_{r-2}$ of $r-2$ points in $r-2$ distinct divisors of g^1_m. Also assume that $g^r_{3r-1} - D_{r-2}$ $(= g^2_{2r+1})$ is simple and without fixed points. g^2_{2r+1} has $r-2$ singularities of multiplicity $m-1$ corresponding to the $r-2$ divisors of g^1_m determined by D_{r-2}. Also g^1_m imposes two conditions on g^2_{2r+1}, so g^2_{2r+1} has a singularity of multiplicity $2r+1-m$. If D_{r-2} is chosen in a general manner all these singularities are disjoint and so contribute $(r-2)(m-1)(m-2)/2 + (2r+1-m)(2r-m)/2$ to the double points of the plane curve determined by g^2_{2r+1}. Consequently

$$p = 3r = (2r)(2r-1)/2 - (r-2)(m-1)(m-2)/2 -$$
$$(2r+1-m)(2r-m)/2 - d'$$

where d' is the contributions of the other double points of the curve. Thus $2d' = -m^2(r-1) + m(7r-5) - 12r + 4 = -(m-4)((r-1)m - (3r-1))$. If $m \geq 5$ we see that the right-hand side of this equation is negative which is a contradiction. Thus $m = 4$ and $d' = 0$. q.e.d.

THEOREM 3.2. *Let* W_{3r} *be a Riemann surface of genus* $3r(r \geq 4, r \neq 5)$ *admitting a simple* g^r_{3r-1}. *Then* W_{3r} *admits a plane model as a curve of degree* $2r + 1$ *with* $r - 2$ *singularities of multiplicity* 3 *and one singularity of multiplicity* $2r - 3$.

Denote the plane curve of the theorem $\mathcal{C}_{2r+1}$.

Now suppose that the $r-1$ singularities of $\mathcal{C}_{2r+1}$ are in general position; that is, no $(\ell+1)(\ell+2)/2$ of them lie on a curve of degree ℓ. (This situation is to be expected, in general, but there are examples where it is not the case.) We can simplify the model by successive quadratic transformations. First transform with the $(2r-3)$-fold point and two triple points as fundamental points. Such a transformation transforms $\mathcal{C}_{2r+1}$ into a curve of degree $2r-1$, $\mathcal{C}_{2r-1}$, with $r-4$ triple points and one $(2r-5)$-fold point. Second transform $\mathcal{C}_{2r-1}$ with the $(2r-5)$-fold point and two triple points as fundamental points to obtain a $\mathcal{C}_{2r-3}$ with a $(2r-7)$-fold point and $r-6$ triple points. Continue.

If r is even, $(r-2)/2$ such transformations yield a curve of degree $r+3$, $\mathcal{C}_{r+3}$, with a single singularity of degree $r-1$. The g^r_{3r-1} is cut out by curves of degree $r/2$ with a $((r-2)/2)$-fold singularity at the $(r-1)$-fold singularity of $\mathcal{C}_{r+3}$.

If r is odd, $(r-3)/2$ such transformations yield a $\mathcal{C}_{r+4}$ with a 3-fold singularity and a r-fold singularity. g^r_{3r-1} is cut out by curves of degree $(r+1)/2$ with a $((r-1)/2)$-fold singularity at the r-fold singularity of $\mathcal{C}_{r+4}$ and also passing through the triple point. It is worth remarking that in this case of r odd, these curves are the examples of Riemann surfaces for which equality is attained in the following classical inequality:

$$p \leq (n-1)(m-1)$$

for W_p admitting a g^1_n and g^1_m ($n = 4$ and $m = r+1$). In fact, in this case of r odd,

$$g^r_{3r-1} = g^1_{r+1} + ((r-1)/2)\, g^1_4$$

where the lines through the triple point of C_{r+4} cut out the g^1_{r+1}.

Since these curves C_{r+3} (r even) and C_{r+4} (r odd) are easily constructed, the question of the existence of W_{3r}'s with simple half-canonical g^r_{3r-1}'s is settled. In fact, the dimension of such surfaces in Teichmüller space can be easily computed using formula (*) of Section 2. Remembering the six-dimensional space of plane collineations which leave a point fixed, the dimension is seen to be $5r + 3$ for r even, $r \geq 4$. The same answer for r odd is derived by taking into account the one-dimensional freedom in choosing the point to lie on the line connecting the two singular points of C_{r+4}.

4. Four groups of automorphisms

We know that if W_{3r} admits a simple g^r_{3r-1} then it is unique. We now consider other possible half-canonical series on such a W_{3r}.

LEMMA 4.1. *Suppose W_{3r} admits a simple g^r_{3r-1}. Then W_{3r} cannot admit a complete half-canonical linear series h^{r-1}_{3r-1}.*

Proof. Suppose W_{3r} admits a simple g^r_{3r-1} and a complete half-canonical h^{r-1}_{3r-1}. Since $2g^r_{3r-1} \equiv 2h^{r-1}_{3r-1}$ there is a smooth two-sheeted cover W_{6r-1} of W_{3r} on which the lifts of the two series are equivalent. This gives rise to a g^{2r}_{6r-2} on W_{6r-1} [2, p. 74] which must be composite by Castelnuovo's theorem. Since g^r_{3r-1} is simple we arrive at a contradiction. q.e.d.

Now suppose that W_{3r} admits a simple g^r_{3r-1} and a g^1_4. We can construct complete half-canonical h^{r-2}_{3r-1}'s as follows. Suppose there are two integral divisors of degree two, P_1 and P_2, so that $|2P_1| = |2P_2| = g^1_4$. Since g^1_4 imposes two conditions on g^r_{3r-1} we see that

$g^r_{3r-1} - g^1_4 = g^{r-2}_{3r-5}$. The latter series is without fixed points and simple if $r \geq 4$ by Lemma 2.8. Let $h^{r-2}_{3r-1} = g^{r-2}_{3r-5} + P_1 + P_2$. Then h^{r-2}_{3r-1} is complete and half-canonical. We now show that this is the only way in which W_{3r} can admit a complete half-canonical h^{r-2}_{3r-1}, if r is large enough.

THEOREM 4.2. *Suppose* W_{3r} $(r \geq 12)$ *admits a simple* g^r_{3r-1} *and a complete* h^{r-2}_{3r-1}. *Then* $h^{r-2}_{3r-1} \equiv g^r_{3r-1} - g^1_4 + P_1 + P_2$ *where* P_1 *and* P_2 *are disjoint integral divisors of degree two whose doubles are in* g^1_4.

Proof. Since g^1_4 imposes at most two conditions on h^{r-2}_{3r-1} and $r \geq 8$, Lemma 2.8 assures us that h^{r-2}_{3r-1} is simple. We now show that h^{r-2}_{3r-1} must have precisely four fixed points by eliminating the other possibilities, case by case.

Case (0): Suppose h^{r-2}_{3r-1} is without fixed points. Then $h^{r-2}_{3r-1} - g^1_4 = h^{r-4}_{3r-5}$ is also without fixed points and is simple by Lemma 2.8 since $r \geq 12$. By Theorem 2.5 h^{r-4}_{3r-5} imposes r conditions on g^r_{3r-1} and so $h^{r-4}_{3r-5} + h^0_4 = g^r_{3r-1}$ where h^0_4 is complete. Since $h^{r-2}_{3r-1} = h^{r-4}_{3r-5} + g^1_4$ we see that $2h^0_4 = 2g^1_4$. By Lemma 2.9 it follows that $h^0_4 = P + Q$ where $|2P| = |2Q| = g^1_4$. But h^0_4 imposes four conditions on g^r_{3r-1}. Therefore, P imposes two conditions on g^r_{3r-1}. Since $2P = g^1_4$ we see that $g^r_{3r-1} - P$ has P as a fixed divisor. Consequently h^{r-4}_{3r-5} must have P as a fixed divisor. This is the desired contradiction.

Case (1): $h^{r-2}_{3r-1} = h^{r-2}_{3r-2} + x$ where h^{r-2}_{3r-2} is without fixed points. h^{r-2}_{3r-2} imposes r conditions on g^r_{3r-1} and so $g^r_{3r-1} = h^{r-2}_{3r-2} + y$. This is a contradiction since one point y cannot impose two conditions on g^r_{3r-1}.

Case (2): $h^{r-2}_{3r-1} = h^{r-2}_{3r-3} + P$ where h^{r-2}_{3r-3} is without fixed points. Then $g^r_{3r-1} \equiv h^{r-2}_{3r-3} + Q$ and $|2P| = |2Q| = g^1_4$. But again Q imposes two conditions on g^r_{3r-1} and so Q must be a fixed divisor for h^{r-2}_{3r-3}. Contradiction.

Case (3): $h^{r-2}_{3r-1} = h^{r-2}_{3r-4} + P$ where degree $P = 3$ and h^{r-2}_{3r-4} is without fixed points. Then h^{r-2}_{3r-4} imposes $r-1$ conditions on g^r_{3r-1} and so $g^r_{3r-1} \equiv h^{r-2}_{3r-4} + g^1_3$. Contradiction.

Case (≥ 5): $h^{r-2}_{3r-1} = h^{r-2}_{3r-6-k} + D$ where degree $D = 5 + k$, $k \geq 0$. Then by Castelnuovo's inequality, Theorem 2.1, h^{r-2}_{3r-6-k} must be composite.

Case (4): $h^{r-2}_{3r-1} = h^{r-2}_{3r-5} + h^0_4$. h^{r-2}_{3r-5} imposes $r-1$ conditions on g^r_{3r-1} and consequently $g^r_{3r-1} \equiv h^{r-2}_{3r-5} + g^1_4$. Thus $2h^0_4 \equiv 2g^1_4$ and the conclusion follows by Lemma 2.9. q.e.d.

The restriction $r \geq 12$ is unnecessary. Some lower cases of r can be handled by more special arguments.

Now we apply this theorem to the question of automorphisms on W_{3r}.

THEOREM 4.3. *Suppose* W_{3r} $(r \geq 12)$ *admits a simple* g^r_{3r-1} *and* $\binom{3r+3}{2}$ *half-canonical* h^{r-2}_{3r-1}*'s . Then* W_{3r} *admits a non-cyclic group of automorphisms* G *of order four whose orbits are the divisors of* g^1_4. *The three subgroups of* G *of order two have orbit spaces which are Riemann surfaces of genus* r.

Proof. W_{3r} admits a non-cyclic automorphism group G of order four whose orbits are g^1_4 if and only if each ramified divisor in g^1_4 is of the form $2P$ where P is an integral divisor with two distinct points. By the Riemann-Hurwitz formula there are at most $3r + 3$ such divisors in g^1_4 since each such divisor contributes two to the ramification of the cover of P^1 given by g^1_4. If such a G exists there are $3r + 3$ such pairs P_i and so $\binom{3r+3}{2}$ half-canonical h^{r-2}_{3r-1}'s which are $g^{r-2}_{3r-5} + P_i + P_j$. Conversely, if there are $\binom{3r+3}{2}$ half-canonical h^{r-2}_{3r-1}'s then by the previous theorem there are at least $3r + 3$ pairs P_i by elementary counting. Since each pair contributes at least two to the ramification and the total ramification is $6r + 6$, each pair must contribute precisely two to the ramification. Consequently each pair must consist of two distinct points.

It remains to prove the last statement of the theorem. By [1] we have $3r = p_1 + p_2 + p_3$ where p_1, p_2, p_3 are the genera of the quotient surfaces in question. But by Lemma 2.7 each $p_j \geq r$. Consequently each $p_j = r$. q.e.d.

To exhibit a general example of a W_{3r} with a simple g^r_{3r-1} and such a four group of automorphisms G seems difficult. The W_{3r}'s admitting a g^r_{3r-1} have dimension $5r + 3$. To admit such a G seems to impose $3r + 3$ additional conditions, so one would guess that such W_{3r}'s have dimension $2r$ in Teichmüller space. W_{3r}'s simply admitting such a four-group have dimension $3r$ so there are further conditions on the latter W_{3r}'s which insure the existence of a g^r_{3r-1}. If r is odd, however, the existence of a g^1_{r+1} does insure the existence of a g^r_{3r-1} ($= g^1_{r+1} + ((r-1)/2)\,g^1_4$).

Our example will be a W_{3r}, r odd, admitting a group of automorphisms H isomorphic to $Z_2 \times Z_2 \times Z_{r+1}$. Let the group be generated by a, b, and c where $a^2 = b^2 = c^{r+1} = e$. Let a representation of the fundamental group of the five-fold punctured sphere into H be determined where the five paths "circling" the punctures are mapped by the representation onto a, b, ab, c, c^{-1} respectively. W_{3r} is the closure of the covering surface corresponding to the kernel of the representation. One sees that the orbit spaces of $Z_2 \times Z_2$ and Z_{r+1} both have genus zero. Our surface admits a g^1_4 whose divisors are the orbits of this four-group and it also admits a g^1_{r+1}, the orbits of Z_{r+1}.

REFERENCES

[1] Accola, R. D. M. "Riemann surfaces with automorphism groups admitting partitions." *Proceedings of the American Mathematical Society*, Vol. 21 (1969), pp. 477-482.

[2] ________. *Riemann surfaces, theta functions, and abelian automorphism groups*. Lecture Notes in Mathematics 483, Springer Verlag, 1975.

[3] Andreotti, A., and Mayer, A. L. "On period relations for abelian integrals on algebraic curves." *Annali della Scuola Normale Superiore di Pisa*, Vol. 21 (1967), pp. 189-238.

[4] Castelnuovo, G. "Ricerche di geometria sulle curve algebriche."
Atti della R. Accademia della Scienze di Torino, Vol. 24 (1889).
(*Memorie Scelte*, pp. 19-43.)

[5] __________. "Sui multipli du una serie lineare di gruppi di punti, etc."
Rendiconti del Circolo Matematico di Palermo, Vol. 7 (1893), pp. 89-
110. (*Memorie Scelte*, pp. 95-113.)

[6] Coolidge, J. L. *Algebraic Plane Curves*. Oxford 1931. (Dover 1959)

[7] Walker, R. *Algebraic Curves*. Dover, 1962.

NONTRIVIALITY OF TEICHMÜLLER SPACE
FOR KLEINIAN GROUP IN SPACE

B. N. Apanasov

1. *Introduction*

Let $\mathfrak{M}_n$ be the group of all Möbius transformations in $\overline{R^n} = R^n \cup \{\infty\}$ and $G \subset \mathfrak{M}_n$ be a discontinuous (Kleinian) subgroup. For simplicity we consider the Kleinian group G to act invariantly on a ball B^n, that is, to be a Fuchsian group. Having introduced the hyperbolic metric on B^n, we turn the group into the discrete group of hyperbolic space isometries, and the natural projection $\pi: B^n \to B^n/G$ inserts a conformal structure on the factor set B^n/G and turns it into a Riemannian manifold of constant negative curvature.

In this class of manifolds the amazingly strong and elegant theorem takes place:

THEOREM 1.1 (G. D. Mostow [10]). *Let Y and Y' be n-dimensional complete Riemannian manifolds of the same constant negative curvature and having finite volume, and let there exist the quasiconformal homeomorphism Y on Y'. Then there exists an isometry of Y on Y' which induces the same isomorphism of the fundamental groups.*

Or, in other words, any isomorphism of the Fuchsian groups G and G', $V(B^n/G) < \infty$, $V(B^n/G') < \infty$ of the form $g \mapsto fgf^{-1} = g'$, where

---

$f : B^n \to B^n$ is a quasiconformal automorphism of the ball, can be represented in the form $g \mapsto AgA^{-1} = g'$, where $A \in \mathfrak{M}_n$, $A(B^n) = B^n$.

In the planer case $(n{=}2)$ the described situation does not occur. Moreover, it is well known that the set of all quasiconformal automorphisms of the disk B^2 compatible with a Fuchsian group G (identifying the conformally equivalent ones) forms the Teichmüller space of the group G and has complex dimension $3p-3+m$, where p is the genus of the surface B^2/G, and m is the number of punctures of B^2/G.

As it may seem, after Mostow's result, that for $n \geq 3$ the Teichmüller space of the Fuchsian group G, $V(B^n/G) < \infty$, degenerates into a point.

However, our main result can be formulated in the following form.

THEOREM A. *For any Fuchsian group* G *in* $\overline{R^n}$, $n \geq 2$ *with compact fundamental polyhedron* B^n/G *such that one side is orthogonal to all other sides that it meets there exists a homeomorphism*

$$\phi : I^m \to T(G)$$

of the open m-dimensional cube, $m \geq 1$, *into Teichmüller space for the group* G.

We also enlist here some applications of this result connected with spatial quasiconformal mappings and Mostow's rigidity theorem to prove the following statement.

THEOREM B. *The boundary of Teichmüller space* $T(G)$ *for the Fuchsian group* $G \subset \mathfrak{M}_3$ *(where* G *is the same as in Theorem A), contains the Kleinian groups* Γ *with the following properties:*

    1) $\mathfrak{L}(\Gamma)$ *is connected;*

    2) $\Omega(\Gamma) = \Omega_0 \cup \Omega_1$, *where* Ω_0 *and* Ω_1 *are the connectivity components;*

    3) $\Gamma(\Omega_0) = \Omega_0$, $\Gamma(\Omega_1) \neq \Omega_1$;

    4) Ω_0 *is a quasiconformal image of the ball;*

    5) *the fundamental group* $\pi_1(\Omega_1)$ *is a free infinitely-generated group.*

This theorem shows once more the essential difference between the spatial case and the planar one (cf. [1]).

2. *Notation and terminology*

For $x \in R^n$ we use the representation $x = \Sigma x_i e_i = (x_1, \cdots, x_n)$ where $e_1, \cdots, e_n$ is an orthogonal basis. By $B^n(x, r)$ we denote the open ball with center $x \in R^n$ and radius $r > 0$, $S^{n-1}(x, r) = \partial B^n(x, r)$, $B(x, r) = B^3(x, r)$, $S(x, r) = S^2(x, r)$, $H^n = \{x \in R^n : x_n > 0\}$. The hyperbolic measure of a set M in B^n or in H^n is denoted by $V(M)$. The closure $\overline{M}$, the boundary ∂M of sets $M \subset \overline{R^n}$ will always be in $\overline{R^n}$.

A Möbius group $G \subset \mathfrak{M}_n$ is said to be discrete if no sequence of distinct elements of G converges (pointwise) to 1, or, equivalently, to any $A \in \mathfrak{M}_n$. The limit set of the group $G \subset \mathfrak{M}_n$ is denoted by $\mathfrak{L}(G)$ and consists of the points of accumulation of the sets

$$G(y) = \{g(y) : g \in G\}, \quad y \in \overline{R^n}. \tag{2.1}$$

Its complement $\Omega(G) = \overline{R^n} \setminus \mathfrak{L}(G)$ is called the discontinuity set. The group G is Kleinian if $\Omega(G) \neq \emptyset$. In this case one can define the so-called fundamental domain $F(G)$ which contains a representative from each orbit $G(y)$ (2.1). The Kleinian group $G \subset \mathfrak{M}_n$ is called quasifuchsian if the Jordan surface $\mathfrak{L}(G)$ breaks $\Omega(G)$ into two connected and simply connected components.

A quasiconformal automorphism f of the space $\overline{R^n}$ is compatible with the Kleinian group G (f is a quasiconformal deformation of the group G), if

$$fgf^{-1} = \{fgf^{-1}, g \in G\} \subset \mathfrak{M}_n. \tag{2.2}$$

Two quasiconformal deformations f and f_1 of the group G are equivalent if there exists $A \in \mathfrak{M}_n$ such that the equalities:

$$f_1 g f_1^{-1} = A f g f^{-1} A^{-1}, \quad g \in G$$

hold. We factor the set of quasiconformal deformations of G with respect

to this equivalence relation to obtain the space of quasiconformal deformations of the Kleinian group G. The space $T(G)$ of all the quasiconformal images of the group G (with the analogous equivalence relation)
is naturally associated with it. The convergence of the group can be defined via the coefficients of their matrix representations in the Lorentz
group [3]. The topology thus introduced is equivalent to the one determined by the usual Teichmüller metric [8, 12]—the dilatation logarithm of
the quasiconformal automorphism f.

3. *Preliminary results*

The following statement is obtained from the well-known results [2] on
the groups generated by reflections in hyperbolic space.

LEMMA 3.1. *Let the group* $G \subset \mathfrak{M}_n$ *be generated by a finite number of
involutions* $T_i = \mathfrak{I}_i \circ \mathcal{O}_i$, *where* $\mathfrak{I}_i$ *is inversion with respect to the
sphere* $S^{n-1}(x^i, r_i)$, $x_n^i = 0$, *and* $\mathcal{O}_i$ *is reflection with respect to the
hyperplane* L_i *which passes through the point* x^i, *is orthogonal to the
subspace* R^{n-1} *and is the plane of symmetry of the polyhedron* $P(G) \subset H^n$
bounded by the spheres $S^{n-1}(x^i, r_i)$ *and, perhaps, by the plane*
$\{x : x_n = 0\} = R^{n-1}$.

Then G *is discrete iff all the dihedral angles of the polyhedron* $P(G)$
are integral parts of π.

3.2. Under the extension of the group $G \subset \mathfrak{M}_n$ into the half-space H^{n+1}
we understand the following. Each element $g \in G$ is the superposition of
reflections with respect to lower-dimensional planes $L_i \subset R^n$ and, perhaps, inversion with respect to some sphere $S^{n-1}(x_g, r_g)$ [4]. Consider
n-planes $\tilde{L}_i \subset R^{n+1}$, $\tilde{L}_i$ R^n, $\tilde{L}_i \cap R^n = L_i$, the spheres $S^n(x_g, r_g)$ and
the mappings $\tilde{g} \in \mathfrak{M}_{n+1}$ which are superpositions of reflections with respect to $\tilde{L}_i$ and, perhaps, of inversions with respect to $S^n(x_g, r_g)$. We
obtain the group $\tilde{G} \subset \mathfrak{M}_{n+1}$ which leaves the half-space H^{n+1} invariant
and the restriction of which to the subspace R^n coincides with the
group G.

If now the group G is discrete, then its extension $\tilde{G}$ acts discontinuously in H^{n+1} and the interior of the intersection $\mathrm{int}(\overline{P} \cap \overline{R^n}) = F(G)$ where $P \subset H^{n+1}$ is some hyperbolically convex fundamental polyhedron of the group $\tilde{G}$ is (in case $F(G) \neq \emptyset$) the fundamental domain of the group G.

LEMMA 3.3. *Let* $G \subset \mathfrak{M}_n$ *be a Kleinian group and* F_0 *be the component of its fundamental domain which contains a point* x_0. *If for every generator* A *of the group* G, *the points* x_0 *and* $A(x_0)$ *can be connected by a curve lying in* $\Omega(G)$, *then the component* $\Omega_0 \supset F_0$ *of the discontinuity set* $\Omega(G)$ *is invariant.*

The proof of this lemma is simple and can be done in analogy with the planar case [1, 15].

4. *Proofs of Theorems A and B*

Though our method is independent of the space dimension we, for simplicity, show it in the case $n = 3$, and by the example of a definite Fuchsian group. It will be clear from the proof how to modify it in the general case.

4.1. Let $Q = \{x \in R^3 : |x_i| \subset \frac{1}{2}\}$ be the unit cube. Circumscribe the spheres S_i, $1 \leq i \leq 8$, of radius $\sqrt{3}/3$ from its vertices. They intersect at the angles $\pi/3$, and are orthogonal to the sphere $S_0 = S(0, \sqrt{15}/6)$. Construct six more spheres S_i, $9 \leq i \leq 14$:

$$S_9 = S(re_1, \rho), \ S_{10} = S(-re_1, \rho), \ S_{11} = S(re_2, \rho),$$

$$S_{12} = S(-re_2, \rho), S_{13} = S(re_3, \rho), \ S_{14} = S(-re_3, \rho), \qquad (4.2)$$

where $r = 5/6$, $\rho = \sqrt{10}/6$.

Denote as L_i, $1 \leq i \leq 8$, the planes passing through the points 0, e_3 and the center of the sphere S_i. Further, put

$$L_9 = L_{10} = \{x : x_3 = 0\}, L_{11} = L_{12} = L_{13} = L_{14} = \{x : x_1 = 0\}. \qquad (4.3)$$

Consider the group $G \subset \mathfrak{M}_3$ generated by the involutions A_i, $1 \leq i \leq 14$, each of them being the superposition of inversion with respect to the sphere S_i and of reflection with respect to the plane L_i. The group G leaves the ball $B(O, \sqrt{15}/6)$ invariant. It follows from Lemma 3.1 that G is a Fuchsian group of the first kind with the compact fundamental domain. Consider also the system of spheres $\{\sigma_i(t)\}_{1 \leq i \leq 14}$, which is obtained from the system $\{S_i\}$ by the substitution of the spheres S_{13} and S_{14} (with the centers $\pm \frac{5}{6} e_3$) for the spheres $\sigma_{13}(t)$ and $\sigma_{14}(t)$ with the centers $\pm t \cdot e_3$ and orthogonal to the corresponding spheres $\sigma_i(t) = S_i$, $1 \leq i \leq 12$ (and, consequently, having radii $\sqrt{t^2 - t + 5/12}$).

The group $G(t) \subset \mathfrak{M}_3$, $t_0 = 5/12 < t < t_1 = 5(1 + \sqrt{5/6})$ generated by the involutions A_i^t which correspond to the spheres $\sigma_i(t)$, $1 \leq i \leq 14$, is a Kleinian one. It directly follows from consideration of the extended (see 3.2) group $\tilde{G}(t)$ and from an application of Lemma 3.1. Hence, we obtain that the fundamental domain of the group $G(t)$ has the form:

$$F^t = \bigcap_{1 \leq i \leq 14} \text{ext } \sigma_i(t) \tag{4.4}$$

and decomposes into two components—the bounded one F_0^t and the unbounded one F_1^t.

The set of discontinuity $\Omega = \Omega(G(t))$ decomposes into two invariant components Ω_0 and Ω_1, both homeomorphic to a ball. This follows from Lemma 3.3 and properties of F^t (see [6]).

4.5. Let F_0 and F_1 denote the polyhedra F_0^t and F_1^t with $t = 5/6$ (that corresponds to Fuchsian G). Using the uniform expansions along two families of circles, orthogonal to the spheres $\sigma_{13}(t)$ and $\sigma_{14}(t)$ correspondingly, one can construct $q(t)$-quasiconformal mappings

$$f_t^0 : F_0 \to F_0^t \quad \text{and} \quad f_t^i : F_1 \to F_i^t \tag{4.6}$$

that map the sides of F_t onto the corresponding sides of F_i^t, $i = 0, 1$.

Here, $q(t)$ is continuous and

$$\lim_{t \to 5/6} q(t) = 1 .$$

Using well-known facts concerning quasiconformal mappings ([9], [13], [14]), we can extend f_t^0 and f_t^i periodically until we obtain the mappings

$$\Phi_0^t : \text{int } S_0 \to \Omega_0(t) \quad \text{and} \quad \Phi_i^t : \text{ext } S_0 \to \Omega_1(t) \tag{4.8}$$

which, being compatible with G , will be equal to a restriction of a $q(t)$-quasiconformal deformation Φ_t of the Fuchsian group G . It follows from the properties of $\mathcal{L}(G)$ and $\mathcal{L}(G(t))$ that for each t' , t'' , $t_0 < t' < t'' < t_1$, deformations $\Phi_{t'}$ and $\Phi_{t''}$ are not conjugate in the group $\mathfrak{M}_n$. Nevertheless, when $|t' - t''| < \varepsilon$ we can conjugate them with the help of a $a(\varepsilon)$-quasiconformal deformation which we can construct in the same way as above. Here we also have

$$\lim_{\varepsilon \to 0} a(\varepsilon) = 1 .$$

Thus, the mapping

$$\phi : t \to \Phi_t$$

is a homeomorphic imbedding of the interval (t_0, t_1) into the space $T(G)$.
     Theorem A is proved.

4.9. Our proof of Theorem B will be reduced to the clearing up of the structure of the groups $G(t)$ (constructed in Section 4.4) for the parameter values $t = t_0 = 5/12$ and $t = t_1 = 5(1 + \sqrt{5/6})$. We consider the case of the group $G(t_1)$. In case $t = t_0$ the situation is analogous and even slightly simpler.

     First we note that the group $G(t_1)$ is Kleinian. Its fundamental poly-hedron has two components—the bounded $F_0^{t_1}$ and the unbounded $F_1^{t_1}$. The polyhedron $F_0^{t_1}$, like all F_0^t , $t_0 < t < t_1$, is connected and simply connected. At the same time the fundamental group $\pi_1(F_1^{t_1})$ is a free group on eight generators. It becomes clear if we note that the spheres $\sigma_{13}(t_1)$ and $\sigma_{14}(t_1)$ touch the corresponding spheres $\sigma_i(t_1) = S_i$, $i = 9, \cdots, 12$, where the points of contact p_i, q_i, $9 \leq i \leq 12$ lie outside the spheres S_j, $1 \leq j \leq 8$, are the unique fixed points of some mappings of parabolic type, and form five cycles. This gives us the direct calculation. The application of Lemma 3.3 gives us the discontinuity set of the group $G(t_1)$ falling into two components $\Omega_0(t_1) \supset F_0^{t_1}$ and $\Omega_1(t_1) \supset F_1^{t_1}$ which are invariant. At the same time, using the density of the fixed points of different classes of transformations from the group $G(t_1)$ in its limit set [5] we obtain that $\mathcal{L}(G(t_1))$ is connected.

The quasiconformal equivalence of the domain $\Omega_0(t_1)$ to the ball is proved, as in Section 4.5, by constructing the quasiconformal mapping

$$\Phi^0_{t_1} : \text{int } S_0 \to \Omega_0(t_1) ,$$

compatible with the group G and the last statement of the theorem on the fundamental group $\pi_1(\Omega_1(t_1))$ directly follows from the previous one and from the density of parabolic points in $\mathcal{L}(G(t_1))$ (see [5]). It is also obvious that the group $G(t_1)$ cannot be obtained from G with the help of a quasiconformal automorphism of the space $\overline{R^n}$, since the component $\Omega_1(t_1)$ is not simply-connected. Thus, $G(t_1) \notin T(G)$, and, consequently, lies on the boundary of Teichmüller space of the group G. Theorem B is proved.

5. *Some corollaries and remarks*

REMARK 5.1. From the proofs of Theorems A and B it is clear that the simultaneous deformation of the Fuchsian group G by three parameters: by generating mappings, corresponding to the spheres $\sigma_{13}(t)$ and $\sigma_{14}(t)$, $\sigma_{11}(r)$ and $\sigma_{12}(r)$, $\sigma_9(s)$ and $\sigma_{10}(s)$, the variations corresponding to the parameters r and s, being built over t (see 4.1), is possible. In particular, if we consider the variations corresponding to $s = \frac{5}{6}$, $t > \frac{5}{6}$

and $\tau < \frac{5}{6}$, and consider the boundary group $G\left(\frac{5}{6}, t_1, \tau_0\right)$, then we obtain the following statement.

COROLLARY 5.2. *On the boundary of Teichmüller space* $T(G)$ *of the Fuchsian group* $G \subset \mathfrak{M}_3$ *from Section 4.1 there exist Kleinian groups* Γ *with the following properties:*

1) $\mathfrak{L}(G)$ *is connected;*

2) $\Omega = \Omega_0 \cup \Omega_1$ *where* Ω_0 *and* Ω_1 *are the connectivity components;*

3) $\Gamma(\Omega_i) = \Omega_i$, $i = 0, 1$, *i.e. the components* Ω_0 *and* Ω_1 *are invariant;*

4) *The fundamental groups* $\pi_1(\Omega_0)$ *and* $\pi_1(\Omega_1)$ *of the invariant components* Ω_0 *and* Ω_1 *are free infinitely generated groups.*

COROLLARY 5.3. *The Mostow rigidity theorem* [10] *doesn't take place in the case of discrete groups of isometries of the* n-*dimensional hyperbolic space* B^n, $n \geq 2$, *inducing discontinuous groups on* $\partial B^n = S^{n-1}$.

In other words, for any $n \geq 2$ there exist isomorphisms of Fuchsian groups of the second kind induced by quasiconformal homeomorphisms of $\overline{R^n}$ which cannot be continued up to the inner automorphism of the group $\mathfrak{M}_n$. We recall that the hyperbolic volume for such groups is $V(B^n/G) = \infty$.

This corollary trivially follows from Theorem A (and in case $n = 2, 3$ is known even earlier).

The above-mentioned phenomenon of "non-rigidity" takes place also in the case of discrete groups of isometries of hyperbolic space B^n, $n \geq 3$, whose limit set is ∂B^n. One can find the details in [7].

REMARK 5.4. Due to the fact that the fixed points of loxodromic elements are dense in $\mathfrak{L}(G(t))$, $t_0 < t < t_1$, we obtain that our quasiconformal automorphism Φ_t of the space $\overline{R^n}$ maps the ball B^n onto a domain $\Omega_0(t)$ which has no tangent plane at any of its boundary points (see [11]).

REMARK 5.5. The described method (after some modification) can be applied in the case when all the faces of the fundamental polyhedron

B^n/G of the Fuchsian group $G \subset \mathfrak{M}_n$ of the first kind contain limit vertices. For an infinitely-generated group a result of such type was obtained by A. V. Tetenov, a student of mine (see [6]).

INSTITUTE OF MATHEMATICS
SIBERIAN BRANCH OF THE USSR ACADEMY OF SCIENCES
NOVOSIBIRSK-90,
USSR 630090

REFERENCES

[1] Accola, R.D.M., Invariant domains for Kleinian groups, Amer. J. of Math. <u>88</u> (1966), pp. 329-336.

[2] Александров, А.Д. О заполнении пространства многогранниками, Вестник Ленинградского Гос.Унив., Сер. мат.-физ.-хим., 1954, № 2, стр. 33-43.

[3] Апанасов Б.Н. Об одном аналитическом методе в теории клейновых групп многомерного эвклидова пространства, Доклады АН СССР, <u>222</u> (1975), № I, стр. II-I4.

[4] Апанасов Б.Н. Об одном классе клейновых групп в R^n. Доклады АН СССР, <u>215</u> (1974), № 3, стр. 509-5I0.

[5] Апанасов Б.Н. О клейновых группах в пространстве, Сиб.матем. ж., <u>16</u>, 1975, № 5, стр. 891-898.

[6] Апанасов Б.Н., Тетенов А.В., О существовании нетривиальных квазиконформных деформаций клейновых групп в пространстве, Доклады АН СССР, <u>239</u> (1978), № I, стр. I4-I7.

[7] Апанасов Б.Н., К теореме жесткости Мостова, Доклады АН СССР, <u>243</u>, (1978), № 4, стр. I029-I032.

[8] Bers L., Quasiconformal mappings and Teichmüller's theorem, Analytic functions, Princeton, 1960, 89-120.

[9] Deny J., Lions T.L., Les espaces du type de Beppo Levi, Ann. Inst. Fourier, № 5(1955), pp. 305-370.

[10] Mostow G.D., Quasiconformal mappings in n-space and the rigidity of hyperbolic space forms, Publ.Math. de l'Institut· des Hautes Etudes Scientifiques, № 34, 1968.

[11]   Копылов А.П. Поведение пространственного квазиконформного
       отображения на плоских сечениях области определения. Доклады
       АН СССР, 167, (1966), № 4, стр. 743-746.

[12]   Крушкаль С.Л. Квазиконформные отображения и римановы поверх-
       ности, Наука, Новосибирск, 1975.

[13]   Решетняк Ю.Г. Локальная структура отображений с ограниченным
       искажением, Сиб.матем.ж., 10 (1969), № 6, стр. 1319-1341.

[14]   Решетняк Ю.Г. Пространственные отображения с ограниченным
       искажением, Сиб.матем.ж., 7 (1967), № 3, стр. 629-658.

[15]   Форд Л.Р. Автоморфные функции, ОНТИ НКТП СССР, М.-Л., 1936.

THE ACTION OF THE MODULAR GROUP
ON THE COMPLEX BOUNDARY[*]

Lipman Bers

To Professor S. E. Warschawski, on his 75th birthday.

Introduction

The Teichmüller space $T_{p,n}$ of Riemann surfaces of genus p, compact except for n punctures, has a canonical topology, a canonical metric and a canonical complex structure, as well as a canonical group of automorphisms, the modular group $\text{Mod}_{p,n}$. It has, however, several "natural" compactifications and hence several boundaries.

The present note, which develops the method of [6], deals with the complex boundary obtained by identifying $T_{p,n}$ with the Teichmüller space $T(G)$ of an appropriately (but not canonically) chosen Fuchsian group G. The space $T(G)$ is canonically embedded into the universal Teichmüller space $T(1)$ and $T(1)$ is canonically identified with a bounded domain in a complex Banach space B. The modular group $\text{Mod}_{p,n}$ becomes identified with $\text{Mod}(G)$ which appears as a subgroup of the universal modular group $\text{Mod}(1)$.

The main results of this note are stated and proved in §4. They include the following statements.

There is a large class of points on $\partial T(1)$ on which $\text{Mod}(1)$ acts continuously, and there is a large subgroup of $\text{Mod}(1)$ which acts continuously

[*]Work partially supported by the NSF.

33

on the whole boundary $\partial T(1)$. *If* $\dim T(G) < \infty$, *then* $\mathrm{Mod}(G)$ *acts continuously on almost all points of* $\partial T(G)$, *including all so-called totally degenerate points.* (The latter result uses essentially recent theorems by Sullivan [13] and by Thurston [not published].)

The essential work is done in §§2, 3; some results there may be of independent interest.

§1. *Background and notations*

We begin by introducing some (mostly standard) notations and by recalling some known facts (see [2], [5], [10] and the references given there). The main properties of quasiconformal mappings are assumed (cf. [2], [11]).

We denote by Q the group of all quasiconformal self-mappings of the upper half-plane U; for a $w \in Q$, $[w] = w|\hat{R}$ denotes the continuous extension of w to $\hat{R} = R \cup \{\infty\}$. A $w \in Q$ is called normalized if it (or rather $w|\hat{R}$) leaves $0, 1, \infty$ fixed; the normalized $w \in Q$ form the subgroup Q_n which contains the subgroup Q_0 of those w which leave every $x \in R$ fixed. Every $w \in Q$ can be written uniquely as $w = a \circ \hat{w}$ where $\hat{w} \in Q_n$ and a is a real Möbius transformation, i.e. an element of the group Q_c of conformal self-mappings of U. Note that $Q_c = \mathrm{PSL}(2,R)$.

By a Fuchsian group G we mean, in this paper, a discrete subgroup of Q_c. In particular, $1 = \{\mathrm{id}\}$ is the trivial Fuchsian group. The set $Q(G)$ consists of those $w \in Q$ for which wGw^{-1} is again a Fuchsian group; we write $Q_n(G) = Q(G) \cap Q_n$.

Let $L_\infty(U)_1$ denote the open unit ball in the (complex) Banach space $L_\infty(U)$, and let $L_\infty(U,G)_1$ consist of those $\mu \in L_\infty(U)_1$ for which

$$\mu(g(z)) \overline{g'(z)}/g'(z) = \mu(z), \quad g \in G.$$

Every $w \in Q$ has a *Beltrami coefficient* $\mu \in L_\infty(U)_1$ such that

$$w_{\overline{z}} = \mu w_z.$$

If $w \in Q_n$, it is uniquely determined by its Beltrami coefficient μ; in this case we write

$$w = w_\mu .$$

A $w \epsilon Q$ belongs to $Q(G)$ if and only if its Beltrami coefficient is a Beltrami coefficient for the group G, i.e. belongs to $L_\infty(U, G)_1$.

For $\mu \epsilon L_\infty(U)_1$ we denote by w^μ the unique quasiconformal homeomorphism of C onto itself which leaves $0, 1, \infty$ fixed, has in U the Beltrami coefficient μ, and is conformal in the lower half-plane L. The *Schwarzian derivative* of $w^\mu | L$ will be denoted by ϕ^μ.

The *Teichmüller space* of the Fuchsian group G is the set of all $[w]$ with $w \epsilon Q_n(G)$, or, which is the same, of all $[w_\mu]$, $\mu \epsilon L_\infty(U, G)_1$. The space $T(G)$ is a complete metric space under the Teichmüller distance

$$< [w_\mu], [w_\nu] >_G = \frac{1}{2} \inf \log K(\omega)$$

where

$$\omega \epsilon Q(w_\nu G w_\nu^{-1}) \cap [w_\mu \circ w_\nu^{-1}]$$

and $K(\omega)$ denotes the dilatation of ω. The *universal Teichmüller space* $T(1)$ is a group: $T(1) = Q_n/Q_0$, but it is not a topological group. For two Fuchsian groups $G_1 \subset G_2$ we have $T(G_2) \subset T(G_1)$ and the injection is bicontinuous. In particular, $T(G) \subset T(1)$ for all G.

The group Q acts on Q_n by the rule

$$w \mapsto \omega_*(w) = \alpha \circ w \circ \omega^{-1}$$

where $w \epsilon Q_n$, $\omega \epsilon Q$ and $\alpha \epsilon Q_c$ is chosen so that $\alpha \circ w \circ \omega^{-1} \epsilon Q_n$. It is clear that $[\omega_*(w)]$ depends only on $[\omega]$ and on $[w]$, so that in effect we have the group $Q/Q_0 = \mathrm{Mod}(1)$ acting on $T(1)$; $\mathrm{Mod}(1)$ is called the *universal modular group*. Every element of $\mathrm{Mod}(1)$ is an isometry in the Teichmüller metric.

The subgroup $\mathrm{Mod}_t \subset \mathrm{Mod}(1)$ (of translations) is induced by elements $\omega \epsilon Q_n$; it is the group of (*right*) *translations* in $T(1)$. The subgroup

$\text{Mod}_r \subset \text{Mod}(1)$ of *rotations* is induced by elements of $Q_c = \text{PSL}(2, \mathbf{R})$ and is isomorphic to $\text{PSL}(2, \mathbf{R})$. The group $\text{Mod}(1)$ is the semi-direct product of Mod_t and Mod_r.

Let $\omega \in Q(G)$. Then ω_* maps $T(G)$ onto $T(\omega G \omega^{-1})$ and takes the metric $< , >_G$ into $< , >_{\omega G \omega^{-1}}$. The map $\omega_* : T(G) \to T(\omega G \omega^{-1})$ is called an *allowable map*. If ω normalizes G, then ω_* is a self-mapping of $T(G)$. Such ω_* form the *modular group* $\text{Mod}(G)$ of G. (In [4] the definition of $\text{Mod}(G)$ is somewhat different, for a few groups G; this difference is of no importance for what follows.)

Now let B denote the complex Banach space of holomorphic functions $\phi(z)$, $z \in L$, with

$$\|\phi\| = \sup |y^2 \phi(z)| < +\infty .$$

For every Fuchsian group G, let $B(G)$ be the closed linear subspace of B consisting of those $\phi \in B$ for which

$$\phi(g(z)) g'(z)^2 = \phi(z), \quad g \in G .$$

The map

$$(1.1) \qquad\qquad\qquad [w_\mu] \mapsto \phi^\mu$$

is a homeomorphism of $T(1)$ onto a bounded domain in $B = B(1)$; the restriction of this map to each $T(G)$ is a homeomorphism onto a bounded domain in $B(G)$. By abuse of language we shall identify each $T(G)$ with its image under the mapping (1.1). Thus each Teichmüller space $T(G)$ has a complex structure and also a boundary

$$\partial T(G) \subset B(G) ,$$

and all allowable maps are holomorphic.

With every $\phi \in B$ we shall associate a meromorphic function $W_\phi(z)$, $z \in L$, defined as follows:

$$W_\phi(z) = \eta_1(z)/\eta_2(z)$$

where η_1 and η_2 are solution of the ordinary differential equation

$$2\eta''(z) + \phi(z)\,\eta(z) = 0$$

subject to the initial conditions

$$\eta_1(-i) = \eta_2'(-i) = 1 , \quad \eta_1'(-i) = \eta_2(-i) = 0 .$$

An equivalent definition reads: $W = W_\phi(z)$ has ϕ as its Schwarzian
derivative:

$$W(z) = (z+i)^{-1} + o(1), \quad z \to -i .$$

The dependence of W_ϕ on ϕ is holomorphic.

We denote by $\mathcal{S}$ the set of those $\phi \in B$ for which W_ϕ is schlicht
(i.e., injective); by the Kraus-Nehari inequality $\mathcal{S}$ is bounded. For
$\phi \in \mathcal{S}$ it will be convenient to consider instead of W_ϕ the schlicht
function

$$\Omega_\phi(z) = \frac{W_\phi(-2i) - W_\phi(-3i)}{W_\phi(z) - W_\phi(-3i)}$$

which has the Schwarzian derivative ϕ and takes $(-i, -2i, -3i)$ into
$(0, 1, \infty)$. The dependence of Ω_ϕ on $\phi \in \mathcal{S}$ is *holomorphic*.

The Teichmüller space $T(1)$ consists of those $\phi \in \mathcal{S}$ for which W_ϕ
(and hence also Ω_ϕ) has a quasiconformal extension from L to $\hat{C}$.
This is equivalent to the requirement that $\Omega_\phi(L)$ be bounded by a quasi-
circle, an image of a circle under a quasiconformal mapping. If $\phi \in T(1)$,
then $\phi = \phi^\mu$ for some $\mu \in L_\infty(U)_1$ and

$$\Omega_\phi(z) = \frac{w^\mu(z) - w^\mu(-i)}{w^\mu(z) - w^\mu(-3i)} \cdot \frac{w^\mu(-2i) - w^\mu(-3i)}{w^\mu(-2i) - w^\mu(-i)} .$$

If $\phi \in T(G)$, then the group

$$G^\phi = \Omega_\phi\, G\, \Omega_\phi^{-1}$$

is a quasi-Fuchsian group with fixed curve $\partial\Omega_\phi(L)$.

It is known, and easy to verify, that $T(1) \cup \partial T(1) \subset \mathcal{S}$. Gehring [8], [9] proved that $T(1)$ is the interior of $\mathcal{S}$, and that the complement of $T(1) \cup \partial T(1)$ in $\mathcal{S}$ is non-empty.

The Banach space $B(G)$ and the Teichmüller space $T(G)$ are finite dimensional if and only if G is finitely generated and of the first kind, i.e., if and only if the Poincaré area of U/G is finite, i.e., if and only if $PSL(2, R)/G$ has finite volume. If so, and if G is torsion free, the quotient U/G is a Riemann surface of some finite genus p, compact except for n punctures, and $3p - 3 + n \geq 0$. In this case $T(G)$ can be identified with the Teichmüller space $T_{p,n}$ and $\text{Mod}(G)$ with the Teichmüller modular group $\text{Mod}_{p,n}$ operating on $T_{p,n}$.

§2. *Tame convergence*

A sequence $\{\phi_j\} \subset B$ is said to converge *weakly* to a $\phi \in B$ if $\|\phi_j\| = 0(1)$ and $\lim \phi_j(z) = \phi(z)$ for every $z \in L$. Every bounded sequence in B contains a weakly convergent subsequence. If $\lim \phi_j = \phi$ weakly, then Ω_{ϕ_j} converges to Ω_ϕ normally (uniformly on compact subsets of L). The set $\mathcal{S}$ is closed under weak convergence and $T(1)$ is dense in $\mathcal{S}$ with respect to weak convergence. The proofs of the above statements are obvious.

A sequence $\{\phi_j\}$ in $\mathcal{S}$ will be said to converge *tamely* to a $\phi \in \mathcal{S}$, if $\lim \phi_j = \phi$ weakly and if for almost every $z \notin \Omega_\phi(L)$ there is a J such that $z \notin \Omega_{\phi_j}(L)$ for $j > J$. Note that a convergent (in norm) sequence in $\mathcal{S}$ is always weakly convergent, but may fail to converge tamely.

PROPOSITION 2.1. *If $\phi \in \mathcal{S}$ and*

$$(2.1) \qquad\qquad \text{mes } \partial\Omega_\phi(L) = 0 \,,$$

then there is a sequence $\{\phi_j\} \subset T(1)$ such that $\phi = \lim \phi_j$ tamely.

(Here and hereafter ∂ denotes the set-theoretical boundary.)

Proof. Let Δ_n, $n = 1, 2, \cdots$, denote the disc

$$|z + (n+4)i| < n + 4 - (n+4)^{-1}$$

and $f_n(z)$ the function which maps L conformally onto Δ_n and satisfies the conditions

$$f_n(-3i) = -3i, \qquad f(-\bar{z}) = -\overline{f(z)} .$$

Clearly, $\lim f_n(z) = z$ normally. Set

$$V_j(z) = \frac{\Omega_\phi \circ f_j(z) - \Omega_\phi \circ f_j(-i)}{\Omega_\phi \circ f_j(-2i) - \Omega_\phi \circ f_j(-i)} .$$

Then V_j is meromorphic and schlicht in L and takes $(-i, -2i, -3i)$ into $(0, 1, \infty)$. Hence $V_j = \Omega_{\phi_j}$, where $\phi_j \in \mathcal{S}$ is the Schwarzian derivative of V_j. Since V_j converges to Ω_ϕ normally, ϕ_j converges to ϕ weakly. Since $\partial \Omega_{\phi_j}(L)$ is obtained from the real analytic Jordan curve $\Omega_\phi(\Delta_j)$ by a Möbius transformation, Ω_{ϕ_j} admits a quasiconformal extension to $\hat{C}$. Hence $\phi_j \in T(1)$.

Now let z_0 be a point exterior to $\Omega_\phi(L)$. If j is large enough, the Möbius transformation

$$\zeta \mapsto H_j(\zeta) = \frac{\zeta - \Omega_\phi \circ f_j(-i)}{\Omega_\phi \circ f_j(-2i) - \Omega_\phi \circ f_j(-i)} .$$

is arbitrarily close to the identity, so that

$$z_0 \notin H_j[\Omega_\phi(L) \cup \partial \Omega_\phi(L)]$$

and, a fortiori,

$$z_0 \notin V_j(L) = H_j \circ \Omega_\phi(\Delta_j) .$$

Since the boundary of $\Omega_\phi(L)$ is assumed to have measure zero, the convergence of ϕ_j to ϕ is tame.

QUERY: *Is the conclusion of Proposition 1 valid without the assumption (2.1)?*

The result just proved does not, of course, imply that a $\phi \in \partial T(G)$ is a tame limit of elements in $T(G)$, even if mes $\partial W_\phi(L) = 0$. A special result in this direction is, however, valid.

Let G be finitely generated and of the first kind. Then the boundary points of $T(G)$ are classified as follows (cf. [3], [12]). If $\phi \in \partial T(G)$ and $W_\phi(L)$ is dense in $\hat{C}$, ϕ and the group G^ϕ are called *totally degenerate*. Almost all boundary points are of this nature. If ϕ is not totally degenerate, the region of discontinuity $R(G^\phi)$ of G^ϕ is disconnected and $R(G^\phi)/G^\phi$ is a disjoint union of Riemann surfaces. The group G^ϕ and the boundary point ϕ are called either *regular* or *partially degenerate* according to whether the Poincaré area of $R(G^\phi)/G^\phi$ is equal on less than twice the Poincaré area of U/G. If dim $T(G) > 0$, $\partial T(G)$ contains regular points and if dim $T(G) > 2$, $\partial T(G)$ contains partially degenerate points.

PROPOSITION 2.2. *Let G be finitely generated and of the first kind. Every regular boundary point in $\partial T(G)$ is the tame limit of a sequence in $T(G)$.*

The proof is a modification of an argument due to Abikoff (see [1], pp. 224-231). The details are somewhat lengthy and will be presented elsewhere.

§3. *Formal translations*

The group Q_n/Q_0 operates on $T(1)$ as the group Mod_t of (right) translations. We will associate with every $w \in Q_n$, which has a continuous Beltrami coefficient in U, a map $\mathcal{S} \to \mathcal{S}$, called a *formal translation*, which restricts on $T(1) \subset \mathcal{S}$ to the element of Mod_t induced by w.

Throughout this section we consider a fixed *continuous* $\theta \in L_\infty(U)_1$ and set

$$w = w_\theta.$$

We extend the definitions of w and θ to L by setting

$$w(\overline{z}) = \overline{w(z)}, \qquad \theta(\overline{z}) = \overline{\theta(z)} .$$

The letters ϕ and ϕ_j will denote elements of S.

For every θ and ϕ set

$$(3.1) \qquad \delta_{\theta,\phi}(z) = \theta(\Omega_\phi^{-1}(z)) \frac{\overline{d\Omega_\phi^{-1}(z)/dz}}{d\Omega_\phi^{-1}(z)/dz} \quad \text{if } z \in \Omega_\phi(L) ,$$

$$(3.2) \qquad \delta_{\theta,\phi}(z) = 0 \quad \text{if } z \notin \Omega_\phi(L) .$$

Then $\delta_{\theta,\phi} \in L_\infty(\mathbb{C})$ and

$$\|\delta_{\theta,\phi}\|_\infty = \|\theta\|_\infty < 1$$

(where $\| \ \|_\infty$ denotes the L_∞ norm). Hence there is a unique quasi-conformal self-mapping $A_{\theta,\phi}$ of $\hat{\mathbb{C}}$ which satisfies the Beltrami equation

$$\frac{\partial A_{\theta,\phi}(z)}{\partial \overline{z}} = \delta_{\theta,\phi}(z) \frac{\partial A_{\theta,\phi}(z)}{\partial z}$$

and the normalization conditions

$$A_{\theta,\phi} \circ \Omega_\phi \circ w_\theta^{-1}(-i) = 0, \quad A_{\theta,\phi} \circ \Omega_\phi \circ w_\theta^{-1}(-2i) = 1 ,$$

$$A_{\theta,\phi} \circ \Omega_\phi \circ w_\theta^{-1}(-3i) = \infty .$$

The function $V(z) = A_{\theta,\phi} \circ \Omega_\phi \circ w_\theta^{-1}(z)$, $z \in L$, is injective by construction and takes $(-i, -2i, -3i)$ into $(0, 1, \infty)$. Also, $V(z)$ is meromorphic since $V \circ w_\theta = A_{\theta,\phi} \circ \Omega_\phi$ and one computes from (3.1) that in L the Beltrami coefficient of $A_{\theta,\phi} \circ \Omega_\phi$ is θ. Thus $V = \Omega_\psi$ where ψ is the Schwarzian derivative of V. This $\psi \in S$; we write

$$\psi = \theta_*(\phi)$$

and note that ψ is defined by the relation

$$(3.3) \qquad \Omega_\psi \circ w_\theta = A_{\theta,\phi} \circ \Omega_\phi \quad \text{in } L.$$

We must now verify that the restriction of θ_* to $T(1)$ is the element $(w_\theta)_*$ of Mod_t.

The following statement is a slight modification of a result by Gardiner [7, p. 475].

PROPOSITION 3.1. *Let* $\mu, \nu \in L_\infty(U)_1$ *be such that*

$$w_\nu = w_\mu \circ w_\theta^{-1}$$

and set

$$\phi^\mu = \phi, \quad \phi^\nu = \psi.$$

Then $\psi = \theta_*(\phi)$, *i.e.,* (3.3) *holds.*

Note that the hypotheses of the Proposition mean that $\psi = (w_\theta)_*(\phi)$.

Proof. Since ϕ and ψ are the Schwarzian derivatives of $w^\mu|L$ and $w^\nu|L$, respectively, there are Möbius transformations a_μ, a_ν such that

$$\Omega_\phi = a_\mu \circ w^\mu, \quad \Omega_\psi = a_\nu \circ w^\nu$$

in L. Using these relations we extend the definitions of Ω_ϕ and Ω_ψ to all of $\hat{C}$. Then the Beltrami coefficients of $\Omega_\phi|U$ and $\Omega_\psi|U$ are μ and ν respectively. Let

$$q_\mu : \Omega_\phi(U) \to U, \quad q_\nu : \Omega_\psi(U) \to U$$

be conformal bijections, chosen so that the maps $q_\mu \circ \Omega_\phi$ and $q_\nu \circ \Omega_\psi$ keep $0, 1, \infty$ fixed. Then

$$q_\mu \circ \Omega_\phi = w_\mu, \quad q_\nu \circ \Omega_\psi = w_\nu \quad \text{in } U.$$

We define a homeomorphism a of $\hat{C}$ onto itself by setting

$$(3.4) \qquad a|\Omega_\phi(U \cup \hat{R}) = q_\nu^{-1} \circ q_\mu \,,$$

$$(3.5) \qquad a|\Omega_\phi(L \cup \hat{R}) = \Omega_\psi \circ w_\theta \circ \Omega_\phi^{-1} \,.$$

The definition is legitimate since on $\Omega_\phi(\hat{R})$ we have

$$q_\nu^{-1} \circ q_\mu = \Omega_\psi \circ w_\nu^{-1} \circ w_\mu \circ \Omega_\phi^{-1} = \Omega_\psi (w_\mu \circ w_\theta^{-1})^{-1} \circ w_\mu \circ \Omega_\phi^{-1} = \Omega_\psi \circ w_\theta \circ \Omega_\phi^{-1} \,.$$

The map $z \mapsto a(z)$ is quasiconformal off the quasicircle $\Omega_\phi(\hat{R})$, hence everywhere. A direct calculation shows that the relation

$$\frac{\partial a}{\partial \bar{z}} = \delta_{\theta,\phi} \, \frac{\partial a}{\partial z}$$

holds a.e. Also

$$a \circ \Omega_\phi \circ w_\theta^{-1}|L = \Omega_\psi|L \,,$$

so that a takes the points $\Omega_\phi \circ w_\theta^{-1}(-i)$, $\Omega_\phi \circ w_\theta^{-1}(-2i)$ and $\Omega_\phi \circ w_\theta^{-1}(-3i)$ into $0, 1, \infty$. Hence $a = A_{\theta,\phi}$, and relation (3.5) shows that $\Omega_\psi = A_{\theta,\phi} \circ \Omega_\phi \circ w_\theta^{-1}$ as asserted.

PROPOSITION 3.2. *Set* $\psi_j = \theta_*(\phi_j)$, $j = 1, 2, \cdots$, *and assume that*

$$\lim \phi_j = \phi \ \ weakly, \ \ \lim \psi_j = \psi \ \ weakly \,.$$

Then there is a quasiconformal self-mapping F *of* $\hat{C}$ *such that the Beltrami coefficient* $\hat{\delta} = F_{\bar{z}}/F_z$ *of* F *satisfies*

$$(3.6) \qquad \|\hat{\delta}\|_\infty = \|\theta\|_\infty \,,$$

$$(3.7) \qquad \hat{\delta}|\Omega_\phi(L) = \delta_{\theta,\phi}|\Omega_\phi(L) \,,$$

that

$$(3.8) \qquad F \circ \Omega_\phi \circ w_\theta^{-1}(-i) = 0 \,, \ \ F \circ \Omega_\phi \circ w_\theta^{-1}(-2i) = 1 \,,$$

$$F \circ \Omega_\phi \circ w_\theta^{-1}(-3i) = \infty$$

and

$$(3.9) \qquad \Omega_\psi = F \circ \Omega_\phi \circ w_\theta^{-1} \quad in \ \ L.$$

Proof. We have that

$$\Omega_{\psi_j} \circ w_\theta = A_{\theta,\phi_j} \circ \Omega_{\phi_j} \quad in \ \ L.$$

Also

$$\|\delta_{\theta,\phi_j}\|_\infty = \|\theta\|_\infty$$

and $A_{\theta,\phi_j} \circ \Omega_{\phi_j} \circ w_\theta^{-1}$ takes $(-i, -2i, -3i)$ onto $(0, 1, \infty)$. Selecting if need be a subsequence we may assume that $\{A_{\theta,\phi_j}\}$ converges, uniformly in the spherical metric, to a quasiconformal self-mapping F of $\mathbf{C}$ with a Beltrami coefficient $\hat{\delta}$ satisfying

$$(3.10) \qquad \|\hat{\delta}\|_\infty \leq \|\theta\|_\infty.$$

Since $\lim \Omega_{\phi_j} = \Omega_\phi$ and $\lim \Omega_{\psi_j} = \Omega_\psi$ normally in L, this F satisfies (3.8) and (3.9). Finally, noting the definition (3.1) of $\delta_{\theta,\phi}$ we conclude that

$$(3.11) \qquad \lim_{j \to \infty} \delta_{\theta,\phi_j} = \delta_{\theta,\phi} \ \text{a.e. in } \ \Omega_\phi(L).$$

(Here we use the assumption that θ is continuous in U and hence also in L. In general, θ will be discontinuous on R.)

Now (3.11) implies (3.7) and since

$$\|\delta_{\theta,\phi}|\Omega_\phi(L)\|_\infty = \|\theta\|_\infty,$$

(3.10) and (3.7) imply (3.6).

PROPOSITION 3.3. *Assume that*

$$(3.12) \qquad \lim_{j \to \infty} \phi_j = \phi \ \ tamely.$$

Then

(3.13)
$$\lim_{j \to \infty} \theta_*(\phi_j) = \theta_*(\phi) \quad weakly.$$

Proof. It will suffice to show that (3.13) holds under the additional hypothesis that the sequence $\{\psi_j\} = \{\theta_*(\phi_j)\}$ converges weakly to some ψ. We are now in the situation described by Proposition 3.2. Hypothesis (3.12) implies that for almost every $z \notin \Omega_\phi(L)$ there is a J such that

$$\delta_{\theta,\phi_j}(z) = 0 \quad \text{for} \quad j > J.$$

Together with (3.11) this implies that $\lim \delta_{\theta,\phi_j} = \delta_{\theta,\phi}$ a.e. Hence the F in the statement of Proposition 3.2 coincides with $A_{\theta,\phi}$, so that $\psi = \theta_*(\phi)$.

PROPOSITION 3.4. *Assume that*

(3.14)
$$\text{mes } \partial\Omega_\phi(L) = 0 .$$

Then $\theta_*(\phi)$ depends only on $w_\theta|R$ (and, of course, on ϕ).

Proof. For $\phi \in T(1)$ the conclusion holds, independently of hypothesis (3.14), in view of Proposition 3.1. Therefore, under this hypothesis, the conclusion follows from Propositions 2.1 and 3.3.

QUERY: *Is the conclusion of Proposition 3.4 valid for all $\phi \in \partial T(1)$? For all $\phi \in \delta$?*

PROPOSITION 3.5. *If*

(3.15)
$$\text{mes } [C - \Omega_\phi(L)] = 0 ,$$

then θ_* is continuous at ϕ with respect to weak convergence (i.e., if $\lim \phi_j = \phi$ weakly, then $\lim \theta_*(\phi_j) = \theta_*(\phi)$ weakly).

Proof. Note that, by (3.15), weak convergence to ϕ is tantamount to tame convergence and use Proposition 3.3.

PROPOSITION 3.6. *If θ has compact support in* U *(and hence also in* U $\cup$ L *),* θ_* *is continuous with respect to weak convergence.*

Proof. We may assume that we are in the situation described in Proposition 3.2. We note that Ω_{ϕ_j} converges normally to Ω_ϕ in $\mathcal{L}$. There are open discs Δ_1 and Δ_2 such that

$$\Delta_1 \cup \partial\Delta_1 \subset \Delta_2 \,, \quad \Delta_2 \cup \partial\Delta_2 \subset L \,, \quad \theta|L - \Delta_1 = 0 \,.$$

If j is large enough, then $\Omega_{\phi_j}(\partial\Delta_2)$ is so close to $\Omega_\phi(\partial\Delta_2)$ that $\Omega_\phi(\Delta_1) \subset \Omega_{\phi_j}(L)$. For such j,

$$\delta_{\theta,\phi_j}|\hat{\mathbf{C}} - \Omega_\phi(\Delta_1) = 0 \,,$$

so that $\delta_{\theta,\phi_j}(z) = 0$ for almost all $z \notin \Omega_\phi(L)$. The conclusion now follows as in the proof of Proposition 3.3.

PROPOSITION 3.7. *If* $\Omega_\phi(L)$ *is the union of finitely many quasidiscs (domains bounded by quasicircles),* θ_* *is continuous at* ϕ *with respect to weak convergence.*

The proof is an obvious modification of the argument in [6].

Let G be a Fuchsian group. We set $\mathcal{S}(G) = \mathcal{S} \cap B(G)$.

PROPOSITION 3.8. *Assume, under the hypotheses of Proposition 3.2, that* w_θ *and all* ϕ_j *belong to* T(G) *(so that* $\phi \in \mathcal{S}(G)$*, all* ψ_j *belong to* $T(w_\theta G w_\theta^{-1})$ *and* $\psi \in \mathcal{S}(w_\theta G w_\theta^{-1})$*). Then the map* F *in the conclusion of Proposition 3.2 satisfies*

$$(3.16) \qquad F \circ \Omega_\phi \circ g \circ \Omega_\phi^{-1} \circ F^{-1} = \Omega_\psi \circ w_\theta \circ g \circ w_\theta^{-1} \circ \Omega_\psi^{-1}$$

so that

$$(3.17) \qquad FG^\phi F^{-1} = (w_\theta G w_\theta^{-1})^\psi \,.$$

Proof. Relation (3.16) follows from (3.9) and implies (3.17).

In proving the next two propositions we make use of Sullivan's result (cf. [13]) which states that the limit set of a finitely generated Fuchsian group supports no Beltrami differentials for that group. The proof would be somewhat shortened if we used instead a yet unpublished assertion by Thurston that the limit sets of finitely generated Kleinian groups, which are limits of quasi-Fuchsian groups, have measure 0.

PROPOSITION 3.9. *Let* G *be a finitely generated Fuchsian group of the first kind, and let* $w_\theta \in T(G)$. *Let* ϕ *be a totally degenerate boundary point of* $T(G)$ *and* $\{\phi_j\} \subset T(G)$ *a sequence such that*

$$(3.18) \qquad \lim \phi_j = \phi .$$

Then

$$\lim \theta_*(\phi_j) = \theta_*(\phi)$$

and $\theta_*(\phi) \in \partial T(w_\theta \, Gw_\theta^{-1})$.

Proof. Since $\dim B(G) < \infty$ and also $\dim B(w_\theta \, Gw_\theta^{-1}) < \infty$, there is no difference between weak convergence or convergence in norm in these spaces. Also, since each $\psi_j = \theta_*(\phi_j) \in T(w_\theta \, Gw_\theta^{-1})$, statement (3.18) implies that $\theta_*(\phi) \in \partial T(w_\theta \, Gw_\theta^{-1})$.

The map F in the conclusions of Propositions 3.2 and 3.8 satisfies (3.17) and its Beltrami coefficient, is therefore a Beltrami coefficient for G^ϕ, i.e.

$$\hat\delta(g(z)) \, g'(z)/\overline{g'(z)} = \hat\delta(z) \quad \text{for} \quad g \in G^\phi .$$

Hence, denoting by $\Lambda(G^\phi)$ the limit set of G^ϕ, we have

$$(3.19) \qquad \hat\delta | \Lambda(G^\phi) = 0$$

by Sullivan's theorem. Since G^ϕ is totally degenerate, with region of discontinuity $\Omega_\phi(L)$, (3.19) and (3.7) imply that $\hat\delta = \delta_{\theta,\phi}$ so that $F = A_{\theta,\phi}$ and $\psi = \theta_*(\phi)$.

PROPOSITION 3.10. *The conclusion of Proposition 3.9 remains valid also if ϕ is regular or partially degenerated, provided that the stabilizer in G^ϕ of any component of $R(G^\phi)$, distinct from the invariant component $\Omega_\phi(L)$ is a triangle group.*

Proof. Let $D_1, D_2, \cdots,$ be all noninvariant components of $R(G^\phi)$, and let F be the map in Propositions 3.2 and 3.8. If Γ_n is the stabilizer of D_n in G^ϕ, then $F\Gamma_n F^{-1}$ is the stabilizer of the component $F(D_n)$ in the group (3.17). Since Γ_n is a triangle group, so is $F\Gamma_n F^{-1}$, both D_n and $F(D_n)$ are circular discs and $F|\partial D_n$ is a Möbius transformation which we call a_n. Let F_0 be the map defined by

$$F_0(z) = F(z) \quad \text{if} \quad z \in \hat{C} - (D_1 \cup D_2 \cup \cdots),$$
$$F_0(z) = a_n(z) \quad \text{if} \quad z \in D_n.$$

Then F_0 is a quasiconformal homeomorphism of C onto itself,

$$F_0 G^\phi F_0^{-1} = F G^\phi F^{-1}$$

and

$$(3.20) \qquad \Omega_\psi = F_0 \circ \Omega_\phi \circ w_\theta^{-1} \quad \text{in} \quad \Omega_\phi(L) .$$

Let $\hat{\delta}_0$ be the Beltrami coefficient of F_0. Then

$$\hat{\delta}_0 | \Lambda(G^\phi) = 0$$

by Sullivan's theorem, as in the preceding proof,

$$\hat{\delta}_0 | D_1 \cup D_2 \cup \cdots = 0$$

since each a_n is conformal, and, as before

$$\hat{\delta}_0 | \Omega_\phi(L) = \delta_{\theta,\phi} | \Omega_\phi(L) .$$

Also, $F_0 \circ \Omega_\phi \circ w_\theta^{-1}$ takes $(-i, -2i, -3i)$ into $(0, 1, \infty)$. We conclude that $F_0 = A_{\theta, \phi}$, so that by (3.20), $\psi = \theta_*(\phi)$.

§4. *Conclusions*

In §2 we tried to extend the definition of w_*, for some $w \in Q_n$, from $T(1)$ to $\mathcal{S}$. For a $w \in Q_c = \mathrm{PSL}(2, \mathbf{R})$ the corresponding problem is trivial, since in this case w_* is the restriction to $T(1)$ of a linear isometry in B. More precisely, the following is known and very easy to prove (see, for instance [6], p. 28).

PROPOSITION 4.1. *If* $\beta \in Q_c$, *the rotation* $(\beta^{-1})_*$ *acts on a* $\phi \in T(1) \subset B$ *by*
$$\phi \mapsto (\phi \circ \beta)(\beta')^2 .$$

We now associate with every $w \in Q$ whose Beltrami coefficient $w_{\bar{z}}/w_z$ is continuous in U a map $w_\# : \mathcal{S} \to \mathcal{S}$: write

$$w = \beta^{-1} \circ w_\theta, \quad \beta \in \mathrm{PSL}(2, \mathbf{R})$$

(this decomposition is unique) and set, for $\phi \in \mathcal{S}$,

$$w_\#(\phi) = (\psi \circ \beta)(\beta')^2 \quad \text{where} \quad \psi = \theta_*(\phi) .$$

We could now rewrite all the propositions in §3 as statements about $w_\#$. For instance, by Proposition 3.1, we have that

$$w_\# | T(1) = w_* .$$

Instead we state explicitly only theorems about boundary actions of modular groups.

In deriving these results we make use of the fact that for every $w \in Q(G)$ there is a $\omega \in Q(G)$ such that $\omega | \mathbf{R} = w | \mathbf{R}$ and ω has in U a continuous Beltrami coefficient. Hence every element $m \in \mathrm{Mod}(G)$ may be written as

$$m = (\beta \circ w_\theta)_*$$

where $\beta \epsilon Q$, and $\theta \epsilon L_\infty(U, G)_1$ is continuous in U. Thus we may use freely the propositions of §3.

THEOREM 1. *Assume that $\{\phi_j\} \subset T(1)$ and $\lim \phi_j = \phi \epsilon \partial T(1)$ tamely. For every $m \epsilon \mathrm{Mod}(1)$, the sequence $\{m(\phi_j)\}$ converges weakly (and hence to a limit depending only on ϕ and m).*

This follows from Proposition 3.3.

THEOREM 2. *Assume that $\phi \epsilon \partial T(1)$ and either the complement of $\Omega_\phi(L)$ has measure 0, or $\Omega_\phi(L)$ is a finite union of quasicircles. If the sequence $\{\phi_j\} \subset T(1)$ converges to $\phi \epsilon \partial T(1)$ weakly, then, for every $m \epsilon \mathrm{Mod}(1)$, the sequence $\{m(\phi_j)\}$ converges weakly (and hence to a limit depending only on m and ϕ).*

Proof. Use Proposition 3.5 and 3.7.

THEOREM 3. *If $m \epsilon \mathrm{Mod}(1)$ is induced by a $w \epsilon Q$ such that*

(4.1) $$w|\hat{R} \ \textit{is real analytic},$$

and $\phi \epsilon \partial T(1)$, then, for every sequence $\{\phi_j\} \subset T(1)$ converging weakly to ϕ, the sequence $\{m(\phi_j)\}$ converges weakly.

Proof. Condition (4.1) means that there exists a real analytic homeomorphism h of the unit circle onto itself such that

$$\frac{w(x) - i}{w(x) + i} = h\left(\frac{x - i}{x + i}\right), \qquad x \epsilon \hat{R} .$$

This h is the restriction to the unit circle of an injective holomorphic mapping of a neighborhood of the unit circle, which we call again h. There exists a diffeomorphism $\hat{h}$ of $|z| \leq 1$ onto itself such that, for some ε, $0 < \varepsilon < 1$, $\hat{h}(z) = h(z)$ for $1 - \varepsilon \leq |z| \leq 1$. Define $\hat{w}$ by the relation

$$\frac{\hat{w}(z) - i}{\hat{w}(z) + i} = \hat{h}\left(\frac{z - i}{z + i}\right), \qquad z \in U .$$

Then $\hat{w} \in Q$ and

$$\hat{w} = \beta^{-1} \circ w_\theta$$

where $\beta \in Q_c$, θ is continuous and has compact support in U, and $\hat{w}|\hat{R} = w|\hat{R}$, so that $\hat{w}$ and w induce the same element $m \in \mathrm{Mod}(1)$. Now apply Proposition 3.6.

Let G be a Fuchsian group with $\dim B(G) < \infty$. A point $\phi \in \partial T(G)$ will be said to have no moduli if it is either totally degenerate or satisfies the condition in Proposition 3.10.

THEOREM 4. *If $\dim T(G) < \infty$, then every $m \in \mathrm{Mod}(G)$ has a limit at every point of $\partial T(G)$ which has no moduli, and takes it into a similar point.*

This follows from Propositions 3.9 and 3.10.

COROLLARY. *If $\dim T(G) = 1$, then $\mathrm{Mod}(G)$ acts continuously on $T(G) \cup \partial T(G)$..*

Proof. Under the hypothesis no boundary point of $T(G)$ has moduli. Note that now $T(G) \subset C$, but (at this writing) it is not known whether $T(G)$ is a Jordan domain.

THEOREM 5. *If $\dim T(G) < \infty$, then every regular boundary point in $\partial T(G)$ is the limit of a tamely convergent sequence in $T(G)$.*

This follows from Proposition 2.2.

Theorem 4, its Corollary, and Theorem 5 can be easily strengthened to corresponding statements about allowable mappings. We leave this to the reader.

REMARK. During the Stony Brook conference, Jørgensen, Kerckhoff, and
Thurston told me that they expect to prove, by geometric methods, that
Mod(G) cannot be continuous at all regular boundary points.

DEPARTMENT OF MATHEMATICS
COLUMBIA UNIVERSITY
NEW YORK, NEW YORK

REFERENCES

[1]   W. Abikoff, "On boundaries of Teichmüller spaces and on Kleinian
groups III," Acta Math., *134*(1975), pp. 211-237.

[2]   L. V. Ahlfors, *Lectures on quasiconformal mappings*, Van Nostrand,
Princeton, (1966).

[3]   L. Bers, "On boundaries of Teichmüller spaces and on Kleinian
groups I," Ann. of Math., *91*(1970), pp. 570-600.

[4]   __________, "Fiber spaces over Teichmüller spaces," Acta Math.,
130(1973), pp. 89-126.

[5]   __________, "Uniformization, moduli and Kleinian groups," Bull.
London Math. Soc., *4*(1972), pp. 257-300.

[6]   __________, "The action of the universal modular group on certain
boundary points," *Contributions to Algebra*, (a collection of papers
dedicated to Ellis Kolchin), Acad. Press (1977), pp. 25-36.

[7]   F. P. Gardiner, "An analysis of the group operation in universal
Teichmüller space," Trans. Amer. Math. Soc., *132*(1968), pp. 471-486.

[8]   F. W. Gehring, "Univalent functions and the Schwarzian derivative,"
Comm. Math. Helv., *52*(1977), pp. 561-572.

[9]   __________, "Spirals and universal Teichmüller space," Acta Math.,
171(1978), pp. 99-113.

[10] W. Harvey, ed., *Discrete groups and automorphic functions*, Acad.
Press. (1977).

[11] O. Lehto and K. I. Virtanen, *Quasikonforme Abbildungen*, Springer-
Verlag, Berlin, (1965).

[12] B. Maskit, "On boundaries of Teichmüller spaces and on Kleinian
groups II," Ann. of Math., *91*(1970), pp. 608-638.

[13] D. Sullivan, "On the ergodic theory at infinity of an arbitrary discrete group of hyperbolic motions," these Proceedings.

SOME REMARKS ON BOUNDED COHOMOLOGY

Robert Brooks[*]

Let M be a topological space, and let $A_*(M; R)$ denote the real singular complex of M. Gromov [3] has made the following observations: If one attempts to classify the chain complex $A_*(M; R)$ up to chain homotopy, then a complete set of invariants is given by the homology $H_*(M; R)$ of this complex. However, $A_*(M; R)$ carries with it a natural norm, defined by

$$\left\| \sum a_i \sigma_i \right\| = \sum |a_i|$$

with respect to which $A_*(M)$ carries the structure of a complex of normed topological vector spaces, and the differential ∂ is a bounded linear map. One may then ask to classify the complex $A_*(M; R)$ up to bounded chain homotopy.

As a first step in this direction, one may complete $A_*(M)$ with respect to the norm $\| \ \|$, to get a complex of Banach spaces $A_*^{\ell_1}(M)$. The dual complex $A_b^*(M)$ is then naturally the complex of bounded cochains; i.e. cochains ϕ satisfying the property that $\phi(\sigma)$ is bounded independently of the simplex σ. The cohomology $H_b^*(M)$ of this complex is the bounded cohomology of M.

---

[*]Partially supported by NSF Grant #MCS 7802679.

There is an entirely group-theoretic analogue of this concept. Let G be a discrete group, and let $A^*_{E-M}(G)$ denote the Eilenberg-Maclane complex:

$$A^k(G) = \text{the set of all functions } f: \overset{\text{k-times}}{G \times \cdots \times G} \to \mathbf{R}$$

$$\delta f(g_1, \cdots, g_{k+1}) = \sum (-1)^i f(g_1, \cdots, g_{i+1} \circ g_i, \cdots, g_{k+1}) \,.$$

The bounded subcomplex $A^*_b(G)$ of $A^*_{E-M}(G)$ is the set of bounded functions $G \times \cdots \times G \to \mathbf{R}$. Then again $A^*_b(G)$ is equipped with a norm which transforms $A^*_b(G)$ into a complex of Banach spaces.

The main result we will prove is that:

THEOREM. $H^*_b(M)$ *depends only on* $\pi_1(M)$. *In fact,* $H^*_b(M) \simeq H^*_b(\pi_1(M))$.

We will assume the result of Gromov [3], which is in fact the lion's share of the theorem, that $H^*_b(M) = 0$ if $\pi_1(M) = 0$. The idea is to re-formulate the notion of bounded cohomology in more categorical terms, in the style of Hochschild-Mostow [2]. In this language, which will be ex-plicated fully in §1, the crucial point becomes the following

LEMMA. *Let* $G \to E$ *be a principal G-bundle. Then* $A^*_b(E)$ *is a bounded*

$$\downarrow$$
$$M$$

injective G-module.

There is some reason to believe that this lemma is of independent interest, if for no other reasons than pedagogical. In different "catego-ries" this lemma is well known (see [13]) although it does not appear as a general technique, say, in [2].

In §1, we recast the ideas of Hochschild-Mostow [2] in the present setting. In §2, we prove the lemma. §3 then contains an assortment of results using the results of §1 and §2.

In §3(a), we construct explicitly an infinite set of linearly independent generators of $H^2_b(\mathbf{Z} * \mathbf{Z})$. By our main lemma, then, this reflects on the geometric structure of the punctured torus. §3(b) contains some technical results needed in the section following. Finally, §3(c) contains an example showing the power of the theory. We prove a result which might be regarded as a homological version of the Mostow Rigidity Theorem, valid for any compact manifold of negative sectional curvature of dimension > 2: the action of $\mathrm{Aut}\,(\pi_1(M))$ on $H^*(M)$ factors through a finite group. The proof is essentially due to Thurston, and I am told that he has extended this result to a large degree through some substantially subtler techniques. At this time, however, the full conjecture that one would expect remains open.

The ideas in this paper have their source in a number of people, and a number of others have been involved in relaying these ideas to me. The notion of bounded cohomology itself stems from the work of Philip Trauber. His ideas were then amplified on and transmitted by Thurston, Sullivan, and Gromov, who is by-and-large responsible for introducing me both to the subject and the kinds of questions considered for the most part in this paper. I would like to extend my appreciation to all of them, and to any others whose contributions may not be properly recognized in this brief acknowledgement.

§1. *The Hochschild-Mostow theory*

Let V be a Banach space, and G a group acting linearly on V. Then V is a bounded G-module if the action of G on V is continuous, so that for any element $g \in G$, the norm

$$\|g\| = \sup_{\|v\|=1} \|g(v)\|$$

is defined, and if the map $G \to \mathbf{R}$ given by $g \to \|g\|$ is bounded. Thus a bounded G-module is given by a uniformly bounded action of G on V.

A bounded injective G-module V is a bounded G-module with the following property:

                    ROBERT BROOKS

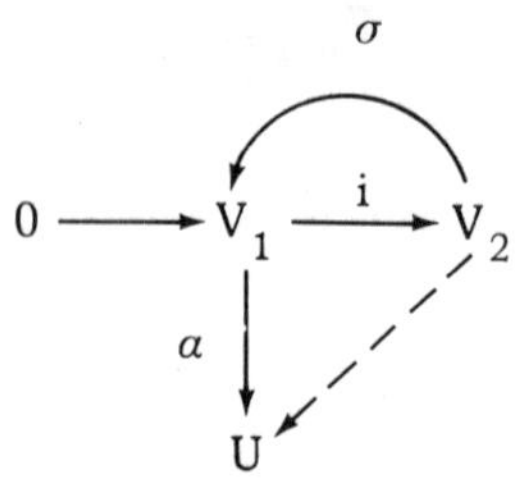

If $i: V_1 \to V_2$ is an injection of bounded G-modules, $\sigma: V_2 \to V_1$ is a splitting of i as topological vector spaces, and $a: V_1 \to U$ a map of bounded G-modules, then there is a G-module map $\beta: V_2 \to U$ such that $a = \beta \circ i$.

If A is a Banach space, let $B(G, A)$ denote the Banach space of bounded maps $f: G \to A$, i.e. maps f such that $g \to \|f(g)\|$ is a bounded function $G \to R$.

LEMMA. $B(G, A)$ *is a bounded injective G-module, where G acts on* $B(G, A)$ *via*
$$(g \cdot f)(x) = f(g^{-1} \cdot x) .$$

Proof. It is evident that $B(G, A)$ is a Banach space, and the action of G on $B(G, A)$ has norm 1, so that $B(G, A)$ is a bounded G-module.

We define β according to the formula $\beta(v)(g) = [a(\sigma(g \cdot v)](e)$ where e denotes the identity element of G. β is plainly a continuous G-module map $V_2 \to U$, and is bounded by $\|a\|$, $\|\sigma\|$, and the bound of G on V_2.

We may now define $H_b^*(G, R)$ as follows:

$$0 \longrightarrow R \xrightarrow{d} V_1 \xrightarrow{d} V_2 \longrightarrow \cdots$$

be a resolution of R by bounded injective G-modules V_i, together with a homotopy operator K such that

$$Kd + dK = \mathrm{id}$$

(K is assumed to be a sequence of maps only of topological vector spaces,

not of G-modules). Then by definition, $H_b^*(G)$ is the cohomology of the
subcomplex

$$(**)\qquad\qquad 0 \to R \to V_1^G \to V_2^G \to \cdots$$

of the G-invariant subspaces of V_i .

Standard homological algebra asserts that $H_b^*(G)$ does not depend on
the bounded injective resolution chosen. Applying the above lemma in-
ductively shows that $B(G\times \overset{k\text{-times}}{\cdots} \times G, R)$ is a bounded injective G-module,
and we recover the definition of bounded cohomology given in the
introduction.

§2. *The main lemma*

We now prove the lemma of the introduction.

Let $G \to E$ be a principal G-bundle.
$$\downarrow$$
$$M$$

We want to show that $A_b^*(E)$ is a bounded injective G-module, where the
action of G on $A_b^*(E)$ is given by deck transformations $g : \phi \to g^{-1^*}(\phi)$.
There is a natural G-module map

$$\iota : A_b^*(E) \to B(G, A_b^*(E))$$

$$\iota(\phi)(h) = h^{-1^*}(\phi) .$$

Clearly $\iota(g^{-1^*}(\phi))(h) = h^{-1^*}(g^{-1^*}(\phi)) = \iota(\phi)(gh))$ and so ι is a
G-module map.

We are done once we find a G-module map $\rho : B(G, A_b^*(E)) \to A_b^*(E)$
such that $\sigma \circ \iota = \mathrm{id}$, for then $A_b^*(E)$ is a G-direct summand of a bounded
injective module. By choosing a triangulation for M , for instance, we
may find a fundamental domain D for the action of G on E . For any
simplex σ , let g_σ be the element of G such that g_σ sends the first
vertex of σ to the domain D ; set

$$\sigma(\phi)(\sigma) = (g_\sigma^{-1})^* \phi(g_\sigma)(\sigma)$$

and extend σ linearly over linear combinations of simplices. Clearly, σ is a bounded linear map, of norm ≤ 1, which satisfies the above conditions. The lemma is then proved.

We can now prove the main theorem. Given a manifold M, then the bounded cochains on M are precisely the $\pi_1(M)$-invariant bounded cochains on $\tilde{M}$. By the main lemma $A_b^*(\tilde{M})$ is a complex of bounded injective $\pi_1(M)$-modules, and by Gromov's theorem, this resolution is acyclic (in fact, his proof constructs such a homotopy operator K). The identification $H_b^*(M) \simeq H_b^*(\pi_1(M))$ then follows from the definitions.

§3. *Some scattered remarks*

(a) We begin by producing infinitely many linearly independent cohomology classes in $H_b^2(Z * Z)$. The construction here is completely algebraic, although, as Gromov has mentioned to me, there is a purely geometric interpretation.

Let w be a word in $Z * Z$ of length ≥ 2.

Set $f_w(g) = $ # of times w occurs in g

$\qquad\qquad - $ # of times w^{-1} occurs in g

where the phrase "w occurs in g" means that w is a subword of the canonical word representing g containing no trivial cancellations.

Clearly f_w is an unbounded 1-cochain on $Z * Z$. However, $\delta f_w(g_1, g_2) = f_w(g_1) + f_w(g_2) - f_w(g_2 g_1)$ is bounded. Roughly speaking, $\delta f_w(g_1, g_2)$ counts the number of times that w occurs in $g_2 g_1$ such that w occurs partly in g_2 and partly in g_1. Clearly δf_w is bounded in terms of the length of the word w.

PROPOSITION. *The bounded 2-cocycles* δf_w *are all non-trivial cocycles, and are all linearly independent.*

Proof. Assume $\sum a_i \delta f_{w_i}$ is zero in $H_b^2(Z * Z)$. Since $\sum a_i \delta f_{w_i} = \delta(\sum a_i f_{w_i})$, and the term on the right is unbounded, then there must be a one-cocycle h such that $\sum a_i f_{w_i} - h$ is a bounded cochain. But h must be of the form $h = b_1 f_{g_1} + b_2 f_{g_2}$, where g_1, g_2 are the two generators of $Z * Z$, since these are the only homomorphisms $Z * Z \to R$. But then $\sum a_i f_{w_i} - b_1 f_{g_1} - b_2 f_{g_2}$ is unbounded, contradiction.

(b) We now turn to the question of isometric structure of the spaces $H_b^*(M)$ and $H_b^*(\pi_1(M))$.

The natural norm on the space $A_b^k(M)$ induces a pseudo-norm on the space $H_b^k(M)$ by the formula:

$$\|a\| = \inf_{[\phi]=a} \|\phi\| .$$

A similar pseudonorm is available for $H_b^k(\pi_1(M))$.

QUESTION: Is the map $H_b^*(M) \to H_b^*(\pi_1(M))$ an isometry? (*)

In this direction, one has the following topological version of the "fundamental principle of hyperbolic geometry":

PROPOSITION 1. *Given a map* $f : M_1 \to M_2$, *the induced map* $f^* : H_b^*(M_2) \to H_b^*(M_1)$ *is of norm* ≤ 1.

Proof. This follows since f^* is of norm 1 on the chain level.

The technique of §2 gives a somewhat sharper estimate.

PROPOSITION 2. *The map* $H_b^k(M) \to H_b^k(\pi_1(M))$ *is of norm* $\leq (k+1)!$

Proof. One first remarks that a contracting operator for $A_b^k(\tilde{M})$ can be constructed of norm ≤ 1. In effect, the usual homotopy operator K of a contracting space is of norm ≤ 1, a simplex σ being sent to the "contractible cone" of σ. The homotopy operator of §2 comes from such geometric construction, together with averaging over amenable groups, so that the bound on K remains unchanged.

(*) NOTE: In [3], Gromov claims an affirmative answer to this question, without proof.

Now $\pi_1(M)$ acts isometrically on the complex $A_b^*(\tilde{M})$, since $\pi_1(M)$ acts by homeomorphisms on $\tilde{M}$. Hence, in the explicit construction of §2,

$$\beta(\phi)(g) = \alpha(\sigma(g \cdot \phi))(e) ,$$

if $\|\alpha\| \leq r$ and $\|\sigma\| \leq 1$, then $\|\beta\| \leq r$. Hence, the norm of the map $A_b^k(\tilde{M}) \to A_b^{k-1}(\pi_1(M))$ is $\leq$ the norm of the map $A_b^{k-1}(\tilde{M}) \to A_b^{k-1}(\pi_1(M))$.

The proposition then follows because $\sigma: A_b^{k-1}(\pi_1(M)) \to A_b^k(\pi_1(M))$ is of norm $\leq k-1$.

(c) In this section, we use the techniques of bounded cohomology to prove a homological version of the Mostow Rigidity Theorem. To state this generalization, we need:

DEFINITION. The bounded part of $H^*(M)$ is the image of the natural map

$$\iota: H_b^*(M) \to H^*(M) .$$

The pseudonorm on $H_b^*(M)$ induces a norm on the bounded part of $H^*(M)$ by the formula:

$$\|\alpha\| = \inf_{\iota(\beta)=\alpha} \|\beta\| .$$

One checks easily that this norm is non-degenerate—if α is any class in $H^*(M)$, and $[u]$ a homology class in $H_*(M)$ such that $\alpha[u] \neq 0$, represent $[u]$ by a chain u in M. Then for any ϕ representing α, we have

$$\|\phi\| \geq \frac{|\phi(u)|}{\|u\|} = \frac{|\alpha(u)|}{\|u\|}$$

so that $\|\alpha\|$ is bounded below.

By an observation of Thurston (see [3]), if M is a compact manifold of negative curvature, then the bounded part of $H^*(M)$ is all of the cohomology of M of dimension ≥ 2. Roughly speaking, there are three components of this observation: (i) a geodesic k-simplex in hyperbolic space (constant curvature -1) has bounded k-dimensional volume, if $k > 1$. (ii) In the universal covering space $\tilde{M}$ of M, the complex of geodesic simplices is (bounded) chain homotopic to the complex of all simplices, so that we may restrict our attention to cochains whose value on any simplex agrees with the value of the corresponding geodesic simplex. (iii) A standard comparison-theorem type argument allows one to conclude that if α is a deRham k-form of M, then its value on any geodesic simplex is bounded in terms of the curvature of M, the volume of a geodesic simplex in hyperbolic space, and, of course, the size of α.

Since M is a $K(\pi_1(M), 1)$, there is a natural map $\phi: \text{Aut}(\pi_1(M)) \to \text{Aut}H^*(M)$. $\text{Ker}(\phi)$ contains the group of inner automorphisms of $\pi_1(M)$, but, perhaps, contains much more.

THEOREM. *Let* M *be a compact manifold of negative curvature,* $\dim(M) > 2$. *Then,*

(a) $\text{Aut}(\pi_1(M))/\text{Ker}(\phi)$ *is a finite group.*

(b) *The order of any element of* $\text{Aut}(\pi_1(M))/\text{Ker}(\phi)$ *is bounded in terms of the Betti numbers of* M.

Proof. Let f be an element of $\text{Aut}(\pi_1(M))$. Then there is a unique homotopy class of homotopy equivalences, which we also denote by f, which realizes the given automorphism of $\pi_1(M)$.

Since both f and f^{-1} are norm-decreasing, f^* must be an isometry of the norm $\| \ \|$. But f^* also preserves the lattice $H^*(M, Z)$ in $H^*(M, R)$.

For $k \geq 2$, let $e_1, \cdots, e_r$ be a basis for the lattice of torsion-free elements of $H^k(M, Z)$. Then since f^* is an isometry, there are only finitely many choices for $f^*(e_1), \cdots, f^*(e_r)$. Also, if f^* is the identity on $H^{n-1}(M)$, by Poincaré duality, it is so also for $H^1(M)$. This then proves (a).

To show (b), we use the fact that f^* is an isometry to show that all eigenvalues are of absolute value 1. Since f^* preserves a lattice, it also follows that all the eigenvalues of f^* are algebraic integers. By Kronecker's theorem, it follows that all the eigenvalues are roots of unity —since each eigenvalue is the solution of an algebraic equation whose degree is bounded by the corresponding Betti number of M, there is a number m which is bounded by the Betti numbers of M such that $(f^m)^*$ is unipotent. But $(f^m)^*$ is an isometry, since f^* is, and these two facts imply that $(f^m)^*$ is the identity, proving (b).

In the case where M is a hyperbolic manifold, the classical Mostow Rigidity Theorem asserts that $\mathrm{Aut}(\pi_1(M))/\mathrm{Inn}(\pi_1(M))$ is finite. Even in this case, I don't know much about the group $\mathrm{Ker}(\phi)/\mathrm{Inn}(\pi_1(M))$.

We may extend this discussion to arbitrary manifolds to some degree, in view of the results of §1.

THEOREM. *Let* M *be an arbitrary manifold, and* $f : M \to M$ *a map which induces an isomorphism on* $\pi_1(M)$. *Then* f^* *restricted to the bounded part of* $H^*(M)$ *has all eigenvalues of absolute value* 1.

Proof. This follows from the fact that $\|f^k\|$ is bounded independently of k, by the result of (b).

One is tempted to apply Kronecker's theorem in this context. The problem is that à priori, the bounded part of $H^*(M)$, need not be a rational subspace of M. Thus, in particular, it may happen that one eigenvalue of f^* may occur in the bounded part of $H^*(M)$, and thus have absolute value 1, while its conjugate with respect to some automorphism of $\overline{\mathbb{Q}}$ over $\mathbb{Q}$ may fail to have absolute value 1.

One can show that this never happens, for instance, when $\pi_1(M)$ has a Jordan-Holder decomposition in which each factor is either an amenable group or the fundamental group of a manifold of negative curvature. Intuitively, one feels that bounded cohomology detects only the "hyperbolic" structure of a group, so that there is at least some hope of the general statement being true. To sum up, we ask:

QUESTION: Is it true, for all groups G, that the bounded part of $H^*(G)$ is a rational subspace of $H^*(G)$?

An affirmative answer to the above question would sharpen the above theorem to read: f^* is of finite order on the bounded part of $H^*(M)$, with the order bounded in terms of the dimension of the bounded part of $H^*(M)$.

A negative answer would, of course, yield interesting examples of groups with non-trivial bounded cohomology.

REFERENCES

[1] Brooks, R., and Trauber, P., "The van Est Theorem for Groups of Diffeomorphisms," *The Hadronic Journal* (1978).

[2] Hochschild and Mostow, "Cohomology of Lie Groups," Ill. J. of Math. (1972).

[3] Gromov, M., "Volume and Bounded Cohomology," preprint.

THE DYNAMICS OF 2-GENERATOR SUBGROUPS
OF PSL(2, C)

Robert Brooks[*] and J. Peter Matelski

A classical result of Shimizu and Leutbecher (see, for instance [6], p. 59) asserts that if $\begin{pmatrix} 1 & 1 \\ 0 & 1 \end{pmatrix}$ and $\begin{pmatrix} a & b \\ c & d \end{pmatrix}$ generate a discrete subgroup of PSL(2, C), then either $c = 0$ or $|c| \geq 1$. This has been strengthened by T. Jørgensen [4] as follows:

JØRGENSEN'S INEQUALITY. If X and Y generate a discrete, non-elementary subgroup of PSL(2, C), then

$$|\mathrm{tr}^2(X) - 4| + |\mathrm{tr}(XYX^{-1}Y^{-1}) - 2| \geq 1 \ .$$

In this paper, we will show the existence of a sequence of inequalities, generalizing Jørgensen's inequality, which X and Y must satisfy in order for $<X, Y>$, the group generated by X and Y, to be discrete. These conditions are mutually independent in the sense that, for given X and Y, at most one can fail to hold. These conditions arise from the Shimizu-Leutbecher process defined below.

For convenience, consider the upper half space model of hyperbolic 3-space. We denote a directed geodesic ℓ by the ordered pair of its end-points; so $\ell = (a, b)$, $a, b \in C$, $a \neq b$. The *complex distance* $\tau = \delta(\ell_1, \ell_2)$

[*]Partially supported by NSF Grant #MCS 7802679.

between two directed geodesics $\ell_1 = (a_1, b_1)$ and $\ell_2 = (a_2, b_2)$ is defined as follows: $\tau \in \mathbf{C}$; $\mathrm{Re}(\tau) \geq 0$ is the hyperbolic distance between the geodesics; $\mathrm{Im}(\tau)$ is the angle made by the geodesics along their common perpendicular and is determined modulo 2π unless $\mathrm{Re}(\tau) = 0$, in which case $\pm \mathrm{Im}(\tau)$ is determined modulo 2π. One may compute the complex distance by the formula:

$$\cosh^2(\tau/2) = (a_1, a_2, b_2, b_1) ,$$

where (z_1, z_2, z_3, z_4) is the usual cross ratio, as can be checked if $\ell_1 = (-1, 1)$ and $\ell_2 = (-e^\tau, e^\tau)$.

Let X be a loxodromic element of $\mathrm{PSL}(2, \mathbf{C})$ and $\mathrm{axis}(X)$ the directed geodesic in hyperbolic space joining the fixed points of X. If ℓ is a perpendicular to $\mathrm{axis}(X)$, then the complex distance τ between ℓ and $X(\ell)$ is called the *complex translation length* of X. In fact X translates $\mathrm{Re}(\tau)$ units along $\mathrm{axis}(X)$ and rotates hyperbolic space by $\mathrm{Im}(\tau)$ about $\mathrm{axis}(X)$. We have

$$\mathrm{tr}^2 X = 4 \cosh^2(\tau/2) ,$$

which makes sense even if X is not loxodromic.

Given X loxodromic with complex translation length τ, and Y in $\mathrm{PSL}(2, \mathbf{C})$, one may check the formula:

$$\mathrm{tr}((YXY^{-1})X^{-1}) - 2 = -(1 - \cosh(\tau))(1 - \cosh(\beta)) ,$$

for β the complex distance from $\mathrm{axis}(X)$ to $\mathrm{axis}(YXY^{-1})$; this follows by normalizing

$$X = \begin{pmatrix} \cosh(\tau/2) & \sinh(\tau/2) \\ \sinh(\tau/2) & \cosh(\tau/2) \end{pmatrix}$$

$$YXY^{-1} = \begin{pmatrix} \cosh(\tau/2) & e^\beta \sinh(\tau/2) \\ e^{-\beta} \sinh(\tau/2) & \cosh(\tau/2) \end{pmatrix} .$$

Given X and Y elements of $PSL(2, \mathbb{C})$ with X loxodromic, we define the Shimizu-Leutbecher sequence inductively by:

$$Y_1 = YXY^{-1}, \qquad Y_{i+1} = Y_i X Y_i^{-1} .$$

Let τ be the complex translation length of X, and let β_i be the complex distance between axis(X) and axis(Y_i). A necessary condition for the group generated by X and Y to be discrete is that the set $\{\cosh(\beta_i)\}$ should form a discrete subset of $\mathbb{C}$.

The following lemma allows one to compute $\cosh(\beta_i)$ inductively:

LEMMA. $\cosh(\beta_{i+1}) = (1 - \cosh(\tau))\cosh^2(\beta_i) + \cosh(\tau)$.

This follows from the hyperbolic law of cosines: if ℓ_0, ℓ_1, ℓ_2 are given, the law of cosines gives a formula for $\omega = \delta(\ell_1, \ell_2)$ in terms of $\tau_1 = \delta(\ell_0, \ell_1)$, $\tau_2 = \delta(\ell_0, \ell_2)$ and α which is the complex distance from the perpendicular between ℓ_0 and ℓ_1 to the perpendicular between ℓ_0 and ℓ_2. The formula is:

$$\cosh(\omega) = \cosh(\tau_1)\cosh(\tau_2) - \cosh(\alpha)\sinh(\tau_1)\sinh(\tau_2) .$$

The lemma follows by setting $\tau_1 = \tau_2 = \beta_i$ and $\alpha = \tau$. One way to check the law of cosines is to normalize so that $\ell_0 = (0, \infty)$, $\ell_1 = (t_1, t_1^{-1})$, and $\ell_2 = (et_2, et_2^{-1})$ where $t_1 = \tanh(\tau_1/2)$, $t_2 = \tanh(\tau_2/2)$, and $e = e^\alpha$; then compute $\cosh^2(\omega/2) = (t_1, et_2, et_2^{-1}, t_1^{-1})$. Note that ℓ_2 does indeed have complex distance τ_2 to ℓ_0 with $(-e^\alpha, e^\alpha)$ as common perpendicular.

Now let $z_i = (1 - \cosh(\tau))(\cosh(\beta_i))$. We may rewrite the above inductive formula as:

$$z_{i+1} = z_i^2 + C, \qquad \text{where} \quad C = (1 - \cosh(\tau))(\cosh \tau) ,$$

and we have that if X and Y generate a discrete group, then $\{z_i\}$ forms a discrete subset of $\mathbb{C}$.

The dynamical behavior of $\mathbb{C}$ under a quadratic polynomial is well understood from the work of Fatou-Julia ([1], [5]; see also [2]). Let

$f^i(z) = f \circ f \cdots \circ f(z)$, where $f(z) = z^2 + C$; a solution ρ of the polynomi-
    i times
al equation $f^i(z) = z$ will be called a stable periodic point of period i
if $\left|\frac{d}{dz} f^i(\rho)\right| < 1$. Then f^i is contracting on any disk $B_\epsilon = \{z : |z-\rho| < \epsilon\}$
on which $\left|\frac{d}{dz} f^i\right| < 1$. The theorem of Fatou-Julia ensures that, for any
choice of C, there is at most one stable periodic orbit. Further results
of Fatou-Julia allow one to draw by computer the region E of C defined
by $E = \{z : f^i(z)$ converges to the stable periodic orbit$\}$ (see Fig. 1), and
the region of C defined by $\{C : z^2 + C$ has a stable periodic orbit$\}$ (see
Fig. 2).

To obtain the above-mentioned inequalities, let p be a stable period-
ic point of f of period n; we may assume that $|p| < 1/2$. Expanding
$$f^n(z) = \sum_{i=0}^{2^n} a_i(z-p)^i \text{ as a Taylor series about } p, \text{ we have}$$

$$|f^n(z)-p| = |z-p| \left| \sum_{i=1}^{2^n} a_i(z-p)^{i-1} \right| \le |z-p|(2^n-1)m \left[\max(1, |z-p|^{2^n-1}\right]$$

where $m = \max(|a_i|)$. Setting $K \le \dfrac{1 - \left|\frac{d}{dz} f^n(p)\right|}{(2^n-1)m}$, we see that on the disk

$|z-p| < \min(K, 1)$, f^n is a contracting map. If also $K < \dfrac{\left|\frac{d}{dz} f^n(p)\right|}{m \cdot (2^n-1)}$, then

$f^n(z) - p$ has no roots other than p in the disk $|z-p| < K$.

In the case $n = 1$, the fixed points of $f(z) = z^2 + (1 - \cosh(\tau))(\cosh(\tau))$
are $\cosh(\tau)$, $1 - \cosh(\tau)$. If $|1 - \cosh(\tau)| < \frac{1}{2}$, we may set $p \le 1 - \cosh(\tau)$,
and $\frac{df}{dz}(p) = 2(1 - \cosh(\tau))$, $m = 1$. We thus have the inequality: If
$0 < |(1 - \cosh(\tau))(\cosh(\beta) - 1)| < \min(1 - 2|\cosh(\tau) - 1|, 2|\cosh(\tau) - 1|)$, then
$\langle X, Y \rangle$ is not discrete. In view of our expressions for $\cosh(\tau)$ and
$\cosh(\beta)$ given above, this becomes Jørgensen's inequality.

In the case $n = 2$, the periodic points of order 2 of $f(z) = z^2 + C$
are $\dfrac{-1 \pm \sqrt{1 - 4(C+1)}}{2}$, and $\frac{d}{dz} f^2(p) = 4(C+1)$. Using the estimates

$|p| < \frac{1}{2}$, $|C| < \frac{5}{4}$, we find $m \leq 4$, so that $0 < |z-p| < \frac{1}{12} \min(1-4|C+1|,$ $4|C+1|)$ implies $<X, Y>$ is not discrete.

In terms of the matrix functions of X and Y, this may be rewritten as:

$$0 < \left| \mathrm{tr}(XYX^{-1}Y^{-1}) - \frac{\mathrm{tr}^2(X)}{2} - \frac{-1 \pm \sqrt{(\mathrm{tr}^2(X) - 5(\mathrm{tr}^2(X) - 1)}}{2} \right|$$

$$< \frac{1}{12} \min \left[(1 - |\mathrm{tr}^4(X) - 6\mathrm{tr}^2(X) - 4|, \; |\mathrm{tr}^4(X) - 6\mathrm{tr}^2(X) + 4|) \right]$$

implies $<X, Y>$ is not discrete.

ACKNOWLEDGEMENTS: We would like to thank Henry Laufer for suggesting and assisting us with the use of the computer, and Bernard Maskit for his advice and encouragement. Also, we want to thank Dusa McDuff for showing us Fatou's paper.

REFERENCES

[1] P. Fatou, *Sur les Equations Fonctionelles*, Bull. Soc. Math. France, 48, (1920), 33-94.

[2] F. Gross, *Factorization of Meromorphic Functions*, Mathematics Research Center, Naval Research Laboratory, Washington, D. C., 1972.

[3] T. Jørgensen, *A Note on Subgroups of SL(2, C)*, Quart. J. Math. Oxford (2), 28 (1977), 209-212.

[4] _________, *On Discrete Groups of Möbius Transformations*, Amer. J. Math., Vol. 98, No. 3, 739-749.

[5] G. Julia, *Memoire Sur l'Iteration des Fonctions Rationelles*, J. Math. 8, 1 (1918).

[6] I. Kra, *Automorphic Forms and Kleinian Groups*, W. A. Benjamin, Reading, Mass., 1972.

[7] J. P. Matelski, *A Compactness Theorem for Fuchsian Groups of the Second Kind*, Duke Math. J. Vol. 43, No. 4, (1976), 829-840.

```
                              (0,1.2)
                                *
                               ***
                                ****  *
    ****        *             *******
    ****   *  ***             *******
    **********         ****     **********
    ***********    *     ****** ****** * *
 ** * ***********  ****   *****************
 **************************  ************************
 ***********************  ************************
 *******  ***********************************************
 ***   * *   *************************************      *
   *****  *******************************   *** ** **
    ****     ****  *******************************
         **    ******************************************
             ***************************************
             **************************** * **
            ****************************
            ****************************
    ****  *  ***************************
    **   *   ***************************
            ****************************
            *************************    *  **
            ************************   *  ****
              ****************** **********
             ***************************
             **************************
    ****  ****************************
         ****************************
 ************************************    **
 *************************************** ****    ****
 ** ** ***  ***********************************  *****
    *           **********************************   * *  ***
             ******************************************* *******
             ************************************************
             ****************  ***************************
             **********   ****  ********** * **
    * * ****** ******      *  *************
    *********    ****          ***********
    ******                 *** * ****
    *******                  *     ****
    * ****
     ***
      *
                              (0,-1.2)
```

Fig. 1. The set E for $f(z) = z^2 + C$ with $C = .1 + .6i$.

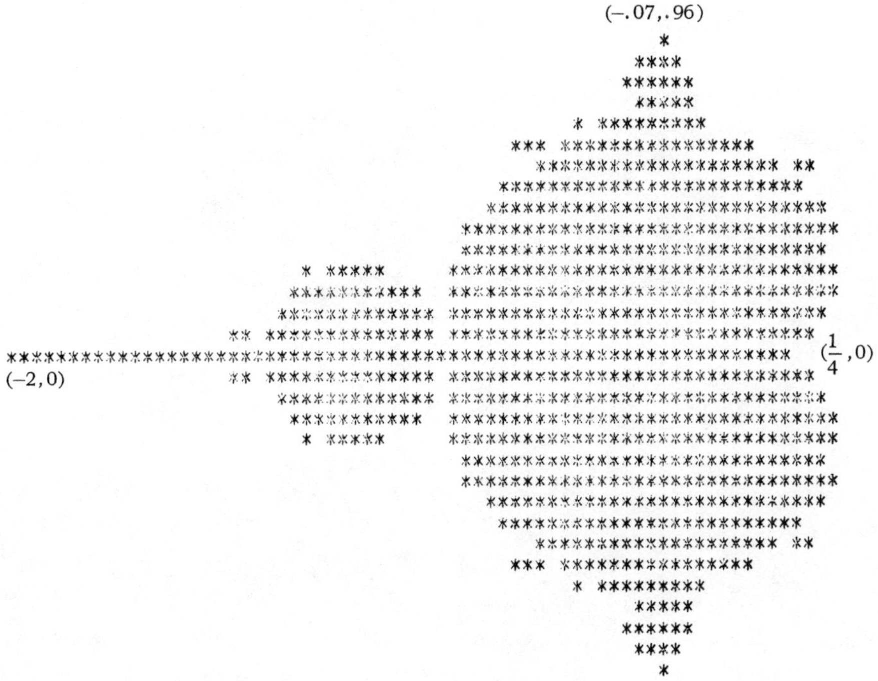

Fig. 2. The set of C's such that $f(z) = z^2 + C$ has a stable periodic orbit.

MINIMAL GEODESICS ON FRICKE'S TORUS-COVERING

Harvey Cohn[*]

Summary

The free group Γ of rank two is representable in $SL_2(R)$ by Fricke's group with torus as fundamental domain. In the lattice covering of the torus, the primitive elements of Γ were previously shown to be precisely the "straightest possible" Poincaré geodesics. Here we show why they are also the "shortest possible." As an application, we can minimize the trace in any coset of Γ/Γ' by using powers of a primitive element. This approach provides a measure of the "nonprimitivity" of a word in Γ and also number-theoretic results on the size of Markoff numbers.

§1. *Introduction*

Robert Fricke's first published work came in two papers ([7] and [8]) from his 1885 Leipzig dissertation. The basic problem involved a transition from the modular group to its commutator subgroup and onward to the second commutator. Topologically, this involved a transition from sphere to torus and onward to the periodic lattice covering of the torus. (Of course, Fricke not only ignored specific references to topology, but he even avoided the word "commutator"!)

---

[*]Research supported by NSF Grant 76-06744, and by the Danish National Research Council.

The prize sought was the first noncongruence subgroup of the modular group (derived analytically by a device described in the title of [7]). Whatever excitement this result generated was surely heightened by a simultaneous discovery by G. Pick [16]. (Felix Klein, as editor of the Annalen, rendered instant mediation by a footnote [7, p. 118] asserting that Fricke's paper had arrived one day earlier.)

In the second paper [8], Fricke considered the image of the modular group in terms of covering operations of the lattice. Yet certain concepts such as the preimages of translation bases were not considered, nor were Poincaré geodesics then in common use for characterizing subgroups of the modular group. It was exactly these concepts which motivated the speaker's earlier work on a modular-theoretic approach to Markoff's forms [1], (generalized by Asmus Schmidt [17]).

For present purposes, we recall results in [2] and [3] to the effect that primitive elements of the commutator group (as a free group of rank two) correspond to the "straight" geodesics in the lattice covering, (see (2.3) below and [4]). Recently, Werner Fenchel communicated privately that "straightness" of geodesics corresponds to a "nonintersection" property established to be a geometrical invariant under automorphisms, (according to an unpublished manuscript [6]).

In this paper, we go further and show a basis for interpreting primitive elements as "short" geodesics. This leads to matrix inequalities which are verifiable numerically:

CONJECTURE 1.1. *Let* $A, B \in SL_2(\mathbf{R})$ *and let* trace $[A, B] = -2$, $|\text{trace } A| > 2$, $|\text{trace } B| > 2$, *(where* $[A, B] = ABA^{-1}B^{-1}$*). Then* $\Gamma = <A, B>$ *is a representation of a free group, (see §3 below). It also follows (see §2 below) that every word* $W \in \Gamma$ *either belongs to* Γ' *(the commutator subgroup of* Γ*) or* $W = M^n X$ *for* $n > 0$, $X \in \Gamma'$, *where* M *is a primitive element of* Γ, *(i.e.,* $\Gamma = <M, M'>$ *for some* M'*). If* $W \in \Gamma$, *write* $M^n = 1$ *for uniformity. Given all this, we define the trace-difference and assert it is nonnegative:*

$$(1.2) \qquad d(W) = |\text{trace } W| - |\text{trace } M^n| \geq 0 .$$

We hope to prove this result on a later occasion. Actually, we establish an overlapping result in §4 which ensures at least partial truthfulness:

THEOREM 1.3. *Conjecture 1.1 surely holds under the symmetry* trace $A =$ trace B *with* $M^n = (AB)^n$. *In any case,* $d(W)$ *exceeds any constant except for at most a finite set of* W *as distinguished to within automorphisms of* Γ *as a free group.*

Thus A and M are equivalent as well as W and $S^{-1}WS$, and likewise A^2B^2 and $ABAB^{-1}$ ($= (AB)^2B^{-2}$), etc. Now Conjecture 1.1 is part of

CONJECTURE 1.4. *The discrete levels of* $d(W)$ *begin as follows (if we identify* W *by automorphisms of* Γ):

$$(1.4a) \qquad d(W) = 0 \qquad \Longleftrightarrow \quad W = A^n \ \text{or} \ [A, B]$$

$$(1.4b) \qquad d(W) = 4 \qquad \Longleftrightarrow \quad W = A^2[A^{-1}, B]$$

$$(1.4c) \qquad d(W) = 2\,|\text{trace } A| \Longleftrightarrow \quad W = A[A, B] .$$

§2. *Measure of power-primitivity*

In the notation of Conjectures 1.1 and 1.4, we set $W = M^nX$ ($X \epsilon \Gamma'$), for M primitive and call W *power-primitive* if $X = 1$. From earlier work [3] we can write M explicitly and compute $d(W)$ algebraically as the measure of the extent to which W fails to be power-primitive. If we abelianize Γ so W becomes $uA + vB$ then

$$(2.1) \qquad W = A^uB^vX_1 \qquad (X_1 \epsilon \Gamma') ,$$

then M is determined by

$$(2.2) \qquad M^n = (A, B)^{u,v}, \qquad n = \gcd(u, v) .$$

Here we use the notation of *step-power*

$$(2.3) \qquad (A, B)^{u,v} = \prod_{j=1}^{u} AB^{s_j}, \qquad s_j = [jv/u] - [(j-1)v/u]$$

for $u, v \in Z$ and $u > 0$, supplemented by

$$(2.4) \qquad (A, B)^{0,v} = B^v, \qquad (A, B)^{-u,-v} = ((A, B)^{u,v})^{-1}.$$

Clearly, then, for $t \in Z$,

$$(2.5) \qquad (A, B)^{tu,tv} = ((A, B)^{u,v})^t.$$

The name "step-power" refers to the increments s_j in the step-function $[jv/u]$ (greatest integral part). Primitivity follows from the relation

$$(2.6) \qquad \Gamma = \langle (A, B)^{u,v}, (A, B)^{u',v'} \rangle, \qquad (uv' - vu' = 1),$$

inductively obtained (see [2]) by the Euclidean algorithm

$$(2.7) \qquad (A, B)^{u,v}(A, B)^{u',v'} = (A, B)^{u+u',v+v'}.$$

We now see how Table I supports Conjecture 1.4. In the left-hand columns several nonprimitive words W are listed with M^n. There is, fortunately, the "trace calculus" of Fricke [9] (also see Horowitz [11]) which permits us to calculate the extent to which W *fails* to be power-primitive,

$$(2.8) \qquad d(W) = |\text{trace } W| - |\text{trace } M^n|,$$

as a polynomial in a, b, c where

$$(2.9) \qquad a = \text{trace } A, \quad b = \text{trace } B, \quad c = \text{trace } AB$$

by the repeated use of the identity

$$(2.10) \quad \text{trace } T \text{ trace } U = \text{trace } TU + \text{trace } TU^{-1}, \quad (T, U \in SL_2(C)).$$

Thus if $W = A^2 B^2$, then $M^2 = (AB)^2$ and we find trace $W = c^2 + 2$, trace $M^2 = c^2 - 2$, and $d(W) = 4$ (as in (1.4b)).

W	M^n	$d(W; a,b,c)$	$d(W; 3,3,3)$	$d(W; \min)$
A^2B^2	$(AB)^2$	4	4	4
A^3B^2	$(ABA^2B)^1$	$4a$	12	8
A^2B^2AB	$(AB)^3$	$4c$	12	8
$A^2B^2(AB)^2$	$(AB)^4$	$4c^2-4$	32	12
A^4B^2	$(A^2B)^2$	$4a^2$	36	16
$AB^2A^{-1}B^2$	B^4	$4b^2$	36	16
A^3BAB^2	$((AB)^2A^2B)^1$	$4ac$	36	32
$(A^2B^2)^2$	$(AB)^4$	$8c^2$	72	32
A^3B^3	$(AB)^3$	$4c+4ab$	48	$46.6\cdots$
$B^3A^2BA^2$	$(AB)^4$	$8b^2+4a^2$	108	48
A^4B^4	$(AB)^4$	$8c^2+4a^2b^2$	396	$364.7\cdots$
$\cdots$	$\cdots$	$\cdots$	$\cdots$	$\cdots$
$A^{m-1}B^{-1}AB$	A^m	$t_m(a)$	$t_m(3)$	$4(m-1)$
$(A^{-1}B^{-1}AB)^mA^{-1}$	$(A^{-1})^1$	$2ma$	$6m$	$4m$
$A^{m+1}B^{-1}A^{-1}B$	A^m	$u_m(a)$	$u_m(3)$	$4m$

Table I. Some values of $d(W\cdots)$ measuring trace-difference between $W(\epsilon\Gamma)$ and its power-primitive component M^n. Here $t_m(a) = 4(\xi^{m-1}-1/\xi^{m-1})/(a^2-4)^{1/2}$, for $\xi = (a+(a^2-4)^{1/2})/2$; and $u_m(a) = 2a(\xi^m-1/\xi^m)/(a^2-4)^{1/2}$. Note $t_m(3)$ and $u_m(3)$ are proportional to alternate Fibonaccian numbers.

Any polynomial relations are valid only modulo the identity

$$(2.11) \qquad a^2 + b^2 + c^2 = abc$$

(see (3.6) below), whereas automorphisms of Γ lead to permutations of a, b, c and of the substitution (from (2.10))

$$(2.12) \qquad c \to c-ab .$$

Therefore, to grasp the relative magnitudes of $d(W)$ for various W, we can substitute $a = b = c = 3$ for the particular group

$$(2.13) \qquad \Gamma_K = \langle A, B \rangle, \qquad A = \begin{pmatrix} 1 & 1 \\ 1 & 2 \end{pmatrix}, \qquad B = \begin{pmatrix} 1 & -1 \\ -1 & 2 \end{pmatrix},$$

reducing $d(W)$ to its minimum under the automorphisms. We could also find the minimum (really the "inf") of $d(W)$ over all possible (a, b, c). This is done in the remaining columns of Table I. (Unfortunately, inequivalent W *can* give equivalent $d(W)$!)

We note Γ_K is precisely Fricke's commutator subgroup of $SL_2(Z)$, (see [1]). If we rewrite it as

$$(2.14) \qquad \Gamma_K = <V_1, V_2>, \qquad V_1 = \begin{pmatrix} 2 & 1 \\ 1 & 1 \end{pmatrix}, \qquad V_2 = \begin{pmatrix} 5 & 2 \\ 2 & 1 \end{pmatrix},$$

$(V_1 = B^{-1}, V_2 = (BAB)^{-1})$, then, by symmetry, all traces of $(A, B)^{u,v}$ occur for $u \geq 0$, $v \geq 0$ as trace $(V_1, V_2)^{u,v}$, just once. These traces are three times the Markoff numbers when $\gcd(u, v) = 1$. Note $V_1 z = 1 + 1/(1+1/z)$ and $V_2 z = 2 + 1/(2+1/z)$. Analogously, A. Schmidt [17] has studied Hecke's modular group

$$(2.15) \qquad \Gamma_H = <G_1, G_2>, \qquad G_1 = \begin{pmatrix} 1 & -2^{\frac{1}{2}} \\ 3 \cdot 2^{\frac{1}{2}} & 5 \end{pmatrix}, \qquad G_2 = \begin{pmatrix} 2^{\frac{1}{2}} & 1 \\ 1 & 2^{\frac{1}{2}} \end{pmatrix}.$$

§3. *Fricke's torus-covering*

For α and $\beta > 0$ define (see [9])

$$(3.1) \qquad A = \begin{pmatrix} \alpha & \alpha \\ \beta^2/\alpha & (\beta^2+1)/\alpha \end{pmatrix}, \qquad B = \begin{pmatrix} \beta & -\alpha^2/\beta \\ -\beta & (\alpha^2+1)/\beta \end{pmatrix}$$

then for $\gamma = (\alpha^2 + \beta^2)(1 + \alpha^2 + \beta^2)/\alpha^2\beta^2$,

$$(3.2) \qquad [A^{-1}, B^{-1}] = A^{-1}B^{-1}AB = \begin{pmatrix} -1 & \gamma \\ 0 & -1 \end{pmatrix}.$$

The matrices A, B generate a free group on two elements. The above parametrization is of the most general type described by

$$(3.3) \quad \Gamma = <A,B> \subset SL_2(R), \quad \text{trace } [A,B] = -2, \quad |\text{trace } A| > 2, \quad |\text{trace } B| > 2$$

at least to within equivalence $(S^{-1}\Gamma S, S \in SL_2(R))$, with a possible sign change of A or B. We abbreviate $C = AB$ and write

(3.4) $a = \text{trace } A, \quad b = \text{trace } B, \quad c = \text{trace } C.$

Then these relations hold:

(3.5) $a = (1+\alpha^2+\beta^2)/\alpha, \quad b = (1+\alpha^2+\beta^2)/\beta, \quad c = (1+\alpha^2+\beta^2)/\alpha\beta$

(3.6) $a^2 + b^2 + c^2 = abc.$

Thus Γ is parametrized by α and β, i.e., $\alpha = b/c$, $\beta = a/c$, (but not uniformly so, see [12]).

The fundamental domain for Γ acting on $\mathcal{H}$, the upper half-plane, is drawn in usual fashion as $\mathcal{D}_{A,B}$ in Figure 1, showing the action of A and B on geodesics with vertices at -1, 0, α^2/β^2, ∞. For our purposes we need to construct $\mathcal{D}_A$ the *special domain* for A.

Let $\mathcal{G}_W$ denote the geodetic arc (ω_1, ω_2), the axis of W, a hyperbolic element of Γ with fixed points ω_1 and ω_2. We assert that for $\mathcal{G}_A = \text{arc}(a_1, a_2)$ the center $a_0 = (a_1+a_2)/2$ satisfies

(3.7) $-1 < a_0 < 0$

if we use some automorphism of Γ to obtain (compare [17])

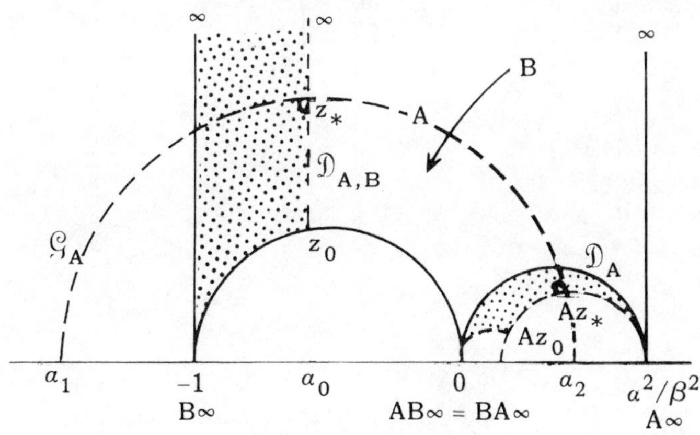

Fig. 1. Formation of the special $\mathcal{D}_A$ from the fundamental domain $\mathcal{D}_{A,B}$ in the upper half z-plane $\mathcal{H}$.

(3.8) $a^2 > |b^2 - c^2|$.

Then the vertical arc (a_0, ∞) has a portion arc (z_0, ∞) which is normal to $\mathcal{G}_A$ at z_*. Likewise its image arc $(Az_0, A\infty)$ is normal to $\mathcal{G}_A$ at Az_*. These arcs, together with arc $(0, Az_0)$ are the new boundary of $\mathcal{D}_A$ as shown in Figure 1.

We next introduce an abelian integral of the first kind for $\mathcal{D}_{A,B}$ (or for $\mathcal{D}_A$), namely u, to transform the z half-plane to cover a lattice in the u-plane. Obviously, the covering is multivalued to the extent that each z has the same image as Xz (for $X \epsilon \Gamma'$). In the classical modular case (2.13), $du = dJ/(J-1)^{1/2}J^{2/3}$ for $J(z)$ the (Klein) modular invariant, (see [1] and [8]). The lattice of the covering is determined by the doubly-periodic images of $z = \infty$.

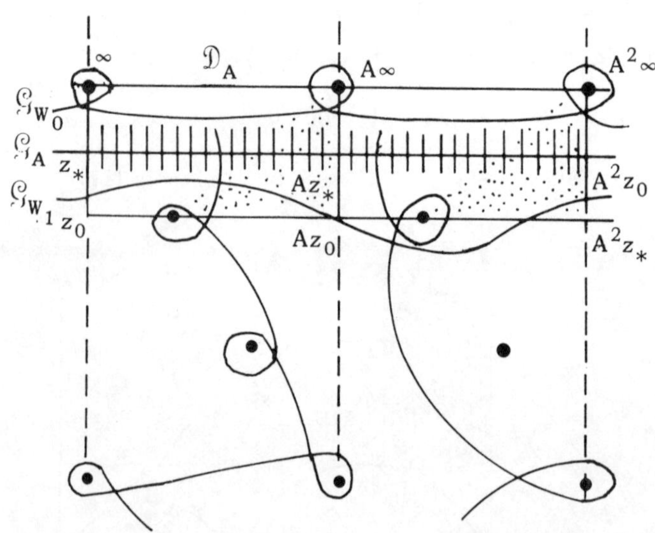

Fig. 2. Lattice covering of torus by $\mathcal{D}_A$ in the u-plane. Labels refer to the z-plane (see Figure 1), and images of $(z=)\infty$ are heavy dots denoting lattice points. The complete transversals are partly dotted but all lines are geodesics. True right angles occur only along $\mathcal{G}_A$. Here $W_0 = [A^{-1}, B^{-1}]A^{-1}$ and $W_1 = AB^{-1}AB$ are nonprimitive, (see §4).

The geodesic arcs normal to $\mathcal{G}_A$ (but lying in $\mathcal{D}_A$) emerge as a *transversal field* uniquely covering the image of $\mathcal{D}_A$. In Figure 2, we label the u-plane by the preimages in the z-plane when multivaluedness does not interfere. We see that these normals extend outside $\mathcal{D}_A$ and there they lose their uniqueness as they wrap themselves about the lattice points (images of $z = \infty$). It is, however, the nature of noneuclidean distance that two nonintersecting, nonparallel geodesics have one common perpendicular at the mutually closest point of approach. Thus the distance in $\mathcal{D}_A$ between arc (z_0, ∞) and arc $(Az_0, \alpha^2/\beta^2)$ is measured along arc (z_*, Az_*). (The metric $ds = |dz|/\mathrm{Im}\, z$ is of course transferred to the u-plane.)

Consider next all geodesics $\mathcal{G}_W$ for $W = AX$ or $A^{-1}X$, $(X \epsilon \Gamma')$. Under the action of W, $\mathcal{G}_W$ has a period which is measured from arc (z_0, ∞) to arc $(Az_0, A\infty)$, using suitable lattice translations in the u-plane (see [3]). Nevertheless unless $W = S^{-1}AS$ or $S^{-1}A^{-1}S$, then $\mathcal{G}_W$ cannot be wholly interior to $\mathcal{D}_A$, by the uniqueness of the geodesics in the field.

There are many ways a geodesic can wind around the lattice points. In Figure 2 we see $\mathcal{G}_{W_0}$ for $W_0 = [A^{-1}, B^{-1}]A^{-1} = A^{-1}B^{-1}ABA^{-1}$. In the u-plane it must intersect itself because in the z-plane it covers the full range $z \to z - \gamma$ of $[A^{-1}, B^{-1}] \epsilon \Gamma'$, (note (3.2)). We also see the geodesic $\mathcal{G}_{W_1}$ for $W_1 = AB^{-1}AB$. It merely covers two periods of $\mathcal{G}_A$ but does not intersect itself. (These cases occur in (1.4c) and (1.4b) above.)

§4. Trace and distance

The geodetic distance is related to trace as follows:

THEOREM 4.1. *Let* A *be hyperbolic in* $SL_2(\mathbf{R})$, *acting on* $\mathcal{H}$, *and let* $\mathcal{G}_A$ *join its fixed points. Let* z *be a point on* $\mathcal{G}_A$. *Then if* s *is the noneuclidean distance on the* $\mathrm{arc}\,(z, AZ)$ *and* $a = |\mathrm{trace}\, A|$,

$$(4.2a) \qquad\qquad\qquad a = 2\cosh s/2$$

$$(4.2b) \qquad s = 2 \log((a + (a^2 - 4)^{1/2})/2) = 2 \cosh^{-1} a/2 \ .$$

If z *is not a point on* $\mathcal{G}_A$, *the corresponding distance* a *on* $\mathrm{arc}(z, Az)$ *satisfies* $a < 2 \cosh s/2$, *or* $s > 2 \cosh^{-1} a/2$.

Conjecture 1.1 then becomes the following (with $M = A$):

CONJECTURE 4.3. *The shortest (periodic) geodesic between* $\mathrm{arc}(z_0, \infty)$ *and* $\mathrm{arc}(A^n z_0, A^n \infty)$, *i.e., through* n *replicas of* $\mathcal{G}_A$ *as laid out horizontally in Figure 2, is the corresponding portion of* $\mathcal{G}_A$.

We do not prove this, but what is apparent from Figure 2 is that only a finite number of homotopy classes of the punctured covering surface can have geodesics of length below a given bound. (One must incidentally check that repeated circuits about ∞ do increase the length unboundedly.) From this, Theorem 1.3 follows. (Note that when $b = c$ there are horizontal and vertical reflectional symmetries in Figure 2, so that if W is in the same Γ/Γ'-coset as A^n, $\mathcal{G}_W$ can be reflected into $\mathcal{D}_A$ for comparison with $\mathcal{G}_A$.)

THEOREM 4.4 (*Triangular inequality*). *Let*

$$(4.5) \qquad\qquad a^2 + b^2 + c^2 = abc$$

for a, b, c *positive, then*

$$(4.6) \qquad \frac{c + (c^2 - 4)^{1/2}}{2} < \frac{a + (a^2 - 4)^{1/2}}{2} \cdot \frac{b + (b^2 - 4)^{1/2}}{2} \ .$$

For proof we construct the corresponding $\Gamma = \langle A, B \rangle$ as in §3, and let z_1 be the intersection of $\mathcal{G}_A$ and $\mathcal{G}_B$ (see A. Schmidt [18] for an elegant construction). Then if $z_0 = B^{-1} z_1$, consider the triangle formed by $\mathrm{arc}(z_0, Bz_0)$ on $\mathcal{G}_B$, $\mathrm{arc}(Bz_0, ABz_0)$ on $\mathcal{G}_A$, and $\mathrm{arc}(z_0, ABz_0)$ which is *not* on $\mathcal{G}_C$, ($C = AB$). Nevertheless, the inequalities are just right for using Theorem 4.1.

§5. *Markoff numbers*

Actually, despite the appearance of Table I, the geodesic for $A^u B^v$ is not much longer than its equivalent for $(A, B)^{u,v}$ even when $|u| + |v|$ is large. This is like saying that the semi-perimeter of a rectangle is of the same order as its diagonal, and indeed the logarithms are asymptotic! This idea tells us the order of magnitude of Markoff numbers.

If we return to (2.14), we write the matrices (see [3])

$$(5.1) \quad M = (V_1, V_2)^{u,v}, \quad M' = (V_1, V_2)^{u',v'}, \quad M'' = (V_1, V_2)^{u+u',v+v'}$$

where $u, u', v, v' \geq 0$ and $uv' - vu' = 1$. Then for the Markoff triple

$$(5.2) \qquad m = (\text{trace } M)/3, \quad m' = (\text{trace } M')/3, \quad m'' = (\text{trace } M'')/3$$

$$(5.3) \qquad\qquad m^2 + m'^2 + m''^2 = 3mm'm''.$$

The triangular inequality (4.6) yields

$$(5.4) \qquad \frac{3m'' + (9m''^2 - 4)^{1/2}}{2} < \frac{3m + (9m^2 - 4)^{1/2}}{2} \cdot \frac{3m' + (9m'^2 - 4)^{1/2}}{2}.$$

By repeated uses (writing M'' now as $M = (V_1, V_2)^{u,v}$), we have

$$(5.5) \qquad \frac{3m + (9m^2 - 4)^{1/2}}{2} < \left(\frac{3 + 5^{1/2}}{2}\right)^u \cdot (3 + 8^{1/2})^v,$$

if we note for V_1 we have $m = 1$ and for V_2 we have $m = 2$. In logarithmic terms, for m large,

$$(5.6) \qquad \log 3m + \sigma(1) \leq u \log(3 + 5^{1/2})/2 + v \log(3 + 8^{1/2}).$$

If we let m_N be the Nth Markoff number, an easy consequence of (5.6) is a count of *relatively prime* lattice points in a triangle as a method of estimating m_N:

$$(5.7) \qquad \limsup \frac{\log 3m_N}{N^{1/2}} \leq \frac{\pi(\log(3 + 8^{1/2})\,\log((3 + 5^{1/2})/2)^{1/2}}{3^{1/2}}.$$

A "lim inf" is also seen to exist by the analogy of the rectangle cited earlier in this section, but we omit details since, recently, Christopher Gurwood [10] showed a *limit* to exist by direct methods. The limit has not been calculated, but the constant in (5.7), namely $2.36247\cdots$, seems close.

MATHEMATICS DEPARTMENT
COLLEGE OF THE CITY OF NEW YORK
NEW YORK, NEW YORK 10031

REFERENCES

[1]   Cohn, H., *Approach to Markoff's minimal forms through modular functions*, Ann. of Math. 61 (1955), pp. 1-12.

[2]   ________, *Representation of Markoff's binary quadratic forms by geodesics on a perforated torus*, Acta. Arith. 18 (1971), pp. 125-136.

[3]   ________, *Markoff forms and primitive words*, Math. Ann. 196 (1972), pp. 8-22.

[4]   ________, *Some direct limits of primitive homotopy words and of Markoff geodesics*, Discontinuous Groups and Riemann Surfaces, Princeton, 1974, pp. 81-98.

[5]   ________, *Ternary forms as invariants of Markoff forms and other $SL_2(Z)$-bundles*, Linear Alg. and Appl., 21 (1978), pp. 3-12.

[6]   Fenchel, W., and Nielsen, J., *Discontinuous Groups*, (to appear).

[7]   Fricke, R., *Über die Substitutionsgruppen, welche zu den aus dem Legendre'schen Integralmodul $k^2(\omega)$ gezogen Wurzeln gehören*, Math. Ann., 28 (1887), pp. 99-119.

[8]   ________, *Die Congruenzgruppen der sechsten Stufe*, Math. Ann., 29 (1887), pp. 97-123.

[9]   ________, *Über die Theorie der automorphen Modulgruppen*, Gött. Nach., (1896), pp. 91-101.

[10] Gurwood, C., *Diophantine approximation and the Markov chain*, Dissertation, N.Y.U, 1976.

[11] Horowitz, R. D., *Characters of free groups represented in the two dimensional special linear group*, Comm. Pure Appl. Math., 25 (1972), pp. 635-649.

[12] Keen, L., *On Fricke moduli*, Advances in the Theory of Riemann Surfaces, Princeton, 1971, pp. 205-209.

[13] Magnus, W., Karrass, A., Solitar, D., Combinatorial Group Theory, New York, 1966.

[14] Markoff, A. A., *Sur les formes binaires indefinies*, *I*, Math. Ann., 15(1879), pp. 381-409; *II*, Math. Ann., 17(1880), pp. 379-400.

[15] McKean, H. P., *Selberg's trace formulas as applied to a compact Riemann surface*, Comm. Pure Appl. Math., 25(1972), pp. 225-246.

[16] Pick, G., *Über gewisse ganzzahlige Substitutionen, welche sich nicht durch algebraische Cohgruenzen erklären lassen*, Math. Ann., 28(1887), pp. 119-124.

[17] Schmidt, A. L., *Minimum of quadratic forms with respect to Fuchsian groups*, *I*, J. reine und angew. Math., 286/287(1976), pp. 341-368.

[18] ————, *Minimum of quadratic forms with respect to Fuchsian groups*, *II*, J. reine und angew. Math., 292(1977), pp. 109-114.

[19] Zieschang, H., Vogt, E., Coldeway, H.-D., Flächen und ebene diskontinuierliche Gruppen, Lecture Notes in Mathematics, #122, Springer-Verlag, Berlin, New York 1970.

Added in proof: Dr. Gerald Myerson kindly communicated a direct proof of Theorem 4.4.

ON VARIATION OF PROJECTIVE STRUCTURES

Clifford J. Earle[*]

1. Introduction

The purpose of this paper is to derive a new variational formula for
the monodromy map associated with (varying) projective structures on
(varying) Riemann surfaces of genus $p \geq 2$.

The monodromy map was investigated in a series of papers by
D. A. Hejhal [6, 7, 8]. Hejhal used a cutting-and-pasting technique in [6]
and [7] to show that the monodromy map defines a local homeomorphism
from the space of projective structures to the space of monodromy groups.
He obtained a variational formula in [8], but that formula is too complicated to be useful for studying the Jacobian of the monodromy map, although it has other interesting applications [9].

In contrast, our formula implies easily that the monodromy map has
nonzero Jacobian and is therefore a local homeomorphism. The formula
also suggests a close relationship between the tangent space to the space
of projective structures and certain Eichler cohomology groups. That relationship is explored more deeply in J. H. Hubbard's paper [10].

The survey article [5] by R. C. Gunning contains much useful information and background material about projective structures. Gunning has

[*]This research was partly supported by a grant from the National Science
Foundation.

independently obtained a variational formula quite similar to ours and observed a relationship with Eichler cohomology (unpublished notes).

2. *Projective structures*

2.1. Let X be a closed Riemann surface of genus $p \geq 2$, and let $\pi : D \to X$ be a holomorphic universal covering of X by the bounded Jordan region $D \subset C$. We shall assume that the group of cover transformations is a quasifuchsian group Γ (with invariant domain D).

As usual, the holomorphic function $\phi : D \to C$ is called a quadratic differential (for Γ) if

$$(2.1) \qquad \phi(\gamma z)\,\gamma'(z)^2 = \phi(z) \quad \text{for all} \quad \gamma \in \Gamma, \ z \in D.$$

The space of all quadratic differentials is denoted by $Q(\Gamma)$.

Given $\phi \in Q(\Gamma)$, let f be any meromorphic solution in D of the Schwarzian differential equation $\{f, z\} = \phi(z)$. Then f is a local homeomorphism from D into the Riemann sphere P_1, and there is a homomorphism ρ from Γ into the group $PL(1, C)$ of Möbius transformations such that

$$(2.2) \qquad f(\gamma z) = \rho(\gamma)\,f(z) \quad \text{for all} \quad \gamma \in \Gamma, \ z \in D.$$

We say that f determines a *projective structure* on X, and we call ρ the *monodromy homomorphism* determined by f.

The meromorphic functions g on D with $\{g, z\} = \phi(z)$ are precisely the functions $g = A \circ f$, $A \in PL(1, C)$. The effect of replacing f by $A \circ f$ is to replace ρ by the homomorphism $\gamma \mapsto A\rho(\gamma)A^{-1}$. The projective structures determined by f and $A \circ f$ are called equivalent; equivalent projective structures determine conjugate monodromy homomorphisms. Thus each $\phi \in Q(\Gamma)$ determines an equivalence class of projective structures and a conjugacy class of monodromy homomorphisms $\rho : \Gamma \to PL(1, C)$.

2.2. In order to study what happens when we change the Riemann surface X, we introduce the Teichmuller space T_p of closed Riemann surfaces of genus $p \geq 2$. T_p is a complex manifold of dimension $d = 3p-3$ and can be embedded in C^d as a bounded contractible domain of holomorphy.

The Bers fiber space (see [1]) over T_p is a region $F_p \subset T_p \times C$ with these properties:

(i) Γ acts freely and properly discontinuously on F_p as a group of biholomorphic maps

$$\gamma(t, z) = (t, \gamma^t(z)) \quad \text{for all} \quad \gamma \in \Gamma, \ (t, z) \in F_p,$$

(ii) $D(t) = \{z \in C; (t, z) \in F_p\}$ is a bounded Jordan region in C for every $t \in T_p$,

(iii) The group $\Gamma(t)$ of self-mappings $z \mapsto \gamma^t(z)$ of $D(t)$ is quasi-fuchsian, and $D(t)/\Gamma(t)$ is the closed Riemann surface corresponding to t,

(iv) If $D^*(t)$ is the exterior of $D(t)$ in P_1, then $D^*(t)/\Gamma(t)$ is a fixed Riemann surface, independent of t.

2.3. There is a complex vector bundle Q over T_p whose fiber over each point t is the space $Q(\Gamma(t))$ of quadratic differentials for $\Gamma(t)$. As before, given $(t, \phi) \in Q$ there exist $f: D(t) \to P_1$ and $\rho: \Gamma \to PL(1, C)$ so that $\{f, z\} = \phi(z)$ and

$$(2.3) \qquad f(\gamma^t z) = \rho(\gamma) f(z) \quad \text{for all} \quad \gamma \in \Gamma, \ z \in D(t).$$

It is the dependence of the conjugacy class of ρ on (t, ϕ) that we wish to study.

2.4. To study the variation of ρ we need local coordinates in T_p. Fix $t_0 \in T_p$ and put $D = D(t_0)$, $\Gamma = \Gamma(t_0)$. Let W be the vector space of functions $\mu: C \to C$ such that

$$(2.4) \qquad\qquad \mu(z) = 0 \quad \text{for all} \quad z \notin D$$

$$(2.5) \qquad \mu(z) = \lambda(z)^{-2} \overline{\phi(z)} \quad \text{for some} \quad \phi \in Q(\Gamma) \text{ and all } z \in D,$$

where $\lambda(z)|dz|$ is the Poincaré metric on D.

Let W_0 be the set of $\mu \epsilon W$ with

$$\sup \{|\mu(z)|; \, z \, \epsilon D\} < 1 .$$

For each $\mu \epsilon W_0$ there is a unique quasiconformal map $w = w^\mu(z)$ of P_1 onto itself satisfying the Beltrami equation

$$w_{\bar{z}} = \mu w_z$$

in C and having the behavior

$$w(z) = z + 0(1/z) \quad \text{as} \quad z \to \infty .$$

For each $\mu \epsilon W_0$, $w^\mu(D)$ is a bounded Jordan region and the group of Mobius transformations $\gamma^\mu = w^\mu \circ \gamma \circ (w^\mu)^{-1}$, $\gamma \epsilon \Gamma$, is quasifuchsian with invariant domain $w^\mu(D)$.

There is a neighborhood W_1 of zero in W_0 which provides a coordinate system at t_0 in such a way that for each $\mu \epsilon W_1$, $D(\mu)$ is the region $w^\mu(D)$ and $\Gamma(\mu)$ is the group of Möbius transformations γ^μ. For $\mu \epsilon W_1$ and $\phi \epsilon Q(\Gamma(\mu))$, (2.3) says that

$$f \circ \gamma^\mu = \rho(\gamma) \circ f$$

in $D(\mu)$ if $\{f, z\} = \phi(z)$. Therefore, in D we have

$$\rho(\gamma) \circ f \circ w^\mu = f \circ \gamma^\mu \circ w^\mu = f \circ w^\mu \circ \gamma .$$

Put $h = f \circ w^\mu : D \to P_1$. Then h is a C^∞ local homeomorphism, and

$$(2.6) \qquad h(\gamma z) = \rho(\gamma) h(z) \quad \text{for all} \quad \gamma \epsilon \Gamma \quad \text{and} \quad z \, \epsilon D .$$

2.5. It will be convenient to normalize so that each $\phi \epsilon Q(\Gamma(\mu))$ determines a unique function $g = g_\phi$ with $\{g, z\} = \phi(z)$ in $D(\mu)$. For this purpose we choose $z_0 \epsilon D$ and choose W_1 so small that $z_0 \epsilon D(\mu)$ for all $\mu \epsilon W_1$. We define $g : D(\mu) \to P_1$ to be the unique meromorphic function satisfying

(2.7) $g(z_0) = g''(z_0) = 0$, $g'(z_0) = 1$, and $\{g,z\} = \phi(z)$ for all $z \in D(\mu)$.

The most general function f with $\{f,z\} = \phi(z)$ is again $f = A \circ g$ for some $A \in PL(1, C)$, and the function h in (2.6) has the form $h = A \circ g \circ w^\mu$. The monodromy homomorphism $\rho : \Gamma \to PL(1, C)$ in (2.6) is thus a function of A, μ, and ϕ.

3. *The variation of* h

3.1. In order to study the dependence of h and ρ on A, μ, and ϕ, we assume that $(A, \mu, \phi) = (A_\tau, \mu_\tau, \phi_\tau)$ depends on a complex parameter τ in such a way that $A \in PL(1, C)$, $\mu \in W_1$, $\phi \in Q(\Gamma(\mu))$ for all τ, and

(3.1) $\qquad \mu = \tau \dot{\mu} + o(\tau)$, $A = A_0 + \tau \dot{A} + o(\tau)$, $\phi = \phi_0 + \tau \dot{\phi} + o(\tau)$.

Our first goal is to find the variation of h.

Now $h = A \circ g \circ w^\mu$, where $g : D(\mu) \to P_1$ satisfies (2.7). The theory of the Beltrami equation (see for instance [12]) gives

(3.2) $\qquad\qquad\qquad w^\mu(z) = z + \tau \dot{w}(z) + o(\tau)$,

where $\dot{w}(z)$ is given explicitly by

(3.3) $\qquad\qquad\qquad \dot{w}(z) = -\frac{1}{\pi} \iint_D \dot{\mu}(\zeta)(\zeta - z)^{-1} d\xi\, d\eta$.

The most important properties of $\dot{w}$ are that $\dot{w}_{\bar{z}} = \dot{\mu}$ and that $\dot{w} = 0$ if and only if $\dot{\mu} = 0$.

Since the solution of the equation $\{g,z\} = \phi(z)$ is the ratio of two independent solutions of the linear equation

$$u''(z) = -\frac{1}{2}\phi(z)u(z),$$

it is evident that

(3.4) $\qquad\qquad\qquad g(z) = g_0(z) + \tau \dot{g}(z) + o(\tau)$, where of course $\{g_0, z\} = \phi_0$. Therefore

$$(3.5) \qquad\qquad h(z) = h_0(z) + \tau\dot{h}(z) + o(\tau) ,$$

with $h_0 = A_0 \circ g_0$ and $\{h_0, z\} = \phi_0(z)$.

    3.2. It will be useful to be slightly more explicit. If we write $f = A \circ g$, then $\{f, z\} = \phi(z)$ in $D(\mu)$, $f_0 = h_0$, and

$$(3.6) \qquad\qquad f(z) = f_0(z) + \tau\dot{f}(z) + o(\tau) ,$$

with

$$(3.7) \qquad\qquad \dot{f}(z) = \dot{A}(g_0(z)) + \dot{g}(z) A_0'(g_0(z)) .$$

The function $\dot{f}$ is meromorphic in D, with possible poles at the poles of f_0, but since all poles of f are simple, the function

$$(3.8) \qquad\qquad f^*(z) = \dot{f}(z) f_0'(z)^{-1} = \dot{f}(z) h_0'(z)^{-1}$$

is holomorphic in D.

    Now $h(z) = f(w(z))$, so (3.2) and (3.6) give

$$(3.9) \qquad\qquad \dot{h}(z) = \dot{f}(z) + h_0'(z)\dot{w}(z) .$$

    3.3. The following lemma summarizes the above observations.

LEMMA 1. *If A, μ, and ϕ satisfy (3.1), then $h = A \circ g_\phi \circ w^\mu$ satisfies (3.5) with $\dot{h}$ given by (3.9). In particular*

$$(3.10) \qquad\qquad h^*(z) = \dot{h}(z) h_0'(z)^{-1} = f^*(z) + \dot{w}(z)$$

is a C^∞ function in D.

    3.4. For our main purpose the explicit form of $\dot{h}$ is unimportant. We need to know merely that h^* is a C^∞ function with $h^*_{\bar{z}} = \dot{w}_{\bar{z}} = \dot{\mu}$, and this additional fact.

LEMMA 2. $h^* = 0$ *in D if and only if* $\dot{A} = \dot{\mu} = \dot{\phi} = 0$.

Proof. If $\dot\mu = 0$, then $\dot w = 0$, and if $\dot\phi = 0$, then $\dot g = 0$. Therefore $\dot h = h^* = 0$ if $\dot A = \dot\mu = \dot\phi = 0$. Conversely, if $h^* = 0$, then (since f^* is holomorphic in D)

$$0 = h^*_{\bar z} = \dot w_{\bar z} = \dot\mu ,$$

so $\dot\mu = \dot w = 0$. Therefore, by (3.8) and (3.10) $\dot f = f^* = 0$, so $\phi(z) = \{f, z\} = \{f_0, z\} + o(\tau)$, and $\dot\phi = 0$. Therefore $\dot g = 0$, and (3.7) gives $\dot A = 0$. The lemma is proved.

3.5. For future reference we wish to give the precise relationship between $\dot\phi$ and f^*.

LEMMA 3. $\dot\phi = (f^*)''' + 2\phi(f^*)' + \phi_0' f^*$.

The proof is merely a computation. One method is to put $\sigma = f''/f' = \sigma_0 + \tau\dot\sigma + o(\tau)$, with $\sigma_0 = f_0''/f_0'$. Then

$$\dot\sigma f_0' = \dot f'' - \sigma_0 \dot f'.$$

Since $\dot f = f_0' f^*$, the derivatives of $\dot f$ can be expressed in terms of f_0' and f^*. That gives

$$\dot\sigma = (f^*)'' - (\sigma_0 f^*)'.$$

Finally, $\phi = \sigma' - \frac{1}{2}\sigma^2$, so $\dot\phi = \dot\sigma' - \sigma_0\dot\sigma$. The lemma follows.

4. The variation of the monodromy map

4.1. Our main result describes the variation of the monodromy homomorphism ρ in terms of h^*.

THEOREM 1. *Let* A, μ, *and* ϕ *satisfy* (3.1), *let* $h = A \circ g_\phi \circ w^\mu$, *and let* $\rho: \Gamma \to PL(1, \mathbb{C})$ *satisfy* (2.6). *Then*

$$(4.1) \qquad \rho(\gamma) = \rho_0(\gamma) + \tau\dot\rho(\gamma) + o(\tau) \qquad \text{for all} \quad \gamma \in \Gamma ,$$

and

$$(4.2) \quad \dot\rho(\gamma)(h_0(z)) = (h_0 \circ \gamma)'(z)[h^*(\gamma z)\gamma'(z)^{-1} - h^*(z)] \quad \text{for all} \quad \gamma \in \Gamma, \, z \in D.$$

Proof. Fix $\gamma \in \Gamma$ and a region $D_0 \subset D$ so that $h(D_0)$ and $h(\gamma(D_0))$ are bounded regions in $\mathbf{C}$ (uniformly in τ near $\tau = 0$). Now in D_0

$$\rho(\gamma)(h(z)) = \rho(\gamma)(h_0(z) + \tau \dot{h}(z) + o(\tau))$$
$$= \rho(\gamma) h_0(z) + \tau \rho(\gamma)'(h_0(z)) \dot{h}(z) + o(\tau) ,$$

and

$$h(\gamma z) = h_0(\gamma z) + \tau \dot{h}(\gamma z) + o(\tau)$$
$$= \rho_0(\gamma) h_0(z) + \tau \dot{h}(\gamma z) + o(\tau) .$$

Therefore, by (2.6),

$$\rho(\gamma)(h_0(z)) = \rho_0(\gamma)(h_0(z)) + \tau[\dot{h}(\gamma z) - \dot{h}(z)\rho(\gamma)'(h_0(z))] + o(\tau) .$$

Since $h_0(D)$ is an open set in $\mathbf{C}$, (4.1) follows, with

$$\dot{\rho}(\gamma)(h_0(z)) = \dot{h}(\gamma z) - \dot{h}(z)\rho_0(\gamma)'(h_0(z))$$
$$= h^*(\gamma z) h_0'(\gamma z) - h^*(z)\rho_0(\gamma)'(h_0(z)) h_0'(z)$$
$$= h^*(\gamma z) h_0'(\gamma z) - h^*(z)(h_0 \circ \gamma)'(z)$$
$$= (h_0 \circ \gamma)'(z)[h^*(\gamma z)/\gamma'(z) - h^*(z)] .$$

That proves (4.2) for $z \in D_0$. Since both sides of (4.2) are meromorphic in D, the theorem is proved.

4.2. The theorem has two important corollaries.

COROLLARY 1. $\dot{\rho}(\gamma) = 0$ *for all* $\gamma \in \Gamma$ *if and only if* $h^* = 0$ *in* D.

Proof. Since $\rho(\gamma) \in PL(1, \mathbf{C})$ for all τ, (4.1) implies that for all $\gamma \in \Gamma$ the function

$$(4.3) \qquad\qquad P_\gamma = \dot{\rho}(\gamma)(\rho(\gamma)')^{-1}$$

is a polynomial of degree at most two. Therefore $\dot{\rho}(\gamma)$ is completely determined by its values on the open set $h_0(D)$ and, by (4.2), $\dot{\rho}(\gamma)$ is zero for all $\gamma \in \Gamma$ if and only if

$$(4.4) \qquad h^*(\gamma z)\gamma'(z)^{-1} = h^*(z) \quad \text{for all} \quad z \in D \quad \text{and} \quad \gamma \in \Gamma .$$

Obviously (4.4) holds if $h^* = 0$ in D. Conversely, if (4.4) holds then h^* defines a C^∞ vector field of type $(1,0)$ on the Riemann surface $X = D/\Gamma$. Therefore for all $\phi \in Q(\Gamma)$ we have

$$0 = \iint_X (\phi(z)h^*(z))_{\bar z}\, dx\, dy = \iint_{D/\Gamma} \phi(z)\, h^*_{\bar z}(z)\, dx\, dy$$

$$= \iint_{D/\Gamma} \phi(z)\, \dot\mu(z)\, dx\, dy .$$

But $\mu \in W_1$ for all τ, so $\mu \in W$ and (2.5) gives

$$\dot\mu(z) = \lambda(z)^{-2}\, \overline{\phi(z)}$$

in D for some $\phi \in Q(\Gamma)$. Therefore

$$0 = \iint_{D/\Gamma} |\phi(z)|^2 \lambda(z)^{-2}\, dx\, dy ,$$

so $\phi = \dot\mu = 0$ in D. Therefore, by (3.3) and (3.10), $\dot w = 0$ and $h^* = f^*$ is holomorphic in D. But all holomorphic vector fields on X are zero, so $h^* = 0$ in D. That proves the corollary.

4.3. A few words of explanation are needed before we state the next corollary. Recall that for each (t, ϕ) in the vector bundle Q there exist $f: D(t) \to P_1$ and $\rho \in \mathrm{Hom}(\Gamma, \mathrm{PL}(1, C))$ so that (2.3) holds. The point (t, ϕ) determines ρ up to conjugation by some $A \in \mathrm{PL}(1, C)$. Hence there is a well defined *monodromy map*

$$Q \to \mathrm{Hom}(\Gamma, \mathrm{PL}(1, C))/\mathrm{PL}(1, C)$$

which associates to each $(t, \phi) \in Q$ the conjugacy class of ρ.

COROLLARY 2. *The monodromy map is a holomorphic local homeomorphism.*

Proof. It is well known (see [4], [5], [7]) that Q is a complex manifold of dimension $6p-6$, that $\mathrm{Hom}(\Gamma, \mathrm{PL}(1, \mathbf{C}))$ is a complex algebraic variety of dimension $6p-3$, and that each monodromy homomorphism ρ is a manifold point of $\mathrm{Hom}(\Gamma, \mathrm{PL}(1, \mathbf{C}))$. Theorem 1 shows that locally the monodromy map is induced by a holomorphic map from $\mathrm{PL}(1, \mathbf{C}) \times Q$ into $\mathrm{Hom}(\Gamma, \mathrm{PL}(1, \mathbf{C}))$. Lemma 2 and Corollary 1 imply that the derivative of that map is injective. Since its domain and range have the same dimension it is a local homeomorphism. Corollary 2 follows because replacing (I, t, ϕ) by (A, t, ϕ) in $\mathrm{PL}(1, \mathbf{C}) \times Q$ has the effect of conjugating ρ by A.

5. An exact sequence in Eichler cohomology

5.1. It is well known that Γ acts on the vector space Π of polynomials of degree at most two by

$$(5.1) \qquad P \cdot \gamma = (P \circ \rho(\gamma))(\rho(\gamma)')^{-1}$$

and that the tangent space at ρ to $\mathrm{Hom}(\Gamma, \mathrm{PL}(1, \mathbf{C}))$ is the space of cocycles $Z^1(\Gamma, \Pi)$ for this action (see [2] and [4]). The map from the tangent space onto $Z^1(\Gamma, \Pi)$ is given by sending $\dot{\rho}$ to the cocycle $\gamma \mapsto P_\gamma$ defined by (4.3). The tangent space to

$$\mathrm{Hom}(\Gamma, \mathrm{PL}(1, \mathbf{C}))/\mathrm{PL}(1, \mathbf{C})$$

at the point determined by ρ is the cohomology group $H^1(\Gamma, \Pi)$. When ρ is the identity map, (5.1) is the Eichler action and $H^1(\Gamma, \Pi)$ is the first Eichler cohomology group.

5.2. Formula (4.2) suggests that we should study the functions

$$(5.2) \qquad \sigma_\gamma(z) = \dot{\rho}(\gamma)(h_0(z))(h_0 \circ \gamma)'(z)^{-1} = P_\gamma(h_0(z)) h_0'(z)^{-1}.$$

They belong to the vector space V of holomorphic functions $\sigma = (P \circ h_0)(h_0')^{-1}$ on D, $P \in \Pi$. V is a vector space of dimension three, since the linear map $P \mapsto (P \circ h_0)(h_0')^{-1}$ from Π to V is bijective. Under that map the action (5.1) of Γ on Π transforms to an action on V given by

$$(5.3) \qquad \sigma \cdot \gamma = (\sigma \circ \gamma)(\gamma')^{-1} \qquad \text{for all} \quad \sigma \in V \quad \text{and} \quad \gamma \in \Gamma .$$

Indeed, if $\sigma = (P \circ h_0)(h_0')^{-1}$, then

$$\begin{aligned}
\sigma \cdot \gamma &= [(P \cdot \gamma) \circ h_0](h_0')^{-1} \\
&= (P \circ \rho(\gamma) \circ h_0)[(\rho(\gamma) \circ h_0)']^{-1} \\
&= (P \circ h_0 \circ \gamma)[(h_0 \circ \gamma)']^{-1} = (\sigma \circ \gamma)(\gamma')^{-1} ,
\end{aligned}$$

by (2.6). We can therefore identify the cohomology group $H^1(\Gamma, V)$ with the tangent space to

$$\mathrm{Hom}(\Gamma, PL(1, \mathbf{C}))/PL(1, \mathbf{C})$$

at the point determined by ρ.

The following description of V is also of interest:

LEMMA 4. *Let σ be a holomorphic function on D. Then $\sigma \in V$ if and only if*

$$(5.4) \qquad\qquad \sigma''' + 2\phi_0 \sigma' + \phi_0' \sigma = 0 ,$$

where $\phi_0(z) = \{h_0, z\}$ in D.

Proof. Since the set of solutions of (5.4) has dimension three, it suffices to prove that every function $\sigma = (P \circ h_0)(h_0')^{-1}$ in V satisfies (5.4). The proof is a simple computation.

5.3. At any point $(t_0, \phi_0) \in Q$, the derivative of the projection $Q \to T_p$ fits into a short exact sequence whose non-trivial terms are the tangent spaces to the fiber $Q(\Gamma)$, to Q, and to T_p. The tangent space to T_p may be identified with the vector space W defined in §2.4 or, more canonically, with the dual space $Q(\Gamma)^*$ of $Q(\Gamma)$. The tangent space to Q is isomorphic to $H^1(\Gamma, V)$ via the monodromy map, and the fiber $Q(\Gamma)$ is its own tangent space. Thus we obtain

THEOREM 2 (Gunning [3]). *There is an exact sequence*

$$(5.5) \qquad 1 \to Q(\Gamma) \to H^1(\Gamma, V) \to Q(\Gamma)^* \to 1 \,.$$

5.4. The results of §§3 and 4 allow us to describe the maps in (5.5) rather explicitly. Let $\mu = \dot{\mu} = 0$, $A = A_0 = I$, and $\phi = \phi_0 + t\dot{\phi}$ in (3.1), so that $\phi \in Q(\Gamma)$ represents a variation along the fiber $Q(\Gamma)$. Then $h^* = f^*$ is holomorphic in D, and

$$(5.6) \qquad \dot{\phi} = (h^*)''' + 2\phi_0 (h^*)' + \phi_0' h^*$$

by Lemma 3. Theorem 1 and (5.2) give

$$(5.7) \qquad \sigma_\gamma(z) = h^*(\gamma z)\gamma'(z)^{-1} - h^*(z) \quad \text{for all } \gamma \in \Gamma \text{ and } z \in D \,.$$

The map from $Q(\Gamma)$ to $H^1(\Gamma, V)$ therefore assigns to each $\dot{\phi} \in Q(\Gamma)$ the cohomology class of the cocycle $\gamma \mapsto \sigma_\gamma$ defined by (5.7), where h^* satisfies (5.6). Notice that Lemma 4 guarantees that two solutions of (5.6) will determine the same cohomology class.

5.5. Next we describe the map from $H^1(\Gamma, V)$ to $Q(\Gamma)^*$. Since the monodromy map determines an isomorphism of tangent spaces, Theorem 1 shows that each cohomology class in $H^1(\Gamma, V)$ comes from a cocycle of the form (5.7), with h^* in C^∞ as in Lemma 1. The corresponding tangent vector to T_p is $\dot{\mu} = h_{\bar{z}}$, and the linear functional ℓ on $Q(\Gamma)$ is

$$(5.8) \qquad \ell(\phi) = \iint_{D/\Gamma} \phi(z) h^*_{\bar{z}}(z)\, dx\, dy \quad \text{for all } \phi \in Q(\Gamma) \,.$$

5.6. The exact sequence (5.5) was described by Gunning in §5 of [3]. It is a generalization to arbitrary projective structures of an exact sequence for Eichler cohomology whose most general form (for finitely generated nonelementary Kleinian groups) was given by Kra [11]. The interpretation of (5.5) as an infinitesimal version of the fibration

$$1 \to Q(\Gamma) \to Q \to T_p \to 1$$

appears to be new.

CORNELL UNIVERSITY
ITHACA, NEW YORK 14853

REFERENCES

[1] Bers, L., Fiber spaces over Teichmüller spaces, *Acta Math.* 130 (1973), 89-126.

[2] Gardiner, F., and Kra, I., Stability of Kleinian groups, *Indiana Univ. Math. J.* 21 (1972), 1037-1059.

[3] Gunning, R. C., Special coordinate coverings of Riemann surfaces, *Math. Ann.* 170 (1967), 67-86.

[4] _________, Analytic structures on the space of flat vector bundles over a compact Riemann surface, in "Several Complex Variables II, Maryland 1970," Springer Lecture Notes 185 (1971), 47-62.

[5] _________, Affine and projective structures on Riemann surfaces, to appear.

[6] Hejhal, D. A., Monodromy groups and linearly polymorphic functions, in "Discontinuous groups and Riemann surfaces" (L. Greenberg, ed.), Ann. of Math. Studies 79 (1974), 247-261.

[7] _________, Monodromy groups and linearly polymorphic functions, *Acta Math.* 135 (1975), 1-55.

[8] _________, The variational theory of linearly polymorphic functions, *J. Analyse Math.* 30 (1976), 215-264.

[9] _________, Monodromy groups and Poincaré series, *Bull. Amer. Math. Soc.* 84 (1978), 339-376.

[10] Hubbard, J. H., The monodromy of projective structures, these proceedings.

[11a] Kra, I., On cohomology of Kleinian groups, *Ann. of Math.* 89 (1969), 533-556.

[11b] _________, On cohomology of Kleinian groups: II, *Ann. of Math.* 90 (1969), 576-590.

[12] Lehto, O., Quasiconformal homeomorphisms and Beltrami equations, in "Discrete groups and automorphic functions," Academic Press, London-New York (1977), 121-142.

VANISHING THETA CONSTANTS

Hershel M. Farkas[*]

Introduction

Several years ago in a paper entitled ''The Vanishing of a Theta Constant is a Peculiar Phenomenon'' [R], H. E. Rauch showed that if we consider the locus $\vartheta[{}^{\varepsilon}_{\varepsilon'}](0, \pi) = 0$ restricted to the Torelli sublocus of $\mathfrak{S}_g$ then the gradient of $\vartheta[{}^{\varepsilon}_{\varepsilon'}](0, \pi)$ with respect to the Torelli coordinates vanishes on this locus. Here $[{}^{\varepsilon}_{\varepsilon'}]$ is an even theta characteristic. For notation and background material the reader is referred to [RF].

Actually, this phenomenon is not at all peculiar but rather one that should be expected. In fact if (z_0, π_0) is a point in $C^g \times \mathfrak{S}_g$ such that $\vartheta(z_0, \pi_0) = 0$ and $\frac{\partial}{\partial z_i} \vartheta(z_0, \pi_0) = 0$ $i = 1, \cdots, g$ with π_0 in the Torelli sublocus of $\mathfrak{S}_g$, then the gradient of $\vartheta(z_0, \pi)$ with respect to the Torelli coordinates vanishes at such a point. In fact an obvious modification of Rauch's proof already shows this and nothing further is required. We however will give a different proof for the generic point (z_0, π_0) in genus 4.

The interesting and perhaps ''peculiar'' thing which happens is that in the former case (vanishing theta constant) not only does the gradient vanish but there is reason to suspect that the second and third order

---

[*]Research partially supported by NSF Grant MCS 7604969A01

derivatives with respect to the Torelli coordinates also vanish. This will not follow however from the results of this note.

I. It is well known that the Torelli sublocus of $\mathfrak{S}_g$ is a complex analytic manifold of dimension $3g - 3 \, (g \geq 2)$. Let us choose $\varepsilon_1, \cdots, \varepsilon_{3g-3}$ as analytic coordinates on the Torelli space so that we also have on the sublocus of $\mathfrak{S}_g$ corresponding to the Torelli space the coordinates π_{ij} $i, j = 1, \cdots, g$ are analytic functions of the $3g - 3$ coordinates $\varepsilon_1, \cdots, \varepsilon_{3g-3}$. The coordinates we choose are those we have called in [F] Patt's coordinates. We recall that $\dfrac{\partial \pi_{ij}}{\partial \varepsilon_k} = \vartheta_i \vartheta_j(P_k)$ where $\vartheta_1, \cdots, \vartheta_g$ are a basis for the holomorphic differentials on the underlying Riemann surface S with the property $\int_{\delta_j} \vartheta_i = \pi_{ij}$ and $P_1 \cdots P_{3g-3}$ are the points of S where the variations are performed [P, FR].

The main idea here is one which has been exploited in [EF, EFMR] and that is the existence of an analytic function $Z(\varepsilon_1, \cdots, \varepsilon_{3g-3})$ which picks out a singular point for the theta function corresponding to $\pi(\varepsilon_1, \cdots, \varepsilon_{3g-3})$. In more fancy language we can think of a fiber space F with base manifold the Torelli space whose fiber over each point in the base is the complex torus or Jacobian variety of the underlying Riemann surface. The existence of such a function is then simply the existence of a local analytic section. The crucial point now is that we have the analytic equations

$$(1) \qquad\qquad\qquad \vartheta(z(\varepsilon),\ \pi(\varepsilon)) \equiv 0$$

$$(2) \qquad\qquad\qquad \frac{\partial \vartheta}{\partial z_x}(z(\varepsilon),\ \pi(\varepsilon)) \equiv 0 \qquad x = 1, \cdots, g$$

where $z(\varepsilon) = z(\varepsilon_1 \cdots \varepsilon_{3g-3})$ and $\pi(\varepsilon) = \pi(\varepsilon_1 \cdots \varepsilon_{3g-3})$.

We now begin by expanding (1) in Taylor series about the point (z_0, π_0). We find by equating the coefficient of ε_r to zero that

$$(3) \quad \sum_{i=1}^{g} \frac{\partial \theta}{\partial z_i}(z_0, \pi_0) \frac{\partial z_i}{\partial \varepsilon_r} + \sum_{i \leq j = 1}^{g} \frac{\partial \vartheta}{\partial \pi_{ij}}(z_0, \pi_0) \frac{\partial \pi_{ij}}{\partial \varepsilon_r} = 0 \quad r = 1, \cdots, 3g-3 .$$

Since by hypothesis $\frac{\partial \vartheta}{\partial z_i}(z_0, \pi_0) = 0$ $i = 1, \cdots, g$ we immediately obtain

$$(4) \quad \sum_{i \leq j = 1}^{g} \frac{\partial \vartheta}{\partial \pi_{ij}}(z_0, \pi_0) \frac{\partial \pi_{ij}}{\partial \varepsilon_r} = 0 \qquad r = 1, \cdots, 3g-3 .$$

Let us now turn our attention to the function $\vartheta(z_0, \pi(\varepsilon))$ and view this as a function on the Torelli space near the point π_0. Expanding this function in Taylor series we find the coefficient of ε_r to be

$$(5) \quad \sum_{i \leq j = 1}^{g} \frac{\partial \vartheta}{\partial \pi_{ij}}(z_0, \pi_0) \frac{\partial \pi_{ij}}{\partial \varepsilon_r} \qquad r = 1, \cdots, 3g-3 .$$

If we compare (5) with (4) we see immediately that the coefficient of ε_r $r = 1, \cdots, 3g-3$ vanishes or that the gradient of $\vartheta(z_0, \pi(\varepsilon))$ vanishes whenever z_0 is a singular point for the associated π.

In particular we see that choosing $z_0 = 0$ we have Rauch's result. Moreover, Rauch, using his trick of expanding on the Riemann surface and then using the heat equations would obtain the same result for $z_0 \neq 0$. All that we have said is valid provided the analytic function $z(\varepsilon)$ exists.

II. If it were true that there always exists a local analytic section we would be finished. This is however a non-trivial matter and we as of yet do not have a complete answer to this question. We shall therefore consider in more detail the case when $g = 4$ and the remainder of this paper is concerned only with the case $g = 4$.

There are actually two cases to consider when $g = 4$. The first one we deal with is the case where the point (z_0, π_0) has the property that the rank of the matrix $\left(\frac{\partial^2 \vartheta}{\partial z_j \partial z_i}(z_0, \pi_0) \right)$ is equal to four. In this case

(which is the generic situation) the existence of a local analytic section
is an immediate consequence of the implicit function theorem and recent
work of A. Beauville [B].

LEMMA 1. *Let* (z_0, π_0) *be a point in* $\mathbf{C}^4 \times \mathfrak{S}_4$ *such that* $\vartheta(z_0, \pi_0) = 0$,
$\frac{\partial \theta}{\partial z_i}(z_0, \pi_0) = 0$ $i = 1, \cdots, 4$ *and rank* $\left(\frac{\partial^2 \vartheta}{\partial z_j \partial z_i}(z_0, \pi_0)\right)$ *equals four. Then
there exists a function* $z(\pi)$ *from* $\mathfrak{S}_4$ *to* $\mathbf{C}^4$ *which is analytic in a suffi-
ciently small neighborhood of* π_0 *with the property* $z(\pi_0) = z_0$ *and*
$\frac{\partial \vartheta}{\partial z_i}(z(\pi), \pi) \equiv 0$ $i = 1, \cdots, 4$.

Proof. The above lemma is just the implicit function theorem.

The above lemma gives us a well-defined analytic function from $\mathfrak{S}_4$
to $\mathbf{C}^4$ which satisfies the requirements for a local analytic section except for the requirement that $\vartheta(z(\pi), \pi) \equiv 0$, and the function is defined
on $\mathfrak{S}_4$ and not on the Torelli sublocus of $\mathfrak{S}_4$. In view of this last remark it is natural to consider the locus $\vartheta(z(\pi), \pi) = 0$ where $z(\pi)$ is
defined by the lemma. Certainly if we restrict ourselves to this locus in
$\mathfrak{S}_4$ then all the requirements for the local analytic section would be
satisfied and the only point at issue would be the domain of definition of
$z(\pi)$.

The locus $\vartheta(z(\pi), \pi) = 0$ is a nine dimensional analytic manifold near
$\pi = \pi_0$. The work of Beauville previously cited [B] shows that each π
sufficiently close to π_0 in this manifold is also in the Torelli sublocus
of $\mathfrak{S}_4$ which is also nine dimensional. Hence the two loci agree set-
theoretically. If we therefore restrict the domain of the function $z(\pi)$ to
the locus $\vartheta(z(\pi), \pi) = 0$ which is also the Torelli sublocus we have produced the local analytic section. We have therefore proven the following:

THEOREM 1. *Let* (z_0, π_0) *be a point in* $\mathbf{C}^4 \times \mathfrak{S}_4$ *such that* π_0 *is in
the Torelli subspace of* $\mathfrak{S}_4$ *and such that* $\vartheta(z_0, \pi_0) = 0$ $\frac{\partial \vartheta}{\partial z_i}(z_0, \pi_0) = 0$
$i = 1, \cdots, 4$. *Furthermore assume that the matrix* $\left(\frac{\partial^2 \vartheta}{\partial z_j \partial z_i}(z_0, \pi_0)\right)$ *has*

rank equal to four. If F is the fiber space with base manifold the Torelli space and fiber over each point the Jacobi variety of the underlying Riemann surface then there exists a local analytic section which chooses a singular point for the associated theta function.

It then follows from the remarks at the beginning of this paper that we also have the following:

COROLLARY. *Let* (z_0, π_0) *satisfy the hypothesis of the theorem. Then the gradient of* $\vartheta(z_0, \pi(\varepsilon))$ *vanishes at* (z_0, π_0).

The second case we need consider is the case $g = 4$ and rank of the matrix $\left(\dfrac{\partial^2 \vartheta}{\partial z_j \partial z_i} (z_0, \pi_0)\right)$ equals three. In this case the argument using the implicit function theorem does not work. Unfortunately the case of $z_0 = 0$ or z_0 a point of order 2 falls into this category so that our argument does not give Rauch's result.

III. The purpose of this note as explained in the introduction has been to show that the vanishing of the gradient of $\vartheta[^{\varepsilon}_{\varepsilon'}](0, \pi(\varepsilon))$ at a point where $\vartheta[^{\varepsilon}_{\varepsilon'}](0, \pi_0) = 0$ is not at all peculiar. The "peculiar" aspect of the situation is that there may be no local analytic section through such a point explained by the fact of the matrix $\left(\dfrac{\partial^2 \theta}{\partial z_j \partial z_i} (z_0, \pi_0)\right)$ having rank three in place of four.

Here we have dealt exclusively with the case of genus 4. We can, in fact show that for any genus $g \geq 4$ a local analytic section exists through a point (z_0, π_0) when the rank of $\left(\dfrac{\partial^2 \theta}{\partial z_j \partial z_i} (z_0, \pi_0)\right)$ equals four. This will be done in another publication. Our interest and motivation for a proof based on the idea of a local analytic section will also be explained there.

We close with the following remark. Even though our result does not give Rauch's result it certainly suggests it very strongly. In a similar fashion it suggests that not only does the gradient vanish when $z_0 = 0$ but that the second and third order derivatives also vanish.

THE HEBREW UNIVERSITY
JERUSALEM, ISRAEL

REFERENCES

[B]   Beauville, A. Prym Varieties and the Schottky Problem, Inventiones (1977), pp. 149-196.

[EF]   Ehrenpreis, L., Farkas, H. M. Some Refinements of the Poincaré Period Relation. Discontinuous Groups and Riemann Surfaces. Annals of Math. Studies, Princeton Univ. Press (1974), pp. 105-120.

[EFMR]   Ehrenpreis, L., Farkas, H. M., Martens, H., Rauch, H. E. On the Poincaré Relation "Contributions to Analysis" Academic Press (1974), pp. 125-132.

[F]   Farkas, H. M. Special Divisors and Analytic Subloci of Teichmüller Space, American Journal of Mathematics (1966), pp. 881-901.

[FR]   Farkas, H. M., Rauch, H. E. Period Relations of Schottky Type on Riemann Surfaces, Annals of Math. (1970), pp. 434-461.

[P]   Patt, C. Variations of Teichmüller and Torelli Surfaces, Journal d'Analyse Math. (1963), pp. 221-247.

[RF]   Rauch, H. E., Farkas, H. M. "Theta Functions with Applications to Riemann Surfaces, Williams & Wilkins (1974).

ANALYTIC TORSION AND PRYM DIFFERENTIALS

John Fay[*]

Introduction

Let M be a compact Riemann surface of genus g given as the quotient of the upper half-plane H by a purely hyperbolic Fuchsian group Γ with fundamental domain $\mathcal{D} = \Gamma \backslash H$. If M is given a marking by a canonical dissection along oriented paths $A_1, B_1, \cdots, A_g B_g \in \pi_1(M, p_0)$ for some basepoint $p_0 \in M$, these paths lifted to H give rise to $2g$ generators in some presentation of Γ, again denoted by $A_1, \cdots, B_g$. If one forms the period matrix τ from the normalized holomorphic differentials $v_1, \cdots, v_g$ for the given marking, then the Jacobi variety $J_0(M) = C^g/(I, \tau)$ can be identified with the set of characters $\chi : \pi_1(M, p_0) \to S^1$ via:

$$
(1) \qquad
\begin{array}{ccc}
\hat{x} - \tau x \in C^g & \xrightarrow{\;\simeq\;} (x, \hat{x}) \in R^{2g} \subset\joinrel\longrightarrow C^{2g} \\[2ex]
\big\downarrow & \big\downarrow \\[2ex]
[\hat{x} - \tau x] \in J_0(M) \xrightarrow{\;\simeq\;} (e^{2\pi i x_j}, e^{2\pi i \hat{x}_j}) \in (S^1)^{2g} \subset\joinrel\longrightarrow (C^*)^{2g}
\end{array}
\left.\begin{array}{l}
\chi(A_j) = e^{2\pi i x_j} \\[2ex]
\chi(B_j) = e^{2\pi i \hat{x}_j}
\end{array}\right\}
$$

Now for any character χ, the Laplace-Beltrami operator $D = 4y^2 \dfrac{\partial^2}{\partial z \, \partial \bar{z}}$ for the Poincaré metric $|dz| = \dfrac{dx \, dy}{y^2}$ on H is self-adjoint on a dense

---

[*]Research partially supported by Sloan Foundation.

subspace of the Hilbert space L^2_χ of all functions $f(z)$ on H satisfying

$$(2) \qquad \iint_{\mathcal{D}} |f|^2 \,\boxed{dz} < \infty \quad \text{and} \quad f(\gamma z) = \chi(\gamma)\,f(z) \quad \text{for all} \quad \gamma \in \Gamma .$$

D has a purely discrete spectrum of eigenvalues $\lambda_n \downarrow -\infty$, and the M-P zeta function associated to D:

$$(3) \qquad \zeta(s, \chi) = \sum_{\lambda_n < 0} |\lambda_n|^{-s} = \frac{1}{\Gamma(s)} \int_0^{+\infty} t^{s-1}(\text{trace } e^{tD} - \delta)\,dt$$

(where $\delta = 1$ if $\chi \equiv 1$ and 0 otherwise) has an analytic continuation to the s-plane with only a simple pole of residue $g-1$ at $s = 1$ [4]. If we define the analytic torsion $T(\chi)$ (here the *square* of that in [6], [7]) by:

$$T(\chi) = T(\overline{\chi}) = \exp\left(-\frac{d\zeta}{ds}(0, \chi)\right) \geq 0$$

then $T(\chi)$ becomes a well-defined real-analytic even function on $J_0(M)$ with a zero of second order at the origin $\chi \equiv I$ [7, p. 172]. Thus $T(\chi)$ is a conformal invariant for any fixed representation χ and so can be viewed as a real-analytic function of the period matrix τ, $\overline{\tau}$ invariant under the change of marking $\begin{pmatrix}\tilde{A}\\\tilde{B}\end{pmatrix} = \begin{pmatrix}dc\\ba\end{pmatrix}\begin{pmatrix}A\\B\end{pmatrix}$, $\begin{pmatrix}ab\\cd\end{pmatrix} \in \text{Sp}(2g, Z)$, when (1) transforms by the rule:

$$(4) \qquad \begin{pmatrix}\tilde{\chi}(A)\\\tilde{\chi}(B)\end{pmatrix} = \begin{pmatrix}d & c\\b & a\end{pmatrix}\begin{pmatrix}\chi(A)\\\chi(B)\end{pmatrix} \quad \text{and} \quad \tilde{\tau} = (a\tau+b)(c\tau+d)^{-1} .$$

    In §1 below we show that $T(\chi)$ has an analytic continuation from $(S^1)^{2g}$ to $(C^*)^{2g}$ in (1), with a divisor of zeroes V determining the period matrix of the given Riemann surface with respect to the marking (1). Thus, by Torelli's theorem, two Riemann surfaces M and M', with canonical markings $\{A_i, B_i\}$ and $\{A'_i, B'_i\}$ will have the same torsion function defined for these markings via (1) if and only if there is a conformal homeomorphism of M onto M' sending $\{A_i, B_i\}$ to $\pm\{A'_i, B'_i\}$. In

§2, using a basic Prym differential of the second kind, we give a higher-genus analogue of the heat equation for $T(\chi)$ as a consequence of the variational formulas for the resolvent of D under quasi-conformal deformations of M.

I am grateful to Professor D. Mumford for emphasizing the importance of non-unitary representations, and for raising a question on the tangent cone to V which led to the conclusion of Theorem 1.3 below.

§1. *Analytic torsion on* $(\mathbf{C}^*)^{2g}$

We begin with the observation of Ray-Singer [6] that for any two non-trivial characters χ, χ',

$$\frac{T(\chi)}{T(\chi')} = \frac{Z(1,\chi)}{Z(1,\chi')}$$

(5)

$$= \exp \sum_{\text{primitive } \gamma} \sum_{n=1}^{+\infty} (\chi'^n(\gamma) - \chi^n(\gamma)) \left(2n \sinh(n\rho_\gamma) e^{(2s-1)n\rho_\gamma}\right)^{-1} \Bigg|_{s=1},$$

where $e^{2\rho_\gamma}$ is the magnification factor for the primitive hyperbolic transformation γ and

$$(6) \qquad Z(s,\chi) = \prod_\gamma \prod_{n=0}^\infty \left(1 - \chi(\gamma) e^{-2\rho_\gamma(n+s)}\right), \quad \mathrm{Re}\, s > 1$$

is the Selberg zeta-function [10]. To give an equivalent formulation of (5), let

$$Q_s(z,z') = \frac{-\Gamma^2(s)}{4\pi\Gamma(2s)} \left(\mathrm{sech}\,\tfrac{r}{2}\right)^{2s} F\left(s, s, 2s; \mathrm{sech}^2\tfrac{r}{2}\right), \quad \frac{z'-z}{z'-\bar{z}} = \mathrm{th}\,\tfrac{r}{2}\, e^{i\theta}$$

be the Green's function for D on H; then for χ unitary, the series

$$(7) \qquad G_s(z,z',\chi) = G_s(z',z,\chi^{-1}) = \sum_{\gamma\epsilon\Gamma} \chi(\gamma)^{-1} Q_s(z, \gamma z')$$

converges uniformly on compact subsets of $\mathrm{Re}\ s > 1$ to the kernel of the resolvent for D on L_χ^2 [9]. Since Q_s is the Laplace transform of the fundamental solution to the heat equation on H, the reasoning in [4, §4] can be applied to give:

$$(8) \quad \frac{1}{1-2s}\frac{Z'(s,\chi)}{Z(s,\chi)} = \mathrm{trace}(G_s - Q_s) = \iint_{\mathcal{D}} \lim_{z' \to z}\ (G_s(z,z',\chi) - Q_s(z,z'))\ \boxed{dz}$$

for all s.

In order to extend (5)-(8) to non-unitary χ, we again denote by L_χ^2 the Hilbert space of all functions f on H satisfying (2) (where the norm is now dependent on $\mathcal{D}$); then D is a non-self adjoint operator on a dense subspace of L_χ^2 and one has:

THEOREM 1.1. *With respect to the marking (1), $T(\chi)$ has a unique analytic continuation from $(S^1)^{2g}$ to a well-defined holomorphic function of $\chi \in (C^*)^{2g}$ satisfying*

$$T(\chi) = T(\chi^{-1}) = \overline{T(\overline{\chi})} \quad \text{for all}\quad \chi\ .$$

On $(C^*)^{2g}$, $T(\chi)$ has a divisor of zeroes V consisting of all representations χ for which 0 is in the (discrete) spectrum of D on L_χ^2.

Proof. Suppose the values of $|\chi(A_j)|^{\pm 1}, |\chi(B_j)|^{\pm 1}$ are bounded by $M_1 \geq 1$, and write any $y \in \Gamma$ in the form $y = \prod_1^N y_i^{n_i}$ with y_i one of the $A_j^{\pm 1}$, $B_j^{\pm 1}$ and with $n_i \geq 0$ and $\sum_1^N n_i$ a minimum. Then an elementary estimate of the length $r(z_0, y z_0)$ of the non-Euclidean segment joining any

basepoint $z_0 \in \mathcal{D}$ to $y z_0$ shows that $r(z_0, y z_0) \geq C_1^{-1} \sum_1^N n_i$ for some $C_1 > 0$ depending only on Γ (see [2] or [5]). Thus

$$|\chi(\gamma)| = \prod_1^N |\chi(\gamma_i')|^{n_i} \le M_1^{C_1 r(z_0, \gamma z_0)} \qquad \text{for any} \quad z_0 \in \mathcal{D} ;$$

and since $|Q_s(z, \gamma z')| = 0(e^{-(\mathrm{Re}\, s) r(z, \gamma z')})$ as $r \to \infty$ uniformly for $\mathrm{Re}\, s \ge \epsilon > 0$, it follows that the series (7) converges uniformly for χ with entries bounded by M_1, for z' bounded away from z in $\mathcal{D}$ and for $\mathrm{Re}\, s > 1 + C_1 \ln M_1$. Now $\mathcal{D}$ is compact and G_s has only the singularity $\frac{1}{2\pi} \ln |z' - z|$ in $\mathcal{D}$; so $G_{s_0}(z, z', \chi)$ is a (generally non-self-adjoint) compact integral operator on L_χ^2 for $s_0 \gg 0$ satisfying

$$(9) \qquad G_s(z, z', \chi) - G_{s_0}(z, z', \chi) =$$

$$(s-s_0)(s+s_0-1) \iint\limits_{\mathcal{D}} G_s(z, z'', \chi) G_{s_0}(z'', z', \chi) \,\boxed{dz''} \quad .$$

Thus from the Fredholm theory [8, Ch. IV] we conclude that $G_s(z, z', \chi)$, the kernel for $[D - s(s-1)]^{-1}$ on L_χ^2, has a simultaneous meromorphic continuation through the s-plane and the representation space $(C^*)^{2g}$. At a pole s_0 of order ν, G_s has a Laurent development

$$G_s(z, z', \chi) = \sum_{k=-\nu}^{+\infty} R_k(z, z')(\lambda_0 - \lambda)^k, \quad \lambda_0 = s_0(s_0 - 1)$$

in $\lambda = s(s-1)$, where the R_k satisfy:

$$(10) \qquad (D - \lambda_0) R_0 + R_{-1} = I \quad \text{and} \quad (D - \lambda_0) R_k = -R_{k-1}, \quad k \ne 0$$

as integral operators on L_χ^2. Choosing any basis $e_1(z, \chi), \cdots, e_m(z, \chi)$ for the (necessarily finite-dimensional) λ_0-subspace, with $(D - \lambda_0) e_j \in$ span $\{e_1, \cdots, e_{j-1}\}$ for each $j \ge 1$, we can write

$$(11) \quad R_{-1}(z, z') = \sum_{i=1}^m \tilde{e}_i(z, \chi^{-1}) e_i(z', \chi), \quad \iint\limits_{\mathcal{D}} \tilde{e}_i(z, \chi^{-1}) e_j(z, \chi) \,\boxed{dz} = \delta_i^j$$

for a dual-basis $\{\tilde{e}_i(z, \chi^{-1})\}$ of the λ_0-subspace of D in $L^2_{\chi^{-1}}$; from (10) and (11) then,

$$(12) \qquad \text{trace } R_{-1}(z, z') = m, \quad \text{trace } R_k(z, z') = 0 \quad \text{for} \quad k < -1.$$

Now let χ' be any fixed non-trivial unitary representation of Γ, and set

$$(13) \qquad \frac{T(\chi)}{T(\chi')} = \exp \int_1^{+\infty} (2s-1) \text{ trace } [G_s(z, z', \chi) - G_s(z, z', \chi')] ds$$

where the integral, taken over the positive real axis say, is convergent near ∞ since

$$\lim_{z' \to z} [G_s(z, z', \chi) - Q_s(z, z')] = O(e^{-(\text{Re } s) r_0}) \quad \text{as} \quad s \to +\infty,$$

with $r_0 = \inf \{r(z, \gamma z) | z \in \mathfrak{D}, \gamma \neq I\}$ the "neck" of $\mathfrak{D}$. Then $T(\chi)$ is well-defined in χ as any point in the discrete spectrum crosses $(1, +\infty)$ since the integral only changes by $2\pi i m$, $m \in \mathbf{Z}$ as in (12). Moreover, if $V \in (C^*)^{2g}$ is the analytic subvariety given by the polar divisor of $G_1(z, z', \chi)$, $T(\chi)$ will by (12) tend to 0 as χ approaches $\chi_0 \in V$ and so, by the Riemann removable singularity theorem, can be made holomorphic near V. Thus, from (5)-(8), (13) gives the unique analytic continuation of torsion from $(S^1)^{2g}$ to $(C^*)^{2g}$; the symmetries of $T(\chi)$ follow from the reflection principle and the fact that $T(\overline{\chi}) = T(\chi) = T(\overline{\chi}^{-1})$ for $\chi \in (S^1)^{2g}$.

If $\chi \in (C^*)^{2g} - V$, $G_s(z, z', \chi)$ is analytic at $s = 1$ and one can write

$$(14) \qquad 4\pi G_1(z, z', \chi) = \ln P(z, z', \chi) \overline{P(z, z', \overline{\chi})}$$

where, for fixed z, the "prime-form" $P(z, z', \chi)$ is analytic in z' with only a simple zero at $z' = z$ in $\mathfrak{D}$. $B(z, z', \chi) = -\frac{1}{\pi} \frac{\partial^2}{\partial \overline{z} \partial z'} \ln P(z, z', \chi)$ is the Bergman kernel for the holomorphic Prym differentials with multipliers χ, while

$$(15) \qquad \Omega(z,z',\chi) = \frac{\partial^2}{\partial z\, \partial z'} \ln P(z,z',\chi) = \frac{1}{(z'-z)^2} + S(z,\chi) + O(z'-z)$$

is orthogonal (under principal value) to the Prym differentials. The Riemann-Roch duality then takes the following form: if $a_i \in C$, $z_1, \cdots, z_m$ are distinct points on M and $\chi \in (C^*)^{2g} - V$,

$$\sum_{j=1}^{m} a_j \Omega(z_j, z, \chi)$$ is an exact differential with multipliers χ and double

poles at $z = z_j$ if and only if $\displaystyle\sum_{j=1}^{m} a_j B(z, z_j, \chi^{-1}) \equiv 0$ for all $z \in \mathcal{D}$.

In the elliptic case $M = C/Z + Z\tau$, $\mathrm{Im}\,\tau > 0$, Kronecker's second limit formula applied to the M–P zeta function for $D = 4\,\dfrac{\partial^2}{\partial z\, \partial \bar{z}}$ on L_χ^2 gives [6]:

$$T(\chi) = |\eta(\tau)|^{-2} \left| \theta \begin{bmatrix} \frac{1}{2} - x \\ \frac{1}{2} + \hat{x} \end{bmatrix}_{\tau} (0) \right|^2, \quad \chi(A) = e^{2\pi i x} \text{ and } \chi(B) = e^{2\pi i \hat{x}}$$

where $\eta(\tau)$ is Dedekind's function, A and B are the transformations $z \to z+1$ and $z \to z+\tau$, and

$$(16) \qquad \theta \begin{bmatrix} \alpha \\ \beta \end{bmatrix}_{\tau} (z) = \sum_{n \in Z^g} \exp\{\pi i (n+a)\tau(n+a)^t + 2\pi i (z+\beta)(n+a)^t\}$$

for $\alpha, \beta \in R^g$, $z \in C^g$ (here $g = 1$). The analytic continuation of $T(\chi)$ to $(C^*)^2$:

$$T(\chi) = |\eta(\tau)|^{-2} e^{-2\pi s^2 \mathrm{Im}\,\tau} \, \theta \begin{bmatrix} \frac{1}{2} \\ \frac{1}{2} \end{bmatrix}_{\tau} (\hat{s} - \tau s)\, \theta \begin{bmatrix} \frac{1}{2} \\ \frac{1}{2} \end{bmatrix}_{-\bar{\tau}} (\hat{s} - \bar{\tau} s)$$

for $(\chi(A), \chi(B)) = (e^{2\pi i s}, e^{2\pi i \hat{s}})$ thus vanishes exactly when $\hat{s} - \tau s \in$

$Z + Z\tau$ or $\hat{s} - \bar{\tau} s \in Z + Z\bar{\tau}$. Likewise, by considering $\exp \displaystyle\int_{z_0}^{z} 2\pi i\, s \cdot v$

for $g > 1$, it is seen that V always contains the (anti-) analytically trivial representations V_0 of the form:

$$(\chi(A_j), \chi(B_j)) = (e^{2\pi i s_j}, e^{2\pi i (\tau s)_j}) \quad \text{or} \quad (e^{2\pi i s_j}, e^{2\pi i (\bar{\tau} s)_j})$$

for some $s \in \mathbb{C}^g$.

PROPOSITION 1.2. $V - V_0$ is the (open) subvariety of all $\chi \in (\mathbb{C}^*)^{2g} - V_0$ for which the cup-product pairing

$$(17) \qquad H^{1,0}(\underline{\chi}) \otimes H^{0,1}(\underline{\chi}^{-1}) \to H^{1,1}(M) \simeq \mathbb{C}$$

is degenerate (here $\underline{\chi}$ is the flat line bundle on M determined from the representation χ).

Proof. Let $\omega_i(z, \chi)$ (resp. $\omega_i(\bar{z}, \chi)$) for $i = 1, \cdots, g{-}1$ be a basis for the Prym differentials $H^{1,0}(\underline{\chi})$ (resp. $H^{0,1}(\underline{\chi})$); by Grauert's Theorem, we may assume that these bases vary local-analytically with $\chi \in (\mathbb{C}^*)^{2g} - V_0$. Now if $\chi \in V - V_0$, there is a single-valued harmonic function $f(z)$ with multipliers χ such that $df = \omega(z, \chi) + \eta(\bar{z}, \chi)$ for $\omega \in H^{1,0}(\underline{\chi})$, $\eta \in H^{0,1}(\underline{\chi})$ and ω, η both non-zero since $\chi \notin V_0$. Thus

$$0 = \int_{\partial \mathfrak{D}} f \omega_j(\bar{z}, \chi^{-1}) = \iint_{\mathfrak{D}} \omega(z, \chi) \wedge \omega_j(\bar{z}, \chi^{-1})$$

for any differential $\omega_j(\bar{z}, \chi^{-1})$ so that

$$(17)' \qquad \det D(\chi) \equiv \det\left(\iint_{\mathfrak{D}} \omega_i(z, \chi) \wedge \omega_j(\bar{z}, \chi^{-1}) \right) = 0$$

in this case. Conversely, if $\chi \notin V$, $G_s(z, z', \chi)$ is analytic at $s = 1$; hence

$$-4 \frac{\partial^2}{\partial \bar{z} \partial z'} G_1(z, z', \chi) = \sum_{1 \leq i, j \leq g-1} a_{ij} \omega_j(\bar{z}, \chi^{-1}) \omega_i(z', \chi)$$

for suitable $a_{ij} \in \mathbb{C}$ will be a reproducing kernel for $H^{1,0}(\underline{\chi})$ and so $\det D(\chi)$ cannot vanish in this case. Thus $\chi \in V - V_0$ if and only if $(17)'$ holds—that is, the pairing (17) degenerates.

From (17)$'$ one can give an equation for V in terms of the Riemann θ-function (16) and the Schottky-Klein prime form $E(z, a)$ used for constructing bases for $H^{1,0}(\underline{\chi})$ over Zariski-open sets in $(C^*)^{2g}$; thus if $b, a_1, \cdots, a_{g-1}$ are generic points on M, V is contained in the subvariety of $e^{2\pi i s_j}, e^{2\pi i \hat{s}_j})$ in $(C^*)^{2g}$ defined by

$$
\det_{1 \le i, j \le g-1} \left\{ \iint_{\mathcal{D}} \theta_\tau\left(\hat{s} - \tau s + \Delta - \sum_{k \ne i} a_k - z\right) \theta_{-\overline{\tau}}\left(\overline{\hat{s} - \overline{\tau}s + \Delta - \sum_{k \ne j} a_k - z}\right) \cdot \right.
$$

$$
\left. \frac{\left| \theta_\tau\left(b + \sum_1^{g-1} a_k - z - \Delta\right) \right|^2}{E(z, b)} \frac{\exp - 4\pi s \cdot \int_{z_0}^z \operatorname{Im} v}{E(z, a_i) \overline{E(z, a_j)}} \, dx \, dy \right\} \equiv 0 .
$$

Here $\Delta = (g-1) p + k^p \epsilon J_{g-1}(M)$ is the divisor class for the Riemann constants k^p for any base point $p \epsilon M$, lifted to C^g so that the above integrand is a well-defined form in $H^{1,1}(M)$; observe that V_0 lies on this variety since $\theta_\tau(\xi - \Delta) \equiv 0$ for any positive divisor ξ of degree $g-1$ by Riemann's theorem.

THEOREM 1.3. *The divisor of zeroes* $V \subset (C^*)^{2g}$ *of* $T(\chi)$ *determines the period matrix of* M *for the marking* (1) *of* M *defining* $T(\chi)$.

Proof. For $s, \hat{s} \epsilon C^g$ corresponding to the representation $(\chi(A), \chi(B)) = (e^{2\pi i s}, e^{2\pi i \hat{s}})$, let $u_i(z, s, \hat{s}) = u_i(z, \chi)$ (resp. $u_i(\overline{z}, s, \hat{s}) = u_i(\overline{z}, \chi)$) for $i = 1, \cdots, g$ be a basis for the meromorphic (resp. conjugate meromorphic) Prym differentials with multipliers χ and with at most a simple pole at some fixed point $p \epsilon M$; by Grauert's theorem, we may assume that this basis varies analytically over all of $(s, \hat{s}) \epsilon C^{2g}$ covering $(C^*)^{2g}$. Now if $\chi \epsilon V$, there is some harmonic function f on $\mathcal{D}$ with multipliers χ and we can write

$$
f(z) = \sum_{i=1}^{g} \left[a_i \int_{z_0}^z u_i(z, \chi) + b_i \int_{z_0}^z u_i(\overline{z}, \chi) \right] + C
$$

for some constants a_i, b_i, C with $\displaystyle\sum_{i=1}^{g} a_i S_i = \sum_{i=1}^{g} b_i \overline{S_i} = 0$,

$$S_i^{(-)} = \sum_{j=1}^{g} \left[(1 - \chi(B_j)) \int_{z_0}^{A_j z_0} u_i(\overset{(-)}{z}, \chi) - (1 - \chi(A_j)) \int_{z_0}^{B_j z_0} u_i(\overset{(-)}{z}, \chi) \right].$$

The dimension of the space of such f is thus the column nullity of the matrix

$$\check{D}(\chi) = \begin{pmatrix} \displaystyle\int_{A_i} u_j(z, \chi) & \displaystyle\int_{A_i} u_j(\overline{z}, \chi) & 1 - \chi(A_i) \\[4mm] \displaystyle\int_{B_i} u_j(z, \chi) & \displaystyle\int_{B_i} u_j(\overline{z}, \chi) & 1 - \chi(B_i) \\[4mm] S_j & 0 & 0 \end{pmatrix}, \quad 1 \le i, \ j \le g$$

(the condition on $\overline{S_i}$ is then redundant here); and so

$$\chi \in V \iff T(\chi) = 0 \iff \det D(\chi) = 0 \ \text{ or } \ \chi \in V_0$$
$$\iff \det \check{D}(\chi) = 0 \iff G_1(z, z', \chi') \ \text{ has a pole at } \ \chi' = \chi \ .$$

Now consider χ near the identity with $1 - \chi(A_i) = \varepsilon_i$, $1 - \chi(B_i) = \hat{\varepsilon}_i$; changing the basis of u_j if necessary so that

$$\left(\int_{A_i} u_j(z, I), \ \int_{B_i} u_j(z, I) \right) = (I, \tau) \ ,$$

the tangent cone to $\det \check{D}(\chi) = 0$ at $\chi = I$ has the equation (up to a constant):

$$\sum_{i,j=1}^{g} \hat{\varepsilon}_i \hat{\varepsilon}_j \pi_{ij} - 2 \sum_{i,k,j=1}^{g} \hat{\varepsilon}_i \varepsilon_k \pi_{ij} \, \mathrm{Re} \, \tau_{jk} + \sum_{i,j,k,\ell=1}^{g} \varepsilon_k \varepsilon_\ell \pi_{ij} \tau_{i\ell} \overline{\tau}_{jk} = 0$$

where π_{ij} is the matrix of cofactors of $\operatorname{Im}\tau$. Thus π_{ij} is determined up to a constant from the coefficients of $\hat{\epsilon}_i\hat{\epsilon}_j$ and then $\operatorname{Re}\tau_{ij}$ is uniquely given from the $\frac{1}{2}\,g(g+1)$ coefficients of $\hat{\epsilon}_i\epsilon_j$ since π_{ij} is a positive definite matrix; the ambiguity in π_{ij} (corresponding to some positive multiple of $\operatorname{Im}\tau_{ij}$) is eliminated by consideration of the coefficients of $\epsilon_k\epsilon_\ell$ and thus the period matrix for the given marking is uniquely determined from the tangent cone to $\det \check{D}(\chi) = 0$ at $\chi = I$.

§2. *Differential of torsion on the moduli space*

Suppose μ is a C^∞-Beltrami differential on $M_0 = M \simeq \Gamma\backslash H$ and

$$w(z) = z + \epsilon\,\phi_\mu(z) + o(\epsilon)$$

is a quasi-conformal mapping of H onto H fixing $0, 1, \infty$ and with dilitation $\epsilon\mu = \dfrac{\partial w}{\partial \bar{z}}\Big/\dfrac{\partial w}{\partial z}$ for ϵ near 0. From the explicit construction of the mapping function [1, p. 105]:

$$\phi_\mu(z) = \overline{\phi_\mu(\bar{z})} = \frac{-1}{\pi}\iint_C \mu(\zeta)\,\frac{z(z-1)}{\zeta(\zeta-1)(\zeta-z)}\,d\xi\,d\eta\,,\quad \zeta = \xi + i\eta$$

so that $\phi_\mu(z)$ is an Eichler integral with $|\phi_\mu(z)| = O(|z|\ln|z|)$ as $z \to \infty$ and with real quadratic period polynomials:

$$(18)\qquad P_\gamma(z) = \phi_\mu(\gamma z)(cz+d)^2 - \phi_\mu(z)\,.$$

If $M_{\epsilon\mu}$ is the compact Riemann surface $\Gamma_{\epsilon\mu}\backslash H$ formed from the Fuchsian group $\Gamma_{\epsilon\mu} = w\Gamma w^{-1}$, then the metric $\dfrac{|dz|}{y}$ on $M_{\epsilon\mu}$ pulled back by w gives rise to the metric on M_0:

$$\frac{|dw|^2}{|\operatorname{Im}w|^2} = \frac{|dz|^2}{y^2} + \epsilon\left\{\frac{2}{y^2}\operatorname{Re}(\bar{\mu}dz^2) + \sigma(z)\frac{|dz|^2}{y^2}\right\} + o(\epsilon)$$

where

$$\sigma(z) = \frac{\partial}{\partial \varepsilon} \left(\boxed{\frac{dw}{\overline{dz}}} \right)_{\varepsilon = 0} = \frac{8y^2}{\pi} \, \mathrm{Re} \iint_H \frac{\mu(\zeta) \, d\xi \, d\eta}{(\zeta - z)^2 (\zeta - \overline{z})^2}$$

is a real automorphic function on M_0. If

$$(19) \qquad D_{\varepsilon\mu} = y^2 \Delta + \varepsilon \Lambda + o(\varepsilon), \quad -\Lambda = 8y^2 \, \mathrm{Re} \, \frac{\partial}{\partial z} \left(\mu \frac{\partial}{\partial z} \right) + \sigma(z) y^2 \Delta$$

is the Laplacian for this metric, then the spectrum of $D_{\varepsilon\mu}$ on L_χ^2 for M_0 is the same as the spectrum of D on L_χ^2 for $M_{\varepsilon\mu}$, and the eigenfunctions of $D_{\varepsilon\mu}$ on M_0 are the pullback under w of the eigenfunctions of D on $M_{\varepsilon\mu}$. So if $\hat{G}_s^{\varepsilon\mu}$ and $G_s^{\varepsilon\mu}$ are the respective resolvents for a representation χ of Γ and $\Gamma_{\varepsilon\mu}$ (compatible with w):

$$(20) \qquad \hat{G}_s^{\varepsilon\mu}(z, z', \chi) = G_s^{\varepsilon\mu}(w(z), w(z'), \chi), \quad z, z' \in \mathcal{D} = \Gamma \backslash H .$$

To describe the perturbation of the spectrum of D we make use of a "weighted-mean" of eigenvalues as in [3]:

THEOREM 2.1. *For fixed $\chi \in (\mathbb{C}^*)^{2g}$, let $\lambda_1^{\varepsilon\mu}, \cdots, \lambda_m^{\varepsilon\mu}$ be the eigenvalues (not necessarily distinct) of $D_{\varepsilon\mu}$ near the eigenvalue λ of D with multiplicity m, and suppose $R_{-1}(z, z')$ is the reproducing kernel (11) for the λ-subspace. Then the weighted mean of eigenvalues has an expansion:*

$$(21) \qquad \frac{1}{m} \sum_{i=1}^{m} \lambda_i^{\varepsilon\mu} = \lambda + \frac{\varepsilon}{m} \left\{ -\lambda \iint_{\mathcal{D}} \sigma(z) R_{-1}(z, z) \, \boxed{dz} \right.$$
$$\left. + 4 \iint_{\mathcal{D}} \left[\mu(z) \frac{\partial^2 R_{-1}}{\partial z \partial z}(z, z') + \overline{\mu(z)} \frac{\partial^2 R_{-1}}{\partial \overline{z} \partial \overline{z}}(z, z') \right]_{z'=z} dx \, dy \right\} + o(\varepsilon) .$$

Proof. For any C^∞-function $f(z)$ automorphic under Γ_0 with multipliers χ^{-1}, (19) gives

$$(D - s(s-1) + \varepsilon \Lambda) \iint_{\mathcal{D}} \hat{G}_s^{\varepsilon\mu}(z, z', \chi) f(z')(1 + \varepsilon \sigma(z')) \, \boxed{dz'} = f(z) + o(\varepsilon)$$

as $\varepsilon \to 0$; expanding this out using Green's theorem and (19) we find

$$H_S(z,z'') \equiv \frac{\partial}{\partial \varepsilon} \hat{G}_S^{\varepsilon\mu}(z,z'',\chi)\big|_{\varepsilon=0} = s(s-1) \iint_{\mathcal{D}} \sigma(z') G_S(z,z',\chi) G_S(z',z'',\chi)\, \boxed{dz'}$$

(22)

$$-4 \iint_{\mathcal{D}} \left[\mu(z') \frac{\partial G_S}{\partial z'}(z,z',\chi) \frac{\partial G_S}{\partial z'}(z',z'',\chi) + \overline{\mu(z')} \frac{\partial G_S}{\partial \bar{z}'}(z,z',\chi) \frac{\partial G_S}{\partial \bar{z}'}(z',z'',\chi) \right] dx'dy'$$

for all $z'' \neq z$ in $\mathcal{D}$. Now if $R_{-1}^{\varepsilon\mu}(z,z')$ is the sum of the reproducing kernels on M_0 for the eigenvalues of $D_{\varepsilon\mu}$ near $\lambda = s(s-1)$,

$$(D_{\varepsilon\mu} - \lambda) R_{-1}^{\varepsilon\mu}(z,z') = -\frac{1}{2\pi i} \int_{|\xi-\lambda|=\delta} (\xi-\lambda) \hat{G}_S^{\varepsilon\mu}(z,z',\chi)\, d\xi$$

for $\xi = s'(s'-1)$ and δ, ε sufficiently near 0. Differentiating this equation with respect to ε, and taking the trace with respect to $\boxed{dz}$, (22) gives

$$\sum_{i=1}^{m} (\lambda_i^{\varepsilon\mu} - \lambda) = -\varepsilon \iint_{\mathcal{D}} \operatorname*{Res}_{s'(s'-1)=\lambda} (s'(s'-1)-\lambda) H_S{}'(z,z'')\big|_{z''=z} \boxed{dz} + o(\varepsilon)$$

$$= \varepsilon \operatorname{trace} \left\{ -\lambda \iint_{\mathcal{D}} \sigma(z') R_{-1}(z,z') R_{-1}(z',z'') \boxed{dz'} \right.$$

$$+ 4 \iint_{\mathcal{D}} \left[\mu(z') \frac{\partial R_{-1}}{\partial z'}(z,z') \frac{\partial R_{-1}}{\partial z'}(z',z'') + \overline{\mu(z')} \frac{\partial R_{-1}}{\partial \bar{z}'}(z,z') \frac{\partial R_{-1}}{\partial \bar{z}'}(z',z'') \right] dx'dy' \left. \right\} + o(\varepsilon)$$

which is the right-hand term in (21); here the higher order terms in the Laurent expansion of G_S do not appear by virtue of the relations (10).

A Poincaré-type series for the variation of the resolvent can be given in terms of the period polynomials (18): thus expressing (22) as an integral over H (or differentiating (7) in ϵ), (20) gives

$$\frac{\partial}{\partial \epsilon} G_s^{\epsilon \mu}(z, z', \chi)\Big|_{\epsilon = 0} = \frac{-1}{y} \sum_{\gamma \in \Gamma} \chi(\gamma)^{-1} \frac{\partial Q_s}{\partial r} (z, \gamma z') \, \text{Im} \, P_{\gamma^{-1}}(z) \, e^{-i\theta(\gamma z', z)}$$

with absolute convergence for $\text{Re } s > 2 + C_1 \ell n \, M_1$ from the growth condition on ϕ_μ (as in the proof of Theorem 1.1); the series has a meromorphic
continuation through the s-plane with poles at those of $G_s(z, z', \chi)$ with
twice the corresponding order. From the (anti-) holomorphic part of (21)
for Beltrami differentials of the form $\mu(z) = y^2 \overline{Q}$, Q a holomorphic quadratic differential, one can note also the following remark. For fixed χ,
let $\lambda_1(r), \cdots, \lambda_p(r)$ be distinct eigenvalues of multiplicity 1 with reproducing kernels $e_i(z, \chi^{-1}) e_i(z', \chi)$, $i = 1, \cdots, p$ on Riemann surfaces M_r
for r in some open set $\mathcal{U}$ in Teichmüller space T_g; then a differentiable function $\phi(\lambda_1, \cdots, \lambda_p)$ is a holomorphic function of moduli on $\mathcal{U}$ if
and only if the Beltrami differential

$$y^2 \sum_{i=1}^{p} \frac{\partial \phi}{\partial \lambda_i} \frac{\partial e_i}{\partial \overline{z}} (z, \chi^{-1}) \frac{\partial e_i}{\partial \overline{z}} (z, \chi)$$

on each surface M_r is orthogonal to the holomorphic quadratic differentials on M_r (a "stationary differential" in the sense of [1]).

    For fixed χ, the differential of torsion $T(\chi)$ on the moduli space
can now be computed from Theorem 2.1, using (21) to give a variation of
the heat-kernel in (3) as in [7, p. 194]; alternatively, proceeding directly
from (22), one has:

THEOREM 2.2. *For* $\chi, \chi' \in (C^*)^{2g} - V$,

$$-2\pi i \sum_{1 \leq i \leq j \leq g} \frac{\partial}{\partial r_{ij}} \ell n \, \frac{T(\chi)}{T(\chi')} \, v_i(z) v_j(z) = \frac{\partial^2 \ell n}{\partial z \, \partial z'} \frac{P(z, z', \chi)}{P(z, z', \chi')}\Big|_{z'=z} = S(z, \chi) - S(z, \chi')$$

(23)

where $P(z, z', \chi)$ and $S(z, \chi)$ are defined by (14)-(15), and r is the
period matrix for the normalized differentials v_i.

Proof. If $Z(s, \chi, \varepsilon)$ is the zeta-function (6) on $M_{\varepsilon\mu}$, (8) gives

$$\frac{\partial}{\partial\varepsilon}\left[\frac{Z'(s,\chi,\varepsilon)}{Z(s,\chi,\varepsilon)} - \frac{Z'(s,\chi',\varepsilon)}{Z(s,\chi',\varepsilon)}\right]_{\varepsilon=0} =$$

$$(1-2s)\,\frac{\partial}{\partial\varepsilon}\iint_{\mathcal{D}} [G_S^{\varepsilon\mu}(w,w',\chi) - G_S^{\varepsilon\mu}(w,w',\chi')]_{w'=w}\,[1+\varepsilon\sigma(z)]\,\boxed{dz}\,\Big|_{\varepsilon=0}.$$

Integrating this equation in s, making use of (20)-(22) and (9) differentiated at $s_0 = s$, one finds

$$\frac{\partial}{\partial\varepsilon}\,\ell n\,\frac{Z(s,\chi,\varepsilon)}{Z(s,\chi',\varepsilon)}\bigg|_{\varepsilon=0} = s(1-s)\iint_{\mathcal{D}} \sigma(z)\,[G_S(z,z',\chi) - G_S(z,z',\chi')]_{z'=z}\,\boxed{dz}$$

$$+4\iint_{\mathcal{D}} \left(\mu(z)\,\frac{\partial^2}{\partial z\,\partial z'} + \overline{\mu(z)}\,\frac{\partial^2}{\partial\overline{z}\,\partial\overline{z'}}\right)(G_S(z,z',\chi) - G_S(z,z',\chi'))_{z'=z}\,dx\,dy$$

for all s. Now let $\mu(z) = 1$ if $|z-\zeta| \le \delta$ and $\mu(z) \equiv 0$ otherwise in $\mathcal{D}$ for fixed $\zeta \in \mathcal{D}$ and δ small; setting $s = 1$, (14)-(15) implies:

$$\frac{\partial}{\partial\varepsilon}\,\ell n\,\frac{Z(1,\chi,\varepsilon)}{Z(1,\chi',\varepsilon)}\bigg|_{\varepsilon=0} = \delta^2[S(\zeta,\chi) - S(\zeta,\chi') + \overline{S(\zeta,\chi)} - \overline{S(\zeta,\chi')}].$$

(23) now follows from (5) and the "Schiffer-variation"

$$\frac{\partial}{\partial\varepsilon}\bigg|_{\varepsilon=0} = -2\pi i\delta^2 \sum_{1\le i\le j\le g} \left(v_i(\zeta)v_j(\zeta)\,\frac{\partial}{\partial\tau_{ij}} - \overline{v_i(\zeta)v_j(\zeta)}\,\frac{\partial}{\partial\overline{\tau}_{ij}}\right)$$

coming from (22) at $s = 1$ for the above choice of μ.

In the elliptic case, the differential (15) can be given explicitly as

$$\Omega(z,z',\chi) = \frac{\partial}{\partial z}\left[e^{2\pi is(z'-z)}\,\frac{\theta[\eta](\hat{s}-\tau s+z-z')\,\theta[\eta]'(0)}{\theta[\eta](\hat{s}-\tau s)\,\theta[\eta](z'-z)}\right]$$

where $[\eta] = \left[\begin{smallmatrix}½\\½\end{smallmatrix}\right]$, $\chi(A) = e^{2\pi is}$ and $\chi(B) = e^{2\pi i\hat{s}}$, and thus (23) reduces to

$$4\pi i \; \frac{\partial}{\partial \tau} \, \ell n \; \frac{\theta[\eta]_\tau(\hat{s}-ts)}{\theta[\eta]_\tau(\hat{s}'-ts')}\bigg|_{t=\tau} = \frac{\theta[\eta]''}{\theta[\eta]}(\hat{s}-\tau s) - \frac{\theta[\eta]''}{\theta[\eta]}(\hat{s}'-\tau s')$$

a consequence of the heat equation on M. However, when $g > 1$ the differentials $\Omega(z, z', \chi)$ and $S(z, \chi)$ are seen from (22) to be non-analytic functions of moduli giving rise to non-zero higher $\bar{\tau}$-derivatives of $\ell n \; T(\chi)/T(\chi')$. Moreover, although $\Omega(z, z', \chi)$ has a Fourier series similar to (5) with coefficients conditionally convergent Poincaré series in z and z', an explicit formula for $\Omega(z, z', \chi)$ on, say, the Jacobi variety independent of the (elements of the) group Γ has yet to be established; likewise, by Propositions 1.2 - 1.3, explicit global bases of the Prym differentials as functions of $\chi \in J_0(M)$ might shed some light on possible explicit formulas for $T(\chi)$ on $J_0(M)$.

REFERENCES

[1]   L. Ahlfors, *Lectures on Quasiconformal Mappings*, Van Nostrand, 1966.

[2]   M. Eichler, *Grenzkreisgruppen und kettenbruchartige Algorithmen*, Acta Arithmetica 11 (1965), pp. 169-180.

[3]   T. Kato, *Perturbation Theory of Linear Operators*, Grund. der Math. Wissen., Band 132 (1966).

[4]   M. McKean, *Selberg's Trace Formula for a Compact Riemann Surface*, Comm. Pure and Applied Math., Vol. 25 (1972), pp. 225-246.

[5]   H. Poincaré, *Mémoire sur les fonctions zétafuchsiennes*, Acta Math. 5 (1884), pp. 209-278.

[6]   Ray-Singer, *Analytic Torsion for Complex Manifolds*, Annals of Math., Vol. 98 (1973), pp. 154-177.

[7]   ________, *R-Torsion and the Laplacian on Riemannian Manifolds*, Advances in Math., Vol. 7 (1971), pp. 145-210.

[8]   Riesz-Sz. Nagy, *Functional Analysis*, Ungar Pub. 1955.

[9]   W. Roelcke, *Das Eigenwertproblem der automorphen Formen in der hyperbolischen Ebene*, Math. Ann. 167-8 (1966-67), pp. 292-337, 261-324.

[10] A Selberg, *Harmonic Analysis and Discontinuous Groups in Weakly Symmetric Riemannian Spaces ...*, Jour. Indian Math. Soc. 20 (1956), pp. 47-87.

ON A NOTION OF QUASICONFORMAL RIGIDITY
FOR RIEMANN SURFACES

Frederick P. Gardiner[*]

Introduction

The motivation for this work came originally from the attempt to understand Schiffer's variational technique from the point of view of quasiconformal mapping. This technique can be used to prove the existence of
quadratic differentials on a Riemann surface with special properties.
Jenkins-Strebel differentials, all of whose non-critical horizontal trajectories are closed, are the prime examples.

It turns out that the special property that these differentials have is
that they are differentials on a subsurface and, in a special sense, it is
impossible to vary the subsurface by a quasiconformal deformation without
varying the larger surface. The special sense is defined in section 3.
Section 1 reviews the necessary features of reduced Teichmüller space.
Section 2 deals with the variational method which is based on the possibility of making one parameter families of Teichmüller trivial deformations.
Methods of this type were first used by Schiffer, [15]. In section 4 numer-
ous examples of quasiconformally rigid surfaces are given and in section 5
all of the quasiconformally rigid subsurfaces of a four-times punctured
sphere are classified.

[*] This research was supported by a grant from the National Science Foundation,
Grant # MCS 77-01711.

1. *Preliminaries*

Let R be a finite Riemann surface type (g, n, k). This means R is
obtained from a compact surface of genus g by deleting n analytic
discs and k points. R is said to have n holes and k punctures. A
marking of R is a selection of generators for the fundamental group of R
of a canonical type. Our objective is to define reduced Teichmüller space,
$T^{\#}(R)$. Intuitively it consists of all marked surfaces of the same type
factored by an equivalence relation. This relation identifies two marked
surfaces if they are conformal and if the conformal mapping between them
preserves the marking.

To define $T^{\#}(R)$ we will take a slightly different approach which is
based on Beltrami coefficients and which is more amenable to describing
its manifold structure. This approach is developed mainly by Ahlfors and
Bers in [2, 3, 5]. It turns out that $T^{\#}(R)$ is a manifold of real dimension

$$(1) \qquad\qquad 6(g-1) + 3n + 2k + \rho$$

where ρ is the real dimension of the continuous group of holomorphic
homeomorphisms of R. The only cases where $\rho > 0$ occur when $g = 1$
or 0, [16].

Let $L(R)$ be the set of all Beltrami differentials on R. By this we
mean that if μ is in $L(R)$ and z is a local parameter for some open set
V in R, then $\mu(z)$ is a measurable function on V, bounded in the sup
norm and the global differential μ behaves in such a way that $\mu(z)\,d\bar{z}/dz$
is invariant. Furthermore, for the function $|\mu|$ on R, we require that the
quantity $\|\mu\|_{\infty}$, which is the essential supremum of $|\mu(z)|$ over all z in
R, be bounded.

Let $M(R) = \{\mu \epsilon L(R); \|\mu\|_{\infty} < 1\}$. Given any μ in $M(R)$ there will be
a Riemann surface R_{μ} and a quasiconformal homeomorphism $w : R \to R_{\mu}$
which satisfies the Beltrami equation

$$(2) \qquad\qquad w_{\bar{z}} = \mu w_{z} .$$

See [3]. We will denote a homeomorphic solution w to (2) by w_{μ}.

Now suppose h is a quasiconformal homeomorphism of R onto R. Let $h^*(\mu)$ be the Beltrami coefficient of $w_\mu \circ h$, that is,

$$(3) \qquad h^*(\mu) = (w_\mu \circ h)_{\bar{z}} / (w_\mu \circ h)_z .$$

If $\sigma(z) = h_{\bar{z}}/h_z$, the explicit formula for $h^*(\mu)$ is

$$(4) \qquad h^*(\mu) = \frac{\sigma(z) + \mu(h(z))\,\theta(z)}{1 + \bar{\sigma}(z)\,\mu(h(z))\,\theta(z)}$$

where $\theta(z) = \overline{(h_z)}/h_z$.

Since σ, μ, and $h^*(\mu)$ are the Beltrami coefficients of quasiconformal mappings with domain R, one sees that they are all elements of $M(R)$ and formula (4) shows how the group of quasiconformal homeomorphisms of R, $D(R)$, acts on $M(R)$.

Let $D_0(R)$ be the group of quasiconformal homeomorphisms of R which are homotopic to the identity. Explicitly, h is homotopic to the identity if there is a continuous mapping $g : R \times I \to R$ such that $g(p, 1) = h(p)$ and $g(p, 0) = p$ for every p in the interior of R. It follows that for such a homeomorphism h, its quasiconformal extension to the boundary of R must preserve boundary components but it may move points along each component.

DEFINITION. *The reduced Teichmüller space of* R, $T^{\#}(R)$, *is the set of orbits in* $M(R)$ *under the action of* $D_0(R)$. *For* μ *in* $M(R)$, *the orbit of* μ *under* $D_0(R)$ *is called the Teichmüller class of* μ *and is denoted by* $[\mu]$.

A theorem of Ahlfors and Bers and extended by Earle shows that $T^{\#}(R)$ is a manifold, [2, 5, 6]. Moreover, the natural mapping

$$(5) \qquad M(R) \xrightarrow{\ \Phi\ } T^{\#}(R)$$

is differentiable and each fiber $\Phi^{-1}(p)$ in $M(R)$ is a closed submanifold of $M(R)$ and at each point μ in $M(R)$ the kernel of the derivative $d\Phi_\mu$ has a closed, complementary subspace in the Banach space $L(R)$.

Let $M_0(R) = \Phi^{-1}(0)$, that is, the orbit of 0 in $M(R)$ under the action of $D_0(R)$. $M_0(R)$ is the trivial Teichmüller class. In most cases $D_0(R)$ and $M_0(R)$ are identified by the mapping which sends h in $D_0(R)$ onto $\mu = h_{\bar{z}}/h_z$ in $M_0(R)$. In fact, if h_i is sent onto μ_i for $i = 1$ and 2 and if $\mu_1 = \mu_2$, then $h_1 \circ h_2^{-1}$ will be a conformal self-mapping of R homotopic to the identity. This will imply $h_1 = h_2$ except in a few cases all of which occur when $g = 1$ or 0. These cases coincide with the cases where the ρ in formula (1) is bigger than 0.

Let $Q(R)$ be the Banach space of holomorphic, quadratic differentials

$$\phi \quad \text{on} \quad R \quad \text{for which} \quad \|\phi\| = \iint_R |\phi|\, dx\, dy \quad \text{is finite and which are real along}$$

the boundary curves of R with respect to boundary uniformizers. A boundary uniformizer is a local parameter z defined in a neighborhood of a point lying on a boundary curve such that z takes the boundary curve into the real axis. With respect to such a parameter the requirement is that $\phi(z)\,dz^2$ should be real for z in $\mathbf{R}$. It is easy to see that this requirement is independent of the choice of boundary uniformizer and that it is enough to force every element of $Q(R)$ to extend to the doubled Riemann surface of R. Moreover, elements of $Q(R)$ may have at most simple poles at the punctures of R because of the condition $\|\phi\| < \infty$. From these remarks it is easy to conclude from the Riemann-Roch theorem that the real dimension of $Q(R)$ is given by formula (1).

$$\text{Let} \quad N = \{\mu \in L(R);\ \operatorname{Re} \iint \mu\phi\, dx\, dy = 0 \quad \text{for all} \quad \phi \quad \text{in} \quad Q(R)\}. \quad \text{Here,}$$

the double integral is taken over the whole Riemann surface R. It turns out that N is the kernel of the derivative of the mapping Φ at the point 0 in $M(R)$. Hence, the tangent space to $T^{\#}(R)$ at the origin is $L(R)/N$ and, under the pairing

$$(6) \qquad\qquad (\mu, \phi) \mapsto \operatorname{Re} \iint \mu\phi\, dx\, dy \ ,$$

the cotangent space is $Q(R)$.

2.   *The variational method*

The importance of the following lemma was first pointed out by Hamilton, [11]. It is a consequence of the differential structure of the mapping (5) and the implicit function theorem.

LEMMA 2.1 (Hamilton). *Let* $\mu \in N$. *Then there exists a function* $\sigma(t,z)$ *which is* C^1 *with respect to* t *uniformly in* z *for sufficiently small* t *and such that* $\sigma(t,z)$ *is in* $M_0(R)$ *for each* t *and* $\sigma(t,z) = t\,\mu(z) + o(t)$ *uniformly in* z.

Lemma 2.1 can be extended in the following way. Let S be a finite Riemann surface and R a nonempty open subset of S and $M_1(R,S)$ be the subset of $M_0(S)$ consisting of elements whose support is contained

in R. Let $N(S) = \{\mu \in L(S) : \iint \mu\,\phi\,dx\,dy = 0$ for all ϕ in $Q(S)\}$. Here

the double integral is taken over S.

LEMMA 2.2. *Let* $\mu \in N(S)$ *and suppose the support of* μ *is contained in* R. *Then there exists a function* $\sigma(t,z)$ *which is* C^1 *with respect to* t *uniformly in* z *for sufficiently small* t *and such that*
   a) $\sigma(t,z)$ *is in* $M_1(R,S)$ *for each* t *and*
   b) $\sigma(t,z) = t\,\mu(z) + o(t)$ *uniformly in* z.

This lemma is proved in [10]. The method is to look at the natural mapping

$$(7) \qquad\qquad M(R) \xrightarrow{\;F\;} T^{\#}(S)$$

where $M(R)$ consists of elements of $M(S)$ whose support is contained in $\bar{R}$. In showing F has surjective derivative at each point, one uses the finite dimensionality of $T^{\#}(S)$. Then one uses the differentiability of (7), as developed by Ahlfors and Bers, and the implicit function theorem.

It is possible to prove a version of Lemma 2.1 by using the main inequality of Reich and Strebel, [14], and this approach will be investigated in later work. Lemma 2.2 seems to defy this approach.

Now let R be connected and of finite type and notice that restrictions of elements of M(S) to R are Beltrami differentials on R . The boundary of R in S may not be analytic. Nonetheless, by the uniformization theorem, it can be realized as a subsurface with analytic boundary curves contained in a larger surface. Thus $T^{\#}(R)$, Q(R), and N can be defined just as in section 2.

Let $f : T^{\#}(R) \to \mathbf{R}$ be a differentiable function. Then $f \circ \Phi : M(R) \to \mathbf{R}$ will be a differentiable function and there will be a continuous linear functional $a : L(R) \to \mathbf{R}$ such that

$$f \circ \Phi(t\mu) = f \circ \Phi(0) + t a(\mu) + o(t) .$$

Since $d\Phi_0$ annihilates elements of N and since, by the chain rule, $a = df_0 \circ d\Phi_0$, one sees that $a(N) = 0$. Hence a factors to a mapping $\tilde{a} : L(R)/N \to \mathbf{R}$ and so $\tilde{a}$ is represented by an element of Q(R). Therefore, one has the following well-known proposition.

PROPOSITION 2.1. *Given a differentiable function* $f : T^{\#}(R) \to \mathbf{R}$, *there exists an element* ϕ *in* Q(R) *such that*

$$(8) \qquad f([t\mu]) = f([0]) + t \, \mathrm{Re} \iint \phi\mu \, dx \, dy + o(t)$$

for small real values of t *and for all* μ *in* L(R).

We will abbreviate (8) by writing $\dot{f} = \phi$.

The basis of the variational method is to consider the natural mapping

$$(9) \qquad M_1(R, S) \xrightarrow{\Psi} T^{\#}(R) .$$

The image of Ψ consists of quasiconformal variations of R which, when viewed as variations of S , do not alter the Teichmüller class of S .

PROPOSITION 2.2. *Suppose* $f : T^{\#}(R) \to \mathbf{R}$ *is a differentiable function* and $\dot{f} = \phi$ in $Q(R)$. *Suppose also that* $f \circ \Psi(\mu) \geq f \circ \Psi(0)$ *for all* μ *in* $M_1(R,S)$. *Then* ϕ *is in* $Q(S)$.

This proposition is proved in [8]. The inequality can be replaced by the reverse inequality and the conclusion still remains true. It is called a globalization principle, because starting with an element of $Q(R)$ it gives a condition under which this element belongs to $Q(S)$.

3. *Poincaré length as a functional on* $T^{\#}(R)$ *and quasiconformal rigidity*

In order to exploit Proposition 2.2 one must look at well-defined differentiable functions on $T^{\#}(R)$, in other words, at conformal invariants. Let R be hyperbolic, which means the universal covering surface for R is U, the upper half-plane, and let λ_R be the Poincaré metric for R, [1]. The Poincaré metric for U is $\lambda_U(z)|dz| = |dz|/y$ and if $\pi: U \to R$ is a universal covering mapping, λ_R is determined by the equation

$$(10) \qquad \lambda_R(\pi(z))|\pi'(z)| = \lambda_U(z) .$$

Let $[\gamma_0]$ be a family of smooth loops on R freely homotopic to a given simple, non-homotopically trivial loop γ_0 on R. Let

$$(11) \qquad L([\gamma_0]) = \inf \int_\gamma \lambda_R(w)|dw|$$

where the infimum is taken over all γ in $[\gamma_0]$.

There are several well-known facts about $L([\gamma_0])$ which we summarize:

i) if γ_0 is not freely homotopic to a puncture of R, then there is a unique Poincaré geodesic γ_1 in $[\gamma_0]$ such that $L([\gamma_0]) = \int_{\gamma_1} \lambda_R(w)|dw|$,

ii) $L([\gamma_0])$ is a conformal invariant and hence determines a well-defined functional $L([\gamma_\mu])$ on $T^{\#}(R)$,

iii) γ_0 determines a conjugacy class in the universal covering group and $L([\gamma_0])$ is the log of the multiplier of any element of that conjugacy class,

iv) $L([\gamma_\mu])$ is a real analytic function on $T^{\#}(R)$ and there is an integral formula for the derivative of $L([\gamma_\mu])$.

One can find a proof of fact i) and iii) in [13], for example. Fact ii) is an obvious consequence of the fact that Poincaré length is a conformal invariant. For μ in $M(R)$, let w_μ be a solution to (2) and $\gamma_\mu = w(\gamma_0)$. So γ_μ will be a simple loop on R_μ. Suppose μ and σ are equivalent by an element h in $D_0(R)$. Since h is homotopic to the identity, h preserves the family $[\gamma_0]$. Hence the mapping $w_\sigma \circ h \circ w_\mu^{-1}$ will take the family $[\gamma_\mu]$ onto the family $[\gamma_\sigma]$ and this mapping is conformal since $w_\sigma \circ h$ and w_μ have the same Beltrami coefficient. This shows that $L([\gamma_\mu]) = L([\gamma_\sigma])$.

Fact iv) is proved in [9]. Let G be the universal covering group of R acting on U with universal covering $\pi : U \to R$ such that $U/G = R$. Under the standard isomorphism between the fundamental group and universal covering group, the class $[\gamma_0]$ will determine a conjugacy class in G. Let A be an element of this class. If we normalize so that ∞ is not a fixed point of A and assume that γ_0 is not homotopic to a puncture, then A is determined by

$$(12) \qquad \frac{Az - b}{Az - a} = M \frac{z - a}{z - b}$$

where $M > 1$ and a and b are attracting and repelling fixed points of A, which lie on the real axis.

Then $\log M = L([\gamma_0])$, (see [9]) and

$$(13) \qquad (\log M)^{\cdot}[\mu] = \mathrm{Re}\left\{ \frac{1}{\pi} \iint_\omega \frac{(a-b)^2}{(z-a)^2 (z-b)^2} \tilde{\mu}(z)\, dx\, dy \right\}$$

where ω is a fundamental domain in U for the cyclic group generated by A and $\tilde{\mu}$ is the lift of μ in $M(R)$ to the upper half plane.

Now let $\mathcal{S}$ be the family of all free homotopy classes of loops on R excluding the trivial homotopy class and any homotopy class of loops homotopic to a puncture of R. Obviously $\mathcal{S}$ is countable. Let $\mathbf{R}_+$ be the set of positive real numbers and $(\mathbf{R}_+)^{\mathcal{S}}$ the product space of functions from $\mathcal{S}$ into $\mathbf{R}_+$ with the product topology. There is a natural mapping

$$(14) \qquad\qquad F : T^{\#}(R) \to (\mathbf{R}_+)^{\mathcal{S}}$$

defined by $F([\mu]) = f_\mu$ where $f_\mu([\gamma]) = L([\gamma_\mu])$.

In [9] the following proposition was proved.

PROPOSITION 3.1. *Given any* $[\mu]$ *in* $T^{\#}(R)$, *there exists a finite set* $\{[\gamma_1], \cdots, [\gamma_n]\} = \mathcal{S}_n \subset \mathcal{S}$ *with* $n = \dim_{\mathbf{R}} T^{\#}(R)$ *such that* $\pi \circ F$ *is a local homeomorphism where* $\pi : (\mathbf{R}_+)^{\mathcal{S}} \to (\mathbf{R}_+)^{\mathcal{S}_n}$ *is the natural projection induced by the inclusion of* $\mathcal{S}_n$ *in* $\mathcal{S}$.

This proposition is proved by showing that the derivatives of the functional $L([\gamma_\mu])$ as $[\gamma]$ varies over $\mathcal{S}$ always generate the vector space $Q(R_\mu)$. It motivates the following definition.

DEFINITION 3.1. *Let* $R \subset S$ *and* R *and* S *be connected, nonempty, finite Riemann surfaces. Then* R *is quasiconformally rigid in* S *if for every* μ *in* $M_1(R, S)$ *and every* $[\gamma]$ *in* $\mathcal{S} = \mathcal{S}(R), L([\gamma_\mu]) \geq L([\gamma])$.

PROPOSITION 3.2. *If* R *is quasiconformally rigid in* S, *then* $Q(R) \subset Q(S)$.

Proof. Let ϕ_γ be the element of $Q(R)$ for which

$$L([\gamma_{t\mu}]) = L([\gamma]) + t \operatorname{Re}\left\{\iint_R \phi_\gamma \mu \, dx \, dy\right\} + o(t).$$

An explicit formula for ϕ_γ can be obtained by taking a relative Poincaré theta series of $(a-b)^2(z-a)^{-2}(z-b)^{-2}$ summed over coset representatives of $G/\langle A \rangle$ where $\langle A \rangle$ denotes the cyclic group generated by A, [9]. The main point is that since $L([\gamma_\mu])$ assumes a minimum at $\mu = 0$ over all μ in $M_1(R,S)$, Proposition 2.2 says that ϕ_γ is in $Q(S)$. Proposition 3.1 says one can pick $\gamma_1, \cdots, \gamma_n$ so that $\phi_{\gamma_1}, \cdots, \phi_{\gamma_n}$ form a basis of $Q(R)$. Therefore, $Q(R) \subset Q(S)$.

This proposition is useful for telling when a subsurface R of S is not rigid in S. For example if R is S with a disc removed, then R cannot be rigid in S since $Q(R)$ has greater dimension than $Q(S)$. The same argument applies if S has genus bigger than one and R is S with a point removed.

It is not known whether $Q(R) \subset Q(S)$ implies R is rigid in S.

4. *Examples of quasiconformally rigid surfaces*

a. *Annuli*

Let S be an annular domain and $f: S \to \{z;\ 1 < |z| < \rho\}$ a biholomorphic mapping. Let R be an annular subset of S, that is, an open subset which is topologically equivalent to the region between two concentric circles in the plane.

PROPOSITION 4.1. *R is quasiconformally rigid in S if, and only if, the restriction of f to R takes R onto the region between two concentric circles centered at the origin.*

Proof. By uniformization, there will be a holomorphic map $g: R \to \{\zeta;\ 1 < |\zeta| < r\}$. The spaces $Q(R)$ and $Q(S)$ are one-dimensional and are spanned by $d\zeta^2/\zeta^2$ and dz^2/z^2, respectively. If R is rigid in S, then $Q(R) \subset Q(S)$ and there exists a real number $c \neq 0$ such that $d\zeta^2/\zeta^2 = c\, dz^2/z^2$. The horizontal (vertical) trajectories of a quadratic differential are curves along which $\phi(z)dz^2 > 0$, $(\phi(z)dz^2 < 0)$. The vertical trajectories of $d\zeta^2/\zeta^2$ are concentric circles centered at the origin in the ζ-plane and the horizontal trajectories are rays emanating from the origin.

The equation $d\zeta^2/\zeta^2 = c\,dz^2/z^2$ says vertical trajectories in the ζ-plane must correspond to either all vertical or all horizontal trajectories in the z-plane. Since R is swept out by vertical trajectories and R is annular, vertical trajectories must correspond to vertical trajectories. It is now easy to see that f takes vertical trajectories of $d\zeta^2/\zeta^2$ on R into concentric circles in $\{z : 1 < |z| < R\}$ and R into the region between two concentric circles centered at the origin.

To prove the converse, let γ be a simple loop in R homotopic to a generator for the fundamental group of R. Let $m(R)$ be the extremal length of the family of smooth curves connecting the two boundary components of R. It is well known and obvious that

$$(15) \qquad\qquad m(R) = \pi/L([\gamma]) \,.$$

Now let R and S both be regions in the z-plane between two concentric circles centered at the origin with $R \subset S$. Thus $S = \{z; 1 < |z| < R\}$ and $R = \{z; r_1 < |z| < r_2\}$ and $1 \le r_1 < r_2 \le R$. Let $S_1 = \{z; 1 < |z| < r_1\}$ and $S_2 = \{z; r_2 < |z| < R\}$. Let $\mu \in M_1(R, S)$ and w^μ be a quasiconformal mapping of S into S_μ satisfying (2) with $\mu \equiv 0$ on $S_1 \cup S_2$. From the comparison principle for extremal length

$$(16) \qquad\qquad m(S_\mu) \ge m(S_{1\mu}) + m(R_\mu) + m(S_{2\mu}) \,.$$

By the hypothesis that μ is in $M_1(R, S)$, $m(S) = m(S_\mu)$ and $m(S_{i\mu}) = m(S_i)$ for $i = 1$ and 2 because w^μ is conformal on $S_1 \cup S_2$. Therefore, $m(S_1) + m(R) + m(S_2) = m(S) = m(S_\mu) \ge m(S_1) + m(R_\mu) + m(S_2)$ and so $m(R) \ge m(R_\mu)$. By using (15), this gives the desired inequality for $L([\gamma_\mu])$.

b. *An annulus in a torus*

Let R be an annular region contained in a torus S. Assume S is realized as C/L where L is a lattice in the z-plane of all points of the form $n + m\tau$ where n and m are integers and $\operatorname{im}\tau > 0$. Let $\pi: \mathbf{C} \to \mathbf{C}/L = S$ be the natural projection.

PROPOSITION 4.2. *R is quasiconformally rigid in* S *if, and only if,* $R = \pi(\text{a strip})$ *where the strip is a region in* $\mathbf{C}$ *bounded by two parallel lines which project to closed curves in* S.

Proof. $Q(S)$ is the complex vector space generated by dz^2. If R is rigid, then a generator for $Q(R)$ lifts to a differential ϕ on $\mathbf{C}$ for which $\phi = c\,dz^2$ for some complex number c. Thus horizontal trajectories for ϕ lift to straight lines in $\mathbf{C}$ and they must be straight lines which project to closed curves in S.

Conversely, suppose $R = \pi(\text{a strip})$. π will take any line in this strip onto a simple curve in R which separates the two boundary components of R. If one translates such a line so that it passes through the origin, it will meet the lattice L at some point not equal to zero since it projects onto a closed curve on S. Let $\tau_0 = m_0 + n_0\tau$ be such a point closest to the origin. Then m_0 and n_0 are relatively prime since the segment from 0 to $m_0 + n_0\tau$ projects to a simple curve.

In any case, one can assume that the preimage of R in a period parallelogram corresponds to one of the following shaded regions.

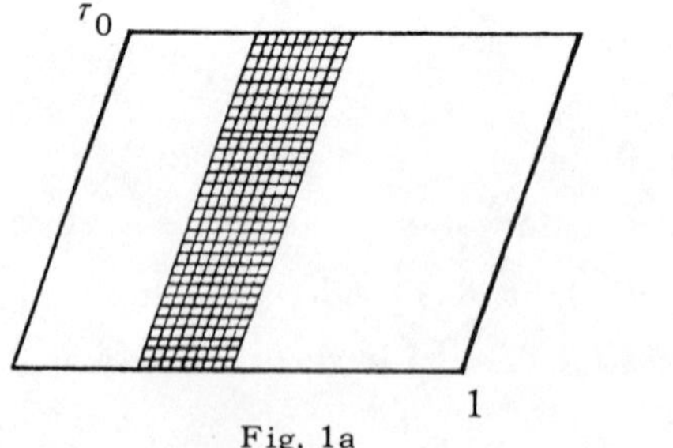
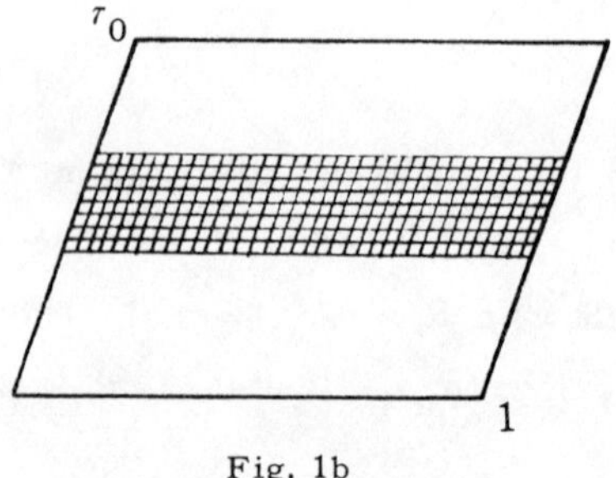

Fig. 1a   Fig. 1b

Let μ be in $M_1(R, S)$ and $\tilde{\mu} = \mu(\pi(z))$. Then $\tilde{\mu}$ will be an automorphic function for the lattice L and the restriction of $\tilde{\mu}$ to the period parallelogram will have support in one of the shaded regions in figure 1. Let w

be a homeomorphic solution to $w_{\bar{z}} = \bar{\mu} w_z$ normalized to fix 0, 1, and ∞. The assumption that μ is in the trivial Teichmüller class of S corresponds to the statement that $w(z+\omega) = w(z)+\omega$ for all ω in L. Now, focus attention only on the case illustrated by figure 1a. Let $m(R)$ be the modulus of the annulus R. Let $\tau_0 = \tau_1 + i\tau_2$ where τ_1 and τ_2 are real. The extremal length of the family of curves joining $[0, \tau_0]$ to $[1, 1+\tau_0]$ and projecting to closed curves on S is $1/\tau_2$. But $w(\tau_0) = \tau_0$ and so τ_2 remains unchanged by the mapping w induced by any μ in $M_1(R, S)$. Now, by an argument analogous to the one given in example 4a, one sees that $m(R_\mu)$ achieves a maximum when $\mu = 0$ for μ in $M_1(R, S)$ and, using (15), this means R is rigid. A similar argument takes care of the case illustrated in figure 1b.

c. *A surface in its double*

Let R be a finite surface with at least one boundary curve and let R^d be the double of R. Let G be a Fuchsian universal covering group of R acting on U. Then G will also act discontinuously on $\bar{U}$, the lower half-plane, and on an open subset of R. The limit set of G will be a subset of R of zero measure. Let $\Omega(G)$ be the full set of discontinuity of G. Then $R^d = \Omega(G)/G$.

PROPOSITION 4.3. R *is quasiconformally rigid in* R^d.

Proof. For any $[\gamma]$ in $\mathcal{S}(R)$, there will be an element A in G such that the conjugacy class of A corresponds to $[\gamma]$ under the isomorphism between fundamental group and universal covering group. Let V be a fundamental domain in $C - \{a, b\}$ for the cyclic group generated by A where a and b are the fixed points of A. (We know $a \neq b$ since γ is assumed to be not homotopic to a puncture of R.) Select V so that it is a ring domain separating the points a and b. Let $\Lambda(V)$ = the extremal length of the family of curves going around V. Let M be the multiplier of A as in equation (12). Then

$$(17) \qquad\qquad L([\gamma]) = \log M = 2\pi/\Lambda(V) .$$

We must look at what happens to all of these quantities where they are deformed by a quasiconformal mapping. Lift the Beltrami differential μ in $M(R)$ to a function $\tilde{\mu}$ on U. Then $\tilde{\mu}$ will satisfy $\tilde{\mu}(A(z))\overline{A'(z)} = \tilde{\mu}(z)A'(z)$ for all A in G. Extend $\tilde{\mu}$ to $\overline{U}$, the lower half-plane, by letting $\tilde{\mu}(\overline{z}) = \overline{\tilde{\mu}(z)}$. Let w_μ be the homeomorphic solution to $w_{\overline{z}} = \tilde{\mu} w_z$ normalized to fix 0, 1, and ∞. Then $A_\mu = w_\mu \circ A \circ w_\mu^{-1}$ will be a linear fractional transformation preserving U and all of the quantities in (17) are transformed. $V_\mu = w_\mu(V)$ will be a region for the cyclic group generated by A_μ. A_μ has a multiplier, M_μ, and $[\gamma_\mu]$ is the free homotopy class of loops in $R_\mu = U/G_\mu$ corresponding to $[\gamma]$ in R under the mapping from R to R_μ induced by w_μ.

Now let $\mu^*(z) = \tilde{\mu}(z)$ for z in U and $\mu^*(z) = 0$ for z in $\overline{U}$ and let w^μ be the solution to $w_{\overline{z}} = \mu^* w_z$ normalized in the usual way. A^μ, V^μ, M^μ are defined by analogy. A basic inequality of Bers [4] says that

$$(18) \qquad\qquad \Lambda(V^\mu) \geq \tfrac{1}{2}\,\Lambda(V) + \tfrac{1}{2}\,\Lambda(V_\mu)\,.$$

This result is nothing more than the comparison principal for extremal length applied to the annulus V^μ and the two subregions $V^\mu \cap w^\mu(U)$ and $V^\mu \cap w^\mu(\overline{U})$. $V^\mu \cap w^\mu(U)$ is biholomorphic to $V_\mu \cap U$ via the holomorphic mapping $w_\mu \circ (w^\mu)^{-1}$ and $V^\mu \cap w^\mu(\overline{U})$ is biholomorphic to $V \cap \overline{U}$ via the mapping w^μ, which is holomorphic in $\overline{U}$.

Suppose $\mu \in M_1(R, R^d)$. Let $\Omega^\mu = \Omega(G^\mu)$ and $G^\mu = w^\mu \circ G \circ (w^\mu)^{-1}$. The fact that μ is in $M_0(R^d)$ implies there is a conformal map $c : \Omega/G \to \Omega^\mu/G^\mu$ homotopic to the map induced by w^μ. Hence, there is a conformal lifting of c, which we denote by f from Ω to Ω^μ for which $fAf^{-1} = A^\mu$ for all A in G. It is not hard to see that f extends continuously to the limit set of G, and that f must be a linear fractional transformation with $f \equiv w^\mu$ on the limit set of G. Therefore, $f =$ the identity and $A = A^\mu$ for all A in G.

From this, one sees that $\Lambda(V^\mu) = \Lambda(V)$. From (18), it follows that $\Lambda(V) \geq \Lambda(V_\mu)$ and from (17), $L([\gamma_\mu]) \geq L([\gamma])$ for all μ in $M_1(R, S)$ and all $[\gamma]$ in $\mathcal{S}$. This is the desired conclusion.

d. *Annuli for Jenkins-Strebel differentials*

Let S be a surface of type (g, o, k) with $g \geq 2$. The maximal number of classes in $\mathcal{S} = \mathcal{S}(S)$ representable by nonintersecting curves is $3g - 3 + k$. Let a_i , $1 \leq i \leq n$, be n positive constants and $n \leq 3g - 3 + k$ and $[\gamma_1], \cdots, [\gamma_n]$ be n classes in $\mathcal{S}$. Recall that $\mathcal{S}$ consists of the set of free homotopy classes of loops in S which are not homotopically trivial and not homotopic to any of the k punctures of S . Let $R_1, \cdots, R_n$ be disjoint subannuli of S such that a generating loop of the fundamental group of R_i is in $[\gamma_i]$. Such a system of annuli $\{R_i\}$ will be called admissible. Form the sum

$$(19) \qquad\qquad P(\{R_i\}) = \sum_{i=1}^{n} a_i^2 \, m(R_i) \; .$$

A theorem of Jenkins [12] says there exists an admissible system $\{R_i\}$ for which the sum (19) achieves its maximum among all admissible systems. Moreover, there is a quadratic differential ϕ in $Q(S)$ such that if z_i is a uniformizing parameter mapping R_i onto the region between two concentric circles centered at the origin in the complex plane, then $\phi = -(a_i^2/8\pi)(dz_i^2/z_i^2)$ in R_i . Each R_i is swept out by freely homotopic closed horizontal trajectories of ϕ and $\bigcup_{i=1}^{n} \bar{R}_i = S$. The annuli R_i are called characteristic annuli for the Jenkins-Strebel differential ϕ , see [17].

PROPOSITION 4.4. *A characteristic annulus* R_1 *of any Jenkins-Strebel differential* ϕ *is quasiconformally rigid.*

Proof. Let $[\gamma]$ be the class of all loops in R_1 which separate the two boundary components of R_1 . Let $\mu \in M_1(R_1, S)$. From Jenkins' observation that (19) is maximized by a Jenkins-Strebel differential and from the fact that w^μ does not alter any of the quantities $m(R_i)$ for $i > 1$, one sees that $m(R_1(\mu)) \leq m(R_1)$. Recalling (15), this implies $\Lambda([\gamma_u]) \geq \Lambda([\gamma])$.

5. *Rigid subsurfaces of a four-times punctured sphere*

Let S be the complex plane C with three points e_1, e_2, and e_3 removed. S can be viewed as a four-times punctured sphere; there is also a puncture at ∞. $Q(S)$ consists of all complex multiples of the differential $d\zeta^2/(\zeta-e_1)(\zeta-e_2)(\zeta-e_3)$, a differential with simple poles at each of the four punctures. To classify rigid subsurfaces R of S we look at the cases where the real dimension of R is either 1 or 2, since by Proposition 3.2, $Q(R) \subset Q(S)$. The cases where $Q(R)$ has dimension zero are dismissed as uninteresting since the conditions for quasiconformal rigidity are then vacuously satisfied.

When $\dim Q(R) = 1$, R must be a) an annulus or b) a disc with two punctures. When $\dim Q(R) = 2$, R must be either c) an annulus with one puncture of d) a four-times punctured sphere. We now examine each of these cases.

a. *R is an annulus*

PROPOSITION 5.1. *Every concentric subannulus R of a Jenkins-Strebel annulus for S is rigid and, conversely, every rigid subannulus R of S is a concentric subannulus of a Jenkins-Strebel annulus on S. A simple loop γ separating the boundary components will contain exactly two of the points e_1, e_2, e_3 in its interior. To every loop there will correspond exactly one Jenkins-Strebel differential.*

Figure 2 shows an example of such a loop on a marked four-times punctured sphere.

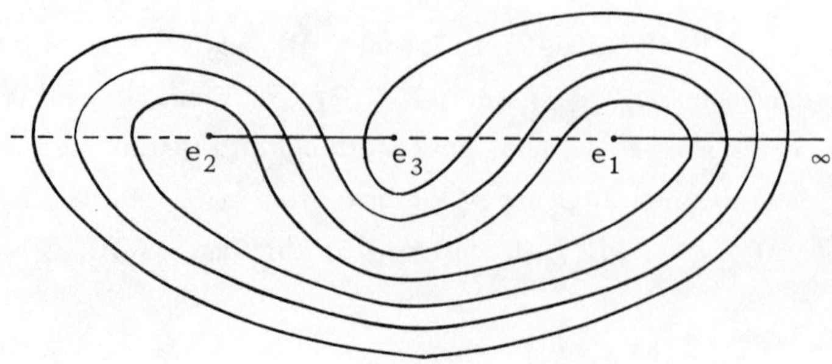

Fig. 2

The loop in figure 2 contains e_2 and e_3 in its interior while e_1 is in its exterior. From this picture it is obvious that there are infinitely many homotopy classes of such loops. We will see that these classes are in one-to-one correspondence with primitive lattice points in an elliptic covering of this sphere.

The first part of the proposition, the statement that concentric subannuli of Jenkins-Strebel annuli are rigid, is essentially contained in Proposition 4.4. If ζ is a uniformizing parameter taking such an annulus onto $\{\zeta; 1<|\zeta|<\rho\}$, by a concentric subannulus one obviously means the preimage of a set of the form $\{\zeta; r_1<|\zeta|<r_2\}$. The proof of Propositions 4.4 and 4.1 obviously take care of this situation. To take care of the second part of the proposition, we will find the subannuli R of S for which $Q(R) \subset Q(S)$. Let $p(z)$ be the Weierstrass p-function,

$$(20) \qquad p(z) = \frac{1}{z^2} + \sum \frac{1}{(z-\omega)^2} - \frac{1}{\omega^2}$$

where the sum is over all non-zero elements ω of the lattice L generated by 1 and τ where $\mathrm{im}\,\tau > 0$. The p-function restricted to the period parallelogram is a double covering of the sphere with branch points of order two at the points 0, $\frac{1}{2}$, $\frac{1+\tau}{2}$, and $\frac{\tau}{2}$. Let $e_1 = p\left(\frac{1}{2}\right)$, $e_2 = p\left(\frac{1+\tau}{2}\right)$, $e_3 = p\left(\frac{\tau}{2}\right)$. Of course, $\infty = p(0)$. It is well known [2] that given any four-times punctured sphere, there exists τ such that $C - \{e_1,e_2,e_3\}$ is biholomorphic to the given sphere. Notice that $p^{-1}(\{e_1,e_2,e_3,\infty\})$ is the lattice of half periods, $L/2$. p restricted to $C - L/2$ is an unramified, planar covering of the given four-times punctured sphere. From now on we will think of p as being so restricted. Under the mapping p, the quadratic differentials for the torus C/L must be transformed into the quadratic differentials for the sphere. In other words, $\zeta = p(z)$ and $c\,dz^2 = d\zeta^2/(\zeta-e_1)(\zeta-e_2)(\zeta-e_3)$.

In fact, from evaluating the leading terms of $p(z)$ and $p'(z)$ at $z = 0$, one sees that

$$(21) \qquad 4\,dz^2 = d\zeta^2/(\zeta-e_1)(\zeta-e_2)(\zeta-e_3).$$

This equation is simply the well-known relation, $p'(z)^2 = 4(p(z)-e_1)(p(z)-e_2)(p(z)-e_3)$. If $Q(R) \subset Q(S)$, then the preimages under p of horizontal trajectories for a generating element of $Q(R)$ will go over into parallel straight lines in $\mathbb{C}$. Since these lines must correspond to closed trajectories, the parallel line which passes through the origin must meet L at some point. Let τ_0 be such a point nearest to the origin. Then images of parallel translates of the segment $[0, \tau_0/2]$ will sweep out horizontal trajectories in R. Let $\tilde{R}$ be the union of the images of the parallel translates of the segment $[0, \tau_0/2]$ excluding those translates which meet the lattice. It is clear that $\tilde{R}$ forms the interior of a Jenkins-Strebel annulus on S and that R is a concentric subannulus of $\tilde{R}$.

It is easy to see from the argument principle that these trajectories enclose exactly two of the points $\{e_1, e_2, e_3\}$. A differential $\phi(\zeta)d\zeta^2$ of $Q(S)$ is a multiple of $d\zeta^2/(\zeta-e_1)(\zeta-e_2)(\zeta-e_3)$ and so it has simple poles at e_1, e_2, e_3 and ∞. Along a trajectory, $\phi(\zeta)d\zeta^2 > 0$ so $\arg \phi(\zeta) = -2 \arg d\zeta$ and the number of zeroes enclosed by a trajectory γ of $\phi(\zeta)d\zeta^2$ is

$$\frac{1}{2\pi} \int d(\arg \phi) = -2 \, \frac{1}{2\pi} \, (2\pi) = -2 \, .$$

This means γ encloses exactly two of the poles.

The uniqueness results of [12] or [17] imply, in this special case, that the homotopy class of γ uniquely determines a Jenkins-Strebel differential with horizontal trajectories freely homotopic to γ. Thus, the homotopy class of γ must determine a complex number c, up to positive real multiple, such that $cd\zeta^2/(\zeta-e_1)(\zeta-e_2)(\zeta-e_3)$ is the associated Jenkins-Strebel differential. From (21) one sees that the line through 0 and $\sqrt{c}$ must point in the direction of one of the points of L.

Given such a line, it is possible to give an algorithm for finding the homotopy class of the simple loop on the marked surface S and, conversely, given the homotopy class, you can easily find the line. Let $[z, w]$ denote the straight line segment from z to w. In figure 2, the

solid line from e_2 to e_3 is meant to be $p\left(\left[\frac{\tau}{2}, \frac{\tau+1}{2}\right]\right)$, the dotted segment from e_3 to e_1 is $p\left(\left[\frac{1}{2}, \frac{\tau+1}{2}\right]\right)$, the solid segment from e_1 to ∞
is $p\left(\left[0, \frac{1}{2}\right]\right)$, and the dotted segment from $-\infty$ to e_2 is $p\left(\left[0, \frac{\tau}{2}\right]\right)$.
Let γ be a representative of a homotopy class transversal to these seg-
ments and crossing them the minimal number of times. Let y be the
number of times γ crosses the solid segments and x the number of
times γ crosses the dotted segments. Always count y positively, but
count x positively or negatively according to the following rule. Move
along with positive orientation starting from any point on the segment
from $-\infty$ to e_2. Look to see which segment is encountered first, the
segment from e_1 to ∞ or the segment from e_2 to e_3. If the segment
from e_2 to e_3 is encountered first, count x positively. If the segment
from e_1 to ∞ is encountered first, count x negatively. Then, taking
c equal to any positive multiple of $(x+y\bar{\tau})^2$, you get the Jenkins-Strebel
differential associated to the class of γ. This formula applies even
when x or y is zero.

b. R *is a disc with two punctures*
Let $R = \{z;\ |z| < 1\ \text{and}\ z \neq p\ \text{and}\ z \neq q\}$. Then $Q(R)$ is generated
by

$$(22) \qquad\qquad -dz^2/(z-p)(1-\bar{p}z)(z-q)(1-\bar{q}z) .$$

This differential has closed horizontal trajectories leaving the points p
and q in their interiors. An arc joining p to q is a critical trajectory.
It is a Jenkins-Strebel differential on R. For fixed p and q, let $g_R(p, q)$
be the Green's function for the disc $R \cup \{p, q\}$ evaluated at p and q.
In [8] it is shown that it is possible to find an $R \subset S$ for which $g_R(p, q)$
is maximized among all R in a given homotopy class. Furthermore, for
this R, (22) becomes a global differential on S. Let $\Lambda_R(p, q)$ be the
extremal length of the family of loops in R containing p and q in their
interiors. It is obvious from the comparison principle for extremal length

that $\Lambda_R(p, q)$ is a monotonic decreasing function of $g_R(p, q)$. Hence, the R which maximizes the latter will minimize the former. In this way, using the uniqueness of Jenkins-Strebel differentials, one sees that (22) becomes a Jenkins-Strebel differential on S. Therefore, the subsurfaces R of S for which R is rigid are just the one of Jenkins-Strebel type described in Proposition 5.1.

c. R *is an annulus with one puncture*

No such R can be rigid for the condition $Q(R) \subset Q(S)$ implies that the puncture of R must coincide with one of the punctures of S. This is so because one of the differentials in R has a pole at the puncture of R and all elements of Q(S) have simple poles at e_1, e_2, e_3 and ∞. Another element of Q(R) has closed trajectories which sweep out R. If this element is a global differential in S, its trajectories must each enclose two punctures of S, an impossibility.

d. R *is a four-times punctured sphere*

The same sort of arguments show that R must equal S and obviously this is a rigid configuration.

DEPARTMENT OF MATHEMATICS
BROOKLYN COLLEGE, CUNY

REFERENCES

[1]   Ahlfors, L. V., *Conformal Invariants; Topics in Geometric Function Theory*, McGraw-Hill, 1973.

[2]   ________, *Lectures on Quasiconformal Mappings*, Van Nostrand, 1966.

[3]   Ahlfors, L. V. and Bers, L., "Riemann's mapping theorem for variable metrics," Ann. of Math., vol. 72, (1960), 385-404.

[4]   Bers, L., "On the boundaries of Teichmüller spaces and on Kleinian groups, I", Ann. of Math. vol. 91, (1970), 570-600.

[5]   ________, "On moduli of Riemann surfaces," Lectures at Forschunginstitut fur Mathematik, Eidgenossiche Technische Hochschule, Zurich, summer, (1964).

[6]   Earle, C. J., "Teichmüller spaces of groups of the second kind," Acta Math. vol. 112, (1964), 91-97.

[7]   Earle, C. J., and Schatz, A., "Teichmüller spaces for surfaces with boundary," J. of Diff. Geo., vol. 4, no. 2 (1970), 169-185.

[8]   Gardiner, F. P., "On relative Teichmüller spaces and a globalization principle in Riemann surface theory," (to appear).

[9]   ________, "Trace moduli for Teichmüller spaces of Kleinian groups," J. D'Analyse, vol. 32, (1978), 212-221.

[10]   ________, "The existence of Jenkins-Strebel differentials from Teichmüller theory," Ann. J. of Math., vol. 99, no. 4 (1977), 1097-1104.

[11]   Hamilton, R. S., "Extremal quasiconformal mappings with prescribed boundary values," Trans. Amer. Math. Soc. 138 (1969), 399-406.

[12]   Jenkins, J. A., "On the existence of certain general extremal metrics," Ann. of Math., 66 (1957), 440-453.

[13]   McKean, H. P., "Selberg's trace formula as applied to a compact Riemann surface," Communications in Pure and Appl. Math., vol. XXV, (1972), 225-246.

[14]   Reich, E., and Strebel, K., "Extremal quasiconformal mappings with given boundary values," Contributions to Analysis, A collection of papers dedicated to Lipman Bers, Academic Press, 1974, 375-391.

[15]   Schiffer, M., and Spencer, D. C., *Functionals of Finite Riemann Surfaces*, Princeton University Press, Princeton, N. J., 1954.

[16]   Springer, G., *Introduction to Riemann Surfaces*, Addison-Wesley, Reading, Mass., 1957.

[17]   Strebel, K., "On quadratic differentials and extremal quasiconformal mappings," Lecture Notes, University of Minnesota, spring, 1967.

SPIRALS AND THE UNIVERSAL TEICHMÜLLER SPACE

F. W. Gehring[*]

1. Introduction.

Suppose that D is a simply connected domain of hyperbolic type in the extended complex plane $\overline{C} = C \cup \{\infty\}$. Then the hyperbolic or non-Euclidean metric ρ_D in D is given by

$$\rho_D(z) = (1 - |g(z)|^2)^{-1} |g'(z)| \, ,$$

where g is any conformal mapping of D onto the unit disk $\{z : |z| < 1\}$. For each function ϕ defined in D we introduce the norm

$$\|\phi\|_D = \sup_{z \in D} |\phi(z)| \, \rho_D(z)^{-2} \, .$$

Next for each function f which is locally univalent and meromorphic in D we let S_f denote the Schwarzian derivative of f.

Now let L denote the lower half-plane, $L = \{z = x+iy : y < 0\}$, and let $B_2 = B_2(L, 1)$ denote the complex Banach space of functions ϕ analytic in L with the norm

$$\|\phi\| = \|\phi\|_L = \sup_{z \in L} 4y^2 |\phi(z)| < \infty \, .$$

[*]This research was supported in part by a grant from the U. S. National Science Foundation, Grant MCS-77-02842.

Next let S denote the family of functions $\phi = S_g$ where g is conformal in L, and let $T = T(1)$ denote the subfamily of those $\phi = S_g$ where g has a quasiconformal extension to $\overline{C}$. Then $\|\phi\| \leq 6$ for all $\phi \in S$ by [8], and hence $T \subset S \subset B_2$. The set T is called the universal Teichmüller space. See [2], [3], [4], [5].

In a recent paper [6], the author established a result, which when combined with an extension theorem of Ahlfors [1], yields the following characterization of T.

THEOREM 1. T *is the interior of* S.

Theorem 1 is closely related to the following interesting open problem raised by Bers in [2], [3], [4], [5].

QUESTION. *Is* S *the closure of* T?

In this talk we answer this question in the negative by sketching a proof of the following result.

THEOREM 2. *There exists a simply connected domain* D *of hyperbolic type and a positive constant* δ *with the following property. If* f *is conformal in* D *and if* $\|S_f\|_D < \delta$, *then* $f(D)$ *is not a Jordan domain.*

COROLLARY. *There exists a* ϕ *in* S *which does not lie in the closure of* T.

The domain D in Theorem 2 can be described in a very explicit manner. Namely, $D = \overline{C} - \gamma$, where γ is the arc

$$\gamma = \{z = \pm i e^{(-a+i)t} : t \in [0, \infty)\} \cup \{0\}$$

and $a \in \left(0, \dfrac{1}{8\pi}\right)$. Hence it is not difficult to derive an analytic expression for the conformal mapping g of L onto D, and $\phi = S_g$ turns out to be a rational function.

The idea behind the proof of Theorem 2 is quite simple. For $a \in (0, \infty)$ let

$$a_1 = \{z = e^{(-a+i)t} : t \in (0, \infty)\}, \quad a_2 = \{z : -z \in a_1\}.$$

Then a_1 and a_2 are logarithmic spirals in D which converge onto the point 0 from opposite sides of ∂D. Next suppose that f is any conformal mapping of D which fixes the points 1, -1, ∞. As $\|S_f\|_D$ approaches 0, f converges to the identity in D. Hence for $\|S_f\|_D$ small, f maps a_1, a_2 onto a pair of disjoint open arcs a_1^*, a_2^* which spiral onto $f_1(0)$, $f_2(0)$, the points which $f(z)$ approaches as $z \to 0$ from the two sides of ∂D.

Now the rate at which a_1 and a_2, and hence a_1^* and a_2^*, spiral depends on a. If a is sufficiently small, then a_1^*, a_2^* will spiral very slowly onto $f_1(0)$, $f_2(0)$. Since a_1^*, a_2^* are disjoint, the points $f_1(0)$, $f_2(0)$ will either coincide or be separated by a distance greater than a positive constant d.

Finally if we make $\|S_f\|_D$ still smaller, we can arrange that $f_1(0)$, $f_2(0)$ lie near 0 and hence within distance d of each other. Then $f_1(0)$ and $f_2(0)$ will coincide and $f(D)$ will not be a Jordan domain.

Complete proofs for Theorem 2 and its corollary are given in [7].

UNIVERSITY OF MICHIGAN
ANN ARBOR, MICHIGAN

REFERENCES

[1] L. V. Ahlfors, Quasiconformal reflections, *Acta Math.* 109 (1963), pp. 291-301.

[2] L. Bers, On boundaries of Teichmüller spaces and on Kleinian groups I, *Ann. of Math.* 91 (1970), pp. 570-600.

[3] _________, Universal Teichmüller space, *Analytic methods in mathematical physics*, Gordon and Breach (1970), pp. 65-83.

[4] _________, Uniformization, moduli, and Kleinian groups, *Bull. London Math. Soc.* 4 (1972), pp. 257-300.

[5] _________, Quasiconformal mappings, with applications to differential equations, function theory and topology, *Bull. Amer. Math. Soc.* 83 (1977), pp. 1083-1100.

[6] F. W. Gehring, Univalent functions and the Schwarzian derivative, *Comm. Math. Helv.* 52 (1977), pp. 561-572.

[7] ————, Spirals and the universal Teichmüller space, *Acta Math.* 141 (1978), pp. 99-113.

[8] Z. Nehari, The Schwarzian derivative and schlicht functions, *Bull. Amer. Math. Soc.* 55 (1949), pp. 545-551.

INTERSECTION MATRICES FOR BASES ADAPTED TO AUTOMORPHISMS OF A COMPACT RIEMANN SURFACE

Jane Gilman and David Patterson

1. Introduction

Consider a conformal automorphism h of prime order p with t fixed points on a compact Riemann surface of genus g $(g \geq 2)$. In [2] J. Gilman describes an integral homology basis (called the basis adapted to h) with respect to which the matrix of the automorphism h has a particularly nice form. For many applications we are interested not only in this homology basis, but also in the intersection matrix of this basis. Whereas the matrix representation for h depends only on the numbers p , g , and t , the intersection matrix also depends on the conjugacy class of h in the mapping class group of the surface. This conjugacy class is determined by a (p–1)-tuple of integers.

For an automorphism with fixed points, only part of this intersection matrix was obtained in [2]. The purpose of this paper is to obtain the complete intersection matrix. In recent discussions with M. Tretkoff, the authors have discovered that he has devised a similar method for constructing a basis for surfaces that are branched coverings of the sphere and for computing intersection numbers of such curves [6].

2. Notation

We now fix some notation and terminology to be used throughout the paper. For h , p , g and t as above, let H be the cyclic group generated by h. The quotient surface W/H will be denoted by W_0 and its genus by g_0. We assume that $t > 0$. At each fixed point P_k $(1 \le k \le t)$ of h, we can assign a rotation number q_k $(1 \le q_k \le p-1)$. The rotation number is characterized by the fact that there is a local coordinate system z near P_k such that the action of h is $z \mapsto \exp(2\pi i q_k/p)z$. (Here we adopt the convention that a positive rotation is one in the counterclockwise direction.) Associated with each rotation number q_k, there is a complementary rotation number s_k $(1 \le s_k \le p-1)$ satisfying $q_k \cdot s_k \equiv 1$ (mod p). For $1 \le s \le p-1$, let n_s be the number of s_k is equal to s. It is known [4], [1] that the $(p-1)$-tuple of integers $(n_1, \cdots, n_{p-1})$ determines the conjugacy class of h in the mapping class group of the surface W.

We introduce further notation that will simplify many of the formulae that follow. Let $\hat{s}$ be the smallest integer s such that $n_s \ne 0$ and let $\hat{q}$ be the integer satisfying $1 \le \hat{q} \le p-1$ and $\hat{q} \cdot \hat{s} \equiv 1$ (mod p). We may assume that the fixed points are ordered so that the complementary rotation numbers are in increasing order. In this case we have $\hat{s} = s_1$ and $\hat{q} = q_1$. For any integer m, let $[m]$ denote the least nonnegative residue of $\hat{q} \cdot m$ mod p. Thus the integer $[m]$ satisfies $0 \le [m] \le p-1$ and $\hat{s} \cdot [m] \equiv m$ (mod p).

3. *The main result*

Before we state the main result we shall describe the basis adapted to h. The following result was suggested by the work of J. Nielsen [4].

THEOREM 1 (J. Gilman [2]). *The surface W has an integral homology basis consisting of the homology classes:*

    (1) $h^j(A_w), h^j(B_w)$ $(1 \le w \le g_0, 0 \le j \le p-1)$

    (2) $h^k(X_i)$ $(3 \le i \le t, 0 \le k \le p-2)$.

The elements of the first set are permuted cyclically by h:

    $h(h^j(A_w)) \cong h^{j+1}(A_w)$ $(1 \le j \le p-2)$

    $h(h^{p-1}(A_w)) \cong h^0(A_w)$.

(with similar equations for the classes involving the B_w's). *The elements of the second set are mapped as follows:*

$$h(h^k(X_i)) \cong h^{k+1}(X_i) \quad (1 \le k \le p-3)$$

$$h(h^{p-2}(X_i)) \cong - \sum_{j=0}^{p-2} h^j(X_i). \quad \textit{(Here} \cong \textit{denotes homology.)}$$

It is known that each of the curves X_i corresponds to a fixed point of h. (See §4.) The basis adapted to h contains $t-2$ of these curves (and their translates). In order to have the basis reflect the conjugacy class of h, we shall use double subscripts on these curves as follows. For $1 \le m_s \le n_s$, let X_{s,m_s} denote the curve (one of the X_i's) corresponding to the $m_s{}^{th}$ fixed point with complementary rotation number s. The t subscript pairs of the form (s, m_s) can be ordered lexicographically: $(r, m_r) < (s, m_s)$ if and only if $r < s$ or $r = s$ and $m_r < m_s$. The $t-2$ curves X_{s,m_s} with the largest subscript pairs are to be included in the homology basis. For ease of reference, we restate Theorem 1 using this new notation.

THEOREM 1'. *If* $(n_1, \cdots, n_{p-1})$ *determines the conjugacy class of* h, *the surface* W *has a homology basis consisting of*:

    (1) $h^j(A_w), h^j(B_w)$ $(1 \le w \le g_0, 0 \le j \le p-1)$

    (2) $h^k(X_{s,m_s})$ $(0 \le k \le p-2$ *and for all pairs* (s, m_s) *with*

                              $1 \le s \le p-1$, $1 \le m_s \le n_s$ *except that the two*

                              smallest pairs are omitted.)

The elements of the first set are permuted cyclically by h:

    $h(h^j(A_w)) \cong h^{j+1}(A_w) \quad (1 \le j \le p-2)$

    $h(h^{p-1}(A_w)) \cong h^0(A_w)$

(with similar equations for the B_w's). *The elements of the second set are mapped as follows:*

    $h(h^k(X_{s,m_s})) \cong h^{k+1}(X_{s,m_s}) \quad (1 \le k \le p-3)$

$$h(h^{p-2}(X_{s,m_s})) \cong - \sum_{j=0}^{p-2} h^j(X_{s,m_s}) .$$

The intersection numbers for this basis are as follows:

THEOREM 2.

    1. $h^j(A_w) \times h^j(B_w) = +1$.

    2.   *If* $(r, m_r) < (s, m_s)$, *then*

$$h^0(X_{r,m_r}) \times h^k(X_{s,m_s}) = \begin{cases} +1 & \text{if} & [k] < [r] \leq [k+s] \\ -1 & \text{if} & [k+s] < [r] \leq [k] \end{cases}$$

    3. $h^0(X_{s,m_s}) \times h^k(X_{s,m_s}) = \begin{cases} +1 & \text{if} & [k] \leq [s] < [k+s] \\ -1 & \text{if} & [k+s] < [s] < [k] \end{cases}$.

    4.   *All other intersection numbers are* 0 *except for those that follow from parts 1, 2, and 3 by applying the identities below to arbitrary classes* C *and* D.

    (a) $C \times D = -D \times C$

    (b) $h^j(C) \times h^k(D) = h^0(C) \times h^{k-j}(D)$

                 ($k-j$ *reduced* mod p).

4.   *Background*

Before we prove Theorem 2, we briefly review the construction of the basis adopted to h. (See the proof of Theorem 1 in [2]. To aid the reader, we have tried to conform to the notation in [2].) Let G be a Fuchsian group with presentation

$$\left\langle a_1, \cdots, a_{g_0}, b_1, \cdots, b_{g_0}, x_1, \cdots, x_t \,;\, x_i^{p} = 1, \, x_1 \cdots x_t \prod_{w=1}^{g_0} [a_w, b_w] = 1 \right\rangle .$$

There is a homomorphism $\phi : G \to H$ such that $\phi(x_i) = h^{s_i}$ $(1 \leq i \leq t)$ and $\phi(a_w) = \phi(b_w) = 1$ $(1 \leq w \leq g_0)$. Here the s_i's have the same meaning as before; they are the complementary rotation numbers. The connection between the rotation numbers and the homomorphism $\phi : G \to H$ is explained in [3, Theorem 7]. Note that in terms of the Fuchsian group G, the number n_s is the number of elliptic generators of G that are mapped onto h^s.

The kernel of ϕ is (isomorphic to) the fundamental group of W. We obtain the basis adapted to h by applying a Schreier-Reidemeister re-writing process to G using the powers of x_1 as coset representatives of G/H. The automorphism h acts on the kernel of ϕ by conjugation by $x_1^{\hat{q}}$. The fundamental group of W is generated by $A_w = a_w$, $B_w = b_w$, and $X_i = x_i x_1^{-[s_i]}$ $(1 \le w \le g_0, 1 \le i \le t)$ and their translates under h:

$$h^k(A_w) = x_1^{[k]} a_w x_1^{-[k]}$$

$$h^k(B_w) = x_1^{[k]} b_w x_1^{-[k]}$$

and

$$h^k(X_i) = x_1^{[k]} x_i x_1^{-[k+s_i]} \quad (1 \le k \le p-1).$$

During the rewriting process some of these generators are eliminated leaving only those listed in Theorem 1. These elements (or, more pre-cisely, their homology classes) form a basis for the homology group. To obtain the basis of Theorem 1$'$ we have replaced X_i by X_{s,m_s} when $\phi(x_i) = h^s$ and $i = n_1 + n_2 + \cdots + n_{s-1} + m_s$.

These elements have a simple geometric description. Let $\hat{W}$ denote the covering surface with the fixed points of h removed and $\hat{W}_0$ the quotient surface with the images of the branch points removed. The fundamental group of $\hat{W}_0$ has a presentation

$$\left\langle a_1, \cdots, a_{g_0}, \beta_1, \cdots, \beta_{g_0}, \delta_1, \cdots, \delta_t; \delta_1 \cdots \delta_t \prod_{w=1}^{g_0} [\alpha, \beta] = 1 \right\rangle.$$

The group G is the quotient of this fundamental group under the homo-morphism mapping $\alpha_w \mapsto a_w$, $\beta_w \mapsto b_w$, and $\delta_i \mapsto x_i$. By relating the unbranched covering to the branched covering and by using the well-known

isomorphism between the defining subgroup of the unbranched covering
and the fundamental group of $\hat{W}$, we see that the elements $h^k(A_w)$,
$h^k(B_w)$, and $h^k(X_i)$ are the images on W of the lifts to the covering
surface $\hat{W}$ of the curves $\delta_1^{[k]} a_w \delta_1^{-[k]}$, $\delta_1^{[k]} \beta_w \delta_1^{-[k]}$ and
$\delta_1^{[k]} \delta_i \delta_1^{-[k+s_i]}$ respectively. We shall compute the intersection numbers
using this description of the curves as the lifts of curves from the quotient
surface.

REMARK. To simplify the calculations in the rest of the paper, we use
the single subscript notation for X_i and its rotation number s_i rather
than the more cumbersome double subscripts for X_{s,m_s} and its rotation
number s. It is then an easy matter to translate the final results into the
double subscript notation of Theorem 2.

5. *A standard topological representative for each basis curve*

The technique we shall use to compute intersection numbers is to
draw a simple picture of each curve and then to read off the intersection
number. Since the intersection number only depends on the homology
class of the curve, we may simplify curves by deforming them using free
homotopies. Each of the curves δ_i in the fundamental group of $\hat{W}_0$ is a
loop about the ith puncture u_i (where u_i is the image of ith fixed
point P_i). Represent δ_i as $l_i^{-1} c_i l_i$ where c_i is a small positively
oriented circle about u_i and l_i is a simple arc from c_i to the base
point of $\hat{W}_0$. (See Figure 1.)

Now consider the lifts of δ_i to the covering surface. In a neighborhood of P_i the projection map is p-to-1; in fact, in terms of local coordinate systems z near P_i and ζ near u_i the projection map is $\zeta = z^p$.
If we consider all possible lifts of δ_i to the covering surface, we get the
picture shown in Figure 1. The p lifts of c_i are simple arcs that fit
together to form a full circle C_i about P_i. The p lifts of l_i are arcs
or rays radiating out from the circle C_i; the end points of these rays are
the p lifts of the base point of $\hat{W}_0$. The picture in an accurate topologi-

cal description of the surface in a neighborhood of the circle C_i and its interior and in a neighborhood of the rays. Note that two distinct rays emanating from P_i do not intersect and that two rays that emanate from different P_i's are either disjoint or intersect only at a common end point.

What are the lifts of δ_i and its powers? Referring to Figure 1, we see that a lift of δ_i is an arc that goes in along a ray, then $(1/p)^{th}$ of the way around C_i in the positive direction, then out along the next ray. A lift of δ_i^2 is a lift of δ_i followed by the lift of δ_i that begins at the end point of the first lift. Note that the portion of this curve that goes out, then in, along the intermediate ray is homotopic to a point and can be eliminated. More generally, a lift of δ_i^k is (homotopic to) an arc that goes in along a ray, then moves the fraction k/p of the way around C_i, then goes out along the kth ray after the beginning ray.

We adopt the following conventions. Choose one of the lifts of the base point of $\widehat{W}_0$ and label it $\underline{0}$. This point is the distinguished base point of the covering surface. The image of the point $\underline{0}$ under h^k is labeled $\underline{k}$. In addition, we label the p rays around each branch point as follows. For the branch point P_i, the rays are labeled $0, 1, \cdots, p-1$ in order as we go around P_i in the positive direction starting with the ray ending at the base point $\underline{0}$. Near P_i, the automorphism h acts as a rotation through an angle of $2\pi\, q_i/p$. The automorphism h^k maps the point $\underline{0}$ to the point $\underline{k}$ and the ray labeled 0 to the ray labeled $kq_i \pmod p$; thus the ray labeled kq_i ends at the point $\underline{k}$. Stated in another way, the ray labeled k ends at the point $\underline{ks_i} \pmod p$. Near P_1, we can use the bracket notation introduced earlier: the ray labeled $[k]$ ends at the point $\underline{k}$.

Recall that $h^0(X_i) = X_i$ is the lift of $\delta_i \delta_1^{-[s_i]}$ starting at the base point $\underline{0}$. Thus we can represent the curve X_i as follows. First take the lift of δ_i that goes $(1/p)^{th}$ of the way around P_i with initial point $\underline{0}$ and end point $\underline{s_i}$, then follow this arc by lift of $\delta_1^{-[s_i]}$ that goes the fraction $[s_i]/p$ of the way around P_1 in the negative direction from the initial point $\underline{s_i}$ to the end point $\underline{0}$. (See Figure 2a.) Now consider the curve $h^k(X_j)$. The curve X_j will consist of two arcs as above, one near P_j and one near P_1, each connecting the point $\underline{0}$ and the point $\underline{s_j}$. The automorphism h^k moves the point $\underline{0}$ to the point $\underline{k}$ and the point $\underline{s_j}$ to the point $\underline{k+s_j}$; thus the curve $h^k(X_j)$ consists of two arcs, one near P_j and one near P_1, each connecting the point $\underline{k}$ and the point $\underline{k+s_j}$. Note that we can also obtain $h^k(X_j)$ as the lift of

$$\delta_1^{[k]} \delta_i \delta_1^{-[k+s_i]}.$$ The curve obtained in this way is not identical on $\widehat{W}$ to the one shown in Figure 2b, but it is freely homotopic on W to the one in Figure 2b.

As a matter of convenience, we now introduce certain standard representatives of the homology classes of $h^0(X_i)$ and $h^k(X_j)$. As a standard representative of $h^0(X_i)$ we take the curve obtained above that lies on the circles C_1 and C_i and on the rays emanating from these circles. On the other hand, the homology class of $h^k(X_j)$ will be represented, not by the original curve shown in Figure 2, but by a deformation of this curve that runs along, but not on, the circles C_1 and C_j and the rays emanating from these circles. (This new curve is indicated by the dotted line in Figure 2b.)

The new curve is constructed as follows. First, those portions of the original curve that lie on the circles C_1 and C_j are to be moved inward toward the points P_1 and P_j so as to lie on the arcs of circles C_1' and C_j' that lie between C_1 and P_1 and between C_j and P_j, respectively. (Of course, the rays must be extended inward to these circles as well.) Second, the two portions that lie on the rays are to be moved slightly to one side of the rays. The side on which we draw the new curve is determined by the following rule: if we transverse a ray in the direction away from P_1 and towards P_j, then the new curve always lies to the left of the ray. (This, of course, forces the arcs of the circles to be shifted slightly.)

These standard representatives of $h^0(X_i)$ and $h^k(X_j)$ have the property that they intersect transversely in a finite number of points. Moreover, if we assume that $i \leq j$, then the only points of intersection of these standard curves lie on the intersection of the circles C_1, C_i, and C_j with a portion of $h^k(X_j)$ that runs along a ray. We will explain this latter point in some detail. Consider an arc α of $h^0(X_i)$ connecting the circle C_1 and the circle C_i. The arc α consists of a ray emanating from C_1 and a ray emanating from C_i joined at a common point labeled $\underline{m}$. Consider also an arc β of $h^k(X_j)$ connecting C_1 and C_j. The arc β runs along a ray from C_1 and along a ray from C_j that are joined at a common endpoint labeled $\underline{n}$. Our claim is that two such arcs never intersect if we use the standard representatives of the curves.

If $\underline{m} \neq \underline{n}$, then there is no problem for the rays that determine α are completely disjoint from the rays that determine β. (Recall that two distinct rays are either disjoint or intersect only at their common endpoint.) Assume now that $\underline{m} = \underline{n}$. In Figure 3 we have drawn the arcs α and β and the rays form C_1, C_i, and C_j that are joined at the common endpoint $\underline{m}$ ($=$ point $\underline{n}$). An important point to notice is that the projection map from W to W_0 is an orientation-preserving homeomorphism of a neighborhood of the common endpoint onto a neighborhood of the base point of $\hat{W}_0$; therefore the order in which we encounter the rays as we move around the common endpoint $\underline{m}$ in the positive direction must be the same as the order in which we encounter their projections 1_1, 1_i, and 1_j as we move around the base point of $\hat{W}_0$ in the positive direction. Thus, if we assume that $i < j$ as in Figure 3a, we should first encounter the ray from P_1, then the ray from P_i, and then the ray from P_j as we move in the positive direction. Referring to Figure 3a, we see that the arcs α and β pass near the point $\underline{m}$ but do not intersect. Note, however, that if we assume $i > j$, the arcs must be joined at the common endpoint as shown in Figure 3b. In this case, the arcs must intersect in the vicinity of the point $\underline{m}$. Note also that the curves will not intersect if

$i = j$. (See Figure 3c.) This argument shows that there are no points of intersection in the common exterior of the circles. It follows that the only points of intersection lie on the circles, since only $h^k(X_j)$ lies on the interior of these circles.

6. *The proof of Theorem 2*

First we note that identities in part 4 of Theorem 2 follow from the general properties of intersection numbers. The first identity merely expresses the fact that the operation $\times$ is skew-symmetric. The second identity says that an homeomorphism (here h^{-j}) preserves intersection numbers. The intersection numbers among curves in the first set of basis elements are given in [2]. The intersection number between a translate of A_w or B_w and a translate of X_i is always 0, since it is possible to represent their homology classes by a pair of disjoint curves. Thus the only case that remains is to consider the intersection of X_i and a translate of X_j.

Using the standard topological representatives of $h^0(X_i)$ and $h^k(X_j)$, we can find the intersection numbers in all cases. As shown in section 5, these curves intersect transversely in a finite number of points that lie on the circles C_1, C_i, and C_j; therefore, we can find the intersection number by counting the algebraic number of crossings. The intersection number will depend on the relative positions of the rays on which (or near which) the curves run. These positions are determined by the relative ordering of the labels on the rays.

The proof now consists of drawing a picture of the curves for all possible orderings of the ray labels 0, $[s_i]$, $[k]$, and $[k+s_j]$. For the sake of brevity, we do the computation only for a few cases. First assume that $i < j$. (See Figure 4.) Since no portion of $h^0(X_i)$ lies on the circle C_j, the only points of intersection lie on the circle C_1. Consider the case when $0 < [k] < [s_i] < [k+s_j]$ shown in Figure 4a. (This is the typical or ''generic'' case in which the four ray labels are distinct.)

There is only one point of intersection which lies on C_1 near the ray labeled k. The intersection number in this case is $+1$.

The case when $0 < [k+s_j] < [s_i] = [k]$ is shown in Figure 4b. (Here two of the ray labels coincide.) There is one point of intersection, and the intersection number is -1. In this situation, if we were to allow $[k+s_j] = 0$, then we would move the ray labeled $\underline{k+s_j}$ until it coincided with the ray labeled $\underline{0}$. Note that we would still have one point of intersection; thus we get an intersection number -1 if $0 = [k+s_j] < [s_i] = [k]$. (See Figure 4c.)

Now assume that $i = j$ as in Figure 5. Unlike the case when $i < j$, we may have points of intersection on C_i as well as on C_1. If $0 < [k+s_i] < [s_i] < [k]$, then we have the picture shown in Figure 5a. There is one point of intersection, and the intersection number is -1. If $0 < [k+s_i] < [s_i] = [k]$, then we have the situation shown in Figure 5b. Here there are two points of intersection, one on C_1 and one on C_i, with opposite signs; therefore the intersection number is 0 in this case.

7. *Examples and comments*

We conclude with some miscellaneous remarks and examples.

REMARK 1. Theorem 2 shows that in general (i.e. if $p \geq 5$ or $t \geq 4$) the basis adapted to h (described in Theorems 1 and 1′) will not be a canonical basis (even if the elements of the basis are reordered). As pointed out by M. Tretkoff, since we have found the intersection matrix for this basis, in any particular case we can apply a standard algorithm [5] to obtain a canonical basis. Computation of a few examples suggests that this cannot be done without destroying the nice properties of the adapted basis relative to the automorphism. (See the example in Remark 2 below.) This is, of course, only natural; whereas an adapted basis is determined by the automorphism, a canonical basis is determined only by the skew-symmetric form on the homology group, independent of the automorphism.

REMARK 2. Let f and h both be homeomorphisms of prime order p with t fixed points on a surface of genus g which are not conjugate in the Mapping Class group. Then even though f and h have the same matrix representation with respect to the adapted basis of Theorem 1, Theorem 2 can be used to show that the images of f and h in the Torelli modular group (or equivalently, in the symplectic group, $Sp(2g, Z)$) will not be conjugate in general.

To see this, consider the example $p = 5$, $t = 3$ and $g_0 = 0$ so that $g = 2$. Let $(2, 0, 1, 0)$ and $(0, 0, 2, 1)$ give the conjugacy classes of f and h respectively. Let $X = X_{3,1}$ and $Y = X_{4,1}$ in the notation of Theorem 2. Then the matrix representations of f and h with respect to the adapted bases $\{X, f(X), f^2(x), f^3(X)\}$ or $\{Y, h(Y), h^2(Y), h^3(Y)\}$ are both given by

$$\begin{pmatrix} 0 & 1 & 0 & 0 \\ 0 & 0 & 1 & 0 \\ 0 & 0 & 0 & 1 \\ -1 & -1 & -1 & -1 \end{pmatrix}.$$

Note that h and f^2 are actually conjugate in the Mapping Class group. By Theorem 2, h and f have intersection matrices $I(h)$ and $I(f)$ respectively where

$$I(f) = \begin{pmatrix} 0 & 1 & 0 & 0 \\ -1 & 0 & 1 & 0 \\ 0 & -1 & 0 & 1 \\ 0 & 0 & -1 & 0 \end{pmatrix} \quad \text{and} \quad I(h) = \begin{pmatrix} 0 & 0 & -1 & 1 \\ 0 & 0 & 0 & -1 \\ 1 & 0 & 0 & 0 \\ -1 & 1 & 0 & 0 \end{pmatrix}.$$

Replace the basis adapted to f by $\{X, X + f^2(X), f(X), f^3(X)\}$ and the basis adapted to h by $\{h^2(Y), h^2(Y) + h^3(Y), Y, h(Y)\}$. Then these are canonical bases and with respect to these new bases f and h have matrix representations F and H where

$$F = \begin{pmatrix} 0 & 0 & 1 & 0 \\ 0 & 0 & 1 & 1 \\ -1 & 1 & 0 & 0 \\ 0 & -1 & -1 & -1 \end{pmatrix} \quad \text{and} \quad H = \begin{pmatrix} -1 & 1 & 0 & 0 \\ -1 & 0 & -1 & -1 \\ 0 & 0 & 0 & 1 \\ 1 & 0 & 0 & 0 \end{pmatrix}.$$

A tedious but straight forward matrix calculation shows that F and H are not conjugate in $Sp(4, Z)$.

REMARK 3. It should be possible to reconstruct $(n_1, \cdots, n_{p-1})$ from the adapted basis and the intersection matrix. For example, if one considers the numbers $h^0(X_{r,m_r}) \times h^k(X_{r,m_r})$ as k varies between 0 and $p-1$, then the number of 1's that occur (which is, of course, equal to the number of -1's) is equal to the minimum of $[r]$ and $p-1-[r]$. This can be seen by investigating the number of k such that $[k] < [r] < [k+r]$ and considering various cases such as $[r] < (p-1)/2$, etc.

REMARK 4. The method itself is general in that one would use similar pictures (with the obvious modification) to compute intersection numbers of lifts of curves with known intersection properties on a factor surface whenever one had an automorphism group on a surface and one knew the branch structure.

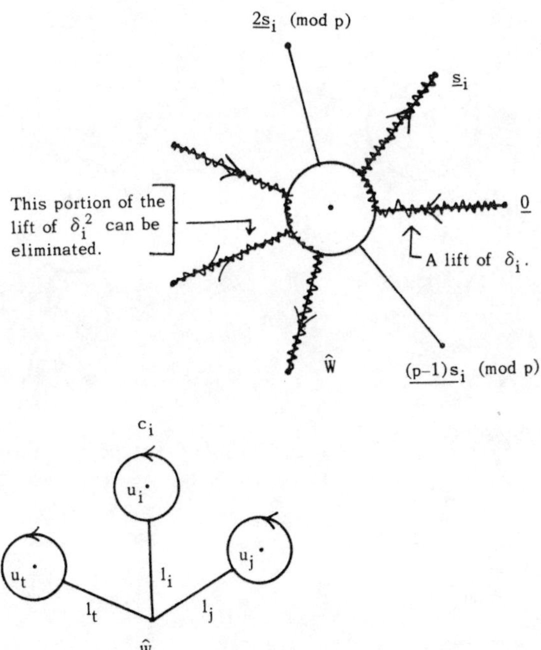

Fig. 1

P_1
P_i
[s_i]
0
s_i
0
s_i
h^0(X_i)-standard representative
(a)

k
[k]
P_1
0
P_j
[k+s_j]
k+s_j
k+s_j
k
h^k(X_j)
h^k(X_j)-standard representative
(b)

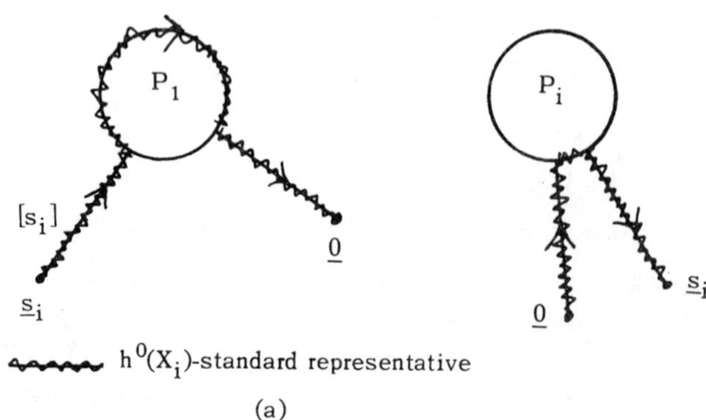

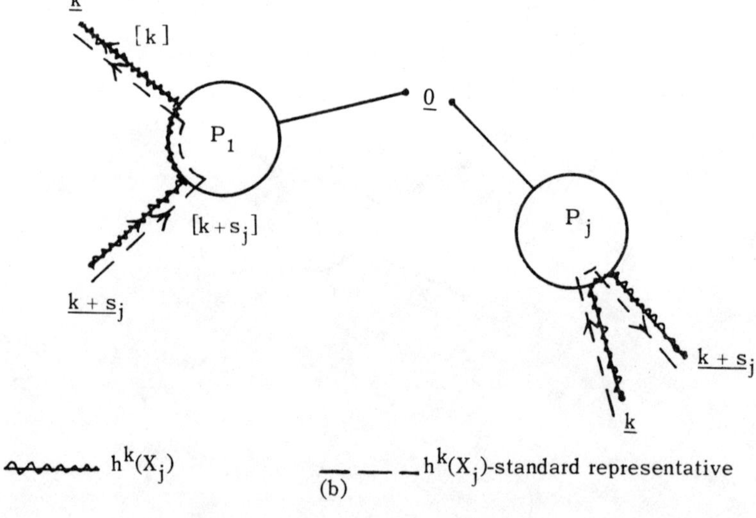

Fig. 2

P_1
P_j
m = n
P_i
(a) i < j
P_1
m = n
P_j
P_i
(b) i > j
P_1
P_i
(c) i = j
α —an arc of h^0(X_i)
β —an arc of h^k(X_j)

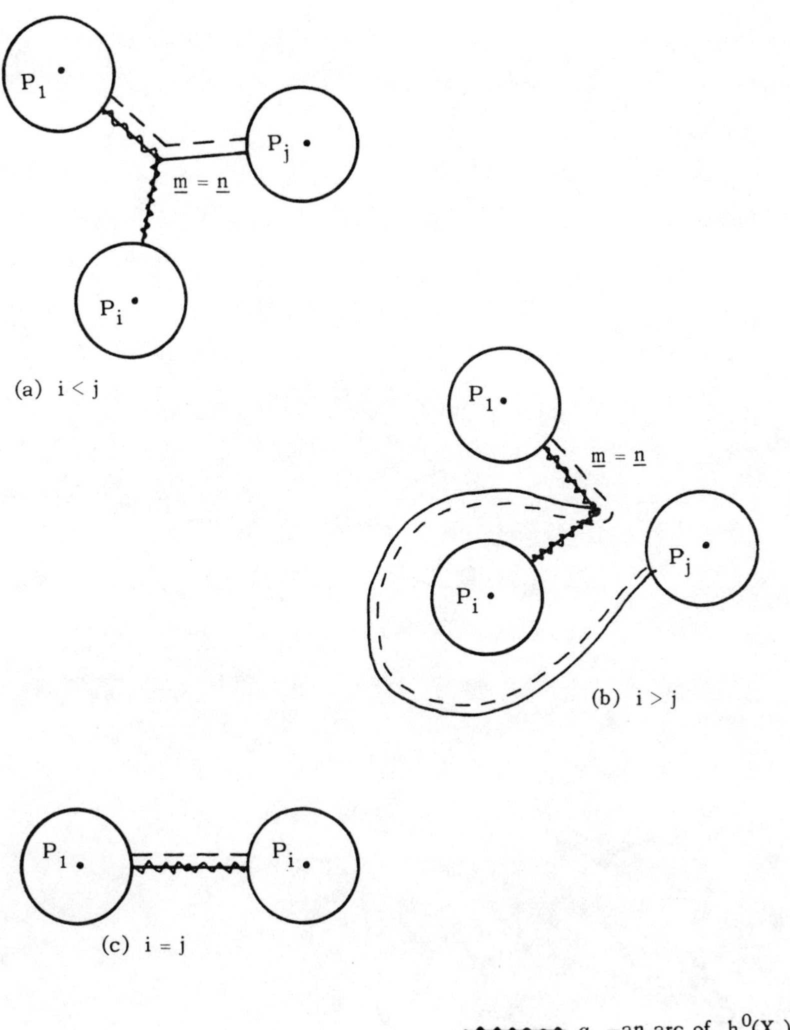

Fig. 3

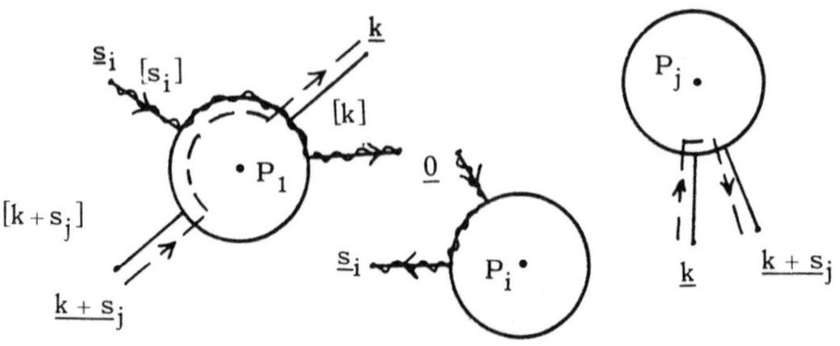

(a) $i < j$ and $0 < [k] < [s_i] < [k+s_j]$

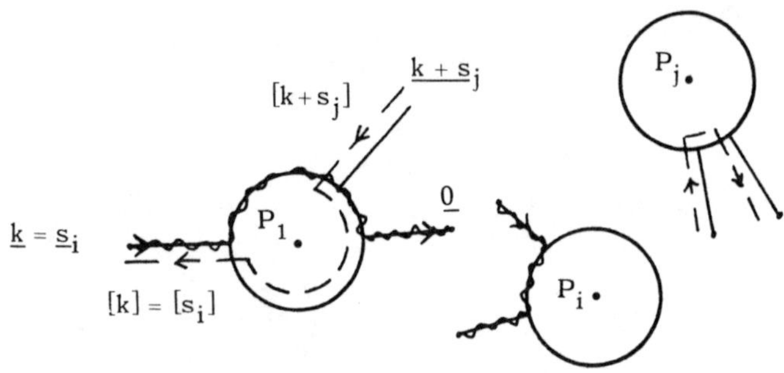

(b) $i < j$ and $0 < [k+s_j] < [s_i] = [k]$

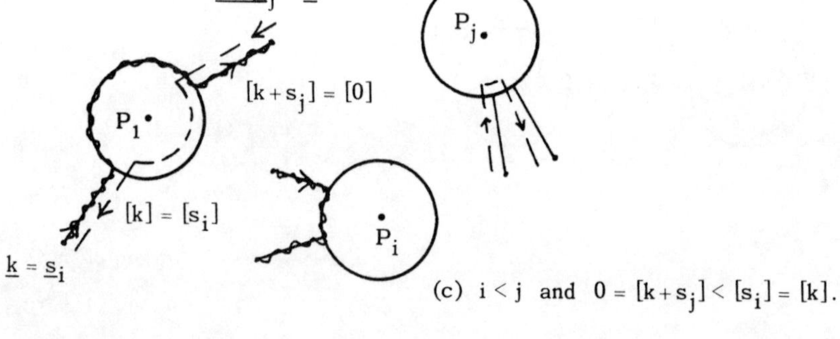

(c) $i < j$ and $0 = [k+s_j] < [s_i] = [k]$.

$\sim\!\!\sim\!\!\sim\!\!\sim\ h^0(X_i)$ $-\ -\ -\ h^k(X_j)$.

Fig. 4

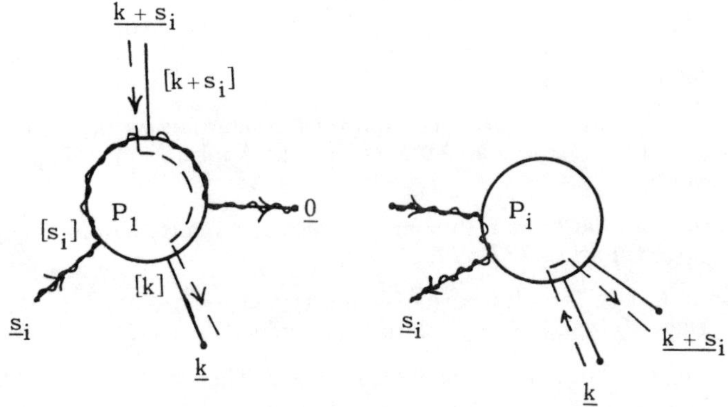

(a) $i = j$ and $0 < [k+s_i] < [s_i] < [k]$.

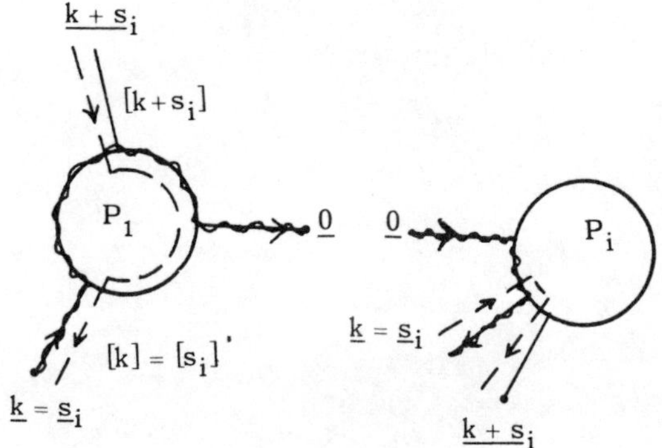

(b) $i = j$ and $0 < [k+s_i] < [s_i] = [k]$

$\text{———}\ h^0(X_i)$

$\text{— — —}\ h^k(X_i)$

Fig. 5

JANE GILMAN
MATHEMATICS DEPARTMENT
RUTGERS UNIVERSITY
NEWARK, NEW JERSEY 07102

DAVID PATTERSON
DIVISION OF MATHEMATICS AND SCIENCE
NOTRE DAME COLLEGE
ST. JOHN'S UNIVERSITY
STATEN ISLAND, NEW YORK 10301

REFERENCES

[1] J. Gilman, On conjugacy classes in the Teichmüller modular group, Mich. Math. J. *23*(1976), pp. 53-63.

[2] ________, A matrix representation for automorphisms of compact Riemann surfaces, Linear Algebra and its Applications *17*(1977), pp. 139-147.

[3] W. J. Harvey, On branch loci in Teichmüller space, Trans. Am. Math. Soc. *153*(1971), pp. 387-399.

[4] J. Nielsen, Die Structur periodischer Transformationen von Flächen, D. K. Dan. Vidensk. Selsk. Math-fys. Medd. XV(1937), pp. 1-77.

[5] C. L. Siegel, Topics in Complex Function Theory, Vol. III, Wiley-Interscience, New York, New York, 1973.

[6] M. Tretkoff, Roadmaps on Riemann Surfaces, to appear.

HOMOMORPHISMS OF TRIANGLE GROUPS INTO PSL(2, C)

Leon Greenberg[*]

In this article, we shall give two methods (one geometric, the other number theoretic) for constructing homomorphisms of Schwarz triangle groups into $PSL(2, \mathbf{C})$. It will be shown that all homomorphisms can be given by the geometric method, while the number theoretic method yields injections. It turns out that a large class of injections are obtained in this way, but we are unable to show that all are. From these considerations we are able to give examples of triangle groups which cannot be injected into $SO(3)$. We can also show that the fundamental group of a closed surface can be injected into $SO(3)$. We conclude with a discussion of the quadratic differentials which induce these homomorphisms.

1. Classification of homomorphisms

In this section, we shall describe all homomorphisms of a triangle group into $PSL(2, \mathbf{C})$. First we consider this kind of question from a more general viewpoint. The following theorem seems to be known.

THEOREM 1. *Let* Γ *be a finitely generated group, and let* A *be a complex algebraic group. Then* Γ *has finitely many normal subgroups* $N_1, N_2, \cdots, N_k,$ *such that*:

[*]This work has been supported by NSF Grant MCS 78-03836.

(1) *Any homomorphism $\Gamma \to A$ factors*:

$$\Gamma \longrightarrow A \qquad (\text{for some } i).$$
$$\Gamma/N_i$$

(2) *For each i , there is an injection*

$$\Gamma/N_i \hookrightarrow A.$$

Proof. Consider a presentation:

$$\Gamma = \langle \gamma_1, \gamma_2, \cdots, \gamma_n; R_\alpha(\gamma_1, \cdots, \gamma_n) = 1, \quad \text{for} \quad \alpha \in S \rangle.$$

For each homomorphism $\theta : \Gamma \to A$, let $x_1 = \theta(\gamma_1), \cdots, x_n = \theta(\gamma_n)$. In this way, a homomorphism θ corresponds to a point $(x_1, x_2, \cdots, x_n) \in A^n$, such that $R_\alpha(x_1, \cdots, x_n) = 1$, for $\alpha \in S$. The set

$$X = \{(x_1, \cdots, x_n) \in A^n | R_\alpha(x_1, \cdots, x_n) = 1, \quad \text{for} \quad \alpha \in S\}$$

is an algebraic set, which can be decomposed into varieties:
$X = X_1 \cup X_2 \cup \cdots \cup X_k$. Let $g_i = (g_{i1}, g_{i2}, \cdots, g_{in})$ be a generic point in X_i . The map $\theta_i(\gamma_t) = g_{it}$ $(1 \leq t \leq n)$ extends to a homomorphism $\theta_i : \Gamma \to A$. Let N_i denote the kernel of θ_i . Thus we have an injection $\Gamma/N_i \hookrightarrow A$.

Now consider any homomorphism $\theta : \Gamma \to A$. Let $x_t = \theta(\gamma_t)$, $1 \leq t \leq n$. The point $x = (x_1, \cdots, x_n) \in X$, so x belongs to some X_i . Since g_i is a generic point of X_i , there is a ring homomorphism $C[g_i] \to C[x]$. Consequently, any polynomial relation satisfied by g_i is also satisfied by x , and so the homomorphism θ factors:

$$\Gamma \xrightarrow{\;\theta\;} A$$
$$\Gamma/N_i$$

q.e.d.

We now return to the case where Γ is a triangle group, and $A = \mathrm{PSL}(2, \mathbb{C})$. Γ arises from a hyperbolic triangle Δ, with angles π/a, π/b, π/c. Let r, s, t be the reflections in the sides of Δ which are respectively opposite these angles, and let $R(a, b, c)$ be the group generated by r, s, t. The triangle group Γ, which we shall now denote $T(a, b, c)$, is the subgroup of index 2 in $R(a, b, c)$ containing the orientation-preserving transformations. $T(a, b, c)$ is generated by $x = st$, $y = tr$, $z = rs$, with defining relations: $x^a = y^b = z^c = xyz = 1$.

We now consider an analogous situation. Let $\overline{\Delta}$ be a triangle in $\mathbb{C}$, whose sides are circular arcs. Suppose that the angles are $\dfrac{\alpha\pi}{a}$, $\dfrac{\beta\pi}{b}$, $\dfrac{\gamma\pi}{c}$ (where α, β, γ are positive integers such that $0 < \dfrac{\alpha}{a}, \dfrac{\beta}{b}, \dfrac{\gamma}{c} < 1$). Let $\overline{r}$, $\overline{s}$, $\overline{t}$ be the reflections in the sides of $\overline{\Delta}$ which are respectively opposite these angles. Let $R\left(\dfrac{\alpha}{a}, \dfrac{\beta}{b}, \dfrac{\gamma}{c}\right)$ be the group generated by $\overline{r}$, $\overline{s}$, t, and let $T\left(\dfrac{\alpha}{a}, \dfrac{\beta}{b}, \dfrac{\gamma}{c}\right)$ be the subgroup of index 2, containing the orientation-preserving transformations. (This is a convenient, although admittedly inconsistent notation.) $T\left(\dfrac{\alpha}{a}, \dfrac{\beta}{b}, \dfrac{\gamma}{c}\right)$ is generated by $\overline{x} = \overline{s}\,\overline{t}$, $\overline{y} = \overline{t}\,\overline{r}$, $\overline{z} = \overline{r}\,\overline{s}$, which satisfy the relations: $\overline{x}^a = \overline{y}^b = \overline{z}^c = \overline{x}\,\overline{y}\,\overline{z} = 1$. (However, these relations may not be defining relations!) Therefore there is a homomorphism $\theta_{\alpha,\beta,\gamma} : T(a, b, c) \to T\left(\dfrac{\alpha}{a}, \dfrac{\beta}{b}, \dfrac{\gamma}{c}\right)$, such that $\theta_{\alpha,\beta,\gamma} : x, y, z \mapsto \overline{x}, \overline{y}, \overline{z}$. We shall now show that any homomorphism $\theta : T(a, b, c) \to \mathrm{PSL}(2, \mathbb{C})$, with non-cyclic image, is of the above form.

LEMMA 1. *The product of two elliptic transformations is non-loxodromic if and only if their fixed points lie on a common circle.*

Proof. This has been proved by Van Vleck [3], Theorem 6.   q.e.d.

THEOREM 2. *Let $\theta : T(a, b, c) \to \mathrm{PSL}(2, \mathbb{C})$ be a homomorphism onto a non-cyclic group. Then there is a circular triangle with angles $\dfrac{\alpha\pi}{a}$, $\dfrac{\beta\pi}{b}$, $\dfrac{\gamma\pi}{c}$, so that $\theta = \theta_{\alpha,\beta,\gamma}$.*

Proof. Let $\bar{x} = \theta(x)$, $\bar{y} = \theta(y)$, $\bar{z} = \theta(z)$. Since $\bar{x}^a = \bar{y}^b = \bar{z}^c = 1$, $\bar{x}$, $\bar{y}$, and $\bar{z}$ are elliptic. Let x_1, x_2 be the fixed points of $\bar{x}$, and y_1, y_2 the fixed points of $\bar{y}$. (Note that $\bar{x}$ and $\bar{y}$ cannot have the same fixed points, since they generate a non-cyclic group. However, they may have a common fixed point.) By Lemma 1, these fixed points lie on a common circle T. Let $\bar{t}$ be the reflection in T. There is a reflection $\bar{s}$ in a circle S, passing through x_1 and x_2, so that $\bar{x} = \bar{s}\bar{t}$. There is a reflection $\bar{r}$ in a circle R, passing through y_1 and y_2, so that $\bar{y} = \bar{t}\bar{r}$. Then $\bar{z} = \bar{y}^{-1}\bar{x}^{-1} = (\bar{r}\bar{t})(\bar{t}\bar{s}) = \bar{r}\bar{s}$. Since z is elliptic, the circles R and S must intersect in two points z_1, z_2, which are the fixed points of $\bar{z}$. Each of the circles R, S, T passes through the fixed points of two of the transformations $\bar{x}$, $\bar{y}$, $\bar{z}$. Let $\bar{\Delta}$ be a triangle whose sides are arcs of the circles R, S and T, and whose vertices are fixed points of $\bar{x}$, $\bar{y}$ and $\bar{z}$. (There are 8 such triangles if $\bar{x}$, $\bar{y}$ have no common fixed points, and 4 such triangles if $\bar{x}$, $\bar{y}$ have a common fixed point.) Since $\bar{x}^a = \bar{y}^b = \bar{z}^c = 1$, $\bar{\Delta}$ must have angles $\frac{a\pi}{a}$, $\frac{\beta\pi}{b}$, $\frac{\gamma\pi}{c}$, for some integers a, β, γ.   q.e.d.

REMARKS:

(1) The homomorphism $\theta_{a,\beta,\gamma}$ depends on the circular triangle $\bar{\Delta}$, and not only on its angles $\frac{a\pi}{a}$, $\frac{\beta\pi}{b}$, $\frac{\gamma\pi}{c}$ (which are determined by a, β, γ, when a, b, c are considered as fixed). However, since two triangles with the same angles are related by a Moebius transformation, $\theta_{a,\beta,\gamma}$ is determined by a, β, γ up to conjugation in $PSL(2, \mathbf{C})$. In this article, we shall consider two such conjugate homomorphisms to be equivalent.

(2) Each circular triangle $\bar{\Delta}$ (or an adjacent triangle in the diagrams below) can be transformed by a Moebius transformation to a spherical, Euclidean or hyperbolic triangle. The group $T\left(\frac{a}{a}, \frac{\beta}{b}, \frac{\gamma}{c}\right)$ is accordingly isomorphic to a subgroup of $SO(3)$, $E(2)$ (the plane Euclidean group), or $PSL(2, \mathbf{R})$. As in the proof of Theorem 2, let x_1, x_2 be

the fixed points of $\bar{x}$, y_1, y_2 the fixed points of $\bar{y}$, and T the circle which passes through these points.

(a) *Spherical* case. The point pairs (x_1, x_2) and (y_1, y_2) separate each other on T.

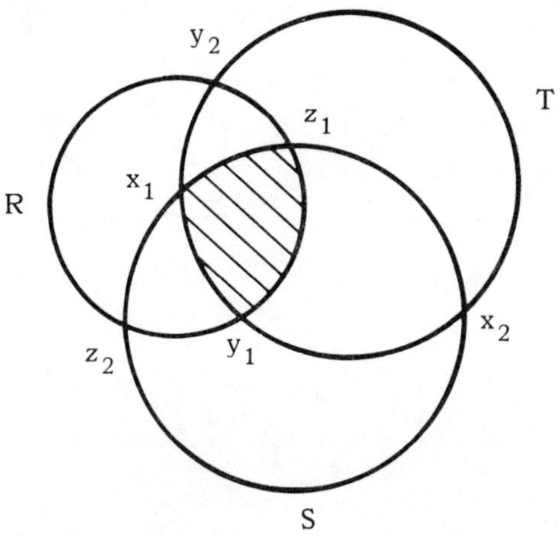

This case is characterized by the following inequalities (which must all be satisfied):

$$\begin{cases} \dfrac{a}{a} + \dfrac{\beta}{b} + \dfrac{\gamma}{c} > 1 \ , \\[2ex] \dfrac{a}{a} + \dfrac{\beta}{b} < 1 + \dfrac{\gamma}{c} \ , \\[2ex] \dfrac{\beta}{b} + \dfrac{\gamma}{c} < 1 + \dfrac{a}{a} \ , \\[2ex] \dfrac{\gamma}{c} + \dfrac{a}{a} < 1 + \dfrac{\beta}{b} \ . \end{cases}$$

(b) *Euclidean* case. The transformations $\bar{x}$ and $\bar{y}$ have a common fixed point (say $x_2 = y_2$).

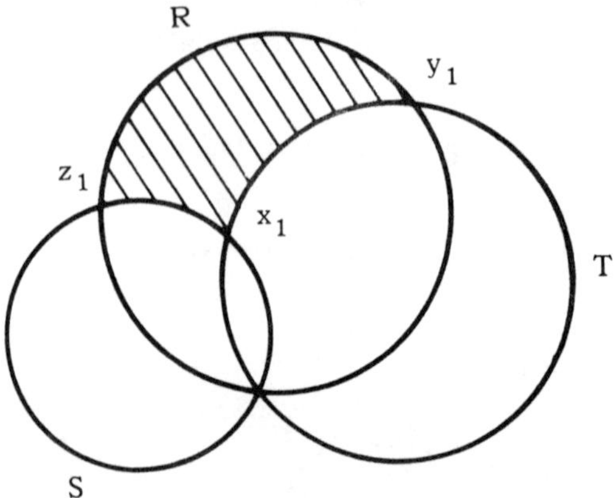

This case is characterized by one of the following equations:

$$\begin{cases} \dfrac{\alpha}{a} + \dfrac{\beta}{b} + \dfrac{\gamma}{c} = 1 \ , \quad \text{or} \\[2ex] \dfrac{\alpha}{a} + \dfrac{\beta}{b} = 1 + \dfrac{\gamma}{c} \ , \quad \text{or} \\[2ex] \dfrac{\beta}{b} + \dfrac{\gamma}{c} = 1 + \dfrac{\alpha}{a} \ , \quad \text{or} \\[2ex] \dfrac{\gamma}{c} + \dfrac{\alpha}{a} = 1 + \dfrac{\beta}{b} \ . \end{cases}$$

(c) *Hyperbolic case.* The point pairs (x_1, x_2) and (y_1, y_2) do not separate each other on T.

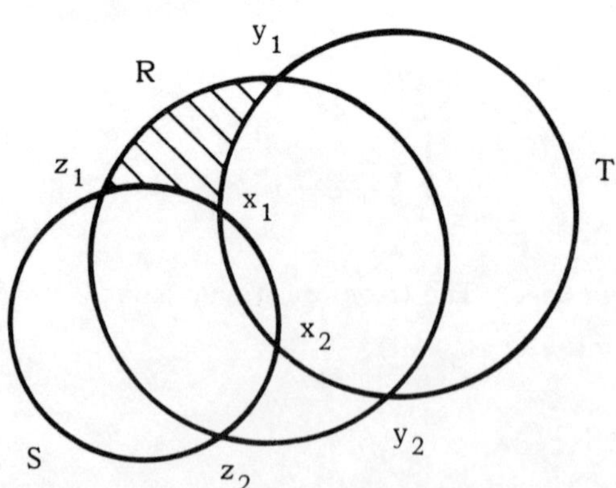

This case is characterized by one of the following inequalities:

$$\left\{ \begin{array}{l} \dfrac{\alpha}{a} + \dfrac{\beta}{b} + \dfrac{\gamma}{c} < 1 \ , \quad \text{or} \\[2ex] \dfrac{\alpha}{a} + \dfrac{\beta}{b} > 1 + \dfrac{\gamma}{c} \ , \quad \text{or} \\[2ex] \dfrac{\beta}{b} + \dfrac{\gamma}{c} > 1 + \dfrac{\alpha}{a} \ , \quad \text{or} \\[2ex] \dfrac{\gamma}{c} + \dfrac{\alpha}{a} > 1 + \dfrac{\beta}{b} \ . \end{array} \right.$$

Theorem 2 has the following consequence.

COROLLARY 1. *The following triangle groups cannot be injected into* SO(3):

$$T(3,6,6) \subset T(2,6,6) \subset T(2,4,6), \quad T(4,6,12), \quad T(2,6,10) \ .$$

Proof. If $\theta_{\alpha,\beta,\gamma} : T(a,b,c) \to T\left(\dfrac{\alpha}{a}, \dfrac{\beta}{b}, \dfrac{\gamma}{c}\right)$ is an isomorphism, then the pairs of integers (α,a), (β,b) and (γ,c) must be relatively prime. But it can be easily seen that in the above cases, no such spherical triangle exists.   q.e.d.

2. *Injections*

We now ask if the homomorphism $\theta_{\alpha,\beta,\gamma} : T(a,b,c) \to T\left(\dfrac{\alpha}{a}, \dfrac{\beta}{b}, \dfrac{\gamma}{c}\right)$ is an isomorphism. In other words, we would like to know if the relations $\overline{x}^a = \overline{y}^b = \overline{z}^c = \overline{x}\,\overline{y}\,\overline{z} = 1$ are defining relations for $T\left(\dfrac{\alpha}{a}, \dfrac{\beta}{b}, \dfrac{\gamma}{c}\right)$. It turns out that this is not always true.

EXAMPLE. $T\left(\dfrac{2}{5}, \dfrac{2}{5}, \dfrac{2}{5}\right)$ is not isomorphic to $T(5,5,5)$. Let $\overline{\Delta}$ be a spherical triangle, whose angles are all equal to $\dfrac{2\pi}{5}$, and let $R\left(\dfrac{2}{5}, \dfrac{2}{5}, \dfrac{2}{5}\right)$ be the group generated by the reflections in the sides of $\overline{\Delta}$. The reflections in the angle bisectors of $\overline{\Delta}$ are contained in $R\left(\dfrac{2}{5}, \dfrac{2}{5}, \dfrac{2}{5}\right)$. The three angle bisectors divide $\overline{\Delta}$ into six triangles with angles $\pi/2$, $\pi/3$, $\pi/5$, and the group $R(2,3,5)$, generated by the reflections in the sides of any of the smaller triangles, coincides with $R\left(\dfrac{2}{5}, \dfrac{2}{5}, \dfrac{2}{5}\right)$. Thus the subgroup $T\left(\dfrac{2}{5}, \dfrac{2}{5}, \dfrac{2}{5}\right) \subset R\left(\dfrac{2}{5}, \dfrac{2}{5}, \dfrac{2}{5}\right)$ coincides with the spherical triangle group $T(2,3,5) \subset R(2,3,5)$, which is a finite group. Since $T(5,5,5)$ is infinite, it cannot be isomorphic to $T\left(\frac{2}{5},\frac{2}{5},\frac{2}{5}\right)$.

Paradoxically, it turns out that $T\left(\frac{3}{5},\frac{3}{5},\frac{3}{5}\right)$ is isomorphic to $T(5,5,5)$. In fact, we will show that $T\left(\frac{\alpha}{a},\frac{\alpha}{b},\frac{\alpha}{c}\right) \approx T(a,b,c)$, if α is odd and relatively prime to a, b and c, while $T\left(\frac{\alpha}{a},\frac{\beta}{b},\frac{\gamma}{c}\right) \not\approx T(a,b,c)$, if α, β and γ are even.

We shall now describe a method of constructing injections $T(a,b,c) \hookrightarrow PSL(2,\mathbb{C})$. We shall use the projective model of the hyperbolic plane H. Let $E^{2,1}$ denote $\mathbb{R}^3$, equipped with the bilinear form $<x,y> = x_1 y_1 + x_2 y_2 - x_3 y_3$, and let $Q(x) = <x,x>$ denote the associated quadratic form. The points of H consist of the 1-dimensional subspaces $P \subset E^{2,1}$, such that the restriction $Q|P$ is negative definite. The lines in H are the 2-dimensional subspaces $L \subset E^{2,1}$, such that the restriction $Q|L$ is indefinite. These subspaces are of the form:

$$L = \{x \in E^{2,1} | <x,e> = 0\}, \quad \text{where} \quad <e,e> = 1.$$

A triangle consists of three 2-dimensional subspaces $L_i = \{x \in E^{2,1} | <x,e_i> = 0\}$ $(i=1,2,3)$, where $<e_i,e_i> = 1$, and e_1, e_2, e_3 are linearly independent. Such a triangle has a Gram matrix $M = (m_{ij})$, where $m_{ij} = <e_i,e_j>$ $(i,j=1,2,3)$. For a triangle with angles π/a, π/b, π/c, the Gram matrix is

$$M = \begin{pmatrix} 1 & -\cos\frac{\pi}{a} & -\cos\frac{\pi}{b} \\ -\cos\frac{\pi}{a} & 1 & -\cos\frac{\pi}{c} \\ -\cos\frac{\pi}{b} & -\cos\frac{\pi}{c} & 1 \end{pmatrix}.$$

In this model, the isometry group of H is the (projectivized) group G of 3×3 real matrices A, such that $A^T M A = M$. $PSL(2,\mathbb{C})$ is isomorphic to the complexification $G_0^{\mathbb{C}}$ of the identity component G_0 of G. The triangle group $T(a,b,c)$ is a group of 3×3 matrices, with coefficients in the field $K = \mathbb{Q}\left(\cos\frac{\pi}{a}, \cos\frac{\pi}{b}, \cos\frac{\pi}{c}\right)$.

Each field isomorphism $\sigma: K \to K^\sigma \subset \mathbf{C}$ induces an isomorphism of $\Gamma = T(a, b, c)$ to a group $\Gamma^\sigma \subset G_0^C$. Thus we obtain an injection of Γ into PSL$(2, \mathbf{C})$. Γ^σ is of hyperbolic or spherical type according as M^σ is indefinite or positive definite.

3. *Field isomorphisms*

Let $t = \ell.\text{c.m.} \ (2a, 2b, 2c)$. Then $K = Q\left(\cos \frac{\pi}{a}, \cos \frac{\pi}{b}, \cos \frac{\pi}{c}\right)$ is a subfield of the cyclotomic field $L = Q\left(e^{\frac{2\pi i}{t}}\right)$. Any isomorphism $\sigma: K \to K^\sigma \subset \mathbf{C}$ can be extended to an isomorphism $\sigma: L \to L$. It is known that the isomorphisms of L are induced by maps $e^{\frac{2\pi i}{t}} \mapsto e^{\frac{2\pi i x}{t}}$, where x is an integer relatively prime to t. Thus, the isomorphisms $\sigma: K \to K^\sigma \subset \mathbf{C}$ are induced by maps:

$$\begin{cases} \cos \dfrac{\pi}{a} \mapsto \cos \dfrac{\pi x}{a} \\[2mm] \cos \dfrac{\pi}{b} \mapsto \cos \dfrac{\pi x}{b} \\[2mm] \cos \dfrac{\pi}{c} \mapsto \cos \dfrac{\pi x}{c} \end{cases}$$

where $(x, 2a) = (x, 2b) = (x, 2c) = 1$.

THEOREM 3. *Let* F_g *denote the fundamental group of a surface of genus* g. *Then there exists an injection* $F_g \hookrightarrow SO(3)$.

Proof. It is well known that $F_2 \supset F_g$, and is easily seen that $F_2 \subset T(2, 8, 8)$. (Consider F_2 as a Fuchsian group, whose fundamental domain D is a regular, non-Euclidean octagon. The lines from the center to the vertices and the perpendicular bisectors of the sides divide D into 16 congruent triangles, with angles $\pi/2$, $\pi/8$, $\pi/8$. The group generated by reflections in the sides of any of these triangles is $R(2, 8, 8)$. From this we see that $F_2 \subset R(2, 8, 8)$, and so $F_2 \subset T(2, 8, 8)$.)

For $T(2, 8, 8)$, the corresponding field is $K = Q\left(\cos \frac{\pi}{8}\right)$, and $t = 16$. If we take $x = 3$, we obtain an isomorphism $\sigma: K \to Q\left(\cos \frac{3\pi}{8}\right)$, and

$$M^{\sigma} = \begin{pmatrix} 1 & -\cos\frac{3\pi}{2} & -\cos\frac{3\pi}{2} \\ -\cos\frac{3\pi}{2} & 1 & -\cos\frac{3\pi}{8} \\ -\cos\frac{3\pi}{8} & -\cos\frac{3\pi}{8} & 1 \end{pmatrix}.$$

Since $(\pi/2, 3\pi/8, 3\pi/8)$ are the angles of a spherical triangle, M^{σ} is positive definite. Therefore $T(2,8,8)$ can be injected into $SO(3)$, and so this is also true for $F_g \subset T(2,8,8)$. q.e.d.

THEOREM 4. *The homomorphism* $\theta_{a,\beta,\gamma} : T(a,b,c) \to T\left(\frac{a}{a}, \frac{\beta}{b}, \frac{\gamma}{c}\right)$ *is induced by a field isomorphism* $\sigma : K \to K^{\sigma}$ *(and consequently* $\theta_{a,\beta,\gamma}$ *is an isomorphism) if and only if the following two conditions are satisfied:*

(1) $(a, 2a) = (\beta, 2b) = (\gamma, 2c) = 1$,

(2) $\begin{cases} a \equiv \beta \mod 2(a,b) \\ \beta \equiv \gamma \mod 2(b,c) \\ \gamma \equiv a \mod 2(c,a), \end{cases}$

or if conditions analogous to (1) and (2) are satisfied when the triple (a, β, γ) *is replaced by* $(a, b-\beta, c-\gamma)$, $(a-a, \beta, c-\gamma)$ *or* $(a-a, b-\beta, \gamma)$.

Proof. Let $\overline{\Delta}$ be the circular triangle with angles $\frac{a\pi}{a}$, $\frac{\beta\pi}{b}$, $\frac{\gamma\pi}{c}$, which determines the homomorphism $\theta_{a,\beta,\gamma}$, and let $\overline{x}$, $\overline{y}$, $\overline{z}$ be the generators of $T\left(\frac{a}{a}, \frac{\beta}{b}, \frac{\gamma}{c}\right)$. If we extend the sides of $\overline{\Delta}$ to complete circles, we obtain a configuration containing several triangles, whose vertices are fixed points of $\overline{x}$, $\overline{y}$ and $\overline{z}$. (There are 8 triangles in the spherical and hyperbolic cases, and 4 triangles in the Euclidean case. See the figures in Remark 2, following Theorem 2.) All of these triangles determine the same homomorphism. ($\theta_{a,\beta,\gamma}$ is really determined by the reflections in the circles, rather than the triangles.) These triangles have angles $\left(\frac{a\pi}{a}, \frac{\beta\pi}{b}, \frac{\gamma\pi}{c}\right)$, $\left(\frac{a\pi}{a}, \frac{(b-\beta)\pi}{b}, \frac{(c-\gamma)\pi}{c}\right)$, $\left(\frac{(a-a)\pi}{a}, \frac{\beta\pi}{b}, \frac{(c-\gamma)\pi}{c}\right)$ and $\left(\frac{(a-a)\pi}{a}, \frac{(b-\beta)\pi}{b}, \frac{\gamma\pi}{c}\right)$. Thus $\theta_{a,\beta,\gamma}$ is induced by a field isomorphism, if and only if one of the following maps σ extends to a field isomorphism of K:

$$\sigma_1\left(\cos\tfrac{\pi}{a},\ \cos\tfrac{\pi}{b},\ \cos\tfrac{\pi}{c}\right) = \left(\cos\tfrac{a\pi}{a},\ \cos\tfrac{\beta\pi}{b},\ \cos\tfrac{\gamma\pi}{c}\right),$$

$$\sigma_2\left(\cos\tfrac{\pi}{a},\ \cos\tfrac{\pi}{b},\ \cos\tfrac{\pi}{c}\right) = \left(\cos\tfrac{a\pi}{a},\ \cos\tfrac{(b-\beta)\pi}{b},\ \cos\tfrac{(c-\gamma)\pi}{c}\right),$$

$$\sigma_3\left(\cos\tfrac{\pi}{a},\ \cos\tfrac{\pi}{b},\ \cos\tfrac{\pi}{c}\right) = \left(\cos\tfrac{(a-a)\pi}{a},\ \cos\tfrac{\beta\pi}{b},\ \cos\tfrac{(c-\gamma)\pi}{c}\right),$$

$$\sigma_4\left(\cos\tfrac{\pi}{a},\ \cos\tfrac{\pi}{b},\ \cos\tfrac{\pi}{c}\right) = \left(\cos\tfrac{(a-a)\pi}{a},\ \cos\tfrac{(b-\beta)\pi}{b},\ \cos\tfrac{\gamma\pi}{c}\right).$$

We consider the case $\sigma = \sigma_1$. From the previous discussion on field isomorphisms, we see that the map $\sigma\left(\cos\tfrac{\pi}{a},\ \cos\tfrac{\pi}{b},\ \cos\tfrac{\pi}{c}\right) = \left(\cos\tfrac{a\pi}{a},\ \cos\tfrac{\beta\pi}{b},\ \cos\tfrac{\gamma\pi}{c}\right)$ extends to a field isomorphism of K, if and only if there is an integer x, such that $(x, 2a) = (x, 2b) = (x, 2c) = 1$, and the congruences

$$\begin{cases} x \equiv a \mod 2a \\ x \equiv \beta \mod 2b \\ x \equiv \gamma \mod 2c \end{cases}$$

admit a solution. By the Chinese remainder theorem, these congruences have a solution if and only if

$$\begin{cases} a \equiv \beta \mod 2(a, b) \\ \beta \equiv \gamma \mod 2(b, c) \\ \gamma \equiv a \mod 2(c, a). \end{cases} \qquad\qquad \text{q.e.d.}$$

THEOREM 5. *If a, b, c are pairwise relatively prime, then the homomorphism $\theta_{a,\beta,\gamma}: T(a, b, c) \to T\left(\tfrac{a}{a}, \tfrac{\beta}{b}, \tfrac{\gamma}{c}\right)$ is an isomorphism only if it is induced by a field isomorphism $\sigma: K \to K^\sigma$.*

Proof. We first note that if $\theta_{a,\beta,\gamma}$ is an isomorphism, then a, β, γ cannot all be even. For consider the spherical, Euclidean or hyperbolic triangle $\bar{\Delta}$ with angles $\tfrac{a\pi}{a}$, $\tfrac{\beta\pi}{b}$, $\tfrac{\gamma\pi}{c}$. Let $R\left(\tfrac{a}{a}, \tfrac{\beta}{b}, \tfrac{\gamma}{c}\right)$ be the group generated by the reflections in the sides of $\bar{\Delta}$. $T\left(\tfrac{a}{a}, \tfrac{\beta}{b}, \tfrac{\gamma}{c}\right)$ is the subgroup of index 2 in $R\left(\tfrac{a}{a}, \tfrac{\beta}{b}, \tfrac{\gamma}{c}\right)$, consisting of the orientation-preserving transformations. Let λ, μ, ν be the angle bisectors in $\bar{\Delta}$, and ℓ, m, n the reflections in λ, μ, ν. If α, β, γ are even, then $\ell, m, n \in R\left(\frac{\alpha}{a}, \frac{\beta}{b}, \frac{\gamma}{c}\right)$, and the products $\ell m, mn, n\ell \in T\left(\frac{\alpha}{a}, \frac{\beta}{b}, \frac{\gamma}{c}\right)$. Since the angle bisectors λ, μ, ν meet at a point, the products ℓm, mn, $n\ell$ have the same fixed points, so they commute. But the corresponding products in $T(a, b, c)$ do not commute, because in this case, the corresponding reflection lines lie outside the triangle, and do not meet at a common point. Thus if α, β, γ are even, then $T\left(\frac{\alpha}{a}, \frac{\beta}{b}, \frac{\gamma}{c}\right)$ has a relation which is not satisfied in $T(a, b, c)$, and $\theta_{\alpha, \beta, \gamma}$ is not an isomorphism.

Since a, b, c are pairwise relatively·prime, at most one of these integers is even. The argument now proceeds by considering various cases where a, b, c, α, β, γ are even or odd. In each case, a suitable replacement of (α, β, γ) by $(\alpha, b-\beta, c-\gamma)$, $(a-\alpha, \beta, c-\gamma)$ or $(a-\alpha, b-\beta, \gamma)$ transforms the situation to one of the cases: (i) α, β, γ are all odd, or (ii) α, β, γ are all even. In case (i), Theorem 4 shows that $\theta_{\alpha, \beta, \gamma}$ is induced by a field isomorphism. In case (ii), $\theta_{\alpha, \beta, \gamma}$ is not an isomorphism. We illustrate the argument with a typical case: Suppose that a, α, β are even, and b, c, γ odd. Replacing (α, β, γ) by $(a-\alpha, \beta, c-\gamma)$, we obtain 3 even integers, so $\theta_{\alpha, \beta, \gamma}$ is not an isomorphism.   q.e.d.

It may be true that Theorem 4 is valid more generally, and that all isomorphisms $\theta_{\alpha, \beta, \gamma}$ are induced by field isomorphisms. There is a good deal of evidence for this. An interesting case study is the group $T(6, 9, 18)$, which has three field-isomorphism classes of non-Euclidean homomorphic images:

$$T_1 \approx T(6, 9, 18) \approx T\left(\frac{1}{6}, \frac{4}{9}, \frac{5}{18}\right) \approx T\left(\frac{1}{6}, \frac{2}{9}, \frac{7}{18}\right),$$

$$T_2 \approx T\left(\frac{1}{6}, \frac{1}{9}, \frac{7}{18}\right) \approx T\left(\frac{1}{6}, \frac{4}{9}, \frac{1}{18}\right) \approx T\left(\frac{1}{6}, \frac{7}{9}, \frac{5}{18}\right),$$

$$T_3 \approx T\left(\frac{1}{6}, \frac{1}{9}, \frac{5}{18}\right) \approx T\left(\frac{1}{6}, \frac{2}{9}, \frac{1}{18}\right) \approx T\left(\frac{1}{6}, \frac{5}{9}, \frac{7}{18}\right).$$

$T\left(\frac{1}{6},\frac{7}{9},\frac{5}{18}\right)$ and $T\left(\frac{1}{6},\frac{5}{9},\frac{7}{18}\right)$ are spherical groups, and the others are hyperbolic. T_1 is not isomorphic to T_2 , because there are 3 reflection lines through the vertices of the $\left(\frac{\pi}{6},\frac{7\pi}{9},\frac{5\pi}{18}\right)$ triangle, which are concurrent. I don't know if T_1 is isomorphic to T_3 .

4. Quadratic differentials

We shall now calculate some quadratic differentials $F(z)dz^2$, which are invariant under $T(a,b,c)$ and induce the homomorphisms $\theta_{\alpha,\beta,\gamma}$.

Let Δ be a hyperbolic triangle (in the unit disc D) with angles $\frac{\pi}{a}, \frac{\pi}{b}, \frac{\pi}{c}$ at its vertices A , B , C . Let $\overline{\Delta}$ be a circular triangle with angles $\frac{\alpha\pi}{a}, \frac{\beta\pi}{b}, \frac{\gamma\pi}{c}$ at its vertices $\overline{A}$, $\overline{B}$, $\overline{C}$. (We assume the triangles have the same orientation.) There exists a conformal map $f : \Delta \to \overline{\Delta}$ such that $f(A) = \overline{A}$, $f(B) = \overline{B}$ and $f(C) = \overline{C}$. The function $f(z) = f_{\alpha,\beta,\gamma}(z)$ can be extended by reflection to a meromorphic function in D , such that $f(\eta z) = \theta_{\alpha,\beta,\gamma}(\eta)f(z)$, for $\eta \in T(a,b,c)$. We shall compute $F(z) = F_{\alpha,\beta,\gamma}(z) = \{f,z\}$.

Let H denote the upper half-plane. We consider conformal maps $w : \Delta \to H$, $v : H \to \Delta$ and $u : H \to \overline{\Delta}$, such that $w(A) = 0$, $w(B) = 1$, $w(C) = \infty$, $u(0) = \overline{A}$, $u(1) = \overline{B}$, $u(\infty) = \overline{C}$, and $v = w^{-1}$. The map $w(z)$ can be extended by reflection to a meromorphic function, defined in D . It then becomes an automorphic function, which generates the field of automorphic functions of $T(a,b,c)$. Geometrically, $w(z)$ is a ramified covering of D over the extended plane $\hat{C}$. It is the projection map from D to the quotient $D/T(a,b,c)$. Any quadratic differential in D , invariant under $T(a,b,c)$, is the w-lift of a quadratic differential on $\hat{C}$, and we shall describe them in this way.

The conformal map $f : \Delta \to \overline{\Delta}$ can be expressed: $f = u \circ w = u \circ v^{-1}$. The Schwarzian $\{f,z\} = \{u,w\}\left(\frac{dw}{dz}\right)^2 + \{w,z\}$. Since $z = v(w(z))$, $0 = \{v,w\}\left(\frac{dw}{dz}\right)^2 + \{w,z\}$, or $\{w,z\} = -\{v,w\}\left(\frac{dw}{dz}\right)^2$. Using the variable $w = w(z)$ in H , so that $\frac{dw}{dz}\cdot\frac{dv}{dw} = 1$, we now have Cayley's identity:

180   LEON GREENBERG

$$\{f, z\} = \frac{\{u, w\} - \{v, w\}}{\left(\dfrac{dv}{dw}\right)^2} \, . \tag{1}$$

Formulas for the Schwarz triangle functions $u(w)$, $v(w)$ are known, and can be found, for example, in Schwarz [2] or Caratheodory [1]. These formulas are the following

$$\{v, w\} = \frac{1 - \dfrac{1}{a^2}}{2w^2} + \frac{1 - \dfrac{1}{b^2}}{2(1-w)^2} + \frac{1 - \dfrac{1}{a^2} - \dfrac{1}{b^2} + \dfrac{1}{c^2}}{2w(1-w)} \, . \tag{2}$$

$$\{u, w\} = \frac{1 - \dfrac{a^2}{a^2}}{2w^2} + \frac{1 - \dfrac{\beta^2}{b^2}}{2(1-w)^2} + \frac{1 - \dfrac{a^2}{a^2} - \dfrac{\beta^2}{b^2} + \dfrac{\gamma^2}{c^2}}{2w(1-w)} \, . \tag{3}$$

When Δ is suitably normalized,

$$\frac{dv}{dw} = \frac{K}{aw^{1-\frac{1}{a}} (1-w)^{1-\frac{1}{b}} F(\ell, m, n; w)^2} \, , \tag{4}$$

where $F(\ell, m, n; w)$ is the hypergeometric function,

$$\ell = \tfrac{1}{2}\left(1 - \tfrac{1}{a} - \tfrac{1}{b} + \tfrac{1}{c}\right) \, , \quad m = \tfrac{1}{2}\left(1 - \tfrac{1}{a} - \tfrac{1}{b} - \tfrac{1}{c}\right) \, , \quad n = 1 - \tfrac{1}{a} \, , \tag{5}$$

$$K = \frac{\Gamma(n)\,\Gamma(1-\ell)\,\Gamma(1-m)}{\Gamma(2-n)\,\Gamma(n-\ell)\,\Gamma(n-m)} \, \sigma \, , \tag{6}$$

$$\sigma = \frac{\sin \frac{\pi}{2}\left(1 - \frac{1}{a} - \frac{1}{b} - \frac{1}{c}\right) \sin \frac{\pi}{2}\left(1 - \frac{1}{a} - \frac{1}{b} + \frac{1}{c}\right)}{\sin \frac{\pi}{2}\left(1 - \frac{1}{a} - \frac{1}{c} + \frac{1}{b}\right) \sin \frac{\pi}{2}\left(1 - \frac{1}{a} + \frac{1}{b} + \frac{1}{c}\right)} \, . \tag{7}$$

Summing up these equations, $F_{a,\beta,\gamma}(z) = \{f, z\}$ satisfies:

$$(8) \qquad F_{\alpha,\beta,\gamma}(z) = \left[\frac{(1-\alpha^2)}{2a^2w^2} + \frac{(1-\beta^2)}{2b^2(1-w)^2} + \frac{\dfrac{1-\alpha^2}{a^2} + \dfrac{1-\beta^2}{b^2} + \dfrac{\gamma^2-1}{c^2}}{2w(1-w)} \right]$$

$$\times \frac{a^2}{k^2} w^{2-\frac{2}{a}}(1-w)^{2-\frac{2}{b}} F(\ell, m, n; w)^4 .$$

The quadratic differential $F_{\alpha,\beta,\gamma}(z)\,dz^2$ is the w-lift of $G_{\alpha,\beta,\gamma}(w)\,dw^2$, where

$$(9) \qquad G_{\alpha,\beta,\gamma}(w) = \frac{(1-\alpha^2)}{2a^2w^2} + \frac{(1-\beta^2)}{2b^2(1-w)^2} + \frac{\left(\dfrac{1-\alpha^2}{a^2} + \dfrac{1-\beta^2}{b^2} + \dfrac{\gamma^2-1}{c^2}\right)}{2w(1-w)} .$$

The homomorphism $\theta_{\alpha,\beta,\gamma} : T(a,b,c) \to T\left(\dfrac{\alpha}{a}, \dfrac{\beta}{b}, \dfrac{\gamma}{c}\right)$ is induced by the quadratic differential $\omega_{\alpha,\beta,\gamma} = F_{\alpha,\beta,\gamma}(z)\,dz^2$.

Summing up our results, we have the following.

THEOREM 6. *The homomorphism* $\theta_{\alpha,\beta,\gamma}$ *is induced by the quadratic differentials* $\omega_{\alpha,\beta,\gamma}$, $\omega_{a,b-\beta,c-\gamma}$, $\omega_{a-\alpha,\beta,c-\gamma}$ *and* $\omega_{a-\alpha,b-\beta,\gamma}$. *These are all obtained from the Schwarzians of conformal maps.*

DEPARTMENT OF MATHEMATICS
UNIVERSITY OF MARYLAND
COLLEGE PARK, MARYLAND 20742

REFERENCES

[1] C. Caratheodory, Theory of Functions of a Complex Variable, vol. 2, Chelsea Publishing Company (1960).

[2] H. A. Schwarz, Ueber diejenigem Falle in welchen die Gaussische hypergeometrische Reihe eine algebraische Function ihres vierten Elementes darstellt, Jour. Reine und Angew. Math. 75, 292-335.

[3] E. G. Van Vleck, On the Combination of Non-loxodromic Substitutions, Trans. A.M.A. 20 (1919), 299-312.

HYPERBOLIC MANIFOLDS, GROUPS AND ACTIONS

M. Gromov

0. Introduction

This lecture gives an outline of basic geometric notions and ideas
associated with the conception of hyperbolicity. Very little is said here
about the hyperbolic space itself (the main source of knowledge is
Thurston's lectures [21]), but it is shown how the phenomenon of hyperbolicity appears in Riemannian geometry, topological dynamics, combinatorial group theory and geometrical theory of mappings. Our presentation
is expository, proofs are only sketched, but constructions and definitions
are illustrated by examples. The comprehensive theory of hyperbolicity
is yet to be built and my purpose here is to provide motivation for future
research. (Questions and conjectures are italicized.)

1. Length Spaces

1.1. *Riemannian spaces with singularities*

Our main example is the following. Take a piecewise smooth polyhedron $X \subset R^N$. The space X carries the metric d induced from R^N.
We get another metric if we use *the length function* on curves in X and
define $R(x,y)$, $x,y \epsilon X$, as the length of the shortest curve in X joining x and y. Such an R is called the *induced length metric*.

We obviously have $R \geq d$ and the equality holds iff X is a convex
set.

© 1980 Princeton University Press
Riemann Surfaces and Related Topics
Proceedings of the 1978 Stony Brook Conference
0-691-08264-2/80/000183-31 $01.55/1 (cloth)
0-691-08267-7/80/000183-31 $01.55/1 (paperback)
For copying information, see copyright page

DIGRESSION. The quantity $\sup(R(x,y) \cdot d^{-1}(x,y))$ measures the distortion of X. One can easily show that the distortion of any topological circle is at least $\pi/2$ and consequently, Distortion$(X) < \pi/2$ implies that X is simply connected. *Probably this inequality implies that X is contractible*, but I can show it only under the stronger assumption Distortion$(X) < \pi/2\sqrt{2}$.* A different approach to the distortion can be found in [10].

DEFINITION. A length structure in a space X is given by a metric and a length function on curves, such that the distance between any two points is equal to the length of the shortest curve (supposed to exist) joining these points. We shall also assume all metric spaces to be complete unless stated otherwise.

FURTHER EXAMPLES. Let V be a Riemannian manifold with boundary. Its length structure is more complicated than that of a manifold due to the fact that the shortest curves between interior points can touch the boundary.

Take a Riemannian manifold and divide it by a group of isometries. When the action is not free then the natural length structure again has new features.

DEFINITIONS. A map from one length space (space with length structure) to another is called *isometric* if it preserves the distance function. Such maps are always embeddings. A map preserving the lengths of the curves is called *path-isometric*. Isometric maps from $[a, b]$, $\mathbf{R}_+$, $\mathbf{R}$ into X are called *straight segment, ray* and *straight line* correspondingly. A locally isometric map $\mathbf{R} \to X$ is called a *geodesic*. There is a natural $\mathbf{R}$-action in the set of all geodesics called the *geodesic flow*.

1.2. *Horofunctions and the ideal boundary*

Let X be a complete metric space. The distance function determines an isometrical embedding $x \to \text{dist}(x,y)$ from X into the space $C(X)$ of

---

* Unpublished.

the continuous functions on X. Consider the factor space $C' = C(X)/(\text{constant functions})$ with the topology of uniform convergence on bounded sets in X. The space X is now embedded into C' and we can define the closure $C\ell(X)$ and the boundary $\partial X = C\ell(X) \setminus X$. The space X is assumed further to be countably compact. In this case, $C\ell(X)$ and $\partial(X)$ are compact.

A function $h \in C(X)$ that projects into a point $b \in \partial X \subset C'$ is called a *horofunction* centered at b. The sets $h^{-1}(-\infty, c) \subset X$ are called (open) *horoballs* centered at c; the levels $h^{-1}(c) \subset X$ are called *horospheres* and the sets $h^{-1}(c, \infty)$ are called *horospaces*.

The *limit set* ∂X_0 (or the set of the limit points) of a closed $X_0 \subset X$ is defined as the intersection of its closure in $C\ell(X_0)$ with ∂X.

When X is a length space, every ray has exactly one limit point. (We identify rays and straight lines with their images.) The corresponding horofunction is called the *ray function* or *the Busemann function*.

1.3. *Isometries*

With an isometry $\gamma: X \to X$, one associates the *displacement function* $\delta_\gamma: X \to R_+$, $\delta_\gamma(x) = \text{dist}(x, \gamma(x))$. An isometry is called *elliptic* if its displacement $X \to R_+$ is proper (i.e., the sets $\delta_\gamma^{-1}[0, a] \subset X$ are compact). The isometry γ is called *hyperbolic* if it generates a free cyclic group Γ, the action of Γ in X is discrete and δ_γ is proper on X/Γ (δ_γ is Γ-invariant function). When γ is elliptic or hyperbolic, δ_γ assumes its minimum.

In a length space, every hyperbolic or a fixed point free elliptic isometry has an invariant geodesic, (an easy argument).

An isometry γ without fixed points is called *parabolic* if each set $\delta_\gamma^{-1}(0, \varepsilon) \subset X$, $\varepsilon > 0$, is unbounded. Every such isometry has, obviously, a fixed point at ∂X. (All isometries can be continuously extended to $C\ell X$.)

1.4. *The word metric*

Consider a group Γ generated by $\beta_1, \cdots, \beta_k$ and denote by $\|\gamma\|$ the length of the minimal word (in β_i and β_i^{-1}) representing γ. Obviously $\|\gamma\| = \|\gamma^{-1}\|$ and $\|\gamma_1\gamma_2\| \leq \|\gamma_1\| + \|\gamma_2\|$. The length function gives rise to the left invariant metric $\|\gamma_1^{-1}\gamma_2\|$ on Γ.

Quasi-isometry. A map f from a metric space X to Y is called quasi-isometric if the ratio $\mathrm{dist}(x_1, x_2)/\mathrm{dist}(f(x_1), f(x_2))$ $x_1, x_2 \,\epsilon\, X$, is pinched between C and C^{-1}. Sometimes one specifies C and calls such an f a C-quasi-isometric map.

Obviously, two different word metrics, corresponding to different choices of generators, are quasi-isometric.

Coarse quasi-isometries. A coarse quasi-isometric map $X \to Y$ is a map f_0 from a subset $X_0 \subset X$ to Y such that X_0 is ϵ-dense[*] in X (i.e., its ϵ-neighborhood coincides with X) and f_0 is C-quasi-isometric. We call f_0 a coarse equivalence if its image is ϵ-dense[*] in Y. This is an equivalence relation if the constants C and ϵ are allowed to vary.

The following obvious fact plays an important role in connecting geometry of a space with its fundamental group.

Let X be a compact length space and $\tilde{X}$ be its universal covering. Then $\tilde{X}$ (with the induced length structure) is coarse equivalent to the fundamental group $\pi_1(X)$. (We suppose here X to be topologically a polyhedron or a more general space admitting covering space theory.)

1.5. *Convex sets*

A set A in a length space is called *convex* if its intersection with every geodesic segment is connected. A set is called *locally* convex if for any $a \,\epsilon\, A$ there is an ϵ such that the intersection of A with any geodesic segment of the length ϵ passing through a is connected.

2. Convex Spaces

2.1. *Definition*

A length space X is called *convex* if the distance function is convex; namely, for any two geodesic segments $x : [a, b] \to X$, $y : [c, d] \to X$ $\mathrm{dist}(x(t_1), y(t_2))$ is convex on $[a, b] \times [c, d]$.

---

[*]In this paper ϵ is positive but not always small.

X is called *locally convex* if it can be covered by open U_i such that dist(x, y) is convex when x, y belong to U_i .

EXAMPLES. A Riemannian manifold is locally convex iff $K \leq 0$ (i.e., sectional curvature ≤ 0). It is convex iff $K \leq 0$ and $\pi_1 = 0$. In the context of length spaces one has by adapting the classical arguments,

THE CARTAN HADAMARD THEOREM. *A simply connected locally convex space is convex. Every convex space is straight (any two points can be joined by the unique geodesic segment) and hence contractible.*

QUESTION. *Are there convex length spaces which are topological manifolds different from* $\mathbf{R}^n$ *?*

2.2. *Manifolds with boundary*

Let V be a Riemannian manifold with smooth boundary B . The boundary is called convex when the second quadratic form is non-negative; it is called concave when this form is non-positive. The boundary is called k-*convex* if the second quadratic form looks infinitesimally as

$$\sum_{1}^{n-1} a_i x_i^2 , \quad n-1 = \dim B ,$$

where among a_i there are at least k non-negative numbers. When k = n–2 , we call B *next-to-convex*. (k = n–1 corresponds to convexity.)

REMARK. *Riemannian manifolds with* $K \leq 0$ *and with next-to-convex boundary* B *have locally convex length structure.* For example, surfaces with boundary are locally convex when $k \leq 0$.

More generally, take a 2-plane τ tangent to B and denote by K'_τ the sectional curvature of the induced metric in B . The following property is necessary and sufficient for local convexity of V (viewed as a length space):

$K \leq 0$ and for any τ where the second quadratic form is negative K'_τ is non-positive. The proofs of all these points are straightforward.

2.3. *Digression*: *k-convex hypersurfaces in* $\mathbf{R}^n$

Let $V \subset \mathbf{R}^n$ be a compact domain with smooth boundary B.

The classical fact stating convexity of a locally convex connected set can be formulated in the following fancy fashion:

B is $(n{-}1)$-convex iff for any straight line $\ell < \mathbf{R}^n$ the homomorphism $H_0(\ell \cap V) \to H_0(V)$ is injective.

By using rudimentary Morse theory one generalizes this theorem:

B is k-convex iff for any $(n{-}k)$-dimensional plane $P \subset \mathbf{R}^n$ the homomorphism $H_{n-k-1}(P \cap V) \to H_{n-k-1}(V)$ is injective. In this case V has the homotopy type of an $(n{-}k)$-dimensional polyhedron.

The most regular behavior is shown by V's with the next-to-convex boundary: Let V have unit volume. Denote by A the $(n{-}1)$-dimensional volume of its boundary and by b_1 the first Betti number of V. *Then* V *contains a ball of radius* $\varepsilon > F(A, b_1, n)$, where for F one can take $\exp(-\exp(A+b_1+n))$.

Sketch of the proof. We actually find a cube C of the size ε contained in V: moving an ε-cube by parallel translation, we first inscribe its 1-skeleton $C^{(1)}$ into V such that the homomorphism $H_1(C^{(1)}) \to H_1(V)$ is trivial. Because V is next-to-convex, the inclusion $C^{(1)} \subset V$ implies $C \subset V$.

2.4. *The double construction*

Take a space X and a closed set $A \subset X$. Take two copies of X and glue them at A.

When X is a Riemannian manifold with $K \leq 0$ and A is a closed locally convex set, then the resulting space Y (with the natural length structure) is locally convex.

The proof is simple.

EXAMPLE. Suppose that the interior $\operatorname{Int} A$ is dense in A. The geodesic flow in $X \setminus \operatorname{Int} A$ with reflection at the boundary is called *Sinai's billiard in X with obstacle* A. Each of the two natural embeddings $X \setminus \operatorname{Int} A \to Y$

sends the orbits of Sinai's billiard into geodesics in Y (this is easy to show). So we have:

The geodesic flow in Y contains the Sinai's billiard as an invariant set.

2.5. *Branched covers*

Take a Riemannian manifold V with $K \leq 0$ and a codimension 2 submanifold $V_0 \subset V$. When V_0 is locally convex (i.e., totally geodesic) any branched covering of V with the branching locus V_0 is a convex length space. (Easy to show.)

Branched torus. Let V be a flat n-dimensional torus and V_0 is a union of flat codimension 2 subtori with transversal intersections. The notion of branched covering still makes sense here, and when all intersections are not only transversal but orthogonal, the branched cover $\tilde{V}$ is a convex length space.

When dim V is odd, a generic linear map $V \to S^1$ gives rise to a fibration $\tilde{V} \to S^1$ where the fiber F is again a branched torus but the subtori forming its branch-locus have non-normal (though transversal) intersections (this was pointed out to me by Thurston).

When dim $V = 3$, for an appropriate $\tilde{V}$ one can change its metric to a Riemannian metric of constant negative curvature (Jorgensen example, see [21], [23]). In this case, F is a surface of genus ≥ 2.

When dim $V \geq 5$, the fiber F does not carry a metric of negative curvature. It follows from the following unpublished theorem of Thurston: *Let F be a compact Riemannian manifold with $K < 0$ and dim $F \geq 3$. Let $a : \pi_1(F) \hookleftarrow$ be an automorphism. Then the orbits of a acting on the conjugacy classes of $\pi_1(F)$ are finite.*

2.6. *Parabolic isometries*

Consider an isometry γ of a length space X with displacement function $\delta_\gamma = \text{dist}(x, \gamma(x))$. Denote by δ_γ^∞ the limit

$$\lim_{n \to \infty} \frac{1}{n} \delta_{\gamma n}(x), \quad x \in X .$$

The limit obviously exists and does not depend on x because

$$\delta_{\gamma^{m+n}}(x) \le \delta_{\gamma^m}(x) + \delta_{\gamma^n}(x) .$$

Let X *be convex and let* γ *have no fixed points. Then*

$$\delta_{\gamma}^{\infty} = \inf_{x \epsilon X} \delta_{\gamma}(x) .$$

Proof. The inequality $\delta_{\gamma}^{\infty} \le \inf_{x \epsilon X} \delta_{\gamma}(x)$ is trivial and it holds for all length spaces. When X is convex, we obviously have:

$$\inf_{x \epsilon X} \delta_{\gamma^2}(x) \ge 2 \inf_{x \epsilon X} \delta_{\gamma}(x) ;$$

this implies the theorem.

COROLLARY. *If* γ *is fixed point free and* $\delta_{\gamma}^{\infty} = 0$, *then* γ *is parabolic.*

2.7. *Straight subgroups and asymptotic torsion*

Let Γ be any group and $\Gamma^0 \subset \Gamma$ be a finitely generated subgroup. We call Γ^0 *straight* in Γ if for any finitely generated $\Gamma' \subset \Gamma$ containing Γ^0, the inclusion $\Gamma^0 \hookrightarrow \Gamma'$ is a quasi-isometric imbedding with respect to the word metrics in Γ^0 and Γ'.

We call a $\gamma \epsilon \Gamma$ an asymptotic torsion element if for some finitely generated Γ' containing γ, we have $\lim_{n \to \infty} \frac{1}{n} \|\gamma^n\| = 0$. This happens exactly when the group generated by γ is finite, or when it is infinite cyclic but not straight.

An example. Let Γ be a nil-potent group without torsion. Then Γ has asymptotic torsion unless it is Abelian.

Combining this with the theorem from the previous section, we conclude.

THEOREM. *Let* Γ *be a group of isometries of a convex space. If* Γ *has no elliptic or parabolic elements, then every nil-potent subgroup in* Γ *is Abelian.*

An analogous argument shows:

If there is an $\varepsilon > 0$ such that $\delta_\gamma(x) \geq \varepsilon$. $x \in X$, $\gamma \in \Gamma^*$ $(= \Gamma \backslash \text{identity})$ then any Abelian subgroup in Γ is straight. Observe also that an Abelian group with $\delta_\gamma(x) \geq \varepsilon$ must be finitely generated and torsion free.

As a corollary, we get a generalization of the Gromoll-Wolf-Lawson-Yau theorem (see [5]):

Let Γ be a soluble group of isometries of a convex space X. If $\delta_\gamma(x) \geq \varepsilon > 0$, $\gamma \in \Gamma^*$, $x \in X$, then Γ contains an Abelian subgroup of finite index.

Proof. If every Abelian subgroup in a soluble group Γ is finitely generated and straight, then Γ obviously contains the required Abelian subgroup.

Observe that for the branched torus the subgroup $\pi_1(F) \subset \pi_1(\tilde{V})$ is usually not straight.

2.8. *The uniformization problems*

Let X be a finite dimensional polyhedron of $K(\pi; 1)$ type. *Does there exist a locally convex space (non-compact but preferably finite dimensional) homotopy equivalent to* X? The Thurston theory [21] says "yes" for a vast class of 3-manifold containing all the known examples.

Let X be any finite dimensional polyhedron. *Does there exist a convex space $\tilde{X}$ and a discrete group (with fixed points) Γ of isometries of $\tilde{X}$ such that $\tilde{X}/\Gamma$ is homeomorphic to* X? When $\dim X = 2$, we have an easy "yes".

3. Strict Convexity

3.1. *Definition*

We shall assume below that our length space X satisfies the following property: for any $x \in X$ there is an ε such that every sphere centered at x of radius $\leq \varepsilon$ contains no geodesic segments. (This is always so for Riemannian manifolds.) When X is convex, the above property implies that large spheres contain no geodesic segments as well.

δ-convexity. We call a convex X *strictly δ-convex* if there is a
positive ν such that for any convex set $A \subset X$ the normal projection
$P : X \to A$ (P sends x to the nearest point from A ; this is a well-
defined map due to the absence of geodesic segments in spheres) satisfies:
for any curve $C \subset X$ with $\mathrm{dist}(C, A) \geq \delta$, we have length $(P(C)) \leq$
$(1-\nu)$ length (C) .

Observe that $(\delta+\delta_1)$-convexity follows from δ-convexity.

X is called *strictly convex* if it is δ-convex for a positive δ .

EXAMPLE. A simply-connected Riemannian manifold (possibly with next-
to-convex boundary) with $K \leq -\kappa$, $\kappa > 0$, is strictly convex (it is
actually δ-convex for all $\delta > 0$).

We call X *locally δ-convex* (strictly convex) if its universal covering
is δ-convex (strictly convex). This is really a "local" notion, because
δ-convexity follows from δ-convexity of the balls of radius δ .

The double construction from section 2.4 gives a strictly convex
space when the underlying manifold has $K \leq -\kappa$, $\kappa > 0$. As a pleasant
example, take a manifold X with $-\kappa_1 \leq K \leq -\kappa_2 < 0$ and with finite total
volume. In this case, one can find a locally convex $A \subset X$ such that the
complement $X \backslash A$ is bounded and homeomorphic to X itself. The double
space Y in this case has the same homotopy type as the usual double of
$X' = X \backslash \mathrm{Int}\, A$. When X has constant curvature, the double of X' can be
itself equipped with a metric of non-positive curvature. This metric can
be chosen real analytic and with negative Ricci curvature (The last possibility was pointed out to me by Ernst Heintze.) Observe that there are
compact manifolds that have C^∞-metrics with $K \leq 0$ but have no real
analytic metrics with $K \leq 0$. The proofs of these facts are easy (see also
[11]).

3.2. Closure

Let X be strictly convex. In this case every horofunction is a ray
function. Moreover, there is a natural one-to-one correspondence between

the points from ∂X and the rays starting from a fixed point $x_0 \epsilon X$.
Intersection between any two horoballs with different centers is compact,
and thus we have the V-*property*: any two distinct points from ∂X can
be joined by a straight line.

For any straight line $\ell \subset X$, the normal projection $X \to \ell$ can be
continuously extended to $C\ell(X)$ and the only fixed points of the extension are the "ends" ℓ^+, $\ell^- \subset \partial X$ of this line.

All these facts are classical for Riemannian manifolds and the classical proofs work in our case with no problems.

When X is the hyperbolic space $(K=-1)$, there is another important
property:

Let $A \subset \partial X$ be a closed set and C be its convex hull. Then the
limit set of C is equal to A.

This feature is probably shared by all strictly convex spaces, but I
could prove it only when X is a Riemannian manifold with $K \leq -1$ and
$A \subset \partial X$ is a finite set. (Recall that the convex hull of A is defined as
the minimal set $C \subset C\ell X$ containing A such that $A \cap X \subset X$ is a
convex set.)

3.3. *Isometries*

Let X be strictly convex and $\gamma: X \to X$ be an isometry. There are
only three possibilities:

a) γ is elliptic. In this case it has a fixed point and the topological
   group generated by γ (in the group $Is(X)$ of all isometries) is
   compact.

b) γ is parabolic. Then γ has no fixed points in X but has a
   unique fixed point x in ∂X. This point is called the center of γ.
   All horospheres centered at x are invariant under γ. The group
   generated by γ is isomorphic to Z and its action in $C\ell(X)\backslash\{x\}$
   is discrete.

c) γ is hyperbolic. In this case γ has two fixed points γ^+ and γ^-
   in ∂X, it generates Z that acts freely and discretely in

$C\ell(X)\backslash\{\gamma^+, \gamma^-\}$. The point γ^+ is an attractor for γ; moreover, for every compact set A, $A \subset C\ell(X)\backslash\{\gamma^-\}$ and any neighborhood $U \subset C\ell(X)$ of γ^+ there is an n such that $\gamma^n(A) \subset U$. The point γ^- is an attractor for γ^{-1}.

This classification is well known for Riemannian manifolds, and the proof depends only on the properties stated in the previous section.

3.4. *Special groups of isometries*

X is again supposed to be strictly convex. Consider a group Γ acting by isometries on X. Γ is called special when one of the following three cases occur:

1) Γ keeps fixed a point $x \in X$. In this case the closure of Γ in the group $\mathrm{Is}(X)$ of all isometries is compact. This is a Lie group when X is a Riemannian manifold (possibly with boundary). When X is a polyhedron, this closure is usually a profinite group; this is always so when X is one-dimensional.

2) There is a straight line $\ell \subset X$ invariant under X. In this case Γ factors as follows: $0 \to \Gamma' \to \Gamma \to \Gamma_\ell \to 0$, where $\Gamma' \subset \Gamma$ is the subgroup keeping ℓ fixed and Γ_ℓ is a subgroup in $\mathrm{Is}(\ell)$.

3) Γ keeps fixed a point $x \in \partial X$. The group Γ can be very complicated even when X is a Riemannian manifold, but when the curvature is pinched, i.e., $-\kappa_1 \leq K(X) \leq -\kappa_2 < 0$, the closure of Γ is isomorphic to an extension of a soluble group by a compact Lie group.

This follows from the Margulis lemma (see [11]).

3.5. *Non-special groups*

Let Γ be a discrete non-special group of isometries of a strictly convex space X.

THEOREM. *There is a closed set* $L \subset C\ell X$ *with the following properties*:

a) L *is an infinite set without isolated points*;

b) *the action of* Γ *in* L *is minimal, i.e., there are no invariant closed subsets*;

c) *the action of* Γ *in* $C\ell(X)\backslash L$ *is discrete.*

Denote by $L^{(2)}$ *the set of distinct pairs* $\ell_1, \ell_2 \in L$. *Such a pair is called axial if there is a hyperbolic isometry from* Γ *keeping this pair fixed.*

d) *Axial pairs are dense in* $L^{(2)}$, *unless all isometries from* Γ *are elliptic.*

Denote by $L^{(3)}$ *the space of pair-wise distinct triples from* L.

e) *The action of* Γ *in* $L^{(3)}$ *is discrete. If the action of* Γ *is uniform in* X *(i.e., the translates of a compact set cover* X) *then* $L = \partial X$ *and* Γ *acts uniformly in* $L^{(3)}$.

All these properties are well known when Γ is a Kleinian group. (Many of them I have learned from Dennis Sullivan.) Classical proofs use only the properties stated in sections 3.3 and 3.4. (See [6] for more information.)

There is another important property of Γ that can be established by Klein's argument:

Let $\gamma_1, \cdots, \gamma_m \subset \Gamma$ be hyperbolic elements where no two of them are powers of a third element from Γ. Then there is a number k such that powers $\gamma_i^{k_i}$ for $k_i \geq k$, $i = 1, \cdots, m$, generate a free group of rank m.

To complete the list of basic properties of Γ, we must mention another one, obvious but quite important:

Every hyperbolic $\gamma \in \Gamma$ can be uniquely written as γ_0^p where γ_0 is not a proper power.

3.6. *Remarks on not strictly convex manifolds*

There are some conditions weaker than strict convexity that lead to the conclusions analogous to the theorem from 3.5.

Consider first the case when X is a Riemannian manifold of non-positive curvature. We suppose that X is simply connected and the group of isometries Is(X) acts on X uniformly, for example, X is the universal covering of a compact manifold. Then X shares all global properties of a strictly convex space (in particular Theorem 3.5) iff it contains no flats. (A flat is a totally geodesic submanifold isometric to R^k, $k \geq 2$.)

As an example, one can take a not flat surface with $K \leq 0$. When this surface has a closed geodesic such that in a neighborhood of this geodesic curvature vanishes, then the universal covering has no strict convexity (because some horospheres contain geodesic segments).

The "no-flats" condition is also satisfied by the universal coverings of branched tori from 2.5 when V_0 intersects all 2-dimensional flat subtori in V. Though $\tilde{V}$ is not a Riemannian manifold, its metric can be smoothed to a Riemannian metric with $K \leq 0$ and with no flats in the universal covering. It is unclear whether in the "no-flats" case $\tilde{V}$ carries a Riemannian metric with $K < 0$.

QUESTION. *Let* V *be a compact Riemannian manifold with* $K \leq 0$ *and its universal covering contains a flat. Does it follow that* $\pi_1(V)$ *contains* $Z + Z$?

When the universal covering of V has no flats, the periodic geodesics are dense in the unit tangent bundle (because Theorem 3.5, in particular, d) is applicable to the universal covering of V), but if there is a flat, it is unclear whether V has more than one simple closed geodesic. The positive answer to the question would provide infinitely many such geodesics.

Of course, there are many compact manifolds with $K \leq 0$ where the density of closed geodesic in the unit tangent bundle is established. This is the case when V is compact and locally symmetric (see [17]).

Another (very easy) example is provided by compact connected manifolds having a tangent vector such that the sectional curvature is negative on all tangent 2-planes containing this vector.

Observe in the end that the whole discussion can be extended to a certain general class of locally convex spaces.

4. Isoperimetric Inequalities

4.1. *Openness at infinity*

Consider a domain Ω in an n-dimensional Riemannian manifold V. We denote by $\mathrm{Vol}(\Omega)$ its volume and by $\mathrm{Vol}(\partial\Omega)$ the (n–1)-dimensional volume of the boundary.

V is called *open at infinity* (see [19]) if there is a constant C such that any $\Omega \subset V$ satisfies $\mathrm{Vol}(\Omega) \leq C\,\mathrm{Vol}(\partial\Omega)$. When V is orientable, openness at infinity is equivalent (see [19]) to any one of the following three conditions:

a) The volume form in V is the differential of a bounded form ω (i.e., $\|\omega\|_V \leq \mathrm{const},\ v \epsilon V$).

b) Any bounded n-form on V is the differential of a bounded form.

c) There is a bounded vector field on V such that its divergence is greater than a fixed positive number.

EXAMPLES. A closed manifold is not open at infinity.

R^n is not open at infinity.

Strictly convex Riemannian manifolds (possibly with boundary) are open at infinity.

THEOREM (Avez [3]). *Let V be a simply connected manifold (without boundary) of non-positive curvature such that the group $\mathrm{Is}(V)$ acts uniformly in V. If V is not open at infinity, then it is isometric to the Euclidean space.*

4.2. *Amenable groups*

A group Γ is called amenable if any action of Γ on a compact space has an invariant measure. Basic properties of such groups can be found in [9]. Here are a few simple facts: A finite extension of a soluble group is amenable. If a group Γ contains a free subgroup of rank ≥ 2, then Γ is non-amenable. Every finitely generated non-amenable group has exponential growth: the number N of words with $\| \ \| \leq R$ satisfies $N \geq (\mathrm{const})^R$.

The following simple fact gives a geometric interpretation of amenability:

Let V *be a compact Riemannian manifold with fundamental group* Γ. *This group is non-amenable iff the universal covering* $\tilde{V}$ *is open at infinity.*

We shall use this fact in a slightly more general situation: take a submanifold $V_0 \subset V$. If $\pi_1(V \setminus V_0)$ is non-amenable, then the universal covering $\widetilde{V \setminus V_0}$ is open at infinity.

4.3. *Quasiconformal maps*

Let V be a complete Riemannian manifold of dimension n. Take a point $v_0 \in V$ and denote by $s(r)$ the $(n-1)$-dimensional volume of the radius r sphere centered at v_0.

AHLFORS' LEMMA. *If the integral* $\displaystyle\int_1^\infty (s(r))^{-\frac{1}{n-1}} \, dr$ *diverges, then* V *cannot be made open at infinity by any conformal change of metric.*

Proof. Take an arbitrary function $f > 0$ on V, multiply the metric by f, and show that there is a ball Ω (with respect to the old metric) such that $\mathrm{Vol}^{\mathrm{new}}(\Omega)/\mathrm{Vol}^{\mathrm{new}}(\partial\Omega)$ becomes arbitrarily large. When $\Omega = \Omega_r$ is a ball centered at v_0 of radius r, we have

$$\mathrm{Vol}^{\mathrm{new}}(\Omega_r) = \int_\Omega f^n = \int_0^r dr \int_{\partial\Omega} f^n$$

$$\geq \int_0^r dr \left(\int_{\partial\Omega} f^{n-1} \right)^{\frac{n}{n-1}} \cdot s(r)^{-\frac{1}{n-1}}$$

$$= \int_0^r dr \, (\mathrm{Vol}^{\mathrm{new}} \partial\Omega_r)^{\frac{n}{n-1}} (s(r))^{-\frac{1}{n-1}} \, ,$$

A straightforward calculation shows now that the divergence of

$$\int_1^\infty (s(r))^{-\frac{1}{n-1}}\, n$$

makes the ratio $\mathrm{Vol}^{\mathrm{new}}(\Omega_r)/\mathrm{Vol}^{\mathrm{new}}(\partial\Omega_r)$ unbounded when

$r \to \infty$.

Let V, W be two orientable Riemannian manifolds of dimension n, where V is complete and W is open at infinity. Let $h : V \to W$ be a map with the following properties generalizing the notion of quasiconformality:

a) The map h factors as

$$V \xrightarrow{\ g\ } V^1 \xrightarrow{\ h'\ } W$$

(with h the composite over the top)

where V^1 is a Riemannian manifold, the map g is a conformal equivalence and h' is a path-quasi-isometric map, i.e., for any curve $\ell \subset V^1$, we have

$$\mathrm{const} \le \frac{\mathrm{length}\ \ell}{\mathrm{length}\ h'(\ell)} \le \mathrm{const}^{-1}\ .$$

b) Consider the balls $\Omega_r \subset V$ as above and denote by w^* the pullback $h^*(w)$ of the volume form in W. Our second condition is the following:

$$\limsup_{r \to \infty} \frac{\displaystyle\int_{\Omega_r} |w^*|}{\displaystyle\int_{\Omega_r} w^*} < \infty\ .$$

This condition is automatically satisfied when h has non-negative Jacobian and such an h is usually called a quasiconformal map.

THEOREM (Picard-Ahlfors [2]). *If* h *satisfies* a); b) *then the integral*

$$\int_1^\infty \mathrm{Vol}(\partial\Omega_r)^{-\frac{1}{n-1}}\, dr$$

converges. In particular, there is no (non-constant) quasiconformal maps from $\mathbf{R}^n$ *into a compact non-flat manifold with* $K \leq 0$ *or into a manifold* W *containing a non-abelian free subgroup in its fundamental group. For example, there is no quasiconformal map from* $\mathbf{R}^2$ *into the 2-sphere minus three points, because the sphere with three punctures has a free fundamental group of rank 2 .*

The proof is obvious in view of the lemma.

Conformal hyperbolicity. There is an analogue of the Kobayashi metric for the quasiconformal situation: this is a metric that must increase no more than k-times under k-quasiconformal maps. No such metric, though, leads to a reasonable length structure, because there are non-Lipschitz quasiconformal maps. (This was pointed out to me by Dennis Sullivan.) One way to start constructing this metric is the following:

For a set $A \subset W$, its conformal capacity $\mathrm{Cap}A$ is defined as

$$\inf_f \int_W \| \mathrm{grad}\, f \|^n \quad \text{where } f \text{ runs over all compactly supported non-negative}$$

functions on W which are equal to 1 on A, $n = \dim W$. For two points $x, y \, \epsilon \, V$, we define $\mathrm{dist}(x,y)$ as the lower bound of capacities of the connected sets containing both points. Of course, such a metric is always zero on a closed manifold, but we use this construction only for open balls, and then we use the functorial Kobayashi-type argument. When W is a closed manifold, the resulting metric vanishes iff there is a (non-constant) quasiconformal map $\mathbf{R}^n \to W$, $n = \dim W$. (Easy to show.)

QUESTION. *Let* W *be a closed simply-connected Riemannian manifold. Does there always exist a non-constant quasiconformal map* $\mathbf{R}^n \to W$?

The locally homeomorphic case. There are relatively few locally homeomorphic quasiconformal maps. For example, for such maps the answer to our question is "No", unless W is homeomorphic to the sphere.

When $W = S^n$ and $n \geq 3$, every locally homeomorphic quasiconformal map $R^n \to S^n$ is injective, and the complement to its image consists of a single point. This is the Zorich Theorem (see [15]).

The Zorich theorem (or rather the proof) has a wider range: *Let* V *be a complete manifold with finite volume and* W *be a simply-connected manifold, both of dimension* $n \geq 3$. *Then any locally homeomorphic quasiconformal map* $V \to W$ *is injective, and the complement to its image is zero-dimensional.*

4.4. *Euclidean inequality*

For an $\Omega \subset R^n$, one classically has $V\ell \leq A_n(S)$ where $V\ell = \mathrm{Vol}(\Omega)$, $S = \mathrm{Vol}(\partial\Omega)$ and A_n denotes the volume of the Euclidean ball with the boundary sphere of volume one.

Let V be a Riemannian manifold, possibly with boundary, such that the length structure is convex. *Does a domain* $\Omega \subset V$ *satisfy the above inequality* $V\ell \leq A_n(S)$? What is easy to prove by standard integral geometry (see [18]) is $V\ell \leq \mathrm{const}_n A_n(S)$ and the problem is to make $\mathrm{const}_n = 1$. That can be done [18] when $n = 2$. Observe also that the inequality $V\ell \leq A_n(S)$ is classically known when V is a simply-connected manifold of constant negative curvature. On the other hand, this inequality is "stable" under direct product. (Easy to show.) As a consequence we get:

If the manifold V metrically splits into the product $V_1 \times \cdots V_i \times \cdots V_k$ such that each V_i is a simply-connected manifold satisfying one of the two conditions: a) V_i has non-positive curvature and $\dim V_i = 2$; b) V_i has constant negative (or zero) curvature, then $V\ell \leq A_n(S)$ for any $\Omega \subset V$.

Intermediate inequalities. Consider a $(k{-}1)$-dimensional cycle C in R^n. Denote by S its volume (mass) and by $V\ell$ the volume of the minimal chain spanning C.

THE FEDERER-FLEMING THEOREM.

$$V\ell^{k-1} \leq \mathrm{const}_n S^k.$$

See [14] for a geometrically oriented proof.

QUESTIONS. *Does the Federer-Fleming inequality hold for all convex manifolds?*

Can one take $\text{const}_n = 1$ or at least replace it by const_k? This is not quite known even for $\mathbf{R}^n$ (see [22] for some information).

We shall see below how to handle $k = 2$.

4.5. *Taut surfaces*

Let X be any length space and D be a 2-dimensional manifold. A map $f: D \to X$ is called *taut* if no compactly supported deformation decreases the induced length in D, i.e., any compactly supported deformation $f_t: D \to V$, $f_0 = f$, satisfying: $\text{length}(f_t(\ell)) \leq \text{length}(f(\ell))$ for all curves $\ell \subset D$, does not change length at all.

When $V = \mathbf{R}^n$, the taut surfaces are essentially the same as saddle-surfaces. In particular, their Gauss curvature is non-positive.

Under very mild restrictions on X, every contractable curve $\ell \subset X$ with $\text{length}(\ell) < \infty$ can be spanned by a taut disk. As a consequence, we have: *if the induced structure in every taut disk $D \to X$ is convex and X is simply connected, then X is convex.*

Taut surfaces allow us to reduce the study of negatively curved spaces to the special case of dimension 2 where much information is available (see [1]). In particular we have:

The Ionin-Burago inequality: Let V be a two-dimensional manifold with boundary. Denote by A its total area; by L the lengths of the boundary, and by $V_+ \subset V$ the set, where the curvature $K(v)$ is greater than -1. Set $\omega_+ = \displaystyle\int_{V_+} (K(v)+1)\,dv$. We have:

$$\tfrac{1}{2}L^2 \geq A\left(\frac{A}{2} - \omega_+ + 2\pi\chi\right),$$

where χ is the Euler characteristic of V.

See [12] for a short proof.

4.6. *Metric inequalities*

The following two theorems link volumes with the distance function.

The Almgren inequality. Let X be an n-dimensional cube with a Riemannian metric. Denote by ℓ_i the length distance between i-th pair of opposite (n–1)-faces. We have:

$$\text{Vol}X \geq \prod_1^n \ell_i$$

(This form of the Almgren inequality was communicated to me by Iu Burago.)

Proof. Let $F_i \subset X$, $i = n$, $i \cdots n$, be (n–1)-faces in X and none of two F_i are opposite. Consider the map $\phi: X \to R^n$ given by the functions $\text{dist}(x, F_1)$, $\text{dist}(x, F_2)$, $\cdots$.

The Jacobian of ϕ does not exceed 1 and, by an obvious topological reason, the image of ϕ contains the direct product $[0, \ell_1] \times [0, \ell_2] \cdots \subset R^n$. Q.E.D.

The simplex inequality. Let X be an n-dimensional simplex with a Riemannian metric. Suppose that for any $x \in \partial X$ the sum of distances from x to all (n–1)-dimensional faces is not less than one, then the volume of X is not less than the volume of the regular Euclidean simplex Δ_1 of unit height.

Proof. Using distances from an $x \in X$ to all (n–1)-faces, we get a map $X \to R^{n+1}$. Use now the projection from R^{n+1} onto the affine hyperplane spanned by the basis vectors. The image of the resulting map $\gamma: X \to R^{n+1}$ contains the standard n-simplex Δ, and the Jacobian of γ does not exceed the ratio $\dfrac{\text{Volume } \Delta}{\text{Volume } \Delta_1}$. Q.E.D.

An application. Let V be a complete simply-connected Riemannian manifold such that every taut disk $D \to V$ satisfies:

$$\text{Area}(D) \leq \text{const} \cdot \text{length}(\partial D)$$

where "const" depends on V but not on $D \to V$.

Let $\ell_1, \ell_2 \subset V$ be straight lines such that

$$\sup_{x \epsilon \ell_1} \ \mathrm{dist}(x, \ell_2) \leq C < \infty \ .$$

Then

$$\inf_{x \epsilon \ell_1} \ \mathrm{dist}(x, \ell_2) \leq \mathrm{const} \ ,$$

and this is the constant from above.

Proof. Take $x, y \ \epsilon \ \ell$ with $\mathrm{dist}(x, y) = d$ and point $x^1, y^1 \ \epsilon \ \ell_2$ with $\mathrm{dist}(x, x^1) \leq C \ \ \mathrm{dist}(y, y^1) \leq C$. Span the geodesic tetragon $x y y^1 x^1$ by a taut disk. Its area A must be less than $\mathrm{const}(2d + 4C)$. Applying the Almgren inequality, we have $A \geq (d - 2C)(\mathrm{Dist}(\ell_1, \ell_2))$. When $d \to \infty$, we get the required estimate.

5. Hyperbolic Systems

5.1. *Shifts*

Let Γ be any countable group. Consider the space A of the functions on Γ with values in a finite set S. (A is equipped with the product topology and homeomorphic to a Cantor set.) The left translations induce an action of Γ in A which is called the *Bernoulli shift*.

An invariant set $B \subset A$ is called *Markov* (or of finite type) if there is a continuous map $\Delta : A \to A$ commuting with the action of Γ such that B consists of all $a \ \epsilon \ A$ satisfying $\Delta(a) = a$. The induced action in such a B is called a *Markov shift*.

Markov shifts and continuous maps commuting with the action of Γ form a nice category. In particular, the product of two Markov shifts with the diagonal action carries a natural structure of a Markov shift.

5.2. *Markov relations*

A Markov relation over B is, by definition, a Markov set $R \subset B \times B$. When R is an equivalence relation (i.e., it is reflexive, symmetric and transitive) we have the quotient B/R with a natural action of Γ. The dynamical systems arising in this way are exactly all *hyperbolic systems* (over Γ).

For example, the basic sets of an axiom A diffeomorphism are hyperbolic with $\Gamma = Z$ (see [4]).

Look now at an example of algebraic origin.

5.3. *Hyperbolic automorphisms*

Let G and Γ be finitely generated groups and let $\{\gamma_1, \cdots, \gamma_k\}$ generate Γ. Suppose that Γ acts on G by automorphisms $g \to g^\gamma$, $g \in G$, $\gamma \in \Gamma$. Consider the space F of all functions $\Gamma \to G$.

A function $f : \Gamma \to G$ is called an *orbit* if $f(\gamma_i \gamma) = (f(\gamma))^{\gamma_i}$, $\gamma \in \Gamma$, $i = 1, \cdots, k$.

A function f is called an ε-orbit, $\varepsilon \geq 0$, if

$$\sup_{\gamma \in \Gamma} \, \text{dist}\,(f(\gamma_i \gamma), (f(\gamma))^{\gamma_i}) \leq \varepsilon \, ,$$

$$i = 1, \cdots, k \, ,$$

where "dist" means a fixed word distance. When ε is not specified, we call an ε-orbit a *pseudo-orbit*.

We say that $f_1, f_2 \in F$ are δ-close if $\text{dist}\,(f_1, f_2) \leq \delta$. When δ is not specified, we just say they are close.

The action is called *hyperbolic* if there are ε_0, $\delta_0 \geq 0$ such that:

a) Every pseudo-orbit is close to an ε_0-orbit;

b) When two ε_0-orbits are close, then they are δ_0-close.

The space F carries the product topology. Take the set F_{ε_0} of all ε_0-orbit and identify close ε_0-orbits. The resulting space we denote by $\overline{G}$. The group G is embedded in F as follows: g goes to the orbit $\{g^\gamma\}$, $\gamma \in \Gamma$, and hence, we get an action of G in $\overline{G}$ coming from multiplication of functions $\Gamma \to G$. The group Γ naturally acts in $\overline{G}/G$ and under the conditions a), b) this action is hyperbolic. (This is easy to check.)

Classical example (Franks [8]). Take $\Gamma = \mathbf{Z}$ and consider an Anosov action of $\mathbf{Z}$ on an infranilmanifold X. Then the induced action of $\mathbf{Z}$ in $G = \pi_1(X)$ is hyperbolic and $\overline{G}/G$ with the action of $\mathbf{Z}$ is canonically isomorphic to the original action of $\mathbf{Z}$ in X. (In particular, X is homeomorphic to $\overline{G}/G$.)

5.4. *The boundary action*

Let Γ act by isometries on a strictly convex space. If the action is discrete, free and uniform, then the associated action on ∂X is hyperbolic. This is an unpublished result of Sullivan (who stated it in a slightly different form). We discuss the proof in the next chapter. Observe that the action in ∂X is minimal (see 3.5), and the dynamical systems that are both hyperbolic and minimal appear seldom. For example, $\mathbf{Z}$ has no such (non-trivial) actions.

6. Coarse Hyperbolicity

6.1. *Coarse invariance of the boundary*

The boundary ∂X of a metric space is naturally invariant under isometries, i.e., an isometry $X \xrightarrow{f} Y$ can be continuously extended to $C\ell(X)$. When f is a quasi-isometry, or more generally a coarse equivalence (see 1.4), such extension is usually impossible. For example, quasi-isometries of $\mathbf{R}^n$, $n \geq 2$, do not generally extend to the $\partial \mathbf{R}^n = S^{n-1}$.

The following fundamental theorem is due essentially to Efremovitch-Tichomirova (see [7]).

If X and Y are strictly convex spaces, then every quasi-isometry (and even a coarse quasi-isometry) $X \xrightarrow{f} Y$ can be continuously extended to a homeomorphism $\partial X \to \partial Y$.

To prove the theorem, we have to provide a coarse invariant construction of the boundary. The following definition is analogous to the one from section 5.3.

6.2. *Coarse boundary*

Let X be a complete metric space. We view a set $F \subset X \times X$ as a set of arrows $x_1 >\!\!\to x_2$ for $(x_1, x_2) \in F$. We call F a λ-field, $\lambda \geq 0$ if

$$\sup_{(x_1, x_2) \in F} \operatorname{dist}\, (x_1, x_2) \leq \lambda \ .$$

We write $x > y + k$, $x, y \in X$, $k = 0, 1, \cdots$, if there are $x_0, x_1, \cdots, x_m \in X$, $m \geq k$, such that $(x_i, x_{i+1}) \in F$ and $x_0 = x$, $x_m = y$.

A field is called straight if for some $\varepsilon < 1$, it satisfies: $x > y + k$ implies $\operatorname{dist}(x, y) \geq (1-\varepsilon) k$.

Denote by $x + \{k, k+1, \cdots\}_F$ the set of all points $z \in X$ satisfying $z > x + k$.

We call a λ-field *compressing* if there are $\delta \geq 10\lambda$ and $p = 0, 1, \cdots$, such that for any $x, y \in X$ with $\operatorname{dist}(x, y) \leq 10\delta$ one can find $i, j \leq p$ such that the Hausdorf distance between $x + \{i, i+1, \cdots\}_F$ and $y + \{j, j+1, \cdots\}_F$ is less than δ.

We call two fields F_1 and F_2 *equivalent* if the Hausdorf distance between sets $x + \{1, 2, \cdots\}_{F_1}$ and $x + \{1, 2, \cdots\}_{F_2}$ is bounded (i.e., finite) for any $x \in X$.

If X is a strictly convex length space, there is a canonical homeomorphism between the boundary ∂X and the set of the equivalence classes of the straight compressing λ-fields. (The last set carries a natural topology.)

Proof. Given a horofunction h on X, we have a λ-field $F \subset X \times X$ defined as follows: $\{x, y\} \in F$ iff $\operatorname{dist}(x, y) \leq \lambda$ and $(h(x) - h(y)) \geq \lambda/10$.

On the other hand, a field F defines a point in ∂X: this is the limit point of the set $x + \{1, 2, \cdots\}_F$. It is easy to check that this correspondence gives the required homeomorphism.

When Γ is a group with the word metric, the set of all λ-fields can be identified with the set of all maps $\Gamma \to S$, where S is the set of all subsets from the radius λ ball in Γ around the identity. When Γ is as in section 5.4, the straight compressing λ-fields (with a fixed λ) form a Markov subset and the equivalence relation of fields is Markov. This explains the Sullivan theorem on hyperbolicity of the action of Γ on ∂X.

As an application of the coarse invariance of ∂X and the action of Γ on ∂X we have:

The geodesic flow on a compact strictly locally convex space $\overline{X}$ is uniquely determined by the fundamental group $\pi_1(\overline{X})$, i.e., if $\pi_1(\overline{X}) = \pi_1(\overline{X}')$ then the (one-dimensional) orbit foliations of the geodesic flows over $\overline{X}$ and $\overline{X}'$ are homeomorphic.

The proof given in [13] for Riemannian manifolds works in the general case.

6.3. *Coarse geometry of symmetric spaces*

We have seen so far that coarse equivalence is strong enough to recover the topology at infinity. The boundary ∂X also carries some geometric structure: a generalized quasiconformal structure (in the terminology of Mostow and Sullivan, see [17]). This structure is best understood in the case of symmetric spaces, in particular for the constant curvature manifolds. In the last case, this is the usual quasiconformal structure on S^{n-1}.

The generalized quasiconformal structure is a coarse invariant. When it comes to symmetric spaces, the Mostov theorem shows [17] that is is a very strong invariant.

If two symmetric spaces (of noncompact type) of rank 1 are coarse equivalent, then they are "essentially" the same (i.e., isometric up to the choice of the normalizing constants).

This result can probably be generalized to more general spaces having large groups of isometries. Here is a sample of such generalization:

Let X_0 be a compact space covered by a convex space X. Suppose that one of the following two conditions is satisfied:

a) The space X (or, equivalently the group $\pi_1(X_0)$) is coarse equivalent to the Lobachevskyi space of dimension n.

b) X is a Riemannian manifold which is quasiconformally equivalent to the standard open ball in $\mathbf{R}^n$.

Then X_0 is homotopically equivalent to an n-dimensional manifold of constant curvature.

Proof. According to Mostow [17], conditions a) and b) make the group $\Gamma = \pi_1(X_0)$ act on S^{n-1} by quasiconformal transformations with uniformly bounded dilatation. Using a remark by Sullivan from [20], we get a measurable conformal structure on S^{n-1} invariant under Γ. When $n = 2$, the proof is concluded (see [20]) by applying the measurable Riemannian mapping theorem. When $n \geq 3$, we observe that the measurable conformal structure is, almost everywhere, close to a flat structure. Using the action of Γ, one can "extend the flatness" to the whole sphere S^n. We shall discuss details at another opportunity.

6.4. *Coarse hyperbolicity*

As we have already seen that the essential features of strictly convex spaces are coarse invariant. That suggests a "coarse" counterpart of the notion of strict convexity. Unfortunately, the needed definitions and the proofs of the basic properties are fairly lengthy though quite trivial. Here we mention only one important property.

We say that a Riemannian manifold X has Is_2-*property* if there is a constant C such that every contractable closed-curve in X of the length ℓ can be spanned by a disk of area $\leq C \circ \ell$.

Observe that this property is a coarse invariant of the universal covering $\tilde{X}$. In particular, if one of two compact manifolds with the same fundamental group satisfies Is_2, so does the other. This allows us to attribute Is_2 to finitely presented groups.

It is not hard to show that fundamental groups of compact spaces with strictly convex universal coverings satisfy Is_2 (compare with Chapter 4). For example, free groups, fundamental groups of compact manifolds of negative curvature, and finite extensions of such groups satisfy Is_2.

An application to complex geometry. Let V be a compact Kähler manifold such that $\pi_1(V)$ satisfies Is_2 and $\pi_2(V) = 0$. Then V is hyperbolic in the sense of Kobayashi, i.e., every holomorphic map $C \to V$ is constant.

Proof. We pass to the universal covering $\tilde{V}$ and use minimality of the images of maps $C \to V$. The rest immediately follows from the Ahlfors Lemma (section 4.3).

6.5. *Attaching cells*

Let X be a Riemannian manifold and $X_0 \subset X$ be a submanifold. We denote by $R(X_0)$ the maximal number ε such that the function $\mathrm{dist}\,(x, X_0)$ is smooth in the open ε-neighborhood of X_0.

Suppose now that $K(X) \leq -1$ and X_0 is the union of disjoint closed geodesics $g_1, \cdots, g_p$ of lengths $\ell_1, \cdots, \ell_p$, and attach 2-disks to each of g_i.

We denote by Y the resulting space and denote by $\overline{Y}$ the manifold with boundary obtained by "thickening" Y, $\dim \overline{Y} = \dim X + 1$. When $\ell_i \geq 10$, $i = 1, \cdots, p$, $R(X_0) \geq 10$, one can construct a Riemannian metric in $\overline{Y}$ such that the universal covering of Y is a strictly convex space. (Easy to show.) This implies that Y itself (i.e., its universal covering) is coarse hyperbolic and, in particular, satisfies Is_2. (Actually, it is much easier to check coarse hyperbolicity than to construct convex length structures. We consider an important example in the next section.)

Using the attaching of 2-cells, one can prove the following:

Let X be a closed manifold of negative curvature. Then the fundamental group $\pi_1(X)$ contains a free non-Abelian normal subgroup.

6.6. *The small cancellation groups*

Let F be a free group freely generated by $f_1, \cdots, f_k$. Consider a system of words $W = \{w_1, \cdots, w_s\}$ with the following properties:

1) if $w \epsilon W$ then $w^{-1} \epsilon W$;

take a $w \epsilon W$ and an arbitrary $w' \epsilon F$ conjugate to w,

2) $w' \epsilon W$ iff $\|w'\| \leq \|w\|$. (Of course, we use the norm associated with the generators f_i.)

We call a $p \epsilon F$ a *common piece* of two words $f_1, f_2 \epsilon F$ if $\|p^{-1}f_1\| = \|f_1\| - \|p\|$ and $\|p^{-1}f_2\| = \|f_2\| - \|p\|$.

The system W is said (see [16]) to satisfy $1/6$-*property* if for any two $w_i, w_j \epsilon W$, $i \neq j$ and their common piece p we have $\|p\| < \frac{1}{6}\|w_i\|$, $\|p\| < \frac{1}{6}\|w_j\|$.

We say that the group presented by W has $1/6$-property if W has this property.

Consider the two-dimensional polyhedron X corresponding to a $1/6$-presentation. It is not hard to show that its universal covering is coarse hyperbolic, in particular, it satisfies Is_2.

Observe, also, that usual algebraic treatment of the small cancellation groups (i.e., $1/6$-group and alike) is based on a combinatorial version of Is_2.

The geometric point of view described here is essentially due to Dehn. As it has been realized (see [16]), his original ideas lead to a better understanding of the small cancellation groups than the algebraic approach.

For example, the word and conjugacy problems (see the next section) or existence of free non-Abelian subgroups are obvious from Dehn's point of view.

6.7. *Algorithmical problems*

Consider a group Γ presented by $\{f_1, \cdots, f_k; w_1, \cdots, w_s\}$ and the presenting homomorphism $F \xrightarrow{\ \square\ } \Gamma$. Denote by $\| \ \|_\square$ the pullback to F of the word norm of Γ. The function $\| \ \|_\square : F \to Z_+$ cannot be, in general, effectively computed. For example, deciding when $\|f\|_\square = 0$ is equivalent to the word problem.

212 M. GROMOV

There is the following geometric interpretation of our problem. Take a finite polyhedron X with a length structure and with $\pi_1(X, x_0) = \Gamma$. Computing $\|y\|$ is essentially the same as estimating the length of the minimal loop at x_0 representing $y \in \Gamma$.

When X is locally convex, the function $\| \ \|$ is effectively computable.

The proof immediately follows from the uniqueness of a closed geodesic loop in each homotopy class.

An analogous argument solves the conjugacy problem as well.

The commutator norm. Let $y \in [\Gamma, \Gamma]$, $\Gamma = \pi_1(X_1)$. Denote by $[y]$ the minimal possible genus of a surface spanning y. When X is a compact manifold of negative curvature, the genus of a minimal surface can be estimated by use of the inequality from section 4.5. That implies computability of the $[\ ]$-norm.

Unfortunately, the extension of those arguments to the coarse hyperbolic case requires more work (or, rather, more terminology) but the basic conclusions are the same.

Observe in the end, that the geometrical interpretation of computability can be used the other way around: when the fundamental group $\pi_1(X)$ is "logically complicated", X has many geodesic loops, closed geodesics, minimal surfaces, etc., (see [13]).

REFERENCES

[1]   A. Alexandrov, V. Zalgallev. *Two-dimensional manifolds of bounded curvature*. Trudy Math. Inst., Steklov. *63*(1967).

[2]   L. Alfors. *Zur Theory der Uberlageruns-floechen*. Acta Math. *65* (1936), 157-194.

[3]   A. Avez. *Varietes riemannienes sans points focaux*. C. R. Ac. Se., Paris, *270*(1970), 188-191.

[4]   R. Bowen. *Markov partitions for axiom-A-diffeomorphisms*. Am. J. Math. *92*(1970), 725-747.

[5]   J. Cheeger and D. Ebin. "Comparison Theorems in Riemannian Geometry," North Holland, 1975.

[6]   P. Ebeolein and B. O'Neill. *Visibility manifolds*. Pac. J. Math. *46* (1973), 45-103.

[7]   V. Efremovitch and E. Tichonirova. *Equimorphisms of hyperbolic spaces*. Isv. Ac. Nauk. *28*(1964), 1139-1144.

[8]   J. Franks. "Anosov Diffeomorphisms, Global Analysis," Proc. Symp. Pure Math. *14*(1970), 61-93.

[9]   F. Greenleaf. Invariant Means on Topological Groups and Their Applications. Van Nostrand, Reinhold C., 1969.

[10] M. Gromov. *Homotopical effects of dilatation*. J. Diff. Geom. (to appear).

[11] _________. *Manifolds of negative curvature*. J. Diff. Geom. (to appear).

[12] _________. *Isometrical immersion, hyperbolic geometry and Burago's isoperimetric inequality*. (Submitted to J. Diff. Geom.)

[13] _________. "Three remarks on geodesic dynamics and fundamental group." Preprint, State Univ. of N. Y. at Stony Brook.

[14] M. Gromov and I. Eliashberg. *Constructing non-singular isoperimetric films*. Trudy, Steklov Inst. *116*(1971), 18-33.

[15] M. Lavrentiev and P. Belinskii. "On locally quasiconformal mappings in space," Contributions to Analysis, Academic Press, 1974.

[16] R. Lynden and P. Schupp. Combinatorial Group Theory, Springer-Verlag, 1977.

[17] D. Mostow. "Rigidity of locally symmetric spaces," Princeton Univ. Press, 1972.

[18] L. A. Santalo, "Introduction to integral geometry," Herman, Paris, 1953.

[19] D. Sullivan. *Cycles for the dynamical study of foliated manifolds and complex manifolds*. Invent. Math. *36*(1975), 225-255.

[20] _________. "On the ergodic theory at infinity of an arbitrary discreet group of hyperbolic motions," These Proceedings.

[21] W. Thurston. Geometry and topology of 3-manifolds, Princeton, 1978.

[22] D. Hoffman and J. Spruck. Sobolev and isoperimetric inequalities for Riemannian submanifolds. *Comm. Pure Appl. Math. 27*(1974), 715-727.

[23] T. Jorgensen, Compact 3-manifolds of constant negative curvature fibering over the circle, Ann. Math., 106, (1977), 61-72.

AUTOMORPHISMS OF COMPACT RIEMANN SURFACES
AND WEIERSTRASS POINTS

Ignacio Guerrero

1. *Introduction*

Let M be a compact Riemann surface of genus $g \geq 2$. We denote by W_q the set of q-fold Weierstrass points on M, i.e. Weierstrass points for holomorphic q-differentials on M (cf. Lewittes [11]). It is well known that W_q is a finite set and that $\underset{q \geq 1}{\cup} W_q$ is dense in M (Olsen [12]). Lewittes [11] showed that if an automorphism of M has five or more fixed points, then each of them is an ordinary Weierstrass point (i.e. is in W_1). Accola [1] has shown that if an automorphism of order N has three or more fixed points, each fixed point is an N-fold Weierstrass point. Farkas and Kra [6] have proved, under the same hypotheses, that each fixed point is a q-fold Weierstrass point for $q \equiv 1 \pmod{N}$, $q > 1$.

The purpose of this paper is to investigate the situation in the remaining cases, namely when the automorphism has one or two fixed points. We will show that if an automorphism has a single fixed point it must be an ordinary Weierstrass point, except for a very special case. A description of the exceptional case is given. It will be seen that in any event the fixed point is in $W_1 \cup W_2$. When the automorphism has two fixed points we will see that it is possible that the fixed points miss the dense set $\underset{q \geq 1}{\cup} W_q$.

The author wishes to express his thanks to Professor R. D. M. Accola for helpful conversations. In particular, the example discussed in section 3.1 was suggested by him.

2. *Automorphisms with one fixed point*

2.1. We first recall that if an automorphism fixes a single point, its order N cannot be a power of a prime. For a proof of this assertion see Guerrero [7]; another proof is obtained as a direct consequence of Theorem 4 in Harvey [8].

Assume now that the only fixed point of $T : M \to M$ is not an ordinary Weierstrass point. Let ε be the rotation constant of T at the fixed point, i.e. locally $T^{-1} : z \mapsto \varepsilon z$. There exists a basis for the space of holomorphic differentials such that the linear map T_1 (we denote by T_q the linear map induced by T on the space of q-differentials) is given by the matrix $\mathrm{diag}(\varepsilon, \varepsilon^2, \cdots, \varepsilon^g)$ (see Lewittes [11]).

The Eichler trace formula (see e.g. Eichler [5] or Guerrero [7]) gives

$$\text{trace } T_1 = 1 + \frac{\varepsilon}{1-\varepsilon} = \frac{1}{1-\varepsilon} .$$

Write $g = g'N + s$, $0 \le s < N$. Note that g' is the dimension of the space of holomorphic differentials on M invariant under T. Thus g' is the genus of the Riemann surface $M/{<}T{>}$ (${<}T{>}$ is the cyclic group generated by T). The trace of T_1 is also given by

$$\begin{aligned}
\text{trace } T_1 &= \varepsilon + \varepsilon^2 + \cdots + \varepsilon^g \\
&= \varepsilon + \varepsilon^2 + \cdots + \varepsilon^s \\
&= \frac{1 - \varepsilon^{s+1}}{1-\varepsilon} - 1 \\
&= \frac{\varepsilon - \varepsilon^{s+1}}{1-\varepsilon} .
\end{aligned}$$

Comparing the two equations for trace T_1 we obtain

$$\varepsilon^{s+1} - \varepsilon + 1 = 0 .$$

Conjugating this equation (recall that ε is a primitive N-th root of 1)

$$\varepsilon^{-(s+1)} - \varepsilon^{-1} + 1 = 0$$

$$\varepsilon^{s+1} - \varepsilon^{s} + 1 = 0 \ .$$

It follows that $\varepsilon^{s} = \varepsilon$ and therefore $s = 1$. Since ε satisfies the equation $\varepsilon^2 - \varepsilon + 1$, ε is a primitive sixth root of 1. Thus we conclude that $N = 6$ and $g = 6g' + 1$.

The Riemann-Hurwitz relation in this case is

$$2g-2 = 6(2g'-2) + \text{Branch number}$$

therefore

$$\text{Branch number} = 12 \ .$$

The Branch number is of the form $5 + 3x + 4y$. We denote by x (resp. y) the number of points in $M/\!<T>$ whose fiber consists of fixed points of T^3 (resp. T^2). The only possibility is that $x = y = 1$.

We summarize our conclusions in the following

PROPOSITION. *If an automorphism* T *of* M *fixes a single point and this point is not an ordinary Weierstrass point, then* T *has order 6 and* $g = 6g' + 1$. *Here* g *and* g' *denote the genera of* M *and* $M/\!<T>$ *respectively. Further, the covering* $M \to M/\!<T>$ *is branched over three points. The fibers over such points consist of fixed points of* T, T^2 *and* T^3 *respectively.*

Suppose now that $T : M \to M$ is an automorphism or order N fixing a unique point p and $g \equiv 1 \pmod{N}$. Assume p is not a q-fold Weierstrass point $(q \geq 2)$. Then there is a basis for the space of holomorphic q-differentials such that the matrix of the induced linear map T_q is $\text{diag}(\varepsilon^q, \varepsilon^{q+1}, \cdots, \varepsilon^{q+d-1})$, $d = (2q-1)(g-1)$ (Lewittes [10]). Therefore

$$\text{trace } T_q = \varepsilon^q(1 + \varepsilon + \cdots + \varepsilon^{d-1}) \ .$$

However, $d = (2q-1)(g-1) \equiv 0 \pmod{N}$, hence

$$\text{trace } T_q = 0 .$$

On the other hand the Eichler trace formula gives

$$\text{trace } T_q = \frac{\varepsilon^q}{1-\varepsilon} \neq 0 .$$

We conclude that the fixed point must be a q-fold Weierstrass point for all $q \geq 2$.

In proposition above we have $N = 6$ and $g \equiv 1 \pmod{6}$. Thus the fixed point is a q-fold Weierstrass point for every $q \geq 2$.

2.2. In the previous section we determined necessary conditions for an automorphism to have a single fixed point which is not an ordinary Weierstrass point. In this section we will show that there exists a unique holomorphic family of Riemann surfaces satisfying such conditions. It is of interest to determine the Riemann surfaces in the family for which the fixed point is not an ordinary Weierstrass point. We give some partial information in this direction.

Theorem 4 in Harvey [8] gives necessary and sufficient conditions for the existence of a Riemann surface M with an automorphism T such that the covering $M \to M/\langle T \rangle$ has a given branch structure. It is very simple to verify that the conditions are satisfied in the situation described in the previous proposition (the Fuchsian group Γ in [8] has signature $(6,3,2; g')$ $g' \geq 1$ in our case).

Denote by $\mathcal{J}$ the relative Teichmüller space determined by such M and $\langle T \rangle$ (see Kuribayashi [10], Earle [3]). $\mathcal{J}$ parametrizes the marked Riemann surfaces having an automorphism topologically equivalent to T. $\mathcal{J}$ is a complex manifold and dimension $\mathcal{J}$ = dimension quadratic differentials on $M/\langle T \rangle -$ Branch set $= 3g'-3+3 = 3g' = (g-1)/2$. There is a holomorphic family $\mathcal{O} \to \mathcal{J}$ such that the fiber over $\tau \in \mathcal{J}$ is a Riemann surface M_τ with an automorphism T_τ topologically equivalent to T (see Earle-Kra [4]).

Next we will show that the family $\mathcal{O} \to \mathcal{J}$ is unique. The relative Teichmüller space $\mathcal{J}$ is biholomorphically equivalent to the fixed point set of $T_* : \mathcal{J}_g \to \mathcal{J}_g$, where T_* is the map induced by T in the Teichmüller space $\mathcal{J}_g$ of Riemann surfaces of genus g. Suppose T and T' are automorphisms of Riemann surfaces M and M' of genus g. Assume further that the coverings $M \to M/\langle T \rangle$ and $M' \to M'/\langle T' \rangle$ satisfy the conditions of the proposition in the previous section. To show that T and T' determine isomorphic families of Riemann surfaces, it is sufficient to show that $\langle T_* \rangle$ is conjugate to $\langle T'_* \rangle$ as a subgroup of the Teichmüller modular group (Earle [3], Harvey [9], Earle-Kra [4]). We have to study the local behavior of T and T' near the fixed points. Let $\mu = \exp(2\pi i/6)$. The rotation constant at the fixed point of T is either μ or μ^5 (primitive sixth roots of 1), the rotation constant at the fixed points of T^2 is either μ^2 or μ^4 (primitive cubic roots of 1), and the rotation constant at the fixed points of T^3 is μ^3 (primitive square root of 1). Topological restrictions (Harvey [9] Lemma 6, Guerrero [7] Lemma 1.2) reduce the possibilities to μ, μ^2, μ^3 and μ^5, μ^4, μ^3 for the rotations at the fixed points of T, T^2 and T^3 respectively. Note that if the rotation constants for T, T^2, T^3 are μ, μ^2, μ^3, then the rotation constants for $V = T^5$, V^2, V^3 are μ^5, μ^4, μ^3 and further $\langle T \rangle = \langle V \rangle$. Of course the same analysis is valid for T'. It follows that $\langle T'_* \rangle = \langle T''_* \rangle$ with T and T'' having the same rotation constants at corresponding fixed points $(T'' = T'$ or $(T')^5)$. We conclude that $\langle T_* \rangle$ and $\langle T'_* \rangle$ are conjugate (see Harvey [9]).

Our family $\mathcal{O} \to \mathcal{J}$ has a holomorphic section $s : \mathcal{J} \to \mathcal{O}$, $s(\tau) =$ the fixed point of $T_\tau : M_\tau \to M_\tau$. Define $\mathcal{W} = \{\tau \in \mathcal{J} \mid s(\tau)$ is an ordinary Weierstrass point$\}$. Locally we can write $\mathcal{W} = \{\tau \in \mathcal{J} \mid w(\tau, s(\tau)) = 0\}$. Here, $w(\tau, z)$ is a holomorphic function determined by the representation in local coordinates of the Wronskian of a basis for the space of holomorphic differentials on M_τ (see Bers [2]). It is clear now that there are three possibilities: (1) $\mathcal{W} = \emptyset$, (2) $\mathcal{W} = \mathcal{J}$, (3) $\mathcal{W}$ is an analytic subset of $\mathcal{J}$ of codimension 1.

Let p be the fixed point of $T: M \to M$ and p' the projection of p on $M/<T>$. If p' is an ordinary Weierstrass point, there is a holomorphic differential ω' with $\mathrm{ord}_{p'}\omega' \geq g'$. The lift ω of ω' satisfies $\mathrm{ord}_p\omega \geq 6g' + 5 > g$, thus p is an ordinary Weierstrass point. If we start with a Riemann surface M' of genus g' and a Weierstrass point p' on it, we can construct a Riemann surface M with an automorphism T such that $M/<T> \simeq M'$ and the covering $M \to M'$ has the same branch structure as the members of our family $\mathcal{O} \to \mathcal{T}$. Further, we require the fixed point of T to be over p'. To construct M, first we uniformize M' by a Fuchsian group Γ of signature $(6, 3, 2; g')$ (the points over which branching occurs can be arbitrarily prescribed) and then apply Harvey's theorem (Harvey [8], Theorem 4). The fixed point of T must be an ordinary Weierstrass point. This shows that $\mathcal{O} \neq \emptyset$.

We have been unable to show that $\mathcal{O} \neq \mathcal{T}$ (we believe that this is the case) i.e. to establish the existence of automorphisms fixing a single point which is not an ordinary Weierstrass point.

3. *Automorphisms with two fixed points*

3.1. Let $T: M \to M$ be an automorphism fixing two points p_1 and p_2. Assume that T has prime order N and that $M/<T>$ has genus 1. The Riemann-Hurwitz relation shows immediately that M has genus $g = N$. Denote by ω the lift to M of a nonzero holomorphic differential on $M/<T>$ (unique up to a constant). The divisor of ω is

$$(\omega) = (N{-}1)p_1 + (N{-}1)p_2$$
$$= (g{-}1)p_1 + (g{-}1)p_2 \, .$$

Suppose p_1 is a q-fold Weierstrass point. By definition, there is a holomorphic q-differential τ whose divisor is of the form

$$(\tau) = [(2q{-}1)(g{-}1)+\alpha]p_1 + \beta p_2 + \delta \, .$$

Here δ is a divisor not containing p_1 or p_2. Of course, $\alpha + \beta + \deg\delta = g{-}1$. We can choose τ to be an eigenvector for T_q, $T_q\tau = \lambda\tau$.

This implies that the divisor δ is invariant under T and therefore $\deg \delta$ is a multiple of $N = g$. Since $\deg \delta < g$ we must have $\delta = 0$. Hence

$$(\tau) = [(2q-1)(g-1)+a]p_1 + \beta p_2 \,, \quad a + \beta = g-1 \,.$$

Now, $f = \tau/\omega^q$ is a meromorphic function on M and

$$\begin{aligned}(f) &= [(q-1)(g-1)+a]p_1 + [\beta - q(g-1)]p_2 \\ &= [(q-1)(g-1)+a](p_1-p_2) \,.\end{aligned}$$

The function f^N is invariant under T, so it projects to a function h on $M/\langle T\rangle$

$$(h) = [(q-1)(g-1)+a](a_1-a_2) \,,$$

where a_i is the image of p_i under the projection $M \to M/\langle T\rangle$. Using Abel's theorem (identify $M/\langle T\rangle$ with its Jacobian) we conclude that a_1-a_2 is a rational point.

Conversely, assume that a_1-a_2 is a nonzero rational point in $M/\langle T\rangle$. Then, there exists a meromorphic function h with $(h)=k(a_1-a_2)$, $k \geq 2$. Let f be the lift of h to M, $(f) = kN(p_1-p_2)$. Consider $\tau = f^{(g-1)}\omega^{kN}$

$$(\tau) = 2kN(g-1)p_1 \,.$$

therefore, p_1 is a kN-fold Weierstrass point.

The above arguments are of course valid for the fixed point p_2. Thus we have shown that p_1 and p_2 are q-fold Weierstrass points for some q if and only if a_1-a_2 is a nonzero rational point on the torus $M/\langle T\rangle$.

Now it is clear how to construct examples of automorphisms fixing two points which are not q-fold Weierstrass points for any $q \geq 1$. Simply construct suitable coverings of a torus with branch points over a_1 and a_2 with a_1-a_2 not rational.

3.2. Let M be the hyperelliptic Riemann surface defined by the equation

$$w^2 = (z^2-a_1^2)(z^2-a_2^2) \cdots (z^2-a_n^2) \,.$$

Choose $n = 2g'+1$, then M has genus $g = 2g'$. Define $T: M \to M$ by $z \to -z$, $w \to w$. T is an involution fixing the two points over 0 and $M/{<}T{>}$ is a Riemann surface of genus g'. Note that the two points over ∞ are not fixed by T, actually the Riemann-Hurwitz relation shows that a Riemann surface of even genus cannot have an involution with four fixed points. We will show that it is possible to choose $a_1, \cdots, a_n$ so that the fixed points are not Weierstrass points of any order.

It is not hard to see that the space of holomorphic q-differentials on M has a basis $\{z^j w^{-q} dz^q, z^k w^{-q+1} dz^q\}$, $j = 0, \cdots, q(g-1)$ and $k = 0, \cdots, (q-1)(g-1)-2$ (except for $q = 1$ and $q = 2$, $g = 2$. Then the differentials $z^j w^{-q} dz^q$ are sufficient).

The points over $z = 0$ are not ordinary Weierstrass points (for any choice of $a_1, \cdots, a_n$), so assume $q \geq 2$. By definition a point over $z = 0$ is a q-fold Weierstrass point if and only if there is a relation

$$P(z) w^{-q}(z) + Q(z) w^{-q+1}(z) = z^d S(z) .$$

Here P and Q are polynomials of degree $s = q(g-1)$ and $t = (q-1)(g-1)-2$ respectively, $S(z)$ is a power series convergent near $z = 0$ and $d = (2g-1)(g-1)$. Note that we have two expressions for $w(z)$, corresponding to the two points over $z = 0$. From now on assume that one of these has been chosen and write

$$w(z) = c_0 + c_1 z + c_2 z^2 + \cdots .$$

We have

$$P(z) + Q(z) w(z) = z^d S(z) w^q(z) .$$

Substituting the expression for $w(z)$ and comparing coefficients one sees that the existence of P and Q (not both zero) is equivalent to the vanishing of

$$W = \begin{vmatrix} c_{s+1} & c_s & \cdots & c_{s-t+1} \\ c_{s+2} & c_{s+1} & \cdots & c_{s-t+2} \\ \cdot & \cdot & & \cdot \\ \cdot & \cdot & & \cdot \\ \cdot & \cdot & & \cdot \\ c_{s+t+1} & c_{s+t} & \cdots & c_{s+1} \end{vmatrix} .$$

Now $w(z) = (z^{2n} - s_1 z^{2n-2} + \cdots + s_{n-1} z^2 - s_n)^{1/2}$ where $s_1, \cdots, s_n$ are the elementary symmetric functions in $a_1^2, \cdots, a_n^2$. The coefficients of the power series for $w(z)$ are of the form

$$s_n^{1/2} F\left(\frac{1}{s_n}, \frac{s_1}{s_n}, \cdots, \frac{s_{n-1}}{s_n}\right)$$

for a polynomial F. We conclude that the vanishing of W is equivalent to the vanishing of a polynomial in $s_1, \cdots, s_n$ or, equivalently, to the vanishing of a polynomial in $a_1, \cdots, a_n$.

A careful analysis of the expansion of $w(z)$ shows that W is not identically zero as a function of $a_1, \cdots, a_n$. In fact, the leading coefficient in s_{n-1} will be a determinant involving binomial coefficients. This determinant can be shown to be nonzero. It will be crucial the fact that g is even.

We have shown that the points over $z = 0$ will be q-fold Weierstrass points when a pair of polynomials $P_q^{\pm}(a_1, \cdots, a_n)$ vanish. Choosing $a_1, \cdots, a_n$ off the zero set of the countable collection of polynomials $\{P_q^{\pm} | q \geq 2\}$ we obtain the example sought.

DEPARTMENT OF MATHEMATICS
UNIVERSITY OF GEORGIA
ATHENS, GEORGIA 30602

REFERENCES

[1]   Accola, R. D. M. On generalized Weierstrass points on Riemann surfaces (to appear).

[2]   Bers, L. Holomorphic differentials as functions of moduli, Bull. Amer. Math. Soc., *67*(1961), 206-210.

[3]   Earle, C. J. Moduli of surfaces with symmetries, Advances in the Theory of Riemann Surfaces, Ann. of Math. Studies *66*(1971), 119-130.

[4]   Earle, C. J. and Kra, I. On sections of some holomorphic families of closed Riemann surfaces, Acta Math., *137*(1976), 49-79.

[5]   Eichler, M. Introduction to the theory of algebraic numbers and functions, Academic Press, New York-London, 1966.

[6]   Farkas, H. and Kra, I. (Forthcoming book on Riemann surfaces.)

[7]   Guerrero, I. On Eichler trace formulas (to appear).

[8] Harvey, W. J. On cyclic groups of automorphisms of a compact Riemann surface, Quart. J. Math. Oxford (2), *17*(1966), 86-97.

[9] ————. On branch loci in Teichmüller space, Trans. Amer. Math. Soc., *153*(1971), 387-399.

[10] Kuribayashi, A. On analytic families of compact Riemann surfaces with non-trivial automorphisms, Nagoya Math. J., *28*(1966), 119-165.

[11] Lewittes, J. Automorphisms of compact Riemann surfaces, Amer. J. of Math., *85*(1963), 732-752.

[12] Olsen, B. On higher order Weierstrass points, Ann. of Math., *95*(1972), 357-364.

AFFINE AND PROJECTIVE STRUCTURES
ON RIEMANN SURFACES

R. C. Gunning

1. Let M be a compact Riemann surface of genus $g > 0$, represented as the quotient of its universal covering space $\tilde{M}$ by the group of covering translations Γ. A *projective structure* on M is described by a complex analytic local homeomorphism $f : \tilde{M} \to P_1$ with the property that for any $T \in \Gamma$,

$$(1) \qquad\qquad f(Tz) = \rho_T(f(z))$$

for some $\rho_T \in PL(1, C)$; here P_1 is the one-dimensional complex projective space (the Riemann sphere), and $PL(1, C)$ is the group of projective transformations (linear fractional transformations) acting on P_1. The mapping f can be viewed as describing a special complex analytic coordinate covering of the Riemann surface M. The coordinate transformations of this coordinate covering are not merely complex analytic mappings but actually projective mappings; so this coordinate covering determines a projective structure on M, in a manner analogous to the determination of a complex structure by a coordinate covering with complex analytic coordinate transformations. If f is any complex analytic local homeomorphism satisfying (1) and σ is any element of $PL(1, C)$ then the composition $f' = \sigma \circ f$ is also a complex analytic local homeomorphism, and satisfies a condition of the form (1) with ρ_T replaced by $\rho'_T = \sigma \circ \rho_T \circ \sigma^{-1}$; the

© 1980 Princeton University Press
Riemann Surfaces and Related Topics
Proceedings of the 1978 Stony Brook Conference
0-691-08264-2/80/000225-20 $01.00/1 (cloth)
0-691-08267-7/80/000225-20 $01.00/1 (paperback)
For copying information, see copyright page

mappings f and f′ are considered as describing *equivalent* projective
structures. It is clear from (1) that the mapping $T \to \rho_T$ is a homomorphism from the group Γ into the group $PL(1, C)$; this homomorphism ρ
is called the *representation* of the projective structure described by the
mapping f. Note that the representations of equivalent projective structures are also equivalent, in the sense of being conjugate representations.

The study of projective structures on Riemann surfaces is of course
closely related to the study of Fuchsian and Kleinian groups, a currently
very active field of research; but it involves a slightly different point of
view, and suggests a different set of problems for investigation. The aim
of this paper is to survey some of these problems and to describe some of
the results known about them. First, however, to indicate what a complete theory of projective structures on Riemann surfaces might look like,
the corresponding theory for the much simpler case of affine structures
will be sketched; things are so simple in that case that very explicit results are easily obtained. Some results about the natural extension to
branched structures will also be included.

2. An affine structure is described by a complex analytic local homeomorphism $f : \tilde{M} \to C$ with the property that for any $T \in \Gamma$ an equation of the
form (1) holds for some $\rho_T \in A(1, C)$; here $A(1, C)$ is the group of affine
transformations on C, so that $\rho_T(z) = a_T z + b_T$ for some complex constants a_T, b_T with $a_T \neq 0$. The notions of equivalent affine structures
and of their representations are the obvious ones.

If M admits an affine structure then it has a coordinate covering for
which the Jacobians of the coordinate transformations are constants,
hence for which the canonical bundle is flat, using the terminology of [4];
and that implies that $g = 1$, which both simplifies and limits the theory
of affine structures. To be more explicit, consider therefore a marked
Riemann surface M of genus $g = 1$; the fundamental group Γ of M is
a free abelian group of rank 2, and a marking is merely a choice of two
free generators A, B for the group Γ. The universal covering space $\tilde{M}$

can be identified with the complex plane C in such a manner that $A(z) = z+1$, $B(z) = z+\omega$, for some complex number ω having positive imaginary part; the Teichmüller space of all such marked Riemann surfaces can be identified with the upper half plane H, with the point $\omega \in H$ representing the given marked Riemann surface. An affine structure on M is described by a complex analytic local homeomorphism $f: C \to C$ such that

$$(2) \qquad f(z+1) = a_A f(z) + b_A, \quad f(z+\omega) = a_B f(z) + b_B.$$

The derivative f' is nowhere zero; and it is apparent from (2) that the quotient function f''/f' is a holomorphic Γ-invariant function on C, and is hence a constant $2\pi i c$. If $c = 0$ then f is itself an affine transformation; and of course any affine transformation f clearly describes an affine structure on M. If $c \neq 0$ then $f(z) = ae^{2\pi i c z} + b$ for some complex constants a, b with $a \neq 0$; and it is clear that any such mapping also describes an affine structure on M. By passing to an equivalent affine structure, by replacing f by the composition $\sigma \circ f$ for some affine transformation σ, it is clear that when $c = 0$ the function f can be reduced to the form $f(z) = z$, while when $c \neq 0$ the function f can be reduced to the form $f(z) = e^{2\pi i c z}$; and no further equivalences are possible. Thus in this manner the set of equivalence classes of affine structures on the marked Riemann surface M can be put into canonical one-to-one correspondence with the complex plane C.

Using this explicit form, note that the mapping $f: C \to C$ describing an affine structure on M is always a covering mapping; indeed, in the normal form chosen above, f is the identity mapping for the affine structure parametrized by $c = 0$, while f is the universal covering of the image $f(C) = C^* = \{z \in C : z \neq 0\}$ for the affine structures parametrized by any $c \neq 0$. The image $\rho(\Gamma)$ of the representation of the affine structure parametrized by c is a group of affine transformations acting on the image $f(C)$; indeed $\rho(\Gamma) = \Gamma$ for $c = 0$, while $\rho(\Gamma)$ is the group of affine transformations on C^* generated by

$$(3) \qquad \rho_A(z) = e^{2\pi i c}\, z, \qquad \rho_B(z) = e^{2\pi i c\omega}\, z$$

for $c \neq 0$. It is easy to verify that the group $\rho(\Gamma)$ is a properly discontinuous group of affine transformations on $f(C)$ if and only if $c = r/(p+q\omega)$ for some integers p, q, r; hence for general parameter values c the group $\rho(\Gamma)$ is not discontinuous. If $\rho(\Gamma)$ is discontinuous the quotient space $f(C)/\rho(\Gamma)$ is also a compact Riemann surface of genus 1, and f induces a covering mapping $f^*: M \to f(C)/\rho(\Gamma)$. The order of this covering mapping f^* is easily seen to be equal to the number of distinct pairs $(\{\nu p/r\}, \{\nu q/r\})$ as ν ranges over all the integers, where $c = r/(p+q\omega)$ and $\{x\} = x - [x]$ is the fractional part of x; thus if p, q, r are coprime the order of this covering is equal to $|r|$ whenever $r \neq 0$.

To describe some further properties of the representations $\rho = \rho_{\omega,c}$ of the affine structures associated to the parameter values $c \in C$, the representations given explicitly by (3), note that the set $\mathrm{Hom}(\Gamma, A(1, C))$ of all homomorphisms from Γ into the affine group $A(1, C)$ can be given a natural complex structure. Indeed viewing Γ as the free abelian group generated by two elements A, B, any $\rho \in \mathrm{Hom}(\Gamma, A(1, C))$ is completely described by the affine transformations $\rho_A(z) = a_A z + b_A$ and $\rho_B(z) = a_B z + b_B$; and these can be any affine transformations satisfying $\rho_A \rho_B = \rho_B \rho_A$. Thus the set $\mathrm{Hom}(\Gamma, A(1, C))$ can be identified with the three-dimensional complex analytic subvariety

$$(4) \qquad V = \{(a_A, a_B, b_A, b_B) \in C^* \times C^* \times C \times C : (a_A-1)b_B = (a_B-1)b_A\}.$$

This subvariety has only one singular point, the point $(1, 1, 0, 0)$ corresponding to the identity representation; the remaining points of V form a connected three-dimensional complex manifold.

Actually of course the principal interest is not so much in the set of all affine structures as in the set of all equivalence classes of affine structures; and hence it is more interesting to consider in place of $\mathrm{Hom}(\Gamma, A(1, C))$ the set of all equivalence classes of affine representations of Γ, the quotient space $\mathrm{Hom}(\Gamma, A(1, C))/A(1, C)$ where $A(1, C)$

acts by conjugation on the representations in $\mathrm{Hom}\,(\Gamma, A(1, C))$. If $\sigma(z) = az + b \,\epsilon\, A(1, C)$ and if $\rho \,\epsilon\, \mathrm{Hom}\,(\Gamma, A(1, C))$ is described by the coordinates (a_A, a_B, b_A, b_B) then the conjugate representation $\rho' = \sigma \circ \rho \circ \sigma^{-1}$ is described by the coordinates

$$(5) \qquad (a'_A, a'_B, b'_A, b'_B) = (a_A, a_B, ab_A - b(a_A - 1), ab_B - b(a_B - 1)) \,.$$

This exhibits $A(1, C)$ as a group of analytic automorphisms of the complex analytic subvariety V, and leads to the question whether the quotient space $W = V/A(1, C)$ can also be given a complex structure. Note first that the singular point $(1, 1, 0, 0) \,\epsilon\, V$ is left fixed by all transformations $\sigma \,\epsilon\, A(1, C)$ as of course it must be. Note next that the orbit under $A(1, C)$ of any point of the form $(1, 1, b_A, b_B)$ is the set of points $\{(1, 1, ab_A, ab_B) : a \,\epsilon\, C^*\}$, hence is an analytic submanifold of V isomorphic to C^*. Note finally that the orbit under $A(1, C)$ of any point of the form (a_A, a_B, b_A, b_B) where $a_A \neq 1$ is the set of points $\{(a_A, a_B, z, (a_B - 1)(a_A - 1)^{-1} z) : z \,\epsilon\, C\}$, hence is an analytic submanifold of V isomorphic to C; and similarly for any point for which $a_B \neq 1$. Thus there are three categories of points of V, corresponding to the three types of orbits under the action of the group $A(1, C)$ on V. However upon restriction to the open subset

$$(6) \qquad V_1 = \{(a_A, a_B, b_A, b_B) \,\epsilon\, V : (b_A, b_B) \neq (0, 0)\} \,,$$

each orbit of $A(1, C)$ in V except for the singular point $(1, 1, 0, 0)$ itself is a submanifold of V_1 isomorphic to C^*. These orbits are moreover just the fibres of the complex analytic mapping $\phi : V_1 \to C^* \times C^* \times P_1$ defined by $\phi(a_A, a_B, b_A, b_B) = (a_A, a_B, [b_A, b_B])$ where $[b_A, b_B]$ is the point of P_1 having homogeneous coordinates (b_A, b_B). Therefore, excluding the singular point of V, the remainder of the quotient space $W = V/A(1, C)$ can be identified with the two-dimensional complex submanifold $W_1 \subset C^* \subset C^* \times P_1$, where

$$(7) \qquad W_1 = \{(a_A, a_B, [b_A, b_B] \,\epsilon\, C^* \times C^* \times P_1 : (a_A - 1)b_B = (a_B - 1)b_A\} \,.$$

Note that the restriction to this submanifold $W_1 \subset C^* \times C^* \times P_1$ of the natural projection $C^* \times C^* \times P_1 \to C^* \times C^*$ is a complex analytic mapping $\psi : W_1 \to C^* \times C^*$ such that $\psi^{-1}(a_A, a_B)$ is a single point whenever $(a_A, a_B) \neq (1, 1)$ but $\psi^{-1}(1, 1) = P_1$; thus W_1 can be described as the complex manifold arising from $C^* \times C^*$ by applying a quadratic transform at the point $(1, 1)$, blowing up the point $(1, 1)$ to P_1.

Now the representation $\rho_{\omega, c} = \rho_\omega(c)$ described by (3) for $c \neq 0$, when considered as a point $\rho_\omega(c) \in V$, has coordinates $(e^{2\pi i c}, e^{2\pi i c\omega}, 0, 0)$; and the orbit of $A(1, C)$ through this point is the subvariety $\{(e^{2\pi i c}, e^{2\pi i c\omega}, b(1 - e^{2\pi i c}), b(1 - e^{2\pi i c\omega})) : b \in C\}$. This orbit does not reduce to the singular point $(1, 1, 0, 0) \in V$, hence represents a point $\rho_\omega(c) \in W_1$; and indeed that point is

$$(8) \quad \rho_\omega(c) = \left(e^{2\pi i c}, e^{2\pi i c\omega}, \left[\frac{e^{2\pi i c} - 1}{c}, \frac{e^{2\pi i c\omega} - 1}{c} \right] \right) \in W_1 \subset C^* \times C^* \times P_1 .$$

On the other hand the representation of the affine structure parametrized by $c = 0$, when considered as a point $\rho_\omega(0) \in V$, clearly has the coordinates $(1, 1, 1, \omega)$. The orbit of $A(1, C)$ through this point does not reduce to the singular point of V either, so represents a point $\rho_\omega(c) \in W_1$; and indeed that point is also given by (8), when extended to the value $c = 0$ by analytic continuation. The mapping $\rho_\omega : C \to W_1$ thus defined is clearly an injective mapping; so distinct equivalence classes of affine structures have distinct representation classes. The mapping ρ_ω is moreover a nonsingular holomorphic mapping; thus the set of representation classes of all affine structures on M is the one-dimensional complex analytic submanifold $\rho_\omega(C) \subset W_1$, isomorphic as a complex manifold to C.

Finally considering the mapping ρ_ω as a function of the point $\omega \in H$, note that whenever ω, ω' are distinct points of H there are infinitely many pairs of parameters (c, c') such that $\rho_\omega(c) = \rho_{\omega'}(c')$; indeed this equality holds precisely when $c = (m\omega' + n)/(\omega' - \omega)$ for some integers m, n, not both of which are zero. Thus although on any fixed Riemann surface distinct affine structures have distinct representation classes, nonetheless for any pair of distinct marked Riemann surfaces there are affine structures having the same representations. The mapping $\rho : H \times C \to W_1$ taking any points $\omega \in H$, $c \in C$ to the point $\rho_\omega(c) \in W_1$ given by (8) is thus not an injective mapping. On the other hand it is easy to see that ρ is a nonsingular holomorphic mapping; thus the image $\rho(H \times C)$ is an open subset of W_1. Whenever $c \neq 0$ the image $\rho_\omega(c) \in W_1$ is a point at which the natural projection $\psi : W_1 \to C^* \times C^*$ is a local homeomorphism; and the composite mapping $\psi \circ \rho : H \times C^* \to C^* \times C^*$ has the explicit form $\psi \circ \rho(\omega, c) = (e^{2\pi i c}, e^{2\pi i c \omega})$. It is easy to see that the image of this mapping is precisely the complement of the subset $\{(x, y) \in C^* \times C^* : |x| = |y| = 1\}$. Indeed whenever $(x, y) \in C^* \times C^*$ write $x = e^{2\pi i \xi}$, $y = e^{2\pi i \eta}$; then there are integers m, n such that $c = \xi + m \neq 0$ and $\omega = (\eta + n)/(\xi + m)$ satisfies $\text{Im}\,\omega > 0$ precisely when not both ξ, η are purely real. On the other hand when $c = 0$ the image $\rho_\omega(c) \in W_1$ is in the subset $\psi^{-1}(1, 1) \subset W_1$, hence can be viewed as a point in P_1; and as such, clearly $\rho_\omega(c) = (1, \omega)$. Thus altogether the image $\rho(H \times C)$ is a proper open subset of W_1, the union of the two sets just described. This provides an explicit determination of precisely which homomorphisms $\rho \in \text{Hom}\,(\Gamma, A(1, C))$ can be the representations of some affine structures on some compact Riemann surfaces of genus 1. Note that this provides a solution to the purely topological problem of determining for which homomorphism $\rho \in \text{Hom}\,(\Gamma, A(1, C))$ there exists a local homeomorphism $f : C \to C$, not necessarily complex analytic, satisfying (2); a more direct, purely topological solution of this problem might be interesting.

3. Turning then to the case of projective structures on compact Riemann surfaces, there is a unique such structure on P_1, and it is easy to verify that any projective structure on a surface of genus 1 is equivalent to an affine structure; hence it can be assumed that M is a marked Riemann surface of genus $g > 1$. The universal covering space $\tilde{M}$ can then be identified with the unit disc Δ in the complex plane; and Γ then becomes a group of projective transformations acting on Δ with $\Delta/\Gamma = M$.

The Schwarzian differential operator $\theta(f) = (f''/f')' - 1/2(f''/f')^2$ plays the role for projective structures that the differential operator f''/f' played for affine structures. The characteristic properties of the Schwarzian differential operator are that $\theta(f) = 0$ precisely when f is a projective transformation, and that $\theta(f \circ g) = \theta(f)(g')^2 + \theta(g)$. From these properties it is easy to see that a complex analytic local homeomorphism f satisfies (1) precisely when $\theta(f)(Tz) \cdot (T'(z))^2 = \theta(f)(z)$, hence precisely when $\phi(z) = \theta(f)dz^2$ represents a quadratic differential on $\Delta/\Gamma = M$; and conversely any quadratic differential on $\Delta/\Gamma = M$ can be written in the form $\theta(f)dz^2$ for some complex analytic local homeomorphism f satisfying (1), the most general such function f being $\sigma \circ f$ for any projective transformation σ. Thus the set of equivalence classes of projective structures on M can be put into one-to-one correspondence with the $(3g-3)$-dimensional space of quadratic differentials on M.

The quadratic differentials $\phi(z)$ on M cannot be considered as being very explicitly known, and solutions f of the Schwarzian differential equation $\theta(f)dz^2 = \phi(z)$ are even less explicitly known; and that complicates the study of projective structures. Nonetheless some interesting general properties of the mappings $f : \Delta \to P_1$ can be established. The mapping f is not always a covering mapping; indeed it is a covering mapping only for a set of projective structures corresponding to a compact subset of the vector space of quadratic differentials, [10], [11]. The following three conditions are equivalent: (i) f is a covering mapping; (ii) the image $f(\Delta)$ is a proper subset of P_1; (iii) the group $\rho(\Gamma)$ is a properly discontinuous group of transformations acting on the image $f(\Delta)$, [5], [10]. As in the affine case, whenever $\rho(\Gamma)$ is a properly discontinuous group of transformations on $f(\Delta)$ the quotient space $f(\Delta)/\rho(\Gamma)$ is a compact Riemann surface, and f induces an analytic mapping $f^* : M \to f(\Delta)/\rho(\Gamma)$; and again this is not necessarily a one-to-one mapping, [11]. Most of the further questions one can ask about these mappings f and f^* remain open.

Turning next to the representations of these projective structures, the set $\mathrm{Hom}\,(\Gamma, \mathrm{PL}(1, \mathrm{C}))$ can be given the structure of a complex analytic variety just as was done for the corresponding set in the case of affine structures. If $A_1, \cdots, A_g$, $B_1, \cdots, B_g$ are the standard generators of Γ, subject to the relation $C_1 \cdots C_g = I$ where $C_j = A_j B_j A_j^{-1} B_j^{-1}$, then an element $\rho \,\epsilon\, \mathrm{Hom}\,(\Gamma, \mathrm{PL}(1, \mathrm{C}))$ is determined by the $2g$ elements $X_j = \rho(A_j)$, $Y_j = \rho(B_j)$ of $\mathrm{PL}(1, \mathrm{C})$; and these can be arbitrary elements of $\mathrm{PL}(1, \mathrm{C})$, subject to the condition imposed by the defining relation. Thus $\mathrm{Hom}\,(\Gamma, \mathrm{PL}(1, \mathrm{C})$ can be identified with the complex analytic subvariety

$$(9) \qquad V = \{(X_j, Y_j) \,\epsilon\, \mathrm{PL}(1, \mathrm{C})^{2g} : X_1 Y_1 X_1^{-1} Y_1^{-1} \cdots X_g Y_g X_g^{-1} Y_g^{-1} = I\}$$

of the complex manifold $\mathrm{PL}(1, \mathrm{C})^{2g}$; this is a subvariety with singularities. The affine transformations $A(1, \mathrm{C})$ can be viewed as forming a complex submanifold of $\mathrm{PL}(1, \mathrm{C})$; and the subset $V_0 = \{Z(X_j, Y_j)Z^{-1} : X_j, Y_j \,\epsilon\, A(1, \mathrm{C}), Z \,\epsilon\, \mathrm{PL}(1, \mathrm{C})\}$ is an analytic subvariety of the product manifold $\mathrm{PL}(1, \mathrm{C})^{2g}$, with the property that the complement $V_1 = V - V_0$ is a $(6g-3)$-dimensional complex analytic manifold. Furthermore the quotient space $W_1 = V_1 / \mathrm{PL}(1, \mathrm{C})$, where $\mathrm{PL}(1, \mathrm{C})$ acts as a group of analytic automorphisms of V_1 by conjugating all the elements X_j, Y_j simultaneously, has the natural structure of a $(6g-6)$-dimensional complex analytic manifold, [6], [7].

Identifying the vector space of quadratic differentials on M with C^{3g-3}, it follows readily from well-known properties of the Schwarzian differential operator that the mapping $\rho_M : \mathrm{C}^{3g-3} \to V$, which associates to each quadratic differential the representation of the associated projective structure, is a complex analytic mapping. The image must lie in the open subset $V_1 \subset V$, since any $\rho \,\epsilon\, V_0$ is equivalent to an affine representation and M admits no affine structures; so this mapping actually extends to a holomorphic mapping $\rho_M : \mathrm{C}^{3g-3} \to W_1$. Using the properties of the Schwarzian differential operator again, it is not difficult to see that this mapping ρ is a nonsingular injective holomorphic mapping, [4], [9].

Thus distinct projective structures on M have distinct representation classes; and the image $\rho(C^{3g-3})$ is probably a complex submanifold of W_1 isomorphic as a complex manifold to C^{3g-3}, although it is not yet known to be a closed subset of W_1.

The set of all marked Riemann surfaces of genus g is parametrized by the Teichmüller space T_g, a $(3g-3)$-dimensional complex manifold; and there is a natural complex analytic vector bundle Q of rank $3g-3$ over T_g such that the quadratic differentials on a marked Riemann surface M can be identified with the fibre of Q over the point representing M, [2]. There is a complex analytic mapping $\rho: Q \to W_1$ such that the restriction of ρ to the fibre over the point representing M is just the mapping ρ_M; and it can be demonstrated that this mapping ρ is a nonsingular complex analytic mapping, [3]. Thus the image $\rho(Q) \subset W_1$, the set of all those equivalence classes of homomorphisms $\rho \in \mathrm{Hom}\,(\Gamma, \mathrm{PL}(1, C))$ which are representations of some projective structures on some compact Riemann surfaces of genus g, is an open subset of W_1. As in the affine case this mapping $\rho: Q \to W_1$ is not injective; perhaps the most interesting illustration of this is provided by the simultaneous uniformizations described by L. Bers, [1].

The problem of determining the image $\rho(Q) \subset W_1$ is really the purely topological one of determining for which homomorphisms $\rho \in \mathrm{Hom}(\Gamma, \mathrm{PL}(1,C))$ there exist local homeomorphisms $f: \Delta \to P_1$, not necessarily complex analytic, satisfying (1); but analytic methods provide several partial results. First, since Q is connected the image $\rho(Q) \subset W_1$ must be contained in a connected component of W_1; and this is a nontrivial result, since W_1 is not connected. To analyze this situation more closely, recall that every projective transformation can be represented by a 2×2 complex matrix of determinant 1, leading to the exact sequence of groups

$$0 \longrightarrow Z/2Z \longrightarrow \mathrm{SL}(2, C) \overset{\phi}{\longrightarrow} \mathrm{PL}(1, C) \longrightarrow 0\ ;$$

and the homomorphism ϕ induces a mapping

$$\phi : \mathrm{Hom}\,(\Gamma, \mathrm{SL}(2, \mathbf{C})) \to \mathrm{Hom}\,(\Gamma, \mathrm{PL}(1, \mathbf{C}))\,.$$

The set $\mathrm{Hom}\,(\Gamma, \mathrm{SL}(2, \mathbf{C}))$ can be given the structure of a complex analytic subvariety $V^* \subset \mathrm{SL}(2, \mathbf{C})^{2g}$, paralleling the imposition of the complex structure V on the set $\mathrm{Hom}\,(\Gamma, \mathrm{PL}(1, \mathbf{C}))$; and in these terms the mapping $\phi : V^* \to V$ is a complex analytic mapping. The inverse image $V_1^* = g^{-1}(V_1)$ is the set of irreducible linear representations in $\mathrm{Hom}\,(\Gamma, \mathrm{SL}(2, \mathbf{C}))$, and it has the structure of a $(6g{-}3)$-dimensional complex manifold. The quotient space $W_1^* = V_1^*/\mathrm{SL}(2, \mathbf{C})$, where $\mathrm{SL}(2, \mathbf{C})$ acts by conjugation, has the structure of a $(6g{-}6)$-dimensional complex manifold, [6], [7]. It is not hard to see that W_1^* is a connected complex manifold; the traces of the appropriate elements give coordinates by means of which this can be demonstrated, following the approach of [8] extended naturally to the complex case. Now using yet again the properties of the Schwarzian differential operator, it can be seen that $\rho(Q) \subset \phi(W_1^*) \subset W_1$, [4], [5]; thus each $\rho \,\epsilon\, \mathrm{Hom}\,(\Gamma, \mathrm{PL}(1, \mathbf{C}))$ that is the representation of a projective structure on M lifts to a representation $\rho^* \epsilon \mathrm{Hom}(\Gamma, \mathrm{SL}(2, \mathbf{C}))$. That is the essence of the connectivity condition.

The condition that each representation of a projective structure on M lifts to a linear representation provides a connection between the projective structures on M and the complex analytic vector bundles over M, since each $\rho^* \,\epsilon\, \mathrm{Hom}\,(\Gamma, \mathrm{SL}(2, \mathbf{C}))$ determines a complex analytic vector bundle of rank 2 over M; and that leads to some interesting further results, the statements of which do not require any knowledge of the properties of complex analytic vector bundles. These results principally involve an analytic invariant of ρ^* called the *divisor order* which can be defined as follows. There are infinitely many pairs of meromorphic functions f_1, f_2 on Δ such that the vector-valued meromorphic function $F = \begin{pmatrix} f_1 \\ f_2 \end{pmatrix}$ satisfies $F(Tz) = \rho^*_T F(z)$ for all $T \,\epsilon\, \Gamma$. For each such function F it is clear that there are complex analytic functions g, h on Δ having no common zeros on Δ, such that gf_1/h, gf_2/h are also complex analytic functions on Δ having no common zeros on Δ; and $\operatorname{order}_M F = \operatorname{order}_M h - \operatorname{order}_M g$ is independent of the choice of these functions g, h, where $\operatorname{order}_M g$ is the total number of zeros of g, counting multiplicities, in a fundamental domain for the action of Γ on Δ. The divisor order of ρ^* is defined to be $\operatorname{div} \rho^* = \max_F \operatorname{order}_M F$; and it can be shown that it is a finite integer, in the range

$$\left[\frac{1-g}{2}\right] \leqq \operatorname{div} \rho^* \leqq g-1 ,$$

where $[x]$ is the greatest integer $\leq x$, although these inequalities are not needed here. Then the representations $\rho^* \in \operatorname{Hom}(\Gamma, SL(2, C))$ corresponding to projective structures on M are precisely those representations for which $\operatorname{div} \rho^* = g-1$, [5]. The explicit determination of the divisor order of ρ^* is as yet generally impossible, since few relations between the divisor order and any other invariants of the representation are known. However it is easy to see that $\operatorname{div} \rho^* \leqq 0$ whenever ρ^* is a unitary representation; and that implies immediately that no unitary representation can correspond to a projective structure on any Riemann surface, [6]. It should be mentioned in passing that the most spectacular result about the divisor order is the rather deep converse assertion that $\operatorname{div} \rho^* \leqq 0$ implies ρ^* is analytically equivalent to a unitary representation, [15].

It is tempting to conjecture that all equivalence classes of representations $\rho^* \in \operatorname{Hom}(\Gamma, SL(2, C))$ correspond to some projective structures on some compact Riemann surfaces of genus g, except for those representations already shown to be excluded; the excluded representations are the reducible representations and the unitary representations, and any representations that become reducible or unitary when restricted to any subgroup of finite index in Γ.

4. The preceding discussion can be extended to cover structures with possible branching. A *branched projective structure* on M is described

by a complex analytic mapping $f : \tilde{M} \to P_1$, not necessarily a local homeo-morphism, satisfying (1); and a *branched affine structure* on M is described correspondingly by a complex analytic mapping $f : \tilde{M} \to C$ satis-fying an equation of the form (1) but with $\rho_T \in A(1, C)$. Thus the only difference between regular and branched structures is that in the latter case the mapping f is allowed to have branch points. The notions of equivalent structures, and of the representations of structures, are intro-duced just as in the case of unbranched structures. These branched struc-tures were studied by Mandelbaum in [12], [13], [14]. (It should be pointed out though that Mandelbaum uses a slightly different terminology when considering affine structures. What are here called branched affine struc-tures he calls regular branched affine structures; his branched affine structures are the more general ones for which f is a mapping to P_1 rather than merely to C.) To discuss the branched affine structures an approach rather different from that used by Mandelbaum will be followed here.

If $f : \tilde{M} \to C$ is a complex analytic mapping such that for any $T \in \Gamma$

$$(10) \qquad f(Tz) = a_T f(z) + b_T$$

for some constants $a_T \in C^*$, $b_T \in C$, then the differential $\phi(z) = df(z)$ is a complex analytic differential form on $\tilde{M}$ such that for any $T \in \Gamma$

$$(11) \qquad \phi(Tz) = a_T \phi(z) \, ;$$

such a differential is called a *Prym differential* on M associated to the representation $a \in \mathrm{Hom}\,(\Gamma, C^*)$, where a assigns to any $T \in \Gamma$ the value $a_T \in C^*$. Conversely if ϕ is a Prym differential on M associated to the representation $a \in \mathrm{Hom}\,(\Gamma, C^*)$ then there are holomorphic functions f on $\tilde{M}$ such that $df = \phi$, since $\tilde{M}$ is simply connected; and any such f will satisfy an equation of the form (10) for some constants b_T. Any two of these integrals f differ by a constant, hence yield equivalent branched affine structures; and more generally, integrals of any two

nonzero scalar multiples of ϕ differ by an affine transformation, hence
yield equivalent branched affine structures. Therefore there is a one-to-
one correspondence between equivalence classes of branched affine structures on M with representations of the form $\rho_T(z) = a_T z + b_T$ for a
fixed $a \in \mathrm{Hom}\,(\Gamma, C^*)$, and points of the projective space associated to
the vector space of Prym differentials for the representation a. The
vector space of these Prym differentials has dimension g or $g-1$,
according as the representation a is or is not analytically trivial, in
the sense of representing an analytically trivial line bundle over M, [4];
hence there exist branched affine structures on any Riemann surface of
genus $g > 0$. The zeros of the Prym differential $\phi = df$ are precisely
the branch points of f, and the total order of the differential ϕ on M
is just the total branching order of the restriction of f to any fundamental
domain for the action of the group Γ on $\tilde{M}$; and that order is equal to
$2g-2$ for any representation $a \in \mathrm{Hom}\,(\Gamma, C^*)$, [4]. Thus when $g = 1$ a
branched affine structure always reduces to an unbranched affine structure;
so further consideration can be limited to Riemann surfaces of genus
$g > 1$. The situation is rather more complicated than that for the case
$g = 1$.

Although the mappings $f : \Delta \to C$ can be somewhat complicated, it is
easy to see that the image $f(\Delta)$ is either all of C or all of C except
for a single point; and this illustrates a recurrent dichotomy in the discussion of these structures. First, if $f(\Delta)$ is all of C except for a single
point then after replacing f by the composition of f with an affine
transformation it can be assumed that $f(\Delta) = C^*$. Since $\rho(\Gamma)$ must preserve C^*, all the transformations ρ_T reduce to the form $\rho_T(z) = a_T z$;
and hence $f(Tz) = a_T f(z)$ for all $T \in \Gamma$. That implies that the representation $a \in \mathrm{Hom}\,(\Gamma, C^*)$ is analytically trivial; and moreover $f(z) =
\exp w(z)$ where $dw(z)$ is an abelian differential on M, viewed as a
Γ-invariant differential form on Δ. Since $w(z)$ is readily seen to take
on all values, it follows that $f(C)$ does equal C^*, so this possibility
does exist. Next, if $f(\Delta)$ omits at least two points of C then the family
of complex analytic functions $\{f(Tz): T \in \Gamma\}$ is a normal family in Δ;

and consequently the image group $\rho(\Gamma)$ is a subgroup of $A(1, C)$ having compact closure. It is easy to verify that the only compact subgroups of $A(1, C)$ are finite cyclic subgroups or are conjugates of the subgroup $\{z \to az : |a| = 1\}$; hence after conjugation all the transformations ρ_T reduce to the form $\rho_T(z) = a_T z$. However this reduces to the case considered before, in which $f(\Delta) = C^*$, contradicting the assumption that $f(\Delta) \neq C^*$. Thus the only case in which $f(\Delta) \neq C$ is that in which $a \in \mathrm{Hom}\,(\Gamma, C^*)$ is analytically trivial and f is equivalent to the function exhibiting this triviality, that is to say, $f(z) = ag(z) + \beta$ where $a \neq 0$ and $g(Tz) = a_T\, g(z)$ for all $T \in \Gamma$; and in this exceptional case $f(\Delta)$ is the complement of a single point in C.

This exceptional case is also the only one for which the branched affine structure is not uniquely determined by its representation class. For if f_1, f_2 are two complex analytic mappings $f_j : \Delta \to C$ such that $f_j(Tz) = a_T f_j(z) + b_T$ then the difference $g = f_1 - f_2$ is a complex analytic function such that $g(Tz) = a_T\, g(z)$; thus $a \in \mathrm{Hom}\,(\Gamma, C^*)$ is analytically trivial and g exhibits this triviality. Conversely whenever $a \in \mathrm{Hom}\,(\Gamma, C^*)$ is analytically trivial and g exhibits this triviality, then the branched affine structure determined by any mapping $f : \Delta \to C$ for which $f(Tz) = a_T f(z) + b_T$ has the same representation as the branched affine structure determined by the mapping $f + g$; but these are not equivalent branched affine structures. Of course it is possible to modify the notion of equivalence to avoid this exception; but the modified notion is less natural while the exceptional case is exceptional in many other ways as well.

Turning next to the question whether $\rho(\Gamma)$ is a properly discontinuous group of transformations, it is convenient to consider three separate cases. The general classification of properly discontinuous groups of complex analytic automorphisms of C or of C^* is simple and will be assumed known; details can be found in L. R. Ford's *Automorphic Functions*, for example. (i) The first case is that in which $f(\Delta) = C$ and $\rho(\Gamma)$ is the lattice subgroup of C generated by the translations $z \to z + 1$, $z \to z + \omega$, where $\mathrm{Im}\,\omega > 0$; the mapping $f : \Delta \to C$ exhibits $M = \Delta/\Gamma$ as a branched analytic covering of the torus $C/\rho(\Gamma)$, and the function f is an abelian integral. Note that in this case the representation $a \in \mathrm{Hom}(\Gamma, C^*)$ is the identity representation. The condition that there exists such a mapping is of course readily expressed in terms of the period matrix of the abelian differentials on M. (ii) The second case is that in which $f(\Delta) = C^*$ and $\rho(\Gamma)$ is a purely multiplicative group, as in the case of unbranched affine structures; again $f : \Delta \to C^*$ exhibits $M = \Delta/\Gamma$ as a branched analytic covering of the torus $C^*/\rho(\Gamma)$, and $f = \exp w$ where w is an abelian integral. Note that in this case the representation $a \in \mathrm{Hom}(\Gamma, C^*)$ is analytically trivial. The condition that there exists such a mapping is again readily expressed in terms of the period matrix of the abelian differentials on M. (iii) The third case is that in which $f(\Delta) = C$ and $\rho(\Gamma)$ is the extension of a lattice subgroup of C by a cyclic group of order ν, where $\nu = 2, 3, 4$, or 6; thus $\rho(\Gamma)$ consists of transformations of the form $z \to \varepsilon^k z + m + n\omega$ for arbitrary integers k, m, n, where $\varepsilon^\nu = 1$ and $\omega = \varepsilon$ if $\nu > 2$. The mapping $f : \Delta \to C$ exhibits $M = \Delta/\Gamma$ as a branched covering of the one-dimensional projective space $P_1 = C/\rho(\Gamma)$. Note that in this case the representation $a \in \mathrm{Hom}(\Gamma, C^*)$ is never analytically trivial, indeed is a homomorphism for which the image $a(\Gamma)$ is a cyclic subgroup of C^* of order ν. There is a subgroup $\Gamma_0 \subset \Gamma$ of finite index for which the restriction $\rho(\Gamma_0)$ is a lattice subgroup, so the induced branched affine structure on $M_0 = \Delta/\Gamma_0$ is of the type considered in case (i); and M is the quotient of M_0 by a cyclic group of order ν. In all other cases, in particular whenever the representation $a \in \mathrm{Hom}(\Gamma, C^*)$ is not analytically trivial and the values a_T are not ν-th roots of unity for $\nu = 2, 3, 4$, or 6, the group $\rho(\Gamma)$ does not act discontinuously on $f(\Delta)$.

Letting A_j, B_j be the canonical generators of Γ corresponding to the marking of the Riemann surface M, the set $\mathrm{Hom}(\Gamma, A(1, C))$ can be given the structure of a complex analytic variety as in the other cases considered before. Indeed associating to any $\rho \in \mathrm{Hom}(\Gamma, A(1, C))$ the coordinates (a_j, a_j', b_j, b_j') where $\rho_{A_j}(z) = a_j z + b_j$, $\rho_{B_j}(z) = a_j' z + b_j'$, establishes a one-to-one correspondence between the set $\mathrm{Hom}\,(\Gamma, A(1, C))$ and the complex analytic subvariety

$$(12) \qquad V = \left\{ (a_j, a_j', b_j, b_j') : \sum_j \,[(a_j-1)b_j' - (a_j'-1)b_j] = 0 \right\}$$

contained in $(C^*)^g \times (C^*)^g \times C^g \times C^g$; this is an irreducible analytic subvariety of dimension $4g-1$, the only singularity being at the point $a_j = a_j' = 1$, $b_j = b_j' = 0$ corresponding to the identity representation. This reduces to the subvariety (4) when $g = 1$. Conjugation of $\mathrm{Hom}(\Gamma, A(1, C))$ by an element $\sigma \,\epsilon\, A(1, C)$ of the form $\sigma(z) = az + \beta$ has the effect of transforming (a_j, a_j', b_j, b_j') to

$$(13) \qquad \sigma(a_j, a_j', b_j, b_j') = (a_j, a_j', ab_j - \beta(a_j-1), ab_j' - \beta(a_j'-1)) ;$$

and this exhibits $A(1, C)$ as a group of complex analytic automorphisms of V. To describe a complex structure on the quotient space it is convenient to introduce the auxiliary $2 \times 2g$ complex matrix

$$(14) \qquad M = \begin{pmatrix} a_1-1, \cdots, a_g-1, \; a_1'-1, \cdots, a_g'-1 \\[2mm] b_1 \;\;, \cdots, b_g \;\;, \;\; b_1' \;\;, \cdots, b_g' \end{pmatrix}$$

and to decompose V into the three subsets $V = V_0 \cup V_1 \cup V_2$ where $V_\nu = \{(a_j, a_j', b_j, b_j') \,\epsilon\, V : \mathrm{rank}\, M = \nu\}$; note that each subset V_ν is mapped to itself under the action of any automorphism $\sigma \,\epsilon\, A(1, C)$. Here V_0 is just the singular point of V, a separate orbit by itself. The discussion of the subset V_1 exactly parallels that in the case $g = 1$; thus the orbit space $W_1 = V_1/A(1, C)$ can be described as the complex manifold of dimension $2g$ arising from $(C^*)^{2g}$ by blowing the point $a_j = a_j' = 1$ up to the projective space P_{2g-1} of dimension $2g-1$. On the subset V_2 each orbit is a complex submanifold of V_2 of the form $\{(a_j, a_j', ab_j - \beta(a_j-1), ab_j' - \beta(a_j'-1) : a \,\epsilon\, C^*, \beta \,\epsilon\, C\}$, hence as a complex manifold is equivalent to $C^* \times C$; the orbit through any point (a_j, a_j', b_j, b_j') is the product of the point $(a_j, a_j') \in (C^*)^{2g}$ with a two-dimensional linear subspace $L \subset C^{2g}$ containing the complex line joining the point $(a_j-1, a_j'-1) \in C^{2g}$ to the origin. Note that when $(a_j-1, a_j'-1) = (1, 0, \cdots, 0) \in C^{2g}$ each such two-dimensional linear space L can be described by a point $(0, z_2, \cdots, z_{2g}) \in L$ for which not all of the coefficients $z_2, \cdots, z_{2g}$ are zero, and two such points $(0, z_2, \cdots, z_{2g})$ and $(0, z_2', \cdots, z_{2g}')$ describe the same linear subspace precisely when $(z_2', \cdots, z_{2g}') = (cz_2, \cdots, cz_{2g})$ for some $c \in C^*$; thus this set of linear subspaces is parametrized by the complex projective space P_{2g-2} of dimension $2g-2$. Since the points $(a_j-1, a_j'-1)$ can be reduced to the form $(1, 0, \cdots, 0)$ by nonsingular linear changes of coordinates in C^{2g}, and these changes can moreover be taken to be complex analytic functions of the variables a_j, a_j' locally, it follows that $W_2 = V_2/\mathrm{Aut}\,(1, C)$ can be described as an analytic subvariety of dimension $4g-3$ contained in a complex analytic P_{2g-2}-bundle over $(C^*)^{2g}$.

The identity homomorphism $\rho \in V_0$ cannot be the representation of any branched affine structure on M, since $C/\rho(\Gamma)$ is not compact. For a homomorphism $\rho \in V_1$ either ρ_T is a pure translation for each $T \in \Gamma$ (in case $a_j = a_j' = 1$), or after suitable conjugation $\rho_T(z) = a_T z$ for each $T \in \Gamma$. In these cases ρ is the representation of a branched affine structure $f: \Delta \to C$ on M only when $f(z) = w(z)$ or $f(z) = \exp w(z)$ for some abelian differential dw on M; thus the possible representations ρ can be determined explicitly from the period matrix of the abelian differentials on M. The only homomorphisms $\rho \in V_1$ that can be the representations of some branched affine structures on some Riemann surfaces M are those for which the quotients $C/\rho(\Gamma)$ or $C^*/\rho(\Gamma)$ are compact; but it is not clear that enough is yet known about the possible period matrices of Riemann surfaces to show that all these homomorphisms can be representations of some branched affine structures. Determining which homomorphisms $\rho \in V_2$ are representations of branched affine structures on M leads to the difficult problem of determining the period classes of Prym differentials on M. Again it is tempting to conjecture that all $\rho \in V_2$ for which $C/\rho(\Gamma)$ is compact are representations of some branched affine structures on some Riemann surfaces.

5. Mandelbaum's approach to these branched structures involves examining the meromorphic functions f''/f' or $\theta(f)$, and leads to interesting results about the branch points of the mapping f; details can be found in his papers [12], [13], [14]. Let it suffice here, in conclusion, to report that he demonstrates that any $\rho \in \mathrm{Hom}\,(\Gamma, PL(1, C))$ is the representation of some branched projective structure on some Riemann surface. What is much more interesting, though, is his result that on a fixed compact Riemann surface M of genus $g > 1$, any irreducible homomorphism $\rho \in \mathrm{Hom}\,(\Gamma, PL(1, C))$ for which $\mathrm{div}\,\rho = k$ is the representation of a branched projective structure with branching order $2g - 2 - 2k$; and conversely, if ρ is irreducible and is the representation of a branched projective structure with branching order $2g - 2 - 2k$ then $\mathrm{div}\,\rho \geqq k$, and this is actually an equality if $k \geq 0$, [14].

REFERENCES

[1]   L. Bers. Simultaneous uniformization. Bull. American Math. Soc. *66*(1960), 94-97.

[2]   ________. Fiber spaces over Teichmüller spaces. Acta Math. *130* (1973), 89-126.

[3]   C. J. Earle. On variation of projective structures. This volume.

[4]   R. C. Gunning. *Lectures on Riemann Surfaces*. Princeton University Press, (Mathematical Notes 2), 1966.

[5]   ________. Special coordinate coverings of Riemann surfaces. Math. Annalen *170*(1967), 67-86.

[6]   ________ . *Lectures on Vector Bundles over Riemann Surfaces*. Princeton University Press, (Mathematical Notes 6), 1967.

[7]   ________ . Analytic structures on the space of flat vector bundles over a compact Riemann surface. *Several Complex Variables II*, Maryland, 1970. Springer Lecture Notes *185*(1971), 47-62.

[8]   H. Helling. Diskrete Untergruppen von $SL_2(R)$. Invent. Math. *17* (1972), 217-229.

[9]   I. Kra. On affine and projective structures on Riemann surfaces. J. d'Analyse Math. *22*(1969), 285-298.

[10] ________ . Deformations of Fuchsian groups, I, II. Duke Math. J. *36* (1969), 537-546 and *38*(1971), 499-508.

[11] I. Kra and B. Maskit. Remarks on projective structures. This volume.

[12] R. Mandelbaum. Branched structures on Riemann surfaces. Trans. American Math. Soc. *163*(1972), 261-275.

[13] __________. Branched structures and affine and projective bundles on Riemann surfaces. Trans. American Math. Soc. *183*(1973), 37-58.

[14] __________. Unstable bundles and branched structures on Riemann surfaces. Math. Annalen *214*(1975), 49-59.

[15] M. S. Narasimhan and C. S. Seshadri. Stable and unitary vector bundles on a compact Riemann surface. Annals of Math. *82*(1965), 540-567.

BOUNDARY STRUCTURE OF THE MODULAR GROUP

W. J. Harvey[*]

§0. *Introduction*

In this note we introduce a simplicial structure for the collection of
simple loops in a surface, with the main purpose in mind being to provide
an appropriate combinatorial framework for studying the geometry of how
the modular group $\Gamma(S)$ of a surface S acts at infinity on the Teichmüller
space $T(S)$. A model for such a study is the paper of Borel and Serre [4],
which analyses the structure of arithmetic groups by adding suitable
boundary components to the relevant homogeneous spaces.

We content ourselves here with a description of the basic facts and
some simple deductions. A detailed study of the implied algebraic structure for the modular group and a related geometric description of the com-
pactified space of moduli will appear elsewhere. The main point which
emerges is the remarkable closeness of the analogy with arithmetic groups;
underneath the definite lack of homogeneity in the complex analytic char-
acter of T_g, there lies concealed a beautiful real analytic structure
which is mirrored in the action of Γ_g.

[*]Preliminary report presented at the Riemann surfaces conference in July 1978
at S.U.N.Y., Stony Brook. Partial support for the work was provided by NSF Grant
MCS 77-18723 A01. The author is grateful for the hospitality of the Institute for
Advanced Study during preparation of this manuscript.

I want to record here my indebtedness to J.-P. Serre, both for rendering palatable to my tender stomach the rich diet of buildings and arithmetic groups and for demonstrating that one should not be put off by locally infinite phenomena. After all, think of the structure of a cusp!

§2. *Partitions of a surface*

Let S be a surface of finite type (g, n). We denote by $\mathscr{P}(S)$ the set of all *partitions* of S. These are systems Λ of disjoint simple loops in S, such that

(i) no loop in Λ bounds either a disc in S or a single boundary component of S;

(ii) no pair of loops form the boundary of an annulus.

The group of all homeomorphisms of S acts naturally on $\mathscr{P}(S)$, and the orbit of Λ consists of all partitions which determine topologically equivalent ways of dissecting the surface. Normally we regard as identical two partitions that are equivalent by a homeomorphism isotopic to the identity, and our aim is to understand how the group $\Gamma(S)$ of isotopy classes of homeomorphisms acts on the classes of $\mathscr{P}(S)$.

We observe that $\mathscr{P}(S)$ is partially ordered by inclusion, so that it is possible to assemble $\mathscr{P}$ as an abstract simplicial complex by taking singleton loops as vertices with edges corresponding to pairs of loops, the edge joining the vertices representing the individual loops of the pair, and so on. For a k-simplex, which represents a partition by $k+1$ loops, there will be $k+1$ faces, each a $(k–1)$-simplex obtained by omission of one loop.

The resulting simplicial complex is denoted $\widetilde{\mathscr{T}}(S)$ or $\mathscr{T}$. It plays in the theory the role of Tits building, although it lacks certain essential features of that object.

PROPOSITION 1. $\widetilde{\mathscr{T}}(S)$ *is a thick chamber complex of dimension* $N–1$ $(N = 3g{+}n–3)$, *on which* $\Gamma(S)$ *operates simplicially. The quotient is a finite complex.*

[A *chamber complex* (see for instance [11]) of dimension m is a simplicial complex with the property that every simplex is a face of some m-simplex. It is termed *thick* if each (m–1)-simplex abutts at least three m-simplices (chambers).]

Any partition extends to a maximal one with $3g+n-3$ loops, which determines a chamber. This can be done in infinitely many ways if Λ has fewer than N loops since some subsurface of $S\backslash\Lambda$ must have negative Euler characteristic and thus contains infinitely many distinct simple loops.

Therefore $\mathcal{T}$ is locally infinite. We shall abuse notation by using the same symbol for the geometric realization of $\mathcal{T}$, which is the space of all functions λ from the vertex set $\mathcal{T}_0$ to $[0, 1]$ whose support is a simplex and which are such that the sum of the λ-values is 1. Equipped with the weak topology, $\mathcal{T}(S)$ is a C–W complex, on which $\Gamma(S)$ acts by the rule

$$\lambda \mapsto \lambda \circ g^{-1} ,$$

for $\lambda \epsilon \mathcal{T}$ and $g \epsilon \Gamma$. This action is certainly simplicial, and a finite connected subcomplex $\mathcal{K} \subseteq \mathcal{T}$ containing a simplex from every Γ-orbit is easily obtained by choosing one partition for each of the finite number of ways in which S can be dissected into a collection of 3-holed spheres. After passing to the barycentric subdivision of $\mathcal{T}$ (and $\mathcal{K}$), one can construct a precise fundamental domain; the quotient $\mathcal{T}/\Gamma$ is therefore a triangulable finite complex.

EXAMPLES. (a) In genus 1 with 1 hole $\mathcal{T}$ is a discrete set of vertices. The quotient is a point. (b) If S has genus 2, there are two orbits of 2-simplices which represent the following partitions:

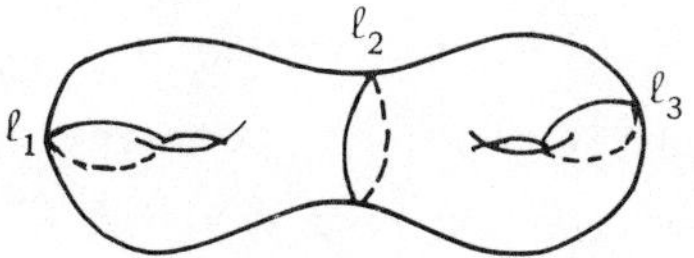

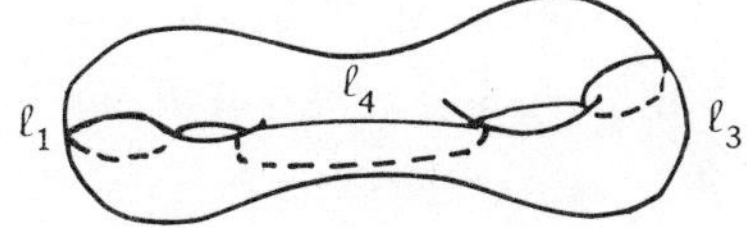

In the barycentric subdivision of $\mathcal{K}$, we find the shaded fundamental region.

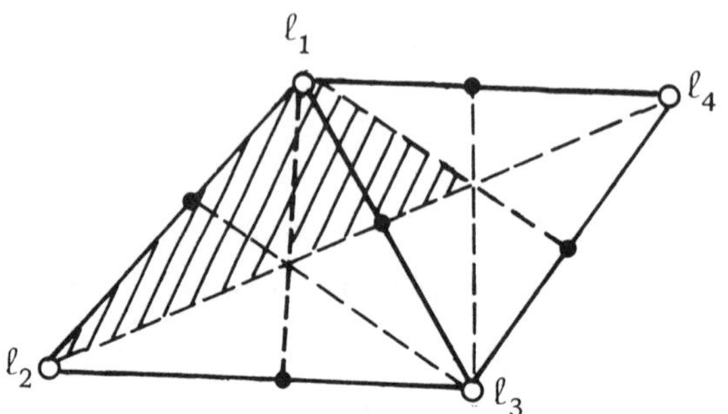

Notice that all the loops are fixed setwise up to homotopy by the hyperelliptic involution.

We end this section by stating the following elementary yet powerful result (cf. Birman's discussion in [D.G] Chapter 6).

PROPOSITION 2. $\mathcal{T}(S)$ *is connected if it has positive dimension.*

The proof consists in determining, for a given pair of loops ℓ, ℓ' in S (with intersection number $m > 0$), how to walk from ℓ to ℓ' by a sequence of loops each disjoint from its predecessor. If $m = 1$, this is two steps when S has type $(g, n) \neq (1, 1)$, and if $m > 1$, an inductive argument due to Lickorish [9] applies.

Since this implies that $\Gamma(S)$ is generated by stability groups of vertices, one obtains by induction on the topological type of S the corollary that $\Gamma(S)$ is generated by a finite number of Dehn twists. A more careful study yields the precise form of Lickorish's result, as sharpened recently by Humphries, that $2g + 1$ twists are sufficient to generate $\Gamma(S_{g,0})$.

§2. *The cuspidal boundary of Teichmüller space*

A method used by several authors ([1], [2], [6], [7]) to construct a compacification of $T(S)/\Gamma(S)$ has important connections with the complex $\mathcal{T}(S)$ of §1. It stems from the observation that to any given partition Λ of S

there is associated a way to degenerate the surface into a singular topological space, known as a *surface with nodes*, or *stable surface*, by
shrinking of the loops in Λ to points. Formally, we make the following
definition.

DEFINITION. The *stable surface* S_Λ of a partition Λ is the topological
space $S/\sim_\Lambda$, obtained from the equivalence relation $\sim_\Lambda$ determined by
the rule

$$x \sim_\Lambda y \iff x \text{ and } y \text{ lie on the same loop of } \Lambda .$$

The projection of loops in Λ to S_Λ determines the set of *nodes* in S_Λ .

Note that homotopically equivalent partitions determine isomorphic
stable surfaces; again we shall abuse notation by identifying these.

The partial ordering on partitions induces a partial ordering on stable
surfaces, according to which there exists a morphism from S_Λ to $S_{\Lambda'}$
whenever $\Lambda \subset \Lambda'$ —here a *morphism* is a continuous map whose fibers are
either points or simple loops, disjoint and homotopically distinct from the
nodes of S_Λ . For our purposes we regard all boundary components of S
as points and require morphisms to fix each one.

Now one associates to each partition Λ of S a *Teichmüller space* of
stable surfaces $T_\Lambda = T(S_\Lambda)$ which classifies marked complex structures
on S_Λ . A natural way to parametrize T_Λ comes from the well-known
Fenchel-Nielsen coordinates, taken with respect to some maximal partition
Λ_0 containing Λ . These determine a homeomorphism from $T(S)$ onto
the product of $N = 3g+n-3$ copies of $\mathfrak{h} = R \times R_+$, well defined after
choice of an ordering $\tilde{\Lambda}_0$, with the R_+-components representing the set
of lengths for the minimal geodesics in the homotopy classes of loops in
Λ_0 , measured in the Riemannian metric on S given by the choice of complex structure $S_x \in T(S)$, and R-components giving the *twist parameters*
which describe the procedure for assembling S_x from the component parts
of $S \backslash \Lambda_0$ in terms of angular deviation from a fixed one. For more details,
the reader is referred to [D.G] Chapter 9 and further references to be found
there.

The various faces of the corner R_+^N obtained by setting Λ-lengths to 0 are now to be viewed as the result of shrinking S by the relation $\sim_\Lambda$ for the various $\Lambda \subset \Lambda_0$ which amounts to identifying a given face with a parametrization of the product of T-spaces for the parts of $S \setminus \Lambda$ *with the Λ-twist parameters added.* In view of the naturality of the procedures and the mutual compatibility of changing to different maximal partitions, it turns out that one can adjoin to $T(S)$ in this fashion the various boundary components ∂T_Λ for all $\Lambda \in \mathscr{P}(S)$, with appropriate identifications arising from the partial ordering corresponding to the procedure of collapsing $T_{\Lambda'}$ onto $T_\Lambda \subset \partial T_{\Lambda'}$ when $\Lambda' \subset \Lambda$. This is the *cuspidal boundary structure* for $T(S)$ which we denote $\partial T(S)$.

THEOREM 1. *The bordified space* $T \cup \partial T$ *is a connected Hausdorff real analytic manifold with boundary, on which the Teichmüller modular group* Γ *acts properly discontinuously.*

THEOREM 2. *There is a weak homotopy equivalence between* $\partial T(S)$ *and the complex* $\mathscr{T}(S)$ *equivariant with respect to the two* $\Gamma(S)$*-actions.*

The boundary structure is closely related to the boundaries studied by Abikoff, by Bers and by Earle and Marden, and all have a common root in Mumford's work on stable curves and the studies of Bers and Maskit [2, 10] on Kleinian groups. The main novelty here is that one blows up one real dimension for each degeneration curve, thereby rendering the action of $\Gamma(S)$ proper, since the stability group of a boundary component $T_\Lambda \times R^{\#(\Lambda)}$ contains the group of Dehn twists about the loops of Λ and these act as translations on the corresponding twist parameter space $R^{\#(\Lambda)}$.

§3. *Comments*

The Fenchel-Nielsen coordinates are in some respects more naturally defined by using the reciprocal of the lengths of Λ-loops as R_+ variables. It is then possible to make canonical the Λ-shrinking process as passage to the cusp at ∞ for the translation group of Dehn twists about Λ-curves, and there is then a natural interpretation in terms of the action of $SL_2(R)^N$ on $\mathfrak{h}^N$.

As a final remark we observe that it is an easy consequence of Theorem 2 that the quotient moduli space $\mathscr{X}(S) = T(S) \cup \partial T(S)/\Gamma(S)$ is a compact real analytic space with boundary. Known properties of $\Gamma(S)$ imply (see e.g. [8]) that there is a finite covering of $\mathscr{X}(S)$ that is a smooth manifold. This result does not extend immediately to the complex analytic Mumford compactification of moduli space.

REFERENCES

[D.G] Discrete Groups and Automorphic Functions (edited by W. J. Harvey), Acad. Press (London), 1977.

[1] W. Abikoff, "Degenerating families of Riemann surfaces," Ann. of Math 105 (1977), 29-44.

[2] L. Bers, "On boundaries of Teichmüller spaces and on Kleinian groups I," Ann. of Math. 91 (1970), 570-600.

[3] ________ , "Spaces of degenerating Riemann surfaces," in Ann. of Math Studies no. 79 (1974), 43-55.

[4] A. Borel and J.-P. Serre, "Corners and Arithmetic Groups," Comment. Math. Helv. 48 (1973), 436-491.

[5] P. Deligne and D. Mumford, "The irreducibility of the space of curves of given genus," Publ. Math. I.H.E.S. 36 (1969).

[6] C. J. Earle and A. Marden, unpublished (but see reference [D.G], Chapter 8).

[7] W. J. Harvey, "Chabauty spaces of discrete groups," in Ann. of Math Study no. 79 (1974), 239-246; see also Chapter 9 of [D.G].

[8] ________ , "Geometric structure of surface mapping-class groups," in Homological Group Theory (ed. C. T. C. Wall), London Mathematical Society Lecture Notes #36, Cambridge University Press (1979), 255-269.

[9] W. B. R. Lickorish, "A finite set of generators for the homeotopy group of a 2-manifold," Proc. Comb. Phil. Soc. 60 (1964), 769-778. (Also corrigendum, *ibid.*, 62 (1966), 679-681.)

[10] B. Maskit, "On boundaries of Teichmüller spaces and on Kleinian groups II," Ann. of Math. 91 (1970), 607-639.

[11] J. Tits, Buildings of spherical type and finite B–N pairs, Lecture Notes in Math *386*, Springer Verlag 1974.

A REALIZATION PROBLEM IN THE
THEORY OF ANALYTIC CURVES

Maurice Heins

1. My talk at the Conference was based on a paper [3] that appeared in
the volume of the Bulletin of the Greek Mathematical Society dedicated
to the memory of Christos Papakyriakopoulos. For this reason, the
present note has the restricted objective of summarizing the paper and of
indicating two further realization problems that may be treated with the
aid of the methods of the paper. We start with these related questions.

2. One of these is the problem of characterizing up to conformal equivalence a Weierstrass class [2] taken with its center map. The answer is
simply stated. The pairs, (Weierstrass class, associated center function),
are conformally equivalent to *exactly* the pairs (S, f), where S is a noncompact Riemann surface and f is a locally simple analytic function on
S. By the Gunning-Narasimhan solution [1] of the problem of Karl Stein
concerning the existence of locally simple analytic functions on noncompact Riemann surfaces, all noncompact Riemann surfaces admit locally
simple analytic functions.

Consequently, there exists a Weierstrass class conformally equivalent
to a given noncompact Riemann surface. Further, for a given pair (S, f)
there exists g analytic on S such that (S, f, g) is equivalent to an

analytic entity ($=$ analytische Gebilde) so that the Weierstrass class induced by (S, f, g) does not admit adjunction of poles or algebraic elements.

On communicating this result to Robert Gunning at the Conference I learned from him that essentially the same question had been proposed to him by Barry Simon of Princeton University. I am indebted to Robert Gunning for this information and for also stating to me a second problem proposed by Barry Simon, namely:

Given a region $\Omega \subset C$, does there exist a holomorphic map of Ω into C^2 whose first projection is the identity map on Ω and which is maximal in the sense that it is not representable as the composition of a (univalent) holomorphic map of a Riemann surface S into C^2 and a proper injective holomorphic map of Ω into S?

This problem has an affirmative answer as does its generalization which replaces Ω by a noncompact Riemann surface and takes the first projection as a given analytic function on S.

It is planned to give a unified account of these questions in the Proceedings of the projected 1979 Durham Instructional Conference (to appear in the L.M.S. Series of the Academic Press).

3. The problem treated in my talk is the following:

Let F be holomorphic on C^2, not the constant 0, but taking the value zero somewhere. Such F will be termed *allowed*. Let A be an analytic entity annihilating F in the sense that with c and v denoting respectively the center and value functions of A the equality $F[c(p), v(p)] = 0$ holds for all $p \in A$ such that $(c(p), v(p)) \in C^2$. One asks for the conformal equivalence classes containing such A. The answer is simple: all. Given a Riemann surface S, there exist an allowed F and an analytic entity A that is conformally equivalent to S and annihilates F.

The facts are classical for compact S. It suffices to refer to the results concerning the problem of Riemann and Klein, cf. [2].

To treat the case where S is not compact we proceed as follows. We construct with the aid of the theorem of Behnke and Stein and Pick-Nevanlinna interpolation theory on Riemann surfaces analytic functions f and g on S which have the following properties:

(1) $f^{-1}(\{|z|<r\})$ is not empty and its components are relatively compact, $0<r<+\infty$.

(2) Let Ω be a component of $f^{-1}(\{|z|<r\})$. Then f has a zero on each component of $S-\overline{\Omega}$.

(3) $0 \notin g(S)$.

(4) For some $a \in C$ such that f has multiplicity one at each point of $f^{-1}(\{a\})$ it is the case that $g|f^{-1}(\{a\})$ is univalent.

(5) The infinite product

$$(3.1) \qquad \prod_{f(p)=z} \left[1 - \frac{w}{g(p)}\right]^{n(p;f)}$$

is uniformly convergent on $\{|z| \leq r\} \times \{|w| \leq r\}$, $0<r<+\infty$.

We see that the infinite product defines an allowed F which is annihilated by the pair (f, g). Further, it is concluded with the aid of (1), (2), and (4) that (S, f, g) is equivalent to an analytic entity taken together with its center and value functions. These facts permit us to conclude that the analytic entity induced by (S, f, g) annihilates F. We conclude that every conformal equivalence class of Riemann surfaces contains an annihilator of an allowed F.

UNIVERSITY OF MARYLAND
COLLEGE PARK, MARYLAND 20742

REFERENCES

[1] Gunning, R. C. and Narasimhan, R., Immersion of open Riemann surfaces. Math. Ann. 174, 103-108 (1967).

[2] Heins, M., Complex Function Theory. New York and London. Academic Press, 1968.

[3] _________, A realization problem in the theory of analytic curves. Bull. Gr. Math. Soc., Papakyriakopoulos Memorial Volume 1978.

THE MONODROMY OF PROJECTIVE STRUCTURES

John H. Hubbard

Introduction

In this paper we shall give a new proof of the result, due to Hejhal [5], that the map associating to an isomorphism class of projective structures its conjugacy class of monodromy homomorphisms is a local homeomorphism.

We shall follow the following plan: show that the domain (Prop. 1) and the range (Prop. 4) are manifolds, identify their tangent spaces (Prop. 2 and 4), and compute the derivative of the map above. It turns out that one space is an Eichler cohomology space and the other is the cohomology of a group; they are canonically isomorphic by a classical theorem of algebraic topology. The derivative is the canonical isomorphism.

The idea of using differential calculus on this problem is not new: both Earle [2] and Gunning [4] have proposed similar proofs; this paper explains the appearance of Eichler cohomology in their computations. Many of the other results I establish in this paper were already known to Hejhal, Kra, Gunning, Maskit, Earle, Weil and no doubt others. The exposition is, I hope, in the spirit of Gunning's book and in fact the paper is largely a matter of putting parameters in arguments appearing there.

I wish to thank Earle, Douady, Kra, and Gunning for helpful conversations, and the N.S.F. for financial support during part of the preparation of this paper.

NOTATION.

$\mathbf{P}^1$ is the complex projective line (the Riemann Sphere)

$G = \mathrm{PGL}_2(\mathbf{C}) = \mathrm{Aut}\ \mathbf{P}^1$; $A \subset G$ is the subgroup of affine maps

$z \mapsto az + b$.

$\mathcal{G} = \mathrm{pgl}_2(\mathbf{C}) = $ Space of analytic vector fields on $\mathbf{P}^1$.

The adjoint action of G on $\mathcal{G}$ corresponds to the direct image of the corresponding vector fields.

We will speak of the fundamental group of a space only after a universal covering space has been chosen; the fundamental group is then the group of automorphisms of the universal covering space. All universal covering maps will be denoted u.

1. *Projective structures*

A *projective atlas* on a Riemann surface X is an open cover U_i of X and analytic maps $a_i : U_i \to P_1$ which are homeomorphisms onto their images such that $a_j \circ a_i^{-1}$ is the restriction to $a_i(U_i \cap U_j)$ of an element of G. Two projective atlases are equivalent if together they form a projective atlas; a *projective structure* on X is an equivalence class of projective atlases.

EXAMPLES. (i) If X is compact of genus ≥ 2 and H is the upper half plane, there is a covering map $u : H \to X$ by the uniformization theorem. Sections of u over simply-connected open subsets of X define a projective atlas.

(ii) Other planar covering spaces of X, such as the Schottky covering space, can be used to describe projective structures.

(iii) If $\Gamma \subset \mathbf{C}$ is a lattice and $X = \mathbf{C}/\Gamma$, appropriate restrictions of the canonical coordinate z of $\mathbf{C}$ define a projective structure on X as above; restrictions of e^{az} also do for any $a \in \mathbf{C} - \{0\}$. In these cases, the changes of coordinates are affine; a projective structure which can be defined by an affine atlas is called an *affine structure*.

Let a be a projective structure on a Riemann surface X, and $\tilde{X}$ be a universal covering space of X, with $u : \tilde{X} \to X$ the covering map.

LEMMA 1. (i) *There exists an analytic map* $f : X \to \mathbf{P}^1$ *such that on any contractible open subset* $U \subset X$ *the composition* $f \circ u^{-1}$ *is a projective chart. Any other such map is of the form* $\sigma \circ f$ *for some* $\sigma \in G$.

(ii) *To every such* f *there corresponds a unique homomorphism* $\rho_f : \pi_1(X) \to G$ *such that* $\rho_f(\gamma) \circ f = f \circ \gamma$, *and* $\rho_{\sigma \circ f} = \sigma \circ \rho_f \circ \sigma^{-1}$.

Proof. Cover X by open subset U_i on which u is injective, and such that there exist projective charts $a_i : u(U_i) \to \mathbf{P}^1$; let $\beta_i = a_i \circ u^{-1}$. Then $\sigma_{ij} = a_j \circ a_i^{-1}$ is a 1-cocycle on X with values in G. Since X is contractible, this cocycle is a coboundary, after refining the cover if necessary, and there exist $\sigma_i \in G$ such that $\sigma_{ij} = \sigma_j^{-1} \circ \sigma_i$. Then on $U_i \cap U_j$, $\sigma_i \circ a_i = \sigma_j \circ a_j$ so all the $\sigma_i \circ a_i$ are restrictions of a global map $f : X \to \mathbf{P}^1$ with the appropriate properties. The second part of (i) is obvious.

In any U_i there is a homomorphism $\rho_{f,i} : \pi_1(X) \to G$ such that $\rho_{f,i}(\gamma) \circ f(x) = f(\gamma(x))$ for $x \in U_i$, since both $f \circ u^{-1}$ and $f \circ \gamma \circ u^{-1}$ are projective coordinates on X. But it is clear from analytic continuation that $\rho_{f,i}(\gamma) \circ f = f \circ \gamma$ on all of $\tilde{X}$, since $\tilde{X}$ is connected and both sides are analytic functions of x.   Q.E.D.

Such a map f is called a developing map of X; ρ_f is the corresponding monodromy homomorphism.

We will need the following fact:

LEMMA 2. *A projective structure on a compact Riemann surface is equivalent to an affine structure if and only if the surface is of genus* 1.

Proof. See [3], p. 173. The result is purely topological, essentially saying that if a surface admits an affine structure, the cotangent bundle is trivial.   Q.E.D.

COROLLARY. *If* X *is a compact Riemann surface of genus* ≥ 2 *and* α *is a projective structure on* X *, then the monodromy homomorphism* ρ_f *for any developing map* f *has non-commutative image.*

Proof. Any commutative subgroup of G is conjugate by an appropriate $\sigma \in G$ to a subgroup of the affine group $A \subset G$. The projective atlas formed by maps of the form $\sigma \circ f \circ u^{-1}$ on contractible open subsets of X is affine.   Q.E.D.

The real interest of projective structures is the geometry of developing maps and the monodromy homomorphism. Beyond this corollary there is little to be said in general; the developing maps may fail to be covering spaces of their images, the monodromy homomorphisms may fail to be isomorphisms, and their images may fail to be discrete. In fact all of these pathologies occur for the family given in example (iii) for appropriate values of a.

2. *The Schwarzian derivative and the affine structure of* P(X)

Let U be a Riemann surface, $x \in U$, and f, g meromorphic functions on U such that $f'(x) \neq 0$, $g'(x) \neq 0$. There exists a unique $\sigma \in G$ such that f and $\sigma \circ g$ agree to order 2 at x. Then $d^3(f - \sigma \circ g)(x)$ is naturally a cubic map $T_x U \to T_{f(x)} P^1$, and $f'(x)^{-1} \circ d^3(f - \sigma \circ g)(x)$ is a cubic map $T_x U \to T_x U$. But for any one dimensional complex vector space V, the cubic maps $V \to V$ correspond naturally to the quadratic maps $V \to C$. Therefore the construction above defines a quadratic form $S(f, g)(x)$ on $T_x U$, and it is easy to see that $S(f, g)$ is a meromorphic quadratic differential form on U, holomorphic at those points x where $f'(x) \neq 0$ and $g'(x) \neq 0$.

If $U \subset C$ and z is the canonical coordinate on C, then $S(f, z) = f'' f' - \frac{3}{2}(f'')^2/(f')^2$, as the classical definition requires.

For any Riemann surface U, let Q(U) be the space of holomorphic quadratic forms on U.

LEMMA 3. *The Schwarzian derivative has the following properties* :

(i) $S(f, g) = S(f, \sigma \circ g) = S(\sigma \circ f, g)$ *for all* $\sigma \in G$.

(ii) $S(f, g) = 0$ *if* $f = \sigma \circ g$, *and conversely* $f = \sigma \circ g$ *if* $S(f, g) = 0$
       and U *is connected.*

(iii) $S(f, g) + S(g, f) = S(f, h)$, *and* $S(f, g) = - S(g, f)$.

(iv) *If* U *is simply connected,* f *is schlicht on* U *and* $q \in Q(U)$,
       there exists a solution g *schlicht on* U *to the equation*
       $S(f, g) = q$.

Proof. Parts i, ii, iii are obvious. Part (iv) is similar to Lemma 1. Q.E.D.

Suppose the projective structures α and β on X are defined by
atlases (U_i, α_i) and (V_j, β_j). Then the quadratic forms $S(\alpha_i, \beta_j)$, defined in $U_i \cap V_j$ coincide on open sets of form $U_{i_1} \cap U_{i_2} \cap V$ so they
are induced by a quadratic form $q = \alpha - \beta$ on X.

Conversely, if $\alpha \in P(X)$ and $q \in Q(X)$, there is a unique projective
structure $\beta \in P(X)$ such that $\beta - \alpha = q$, we shall denote it $q + \alpha$. If
(U_i, α_i) is an atlas defining α and the U_i are simply connected, then β
may be defined by (U_i, β_i) where the β are solutions of $S(\beta_i, \alpha_i) = q$
in U_i, which exist by Lemma 3, (iv).

LEMMA 4. *The map* $Q(X) \times P(X) \to P(X)$ *given by* $(q, a) \mapsto q + a$ *makes*
P(X) *into an affine space under* Q(X).

Proof. All that is left to show is that P(X) is not empty. This follows
from the uniformization theorem, as in the example (i) §1, or from Riemann-
Roch as in [3], p. 172.   Q.E.D.

COROLLARY. *The space* P(X) *is canonically a complex manifold, and*
for all $a \in P(X)$, *we have* $T_\alpha P(X) = Q(X)$.

3. *Relative projective structures*

Let $\pi : X \to S$ be a smooth family of compact Riemann surfaces parametrized by a complex manifold S (i.e. a proper analytic submersion with

fibers $X(s) = \pi^{-1}(s)$ of dimension 1). A relative projective atlas on X is a relative atlas (U_i, a_i) where the U_i form an open cover of X, and the $a_i : U_i \to \mathbf{P}^1$ are analytic maps where restrictions to fibers of π are isomorphisms onto their images, such that over each $s \in S$, the pair $(U_i(s), a_i(s))$ is a projective atlas on $X(s)$.

As above, two relative projective atlases are equivalent if together they still form a projective atlas, and a relative projective structure on X is an equivalence class of relative projective atlases.

REMARK. To say that a family of projective structures $a(s)$ is induced by a relative structure is to say that $a(s)$ depends analytically on s.

EXAMPLES. The family of projective structures obtained by applying the uniformization theorem fiber by fiber *does not* define a relative projective structure: the normalization requiring the images of the universal covering spaces to be the upper half plane cannot be made analytic.

The generalization of the uniformization theorem given by Bers [1] does give relative projective structures on the universal curve over Teichmüller space.

The canonical family of projective structure on $P(X) \times X \to P(X)$ is induced by a relative projective structure.

Let $P_S(X)$ be the set of pairs (s, a) such that $s \in S$ and a is a projective structure on $X(s)$.

PROPOSITION 1. (i) *There is a unique structure of a complex manifold on* $P_S(X)$ *such that the projection* $\rho : P_S(X) \to S$ *given by* $(s, a) \mapsto s$ *is analytic, and analytic families of projective structures on* X *given by section of* p *are induced by relative projective structures on* X.

(ii) *The action of* $Q_S(X)$ *on* $P_S(X)$ *over* S *given by* $((s, q), (s, a)) \mapsto (s, q+a)$ *makes* $P_S(X)$ *into an analytic affine bundle over* S, *under the analytic vector bundle* $Q_S(X)$.

Proof. The proposition is clearly local in S. Suppose a relative projective structure a on X can be found (even locally over small subset of S).

Then the map $(s, q) \to (s, q + a(s))$ is a bijection $Q_S(X) \to P_S(X)$. Give $P_S(X)$ the induced structure; with this structure, $P_S(X)$ clearly satisfies (ii).

To see that it satisfies (i), we need to know that if q is a section of $Q_S(X)$, then the family of projective structures induced by $a(s) + q(s)$ is induced by a relative projective structure.

Let (U_i, a_i) be a relative projective atlas defining a. On $U_i(s)$, the equation $S(a_i(s), \beta_i) = q(s)$ is an analytic differential equation of third order depending analytically on s, whose solutions will exist in $U_i(s)$ by Lemma 3, (iv) if the U_i are sufficiently small, and will depend analytically on s if initial conditions are chosen analytic in s. But this can be done, for instance by picking (locally in S) a section $S \to X$ of π and requiring $\beta_i(s)$ to coincide with $a_i(s)$ to order 2 along the section. Clearly the β_i form a relative projective atlas with the desired properties.

Thus we are left with showing that over sufficiently small open subsets of S, X carries a relative projective structure. This may be shown by appealing to the universal property of Teichmüller space and the simultaneous uniformization theorem of Bers [1]; we shall prove a slightly more general result.

LEMMA 5. *Let* $\pi: X \to S$ *be a proper and smooth family of Riemann surfaces of genus at least* 2, *with* S *a Stein manifold. Then* X *admits relative projective structures.*

Proof. Let (U_i, ϕ_i) be any relative atlas (relative atlases exist by the implicit function theorem). On the $U_i \cap U_j$, consider the section of $\Omega_{X/S}^{\otimes 2}$ given by $S(\phi_i(s), \phi_j(s))$. These define a 1-cochain with values in $\Omega_{X/S}^{\otimes 2}$ for the cover $\{U_i\}$ which is a cocycle by Lemma 2, iii.

But the Leray spectral sequence of π gives an exact sequence

$$0 \to H^0(S, R^1\pi_*\Omega_{X/S}^{\otimes 2}) \to H^1(X, \Omega_{X/S}^{\otimes 2}) \to H^1(S, \pi_*\Omega_{X/S}^{\otimes 2})$$

and the first term is zero because $H^1(X(s), \Omega^{\otimes 2}) = 0$ by Riemann-Roch; the last term is zero because S is Stein.

Therefore refining the cover if necessary, we may assume that there are sections q_i of $\Omega_{X/S}^{\otimes 2}$ over U_i such that

$$q_i - q_j = S(\phi_i, \phi_j) .$$

Solutions a_i of the differential equations $S(a_i, \phi_i) = q_i$ chosen so as to satisfy some analytic initial condition such as to agree with ϕ_i to order 2 along some section $S \to U_i$ of π, will then form a projective atlas for X, perhaps after further refining the cover to make them injective on fibers.

                                        Q.E.D.

COROLLARY. (i) *The family of projective structures on the family of Riemann surfaces* $p^*X \to P_SX$ *which is* a *on the fiber over* $a \in P_SX$ *is induced by a canonical relative projective structure* $a_S(X)$.

(ii) *The space* P_SX *has the following universal property*: *The map which associates to any analytic mapping* $f : T \to P_S(X)$ *the projective structure* $f^*a_S(X)$ *on the family of Riemann surfaces* $(p \circ f)^*X$ *is a bijection of* $\mathrm{Mor}(T, P_S(X))$ *onto the set of relative projective structures on* $(p \circ f)^*X$.

The proof is left to the reader.

4. *Infinitesimal deformations and Eichler cohomology*

In this paragraph, we shall carry out an infinitesimal deformation theory for projective structures analogous to the Kodaira-Spencer theory for complex structures.

Let U be a Riemann surface, a be a projective structure on U and χ an analytic vector field on U. Choose a one-parameter family of maps $\phi_t : U \to U$ with $\phi_0 = \mathrm{id}$ and $\phi_0' = \chi$, and define the Lie derivative of a in the direction χ

$$L_\chi(a) = \lim_{t \to 0} \frac{\phi_t^*(a) - a}{t} .$$

Clearly $L_\chi(a)$ is an analytic quadratic form on U, and we leave it to the reader to prove that if ζ is a projective coordinate on U and $\chi = \chi(\zeta)\frac{\partial}{\partial\zeta}$, then $L_\chi(a) = \chi'''(\zeta)d\zeta^2$. In particular, the Lie derivative does not depend on the family ϕ_t that was chosen.

Let Λ_a be the subsheaf of the sheaf of germs of analytic vector fields which is the kernel of the morphism $\psi \mapsto L_\chi(a)$; we then obtain an exact sequence of sheaves

$$0 \longrightarrow \Lambda_a \longrightarrow TU \xrightarrow{\;L(a)\;} \Omega^{\otimes 2} \longrightarrow 0$$

which will be important; TU stands for the sheaf of germs of vector fields on U (as opposed to the tangent bundle TU), and the Lie derivative is surjective because of the formula that computes it in a projective coordinate.

REMARKS. In a projective coordinate ζ, sections of Λ_a are exactly those vector fields which can be written $p(\zeta)\frac{\partial}{\partial\zeta}$ with p a polynomial of degree at most two; such vector fields are called infinitesimal automorphisms of a because the flows they generate send a to itself.

The sheaf Λ_a is locally constant of rank 3, an example of what topologists call a local system. In particular, it may be thought of as the sheaf of germs of sections of a covering space, with fiber isomorphic to $\mathbf{C}^3$ with the discrete topology.

Let $X \to S$ a family of Riemann surfaces, $s_0 \in S$ and X_0 the surface above it. Suppose a is a relative projective structure on X which induces the projective structure a_0 on X_0. We shall describe a linear map $T_{s_0}S \to H^1(X_0, \Lambda_{a_0})$ which measures the infinitesimal deformation of a.

Let (U_i, a_i) be a relative projective atlas on X; by restricting S and refining the cover $\mathfrak{U} = \{U_i\}$ we may assume that for each i and all $s \in S$ the maps $a_i(s): U_i(s) \to \mathbf{P}^1$ are homeomorphisms onto some open set $V_i \subset \mathbf{P}^1$. Define $\phi_i(s) = a_i(s)^{-1}\circ a_i(s_0)$ and $\phi_{i,j}(s) = \phi_i(s)^{-1}\circ\phi_j(s)$, where $\phi_{i,j}(s)$ is defined in an open subset of $U_i \cap U_j$ which will include any given point for s sufficiently near s_0. The maps $\phi_{i,j}(s)$ satisfy the following two identities:

$$\phi_{i,j}(s) \circ \phi_{j,k}(s) = \phi_{i,k}(s)$$

$$\phi_{i,j}(s)^* \alpha_0 - \alpha_0 = 0 ,$$

the first obviously and the second because (U_i, α_i) is a relative *projective* atlas.

Since $\phi_{i,j}(s_0)$ is the identity map $U_i \cap U_j \to U_i \cap U_j$, the derivative $d_{s_0} \phi_{i,j}(v) = \chi_{i,j}(v)$ is a vector field on $U_i \cap U_j$ for any $v \in T_{s_0} S$, and the derivatives of the identities above give:

$$\chi_{i,j}(v) + \chi_{j,k}(v) = \chi_{i,k}(v)$$

$$L_{\chi_{i,j}}(\alpha_0) = 0 .$$

The first identity says that $\chi(v) = \{\chi_{i,j}(v)\}$ is a cocycle in $C^1(\mathcal{U}, T X_0)$, and the second that it is in fact in $C^1(\mathcal{U}, \Lambda_{\alpha_0})$. Define the *infinitesimal deformation* of a at s_0

$$d_{s_0} a : T_{s_0} S \to H^1(X_0, \Lambda_{\alpha_0})$$

by $d_{s_0}(a)(v) = $ the cohomology class of $\chi(v)$. We leave it to the reader to prove that the class does not depend on the projective atlas that was chosen.

The map $d_{s_0}(a)$ has the following properties:

(i) *It commutes with change of basis*, i.e., if $f : T \to S$ is a map with $f(t_0) = s_0$ and we give $f^* X$ the relative projective structure $f^* a$, then $d_{t_0}(f^* a) = d_{s_0}(a) \circ d_{t_0} f$.

(ii) *The map* $i_* \circ d_{s_0} a : T_{s_0} S \to H^1(X_0, T X_0)$ *obtained by composing* $d_{s_0} a$ *with the map* $H^1(X_0, \Lambda_{\alpha_0}) \to H^1(X_0, T X_0)$ *induced by the inclusion*

$\Lambda_{a_0} \subset TX_0$ *is the Kodaira-Spencer map classifying the deformation of the complex structure of* X *at* x_0.

(iii) Let X_0, a_0 be any compact Riemann surface with a projective structure, and let a be the canonical relative projective structure on $P(X_0) \times X_0 \to P(X_0)$. *Then* $T_{a_0} P(X_0) = Q(X_0)$, *and* $d_{a_0} a : Q(X_0) \to H^1(X_0, \Lambda_{a_0})$ *is the "connecting homomorphism" coming from the long exact sequence associated to the short exact sequence* (1).

Part (i) follows immediately from the construction, (ii) is clear since the cocycle $\chi(v)$ is a definition of the Kodaira-Spencer map [7], (iii) is a computation we shall leave to the reader.

Now let us examine the universal case; let M be a compact surface of genus at least 2, Θ_M the Teichmüller space modelled on M and $\pi : \Xi_M \to \Theta_M$ the universal Teichmüller curve. We shall omit the subscript M in the sequel.

Let $p : P_\Theta \Xi \to \Theta$ be the canonical projection and give $p^* \Xi$ the canonical relative projective structure a given by the corollary to Proposition 1.

PROPOSITION 2. *Let* θ_0 *be a point in* Θ, $X_0 = \pi^{-1}(\theta_0)$ *the Riemann surface above it and* $a_0 \in P(X_0)$. *Then*

$$d_{a_0} a : T_{a_0} P_\Theta \Xi \to H^1(X_0, \Lambda_{a_0})$$

is an isomorphism.

Proof. Consider the diagram

$$
\begin{array}{ccccccccc}
0 & \longrightarrow & T_{a_0}P(X_0) & \longrightarrow & T_{a_0}P_\Theta\Xi & \xrightarrow{\ d_{a_0}p\ } & T_{\theta_0}\Theta & \longrightarrow & 0 \\
 & & \Big\downarrow{\scriptstyle \mathrm{id}} & & \Big\downarrow{\scriptstyle d_{a_0}a} & & \Big\downarrow{\scriptstyle \wr} & & \\
0 & & H^0(X_0, \Omega^{\otimes 2}) & \longrightarrow & H^1(X_0, \Lambda_{a_0}) & \xrightarrow{\ i_*\ } & H^1(X_0, TX_0) & \longrightarrow & 0
\end{array}
$$

268   JOHN H. HUBBARD

The top line is induced by the inclusion of $P(X_0)$ as the fiber of p above θ_0 (cf. Prop. 1), the bottom line is extracted from the long exact sequence associated to the short exact sequence (1), the left vertical map is the isomorphism of the corollary to Lemma 4 and the right vertical map is the Kodaira-Spencer isomorphism. The left-hand square commutes by properties (i) and (iii) of $d_{\alpha_0}\alpha$ and the right-hand square by property (ii). The proposition now follows from the five lemma.   Q.E.D.

REMARK. Proposition 2 identifies the tangent space to the space $P_\Theta\Xi$ of "all projective structures on all Riemann surfaces" exactly in the same sense that the Kodaira-Spencer isomorphism identifies the tangent space to the space Θ of "all Riemann surfaces" as $T_{\theta_0}\Theta = H^1(X_0, TX_0)$.

5. *The space* $\mathrm{Hom}\,(\Gamma, G)$

Let Γ be the fundamental group of a surface, given by generators $\Sigma = \{a_1, \cdots, a_{2g}\}$ subject to the one relation

$$\prod_{i=1}^{g} [a_i, a_{i+g}] = 1 .$$

Clearly the set $\mathrm{Hom}\,(\Gamma, G)$ may be identified with the subset of G^{2g} defined by the analytic equation $f = 1$ where $f : G^{2g} \to G$ is given by $f(\sigma_1, \cdots, \sigma_{2g}) = \prod_{i=1}^{g} [\sigma_i, \sigma_{i+g}]$. This gives $\mathrm{Hom}\,(\Gamma, G)$ the structure of an analytic space.

LEMMA 6. *With this analytic structure,* $\mathrm{Hom}\,(\Gamma, G)$ *has the following universal property: for any analytic space* S, *morphisms* $S \to \mathrm{Hom}\,(\Gamma, G)$ *correspond bijectively to morphisms* $S \times \Gamma \to G$ *which are analytic, and whose restrictions to* $\{s\} \times \Gamma$ *are group homomorphisms for all* $s \in S$.

REMARK. In this lemma Γ is considered as a discrete analytic space. The proof is trivial and left to the reader. In particular, the analytic structure on $\mathrm{Hom}\,(\Gamma, G)$ does not depend on the chosen presentation. We may therefore expect that there is an intrinsic description of the local structure of $\mathrm{Hom}\,(\Gamma, G)$, in particular of its tangent space, etc. The object of this paragraph is to give such a description.

Let $\mathrm{Hom}^*(\Gamma, G) \subset \mathrm{Hom}\,(\Gamma, G)$ be the open set of representations with non-commutative image.

For any representation $\rho : \Gamma \to G$, we may consider $\mathfrak{g}$ as a Γ-module by $\gamma \cdot \xi = \mathrm{Ad}\,\rho(\gamma)(\xi)$; we shall denote this Γ-module $\mathfrak{g}_\rho$. Recall [6] that a derivation $\delta : \Gamma \to \mathfrak{g}_\rho$ is a map satisfying $\delta(\gamma_1\gamma_2) = \delta(\gamma_1) + \gamma_1 \cdot \delta(\gamma_2)$ and that those derivations of the form $\delta_\xi(\gamma) = \xi - \gamma \cdot \xi$ are called principal derivations. Call $\mathrm{Der}\,(\Gamma, \mathfrak{g}_\rho)$ the space of derivations and $\mathrm{IDer}\,(\Gamma, \mathfrak{g}_\rho)$ the subspace of principal derivations. A classical description of $H^1(\Gamma, \mathfrak{g}_\rho)$ is $\mathrm{Der}\,(\Gamma, \mathfrak{g}_\rho)/\mathrm{IDer}\,(\Gamma, \mathfrak{g}_\rho)$; this is the description we shall use.

The tangent space to G at any σ is $\mathfrak{g}$ (in two different ways); we shall use the local chart $\mathfrak{g} \to G$ of G near σ given by $\xi \mapsto \exp(\xi)\sigma$.

PROPOSITION 3. *The space* $\mathrm{Hom}^*(\Gamma, G)$ *is a submanifold of* G^{2g}.

For any $\rho \in \mathrm{Hom}^*(\Gamma, G)$ *the map* $\mathrm{Der}\,(\Gamma, \mathfrak{g}_\rho) \to \mathfrak{g}^{2g}$ *given by* $\delta \mapsto \delta|_\Sigma$ *is an isomorphism of* $\mathrm{Der}\,(\Gamma, \mathfrak{g}_\rho)$ *onto* $T_\rho \mathrm{Hom}\,(\Gamma, G)$.

Proof. A computation which begins

$$\Pi\,[e^{\xi_i}\sigma_i,\, e^{\xi_{i+g}}\sigma_{i+g}] = e^{\xi_1}(\sigma_1 e^{\xi_{1+g}}\sigma_1^{-1})(\sigma_1\sigma_{1+g}\sigma_1^{-1}e^{-\xi_1}\sigma_1\sigma_{1+g}^{-1}\sigma_1^{-1}) \cdots$$
$$= e^{\xi_1}e^{\mathrm{Ad}\,\sigma_1 \cdot \xi_{1+g}}e^{-\mathrm{Ad}(\sigma_1\sigma_{1+g}\sigma_1^{-1})\cdot\xi_1} \cdots$$

and ends using $e^{X_1}e^{X_2} = e^{X_1+X_2} + 0(|X_1|^2 + |X_2|^2)$ shows that the derivative of f at $\boldsymbol{\sigma} = (\sigma_1, \cdots, \sigma_{2g}) \in G^{2g}$ in the direction $\boldsymbol{\xi} = (\xi_1, \cdots, \xi_{2g}) \in \mathfrak{g}^{2g}$ is

$$d_{\boldsymbol{\sigma}}f(\boldsymbol{\xi}) = \sum_{i=1}^{g} \prod_{j=1}^{i-1} [\sigma_j, \sigma_{j+g}] \cdot ((1 - \sigma_i\sigma_{i+g}\sigma_i^{-1}) \cdot \xi_i + (\sigma_i - [\sigma_i, \sigma_{i+g}]) \cdot \xi_{i+g}).$$

If $\sigma = \rho|_\Sigma$ for some $\rho \,\epsilon\, \mathrm{Hom}\,(\Gamma, G)$, essentially the same computation shows that $d_\sigma f(\xi) = 0$ is the necessary and sufficient condition for $\xi : \Sigma \to \mathfrak{g}_\rho$ to extend (obviously uniquely) to a derivation $\Gamma \to \mathfrak{g}_\rho$.

Thus all we need to prove is that $d_\sigma f : \mathfrak{g}_\rho^{2g} \to \mathfrak{g}_\rho$ is surjective if $\rho \,\epsilon\, \mathrm{Hom}^*(\Gamma, G)$. The basic fact (left to the reader) is that if $\tau_1, \tau_2 \,\epsilon\, G$ do not commute, the linear map $\mathfrak{g}_\rho \times \mathfrak{g}_\rho \to \mathfrak{g}_\rho$ given by $(\xi_1, \xi_2) \mapsto (1 - \tau_1)\xi_1 + (1 - \tau_2)\xi_2$ is surjective.

This result gets applied twice. First suppose that for some i, $1 \le i \le g$, σ_i and σ_{i+g} do not commute. Then since

$$(1 - \sigma_i \sigma_{i+g} \sigma_i^{-1})\xi_i + (\sigma_i - [\sigma_i, \sigma_{i+g}])\xi_{i+g}$$

$$= \sigma_i((1 - \sigma_{i+g})\sigma_i^{-1} \cdot \xi_i + (1 - \sigma_{i+g}\sigma_i^{-1}\sigma_{i+g}^{-1})\xi_{i+g})$$

the images of the ξ_i and ξ_{i+g} already fill out the image of $d_\sigma f$.

If each σ_i commutes with σ_{i+g} the expression for the derivative of f simplifies to

$$d_\sigma f(\xi) = \sum_{i=1}^{g} ((1 - \sigma_{i+g}) \cdot \xi_i + (\sigma_i - 1) \cdot \xi_{i+g}))$$

and the result is clear since for some i, j, σ_i and σ_j do not commute.

$$\mathrm{Q.E.D.}$$

REMARK. This result is a special case of the following more general results: If Γ is a group of finite presentation, G is a Lie group with Lie algebra $\mathfrak{g}$, then $\mathrm{Hom}\,(\Gamma, G)$ is an analytic space, its Zariski tangent space at ρ is $\mathrm{Der}\,(\Gamma, \mathfrak{g}_\rho)$ and the equations defining locally $\mathrm{Hom}\,(\Gamma, G)$ in $\mathrm{Der}\,(\Gamma, \mathfrak{g}_\rho)$ may be chosen to have values in $H^2(\Gamma, \mathfrak{g}_\rho)$. In our case, this boils down to the fact that if the image of ρ is not commutative, $H^2(\Gamma, \mathfrak{g}_\rho) = 0$, which may be proved by Poincaré duality.

PROPOSITION 4. (i) *The group* G *acts freely on* $\mathrm{Hom}^*(\Gamma, G)$, *and the quotient* $\mathrm{Hom}^*(\Gamma, G)/G$ *has a unique structure of an analytic manifold such that the projection* $\mathrm{Hom}^*(\Gamma, G) \to \mathrm{Hom}^*(\Gamma, G)/G$ *is analytic.*

(ii) *For any* $\rho \in \mathrm{Hom}^*(\Gamma, G)$, *the derivative of the inclusion* $G \to \mathrm{Hom}^*(\Gamma, G)$ *given by* $\sigma \mapsto \sigma \circ \rho \circ \sigma^{-1}$ *at* $1 \in G$ *is the map* $\mathfrak{g}_\rho \to \mathrm{Der}(\Gamma, \mathfrak{g}_\rho)$ *given by* $\xi \to \delta_\xi$. *In particular the tangent space to* $\mathrm{Hom}^*(\Gamma, G)/G$ *at the image of* ρ *is canonically isomorphic to* $H^1(\Gamma, \mathfrak{g}_\rho)$.

Proof. The fact that G acts freely follows from the fact that commuting is an equivalence relation on non-trivial elements of G. This may for instance be seen by observing that $\sigma_1 \neq 1$ and $\sigma_2 \neq 1$ commute if and only if they have the same fixed points in $\mathbf{P}^1$.

Similarly, if ρ_1 and ρ_2 are not conjugate, they have neighborhoods U_1 and U_2 such that no element of U_1 is conjugate to an element of U_2, since the fixed points of a non-trivial $\gamma \in G$ vary continuously with γ. Therefore the graph of the equivalence relation is closed, and the quotient is Hausdorff. The existence and uniqueness of the analytic structure follows from the analyticity of the action of G, and the derivative in (ii) is computed from $e^\xi \rho e^{-\xi} = e^{\xi - \rho \cdot \xi} \rho + 0(|\xi|^2)$.   Q.E.D.

6. *Hejhal's theorem*

Let $\pi : X \to S$ be a family of Riemann surfaces with a relative projective structure a. Suppose S is contractible and let $\Gamma = \pi_1(X) = \pi_1(X(s))$. If π admits analytic sections, there are relative developing maps $f : \tilde{X} \to \mathbf{P}^1$, i.e., analytic maps which, restricted to $\tilde{X}(s)$, are developing maps of $a(s)$. This is just a matter of picking an analytic normalization, for instance requiring that f should agree to order 2 with a relative analytic chart along a section.

REMARK. The requirement that π admit analytic sections is too stringent. In fact, there are relative developing maps if S is contractible and Stein. Indeed, the space of all developing maps of the $X(s)$ forms a principle analytic bundle under G over S, and so is trivial by Grauert's theorem if S is Stein and contractible. This applies in particular to the universal family over $P_\Theta \Xi$.

Clearly if $f : X \to \mathbf{P}^1$ is a relative developing map, the associated $\rho_f : S \to \mathrm{Hom}\,(\Gamma, G)$ which associates to each $s \in S$ the monodromy homomorphism of $f(s)$, is analytic.

Let $F : P_\Theta \Xi \to \mathrm{Hom}\,(\Gamma, G)/G$ be induced by the above construction, for the universal family of projective structures parametrized by $P_\Theta \Xi$.

REMARK. The global existence of F does not require Grauert's theorem, because we have divided by the action of G. It does require the contractibility of $P_\Theta \Xi$, so that a universal covering space $\widetilde{p^* \Xi}$ induces a universal covering space $p^* \Xi$ over each fiber of $p^* \Xi \to P_\Theta \Xi$.

THEOREM. *The map* F *is an analytic local homeomorphism.*

The fact that F is analytic follows from the fact that F lifts locally (and even globally by Grauert) to an analytic map $P_\Theta \Xi \to \mathrm{Hom}\,(\Gamma, G)$.

By the corollary to Lemma 2, the image of a monodromy homomorphism is never commutative, and so both the range and the image of F are manifolds, whose tangent spaces we know.

Let $a_0 \in P_\Theta \Xi$ be a projective structure on X_0; let $f : X_0 \to \mathbf{P}^1$ be a developing map for a_0 and $\rho : \Gamma \to G$ its monodromy homomorphism. The theorem will now follow from

LEMMA 6. *The derivative* $d_{a_0} F : H^1(X_0, \Lambda_{a_0}) \to H^1(\Gamma, \mathfrak{g}_\rho)$ *is an isomorphism.*

The two tangent spaces look similar; they are in fact canonically isomorphic by the classical theorem of algebraic topology which says that one way to compute the cohomology of a group Γ with values in a Γ-module is to compute the cohomology of a $K(\Gamma, 1)$, with coefficients in the associated local system in the sense of the following lemma.

LEMMA 7. *The group* Γ *acts on* $\tilde{X}_0 \times \mathfrak{g}_\rho$ *by* $\gamma \cdot (x, \xi) = (\gamma \cdot x, \rho(\gamma)_* \xi)$, *and the map* $\tilde{X}_0 \times \mathfrak{g}_\rho \mapsto \Lambda_{a_0}$ *given by* $(x, \xi) \mapsto (u(x), u_* f^* \xi)$ *induces an isomorphism on the quotient.*

The proof is immediate and left to the reader.

We cannot unfortunately use the canonical isomorphism without explicitly constructing it. There are many ways to do this; the one we shall use here is adapted to our knowledge of the two spaces, one via Čech cocycles and the other via derivations.

It is possible to compute Čech cohomology using a generalization of an open cover: an etale cover. The "open sets" are manifolds U_i and immersions $U_i \to X$, whose images are required to cover X. The intersections $U_i \cap U_j$ must be replaced by the fiber products $U_i \times_X U_j$, and similarly for multiple intersections.

Moreover, Leray's theorem still applies: if the U_i as well as all their fiber products are cohomologically trivial (for whatever sheaf we may be considering) the Čech cohomology for that cover is the cohomology of the sheaf (either in the Čech sense of direct limit over all covers, or via resolutions, or whatever, which are all isomorphic).

We shall apply this to the cover consisting of a single open set $\tilde{X} \to X$. The map $\tilde{X} \times_X \tilde{X} \times_X \tilde{X} \times \cdots \times_X \tilde{X} = X \times \Gamma^n$ given by $(x, \gamma_1, \cdots, \gamma_n) \mapsto (x, \gamma_1(x), \cdots, \gamma_n(x))$ and the identification of Lemma 7 give isomorphisms $C^n(X, \tilde{X}; \Lambda_\alpha) = \mathfrak{g}_\rho^{\Gamma^n}$. In fact the complex is the classical inhomogeneous bar complex [6] whose first two differentials are

$$d_0 : \mathfrak{g}_\rho \to \mathfrak{g}_\rho^\Gamma \quad \text{given by} \quad d_0(\xi)(\gamma) = \xi - \gamma \cdot \xi \,;$$

$$d_1 : \mathfrak{g}_\rho^\Gamma \to \mathfrak{g}_\rho^{\Gamma \times \Gamma} \quad \text{given by} \quad d_1 \chi(\gamma_1, \gamma_2) = \chi(\gamma_1) + \gamma_1 \chi(\gamma_2) - \chi(\gamma_1 \gamma_2).$$

In particular, the kernel of d_1 is formed of the derivations $\Gamma \to \mathfrak{g}_p$ and the image of d_0 is formed of the principal derivations.

In our case, X_0 is a $K(\Gamma, 1)$ so Leray's theorem applies to guarantee that the cohomology of the complex is in fact $H^1(X_0, \Lambda_{\alpha_0})$.

LEMMA 6′. *The derivative* $d_{\alpha_0} F$ *is the isomorphism* $H^1(X_0, \Lambda_{\alpha_0}) \to H^1(\Lambda, \mathfrak{g}_\rho)$ *described above.*

Proof. Choose an analytic curve $a(t)$ in $P_\Theta \Xi$; let f_t be a relative developing map and ρ_t the corresponding monodromy homomorphisms. Define (as in the construction of §4) a family of analytic maps $\phi_t : U_t \to X(t)$ which:

    a) are analytic isomorphisms onto their images, and analytic in t;

    b) are defined in subsets $U_t \subset \tilde{X}_0$ which fill out $\tilde{X}_0$ as t becomes small;

    c) satisfy $f_0 = f_t \circ \phi_t$ in U_t, and $\phi_0 =$ identity of $\tilde{X}_0$.

Then $\frac{d}{dt} a(t)\Big|_{t=0}$ is represented by the Čech cocycle for the cover $\tilde{X}_0$ which is, on the component $\tilde{X}_0 \times \{\gamma\}$ of $\tilde{X}_0 \times_{X_0} \tilde{X}_0$, given by

$$\xi_\gamma = \frac{d}{dt} (\phi_t^{-1} \circ \gamma \circ \phi_t)\Big|_{t=0}.$$

Using $f_t \circ \gamma = \rho_t(\gamma) f_t$ the expression above may be written

$$\xi_\gamma = \frac{d}{dt} (f_0^{-1} \circ \rho_t(\gamma) \circ f_0)\Big|_{t=0}, \quad \text{where the entire expression } f_0^{-1} \circ \rho_t(\gamma) \circ f_0 \text{ is}$$

defined in U_t.

If we write $\frac{d}{dt} \rho_t(\gamma)\Big|_{t=0} = \xi'_\gamma$ (it is best to think of ξ'_γ as a vector field on P^1), then differentiating the expression above gives $\xi_\gamma = f_0^* \xi'_\gamma$. This is the identification of Lemma 7.   Q.E.D.

REMARKS. Some obvious questions, unsolved to the author's knowledge, are:

    What is the image of F?

    What do the fibers of F look like, and their projections in Teichmüller space? It is known [8] that F is not injective, but it is injective on fibers [3].

BIBLIOGRAPHY

[1] L. Bers, *Simultaneous Uniformization*, Bull. A.M.S. 66 (1960), 94-97.

[2] C. Earle, *On Variations of Projective Structures*, these proceedings.

[3] Gunning, R. C., *Lectures on Riemann Surfaces*, Princeton University Press, 1966.

[4] ——————, unpublished notes.

[5] Hejhal, D. A., *Monodromy Groups and Poincaré Series*, Bull. A.M.S. 84 (1978), 339-376.

[6] Hilton, P. J. and Stammbach, U., *A Course in Homological Algebra*, Springer-Verlag, 1971.

[7] Kodaira, K. and Spencer, T., *On Deformations of Complex Analytic Structures, I and II*, Ann. Math. 67 (1958), 328-466.

[8] Maskit, B., *One Class of Kleinian Groups*, Ann. Acad. Sci. Fenn. 442 (1969).

HOLOMORPHIC FAMILIES OF RIEMANN SURFACES
AND TEICHMÜLLER SPACES

Yoichi Imayoshi

Introduction

Let $\mathfrak{S}$ be a two dimensional complex manifold. Denote by D the unit disc $|t| < 1$ and by D^* the punctured disc $0 < |t| < 1$ in the complex t-plane. We assume that a proper holomorphic mapping $\pi: \mathfrak{S} \to D^*$ satisfies the following two conditions; for every $t \in D^*$,

i) the fiber $S_t = \pi^{-1}(t)$ over t is a one-dimensional, non-singular irreducible analytic subset of $\mathfrak{S}$, and

ii) the rank of the Jacobian of π is equal to one at each point of S_t. It is well known that the fiber space $(\mathfrak{S}, \pi, D^*)$ is differentiably locally trivial. Therefore, every fiber S_t has the same genus g as a Riemann surface. We shall always assume that $g \geq 2$.

We call the triple $(\mathfrak{S}, \pi, D^*)$ satisfying above conditions a holomorphic family of compact Riemann surfaces of genus g. A fiber S_t satisfying above two conditions i) and ii) is called ordinary. It should be noted that, in general, the fiber over $t = 0$ cannot be considered.

Further, let $\hat{\mathfrak{S}}$ be a two-dimensional complex analytic space and $\hat{\pi}: \hat{\mathfrak{S}} \to D$ be a proper holomorphic mapping. Assume that the fiber $\hat{S}_0 = \hat{\pi}^{-1}(0)$ over $t = 0$ is a one-dimensional compact analytic subset of of $\hat{\mathfrak{S}}$ and that the triple $(\hat{\mathfrak{S}} - \hat{S}_0, \hat{\pi}|\hat{\mathfrak{S}} - \hat{S}_0, D^*)$ is a holomorphic family as stated above. In this case when the fiber $\hat{S}_0$ is not ordinary, we call the triple $(\hat{\mathfrak{S}}, \hat{\pi}, D)$ a holomorphic family of compact Riemann surfaces of genus g with a singular fiber over $t = 0$.

In this paper, given a holomorphic family $(\mathfrak{S}, \pi, D^*)$, we regard the fiber S_t over t $(t \epsilon D^*)$ as a point in a Teichmüller space and we try to construct the fiber over $t = 0$ from the limit point of $\{S_t\}_{t \epsilon D^*}$ in the Teichmüller space when t tends to zero and try to construct a holomorphic family $(\hat{\mathfrak{S}}, \hat{\pi}, D)$ with or without a singular fiber over $t = 0$, which can be considered as a natural extension of $(\mathfrak{S}, \pi, D^*)$. This $(\hat{\mathfrak{S}}, \hat{\pi}, D)$ will be called a completion of $(\mathfrak{S}, \pi, D^*)$. Here the concept of the homotopical monodromy $\tilde{\mathfrak{M}}$ of $(\mathfrak{S}, \pi, D^*)$, which will be defined later, plays an essential role. If $\tilde{\mathfrak{M}}$ is trivial or of finite order, then the fiber over $t = 0$ can be constructed by a compact Riemann surface without nodes of genus g. On the other hand, if $\tilde{\mathfrak{M}}$ is of infinite order, then the fiber over $t = 0$ can be constructed by a compact Riemann surfaces with nodes of genus g. A precise description of the fiber $t = 0$ is given in Theorems 1, 2, and 4 in §§3, 4, and 6. More properties of the completion of $(\mathfrak{S}, \pi, D^*)$ are given in Theorems 5 and 6.

Nishino [16] has already discussed the quite similar problem to construct the fiber over $t = 0$ and obtained the same results as ours without using the theory of Teichmüller spaces.

The author would like to express his hearty gratitude to Professor Kuroda for his constant encouragement and advice and wish to express his thanks to my colleagues Yamamoto and Sekigawa for immeasurably profitable conversations with them. The author is also indebted to Professor Bers for pointing out some errors in the original version of this paper.

§1. *Preliminaries*

First we introduce terminologies and notations which will be used throughout this paper. These are due to Ahlfors [3], [4] and Bers [6], [7], [8].

Let C be the complex plane and let $\hat{C}$ be the extended complex plane. We denote by U and L the upper and the lower half-planes in C, respectively, and by R the real axis in C.

Let $SL'(2; C)$ be the set of all complex Möbius transformations of the form

$$g: z \mapsto \frac{az+b}{cz+d}, \, a, b, c, d \in C \quad \text{and} \quad ad - bc = 1 \, ,$$

and let $SL'(2; R)$ be the set of all $g \in SL'(2; C)$ whose coefficients are all real.

Let us denote by G a Fuchsian group acting on U and by Q_{norm} the group of all quasiconformal automorphisms w of U satisfying normalization conditions $w(0) = 0$, $w(1) = 1$ and $w(\infty) = \infty$. Here we note that any $w \in Q_{norm}$ can be extended to a homeomorphism of $\bar{U}$ onto itself, where $\bar{U}$ is the closure of U in $\hat{C}$. We set

$$Q_{norm}(G) = \{w \in Q_{norm} \mid wGw^{-1} \subset SL'(2; R)\} \, .$$

Let $L^{\infty}(U)$ be the complex Banach space of (equivalence classes of) bounded measurable functions on U and let $L^{\infty}(U, G)$ be the closed linear subspace of $L^{\infty}(U)$ such that every $\mu \in L^{\infty}(U, G)$ satisfies

$$\mu(g(z))\overline{g'(z)}/g'(z) = \mu(z) \quad \text{for every} \quad g \in G \, .$$

We denote by $L^{\infty}(U)_1$ the unit ball in $L^{\infty}(U)$ and put $L^{\infty}(U, G)_1 = L^{\infty}(U, G) \cap L^{\infty}(U)_1$.

Furthermore, we mean by $B_2(L, G)$ the set of all holomorphic functions ϕ defined on L such that

$$\phi(g(z)) g'(z)^2 = \phi(z) \quad \text{for every} \quad g \in G$$

and such that its norm

$$\|\phi\| = \sup_{z \in L} y^2 |\phi(z)|, \quad z = x + iy$$

is bounded. The space $B_2(L, G)$ is the complex Banach space of holomorphic quadratic differentials in L with respect to the Fuchsian group G.

Now we can define Teichmüller spaces in three ways.

1) Given a reference compact Riemann surface S of genus $g(\geq 2)$, a marked Riemann surface with respect to S is a compact Riemann surface S' of genus g with an orientation-preserving topological mapping $f : S \to S'$, which we call the marking on S'. We denote this marked surface by (S, f, S'). We define an equivalence relation between marked surfaces by calling two marked surfaces (S, f, S') and (S, g, S'') homotopically equivalent if and only if there exists a conformal mapping h such that the diagram

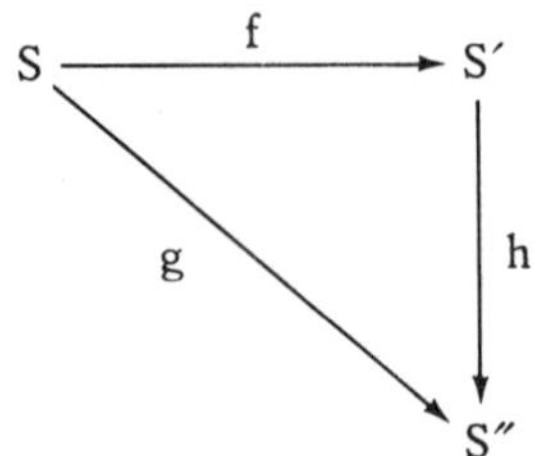

commutes up to homotopy, that is, the self-mapping $g^{-1} \circ h \circ f : S \to S$ is homotopic to the identity. We denote by $[S, f, S']$ an equivalence class of marked surfaces and by $T(S)$ the set of all those equivalence classes, which is the Teichmüller space of S.

2) Let G be the finitely generated Fuchsian group of the first kind acting on U induced by the universal covering group of the compact Riemann surface S of genus g. Two elements w_1 and w_2 in Q_{norm} are called equivalent if $w_1 = w_2$ on R. The equivalence class of $w \in Q_{norm}$ is denoted by $[w]$. The Teichmüller space $T(G)$ of the Fuchsian group G is the set of all equivalence classes $[w]$ of elements $w \in Q_{norm}(G)$.

3) The Teichmüller space $T(G)$ with its complex structure can be regarded canonically as a bounded domain in the complex Banach space $B_2(L, G)$ in the following way: For every $\mu \in L^{\infty}(U)_1$, there is a unique quasiconformal automorphism w of C with $w(0) = 0$, $w(1) = 1$, $w(\infty) = \infty$ such that w has the Beltrami coefficient μ on U and is conformal on L.

We write $w = w^\mu$ and denote by ϕ_μ the Schwarzian derivative of w^μ in L, that is,

$$\phi_\mu = \{w^\mu, z\} = ((w^\mu)''/(w^\mu)')' - \frac{1}{2}((w^\mu)''/(w^\mu)')^2 .$$

Nehari's theorem shows that $\phi_\mu \in B_2(L, I)$ and $\|\phi_\mu\| \leq \frac{3}{2}$, where I is the group consisting of only the identity mapping of $\hat{C}$ onto itself. As is well known, $w^\mu|L, w^\mu|R$ and ϕ_μ depend only on $[w_\mu]$, where w_μ is the quasiconformal automorphism of U with the Beltrami coefficient μ and $w(0) = 0$, $w(1) = 1$, $w(\infty) = \infty$. If $\mu \in L^\infty(U, G)_1$, then $\phi_\mu \in B_2(L, G)$ and the mapping $[w_\mu] \mapsto \phi_\mu$ is a biholomorphic bijection of $T(G)$ onto a holomorphically convex bounded domain in $B_2(L, G)$ containing the open ball of radius $1/2$. From now on, we will identify $T(G)$ with its canonical image in $B_2(L, G)$.

Note that $T(S)$ is canonically biholomorphic to $T(G)$. Therefore, we can also identify $T(S)$ with $T(G)$.

For every holomorphic function ϕ on the lower half-plane L, the Schwarzian differential equation

$$(1.1) \qquad \{w, z\} = (w''/w')' - \frac{1}{2}(w''/w')^2 = \phi$$

has a meromorphic solution defined in L. If $\eta_1(z)$ and $\eta_2(z)$ are two linearly independent solutions of the linear differential equation

$$(1.2) \qquad 2\eta''(z) + \phi(z)\eta(z) = 0 ,$$

then $w(z) = \eta_1(z)/\eta_2(z)$ is a solution of (1.1) and every solution of (1.1) can be obtained in this way.

Let $\phi \in B_2(L, G)$ and let $w_1(z)$ satisfy $\{w, z\} = \phi$ in L. For each $g \in G$, there is a unique $\hat{g} \in SL'(2; C)$ such that $w_1 \circ g = \hat{g} \circ w_1$. The mapping $g \mapsto \hat{g}$ induced by w_1 in this way gives clearly a homomorphism $G \to SL'(2; C)$ and it is determined by ϕ up to a conjugation in $SL'(2; C)$.

Now we associate every $\phi \in B_2(L, I)$ with a solution $W_\phi(z) = \eta_1(z)/\eta_2(z)$ of the equation (1.1), where η_1 and η_2 are solutions of

(1.2) normalized by the conditions

(1.3) $\eta_1 = \eta_2' = 1$ and $\eta_1' = \eta_2 = 0$ at $z = -i$.

This solution $W_\phi(z)$ of (1.1) is determined uniquely. The homomorphism $G \to SL'(2; C)$ induced by W_ϕ, which carries g into $\hat{g}$ in such a way that $W_\phi \circ g = \hat{g} \circ W_\phi$, is denoted by χ_ϕ.

For every $[w_\mu] \in T(G)$, we put $W^\mu = W_{\phi_\mu}$. Then there is a Möbius transformation $\beta_\mu \in SL'(2; C)$ such that $W^\mu = \beta_\mu \circ w^\mu$ in L, so we use this relation to define W^μ for all $z \in C$ and we set $D_{\phi_\mu} = W^\mu(U)$. The definition is legitimate since $W^\mu(U) = D_{\phi_\mu}$ is the complement of the closure of $W^\mu(L)$ and since $W^\mu|L$ depends only on $[w_\mu] = \phi_\mu$. We should notice that, by Koebe's one-quarter theorem, $D_{\phi_\mu} \subset (|w| < 2)$ for every $[w_\mu] = \phi_\mu \in T(G)$.

The set of pairs (ϕ_μ, z) with $\phi_\mu \in T(G)$ and $z \in D_{\phi_\mu}$ is called the fiber space over $T(G)$ and denoted by $F(G)$.

We set $N(G) = \{\omega \in Q | \omega G \omega^{-1} = G\}$ and $Q_0 = \{\omega \in Q | \omega = \text{identity on } R\}$, where Q is the group of all quasiconformal automorphisms of U. Every $\omega \in N(G)$ induces an automorphism ω_* of $Q_{\text{norm}}(G)$ defined as follows: If $w \in Q_{\text{norm}}(G)$, then $\omega_*(w)$ is the uniquely determined element $\omega_*(w) = a \circ w \circ \omega^{-1}$ of $Q_{\text{norm}}(G)$, where $a \in SL'(2; R)$ depends on w. This ω_* is an isometry and a holomorphic mapping of $Q_{\text{norm}}(G)$, and since $[\omega_*(w)]$ in $T(G)$ depends only on $[\omega]$ and $[w]$, ω_* may be considered as an isometric and holomorphic automorphism of $T(G)$ which depends only on $[\omega]$ and maps $[w] \in T(G)$ into $[\omega_*(w)]$. The group of all these automorphisms is called the extended modular group of G and is denoted by $\text{mod}(G)$; it can be identified with the quotient $N(G)/(N(G) \cap Q_0)$.

Each element of $\text{mod}(G)$ induces a holomorphic automorphism of $F(G)$ as follows: If $\omega \in N(G)$, $[w_\mu] \in T(G)$ and $z \in D_{\phi_\mu}$, then

(1.4) $[\omega]_*([w_\mu], z) = ([w_\nu], \hat{z})$, where $w_\nu = \omega_*(w_\mu)$ and $\hat{z} = [\omega]_*(z) = W^\nu \circ \omega \circ (W^\mu)^{-1}(z)$. The mapping $z \mapsto \hat{z}$ defined as above is a conformal bijection of D_{ϕ_μ} onto D_{ϕ_ν} and depends only on the equivalence class $[w_\mu]$ and $[\omega]$.

The modular group $\mathrm{Mod}\,(G)$ of G is defined as the factor group

$$\mathrm{Mod}\,(G) = \mathrm{mod}\,(G)/G \cong (N(G)/(N(G) \cap Q_0))/G \,.$$

The element of $\mathrm{Mod}\,(G)$ induced by $\omega \in N(G)$ is denoted by $<\omega>$. Every element $<\omega>$ of $\mathrm{Mod}\,(G)$ induces an automorphism $[\omega]_*|T(G)$ of $T(G)$.

For an orientation-preserving topological self-mapping $f : S \to S$, we can define an automorphism $f_* : T(S) \to T(S)$ by $f_*([S, g, S']) = [S, g \circ f^{-1}, S']$. The set of all f_* thus obtained is denoted by $\mathrm{Mod}\,(S)$ and is called the group of modular transformations of $T(S)$. The group $\mathrm{Mod}\,(S)$ is canonically isomorphic to $\mathrm{Mod}\,(G)$. Therefore, we will identify $\mathrm{Mod}\,(S)$ with $\mathrm{Mod}\,(G)$.

§2. Analyticity

Let $(\mathfrak{S}, \pi, D^*)$ be a holomorphic family of compact Riemann surfaces of genus $g(\geq 2)$. Let G be a Fuchsian group acting on the upper half-plane U determined by the universal covering group of a compact Riemann surfaces S of genus G.

Let $(a, b) = (a_1, \cdots, a_g, b_1, \cdots, b_g)$ be a canonical system on S. We fix a point t_0 in D^* and also fix a canonical system (a_{t_0}, b_{t_0}) on S_{t_0}. By continuously moving canonical systems starting from (a_{t_0}, b_{t_0}), we have a canonical system (a_t, b_t) on S_t. When t turns around the origin, we suppose that the canonical system on S_t changes from (a_t, b_t) to $(a_t{}', b_t{}')$. If we regard a canonical system as a homology basis, there is an element $\mathfrak{M}$ in the Siegel modular group $S_p(g; \mathbf{Z})$ of degree g such that $\begin{pmatrix} a_t{}' \\ b_t{}' \end{pmatrix} = \mathfrak{M} \begin{pmatrix} a_t \\ b_t \end{pmatrix}$. This element $\mathfrak{M}$ does not depend on t. We call $\mathfrak{M}$ the homological monodromy of $(\mathfrak{S}, \pi, D^*)$. On the other hand, if we regard a canonical system as a homotopy basis, there is an element $\tilde{\mathfrak{M}}$ in $\mathrm{Mod}\,(G)$

such that $[S_t, a_t', b_t'] = \widetilde{\mathfrak{M}}[S_t, a_t, b_t]$, where $[S_t, a_t, b_t]$ denotes the point in $T(S)$ marked by the homotopy basis (a_t, b_t) on S_t. If $[S_t, a_t, b_t]$ and $[S_t, a_t', b_t']$ in $T(S)$ correspond to $[S, f_t, S_t]$ and $[S, f_t', S_t]$ in $T(G)$, respectively, then $\widetilde{\mathfrak{M}}$ is given by $[(f_t')^{-1} \circ f_t]_*$. This element does not depend on t and is called the homotopical monodromy of $(\mathfrak{S}, \pi, D^*)$.

Let $\phi_t \in T(G)$ be the holomorphic quadratic differential corresponding to $[S_t, a_t, b_t]$. Then the following theorem is well known.

THEOREM. *The mapping* $\Phi : D^* \to T(G)$ *defined by* $\Phi(t) = \phi_t$ *is analytic. In general, this mapping* Φ *is not single-valued and it depends on the homotopical monodromy* $\widetilde{\mathfrak{M}}$ *of* $(\mathfrak{S}, \pi, D^*)$.

§3. *The case where* $\widetilde{\mathfrak{M}}$ *is trivial.*

First we prove the following.

THEOREM 1. *If the homotopical monodromy* $\widetilde{\mathfrak{M}}$ *is trivial, then the mapping* $\Phi : D^* \to T(G)$ *is single-valued and it has the holomorphic extension* $\tilde{\Phi} : D \to T(G)$, *so there exists a non-singular compact Riemann surface* S_0 *of genus* g *corresponding to the point* $\tilde{\Phi}(0)$ *of* $T(G)$.

Further, a holomorphic family $(\hat{\mathfrak{S}}, \hat{\pi}, D)$ *of compact Riemann surfaces of genus* g *can be constructed canonically in such a way that*

(i) $\hat{\mathfrak{S}}$ *is a two-dimensional complex manifold,*

(ii) *the fiber* $\hat{S}_t = \hat{\pi}^{-1}(t)$ *is ordinary for every* t *in* D, *and*

(iii) *the fiber* S_t *is conformally equivalent to* $\hat{S}_t$ *for every* t *in* D^* *and* $\mathfrak{S}$ *is naturally isomorphic to* $\hat{\mathfrak{S}} - \hat{S}_0$ *so that the following diagram is commutative:*

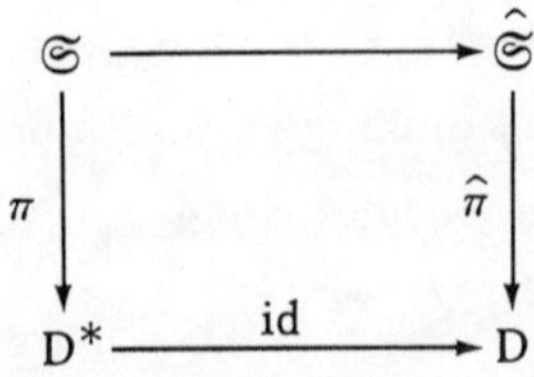

Proof. First, by a result of Earle [10], $T(G)$ is complete with respect to its Carathéodory distance. Therefore, the holomorphic mapping $\Phi : D^* \to T(G)$ has the holomorphic extension $\tilde{\Phi} : D \to T(G)$.

Next we prove the second assertion of our theorem. We set $\Sigma = \{(t,z)\,|\,t \in D, z \in D_{\phi_t}\}$. Denote by $\mathfrak{g}$ the group of analytic automorphisms of Σ consisting of $g(t,z) = (t, (\chi_{\phi_t}(g))(z))$ for all elements g in G. Then the group $\mathfrak{g}$ acts on Σ properly discontinuously and fixed-point freely. Hence, if we set $\hat{\mathfrak{S}} = \Sigma/\mathfrak{g}$, then $\hat{\mathfrak{S}}$ is a two-dimensional complex manifold with the canonical complex structure. Denote by $[t,z]$ the point in $\hat{\mathfrak{S}}$ represented by $(t,z) \in \Sigma$. The mapping $\hat{\pi}: \hat{\mathfrak{S}} \to D$, which carries $[t,z]$ into t, is holomorphic and the rank of Jacobian of $\hat{\pi}$ is equal to one at each point of $\hat{\mathfrak{S}}$. The fiber $\hat{S}_t = \hat{\pi}^{-1}(t)$ over t is ordinary for every $t \in D$ and S_t is conformally equivalent to $\hat{S}_t$ for every $t \in D^*$ by its construction. Furthermore, $\mathfrak{S}$ is naturally isomorphic to $\hat{\mathfrak{S}} - \hat{S}_0$ so that the diagram in Theorem 1 is commutative.

§4. *The case where $\tilde{\mathfrak{M}}$ is of finite order.*

In this section, we prove the following.

THEOREM 2. *If the homotopical monodromy $\tilde{\mathfrak{M}}$ is of finite order, then for the mapping $\Phi: D^* \to T(G)$ defined by $\Phi(t) = \phi_t$, the limit $\phi_0 = \lim\limits_{t \to 0} \phi_t$ exists in $T(G)$ and ϕ_0 is a fixed point of $\tilde{\mathfrak{M}}$.*

Further, a holomorphic family $(\hat{\mathfrak{S}}, \hat{\pi}, D)$ of compact Riemann surfaces of genus g can be constructed in such a way that

   (i) $\hat{\mathfrak{S}}$ is a two-dimensional normal complex space,

   (ii)   if S_0 is the Riemann surface corresponding to ϕ_0 in $T(G)$ and if Γ_0 is an automorphism group of S_0 induced by $\tilde{\mathfrak{M}}$, then $\hat{S}_0 = \hat{\pi}^{-1}(0)$ is isomorphic to S_0/Γ_0, and

   (iii)   the fiber $\hat{S}_t = \hat{\pi}^{-1}(t)$ is ordinary for every t in D^*, the fiber S_t is conformally equivalent to the fiber $\hat{S}_t$ for every t in D^* and $\mathfrak{S}$ is naturally isomorphic to $\hat{\mathfrak{S}} - \hat{S}_0$ such that the following diagram is commutative:

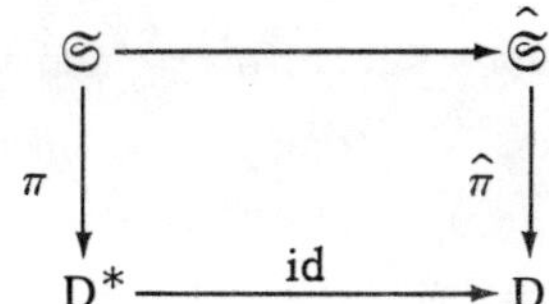

Proof. We suppose the homotopical monodromy $\tilde{\mathfrak{M}}$ has a finite order $\rho\,(\geq 2)$. We set $\Delta = (|\tau| < 1)$ and $\Delta^* = (0 < |\tau| < 1)$ in the complex τ-plane. Let $(\tilde{\mathfrak{S}}, \tilde{\pi}, \Delta^*)$ be a holomorphic family constructed by $(\mathfrak{S}, \pi, D^*)$ from the relation $t = \tau^\rho$. Then the homotopical monodromy of $(\tilde{\mathfrak{S}}, \tilde{\pi}, \Delta^*)$ is trivial. Thus, Theorem 1 implies that the holomorphic mapping $\Psi : \Delta^* \to T(G)$ sending τ into $\Phi(\tau^\rho) = \phi_{\tau^\rho}$ has the analytic extension $\tilde{\Psi} : \Delta \to T(G)$. The quadratic differential $\Psi(\tau)$ in $T(G)$ will be denoted by ψ_τ for a point τ in Δ. Then $\lim_{t\to 0} \phi_t = \lim_{\tau\to 0} \psi_\tau = \psi_0$. Therefore, $\lim_{t\to 0} \phi_t$ exists in $T(G)$ and it coincides with ψ_0. It is obvious that $\phi_0 = \lim_{t\to 0} \phi_t$ is a fixed point of $\tilde{\mathfrak{M}}$ by Lindelöf's theorem.

Next, we will construct a holomorphic family $(\hat{\mathfrak{S}}, \hat{\pi}, D)$ satisfying the second assertion of the theorem. We put

$$\Sigma = \{(\tau, z)| \tau \,\epsilon\, \Delta, \, z \,\epsilon\, D_{\psi_\tau}\}$$

and

$$g_n(\tau, z) = (e_n(\tau), [\omega^n \circ g]_*(z)) ,$$

where $\tau \,\epsilon\, \Delta,\, z \,\epsilon\, D_{\psi_\tau},\, g \,\epsilon\, G,\, n \,\epsilon\, \mathbf{Z},\, e_n(\tau) = \left(\exp\left(\frac{2n\pi i}{\rho}\right)\right)\tau,\, \omega \,\epsilon\, N(G)$ with $\tilde{\mathfrak{M}} = <\omega>$ and $[\omega^n \circ g]_*(z)$ is the conformal bijection of D_{ψ_τ} onto $D_{\psi_{e_n(\tau)}}$ induced by $\omega^n \circ g$ as (1.4). Then g_n is an analytic automorphism of Σ and the group $\mathfrak{g}$ of all these automorphisms acts properly discontinuously on Σ.

Now we recall the canonical ringed structure of the quotient of a bounded domain $\mathfrak{D}$ in $\mathbf{C}^N$ by a discrete subgroup Γ of $\mathrm{Aut}(\mathfrak{D})$, where $\mathrm{Aut}(\mathfrak{D})$ is the group of analytic automorphisms of $\mathfrak{D}$. Then Γ acts on $\mathfrak{D}$ in properly discontinuous fashion and the orbit space $X = \mathfrak{D}/\Gamma$ with the canonical quotient topology is a locally compact Hausdorff space. Let $\pi : \mathfrak{D} \to X$ be the canonical projection. We define a ringed structure $\mathfrak{R}$ on X as follows: If V is an open subset of X and if f is a complex-valued continuous function on V, then $f \,\epsilon\, \mathfrak{R}_V$ if and only if $f \circ \pi$ is holomorphic on $\pi^{-1}(V) \subset \mathcal{D}$. It is well known that the ringed space $(X, \mathcal{R})$ defined as above is a normal complex analytic space.

Therefore, if we introduce the above canonical ringed structure on $\hat{\mathfrak{S}} = \Sigma/\mathfrak{g}$, then $\hat{\mathfrak{S}}$ is a two-dimensional normal complex analytic space. The projection $\hat{\pi}: \hat{\mathfrak{S}} \to D$ sending $[\tau, z]$ into τ^ρ is holomorphic and the fiber $\hat{S}_t = \hat{\pi}^{-1}(t)$ is ordinary and conformally equivalent to S_t for every $t \in D^*$.

If S_0 is the Riemann surface corresponding to the point ϕ_0 in $T(G)$ and if $\mathfrak{g}_0$ is the conformal automorphism group of D_{ϕ_0} induced by $\mathfrak{g}$, then $\mathfrak{g}_0$ induces an automorphism group Γ_0 of S_0 and the fiber $\hat{S}_0 = \hat{\pi}^{-1}(0)$ is isomorphic to S_0/Γ_0.

Moreover, by its construction, $\mathfrak{S}$ is analytically isomorphic to $\hat{\mathfrak{S}} - \hat{S}_0$ such that the diagram in Theorem 2 is commutative.

§5. *Deformation spaces of Riemann surfaces with nodes*

To consider the case where $\tilde{\mathfrak{M}}$ is of infinite order, we will first explain deformation spaces of Riemann surfaces with nodes, which is due to Bers [9].

A compact Riemann surface with nodes of (arithmetic) genus g (≥ 2) is a connected compact one-dimensional complex analytic space S, on which there are $k = k(S)$ $(\leq 3g-3)$ points $P_1, \cdots, P_k$, called nodes, such that (i) every nodes P_j has a neighborhood isomorphic to the analytic set $\{z_1 z_2 = 0 \mid |z_1| < 1, |z_2| < 1\}$ with P_j corresponding to $(0,0)$, (ii) the set $S - \{P_1, \cdots, P_k\}$ has r (≥ 1) components $\Sigma_1, \cdots, \Sigma_r$ called parts of S, where each Σ_i is a Riemann surface of genus g_i, compact except for n_i punctures, with $3g_i - 3 + n_i \geq 0$ and $n_1 + \cdots + n_r = 2k$, and (iii) $g = (g_1 - 1) + \cdots + (g_r - 1) + k + 1$.

The condition (ii) implies that every part carries a Poincaré metric and the condition (iii) is equivalent to the requirement that the total Poincaré area of S equals $4\pi(g-1)$.

From now on, g (≥ 2) is fixed and the letter S, with or without subscripts or superscripts, always denotes a surface with properties (i)-(iii).

If $k(S) = 0$, then S is called non-singular; if $k(S) = 3g-3$, then S is called terminal.

A continuous surjection $a : S' \to S$ is called a deformation if, for every node $P \in S$, $a^{-1}(P)$ is either a node or a Jordan curve avoiding all nodes and if, for every part Σ of S, $a^{-1}{}_{|\Sigma}$ is an orientation-preserving homeomorphism.

Once and for all we choose an integer $\nu (> 3)$ which will be fixed throughout the following discussion.

Two deformations $a : S' \to S_0$ and $\beta : S'' \to S_0$ are called equivalent to each other, if there exists a homeomorphism f of S' onto S'' such that $a = \beta \circ f$ and if f is homotopic to a product of ν-th powers of Dehn twists about Jordan curves mapped by a into nodes, followed by an isomorphism. We denote by $<a> = <S', a, S_0>$ the equivalence class of a deformation $a : S' \to S_0$. Given a compact Riemann surface S_0 with nodes of genus g, the deformation space $X(S_0)$ consists of equivalence classes $<S', a, S_0>$ of all deformations $a : S' \to S_0$.

To every node $P \in S_0$, there belongs a distinguished analytic hypersurface $<P> (\subset X(S_0))$ consisting of all $<a> \in X(S_0)$ such that $a^{-1}(P)$ is a node of $a^{-1}(S_0)$. If $P_1, \cdots, P_k$ are all nodes of S_0, we denote by $X_0(S_0)$ the set $X(S_0) - <P_1> \cup \cdots \cup <P_k>$, that is, $X_0(S_0) = \{<S', a, S_0> | S'$ is non-singular$\}$.

Every deformation $\beta : S_0 \to S_1$ induces an allowable holomorphic mapping $\beta_* : X(S_0) \to X(S_1)$ which sends $<a> \in X(S_0)$ into $<\beta \circ a> \in X(S_1)$. Let $\Gamma(S_0)$ induced by all topological orientation-preserving self-mappings of S_0 and let $\Gamma_0(S_0)$ be its subgroup induced by all automorphisms of S_0. Then the group $\Gamma(S_0)$ is discrete and the subgroup $\Gamma_0(S_0)$ is finite and is the stabilizer of $<\mathrm{id}> \in X(S_0)$ in $\Gamma(S_0)$.

Let R_g be the Riemann's moduli space of non-singular compact Riemann surfaces of genus g and let M_g be the moduli space of compact Riemann surfaces with nodes of genus g. Let $\pi : T(G) \to R_g$ and $\Pi : X(S_0) \to M_g$ be the canonical projections.

Now we can prove the following lemma.

LEMMA 1. *The holomorphic mapping* $J : D^* \to R_g$ *sending* t *into* $[S_t]$ *has a holomorphic extension* $\hat{J} : D \to M_g$.

Proof. To prove this lemma, we use the following theorem (Kobayashi [12], Kobayashi and Ochiai [13]).

THEOREM. *Let* $\mathfrak{D}$ *be a complex space and* Γ *a properly discontinuous group of holomorphic transformations of* $\mathfrak{D}$ *such that* $M = \mathfrak{D}/\Gamma$ *is an open subset of a complex space* Y. *Assume that*

(1) *the pseudo-distance* $d_M{}'$ *is a distance;*

(2) *the closure* $\overline{M}$ *of* M *in* Y *is compact;*

(3) *given a point* $p \in \partial M$ *and a neighborhood* U *of* p *in* Y, *there exists a smaller neighborhood* V *of* p *in* Y *such that*

$$d_M{}'(M \cap (Y-U), M \cap V) > 0 .$$

Let X *be a complex manifold and* A *a locally closed complex submanifold of* X. *Then every locally liftable holomorphic mapping* $f : X - A \to M$ *extends to a holomorphic mapping* $\hat{f} : X \to Y$.

We apply this theorem to the case where $\mathfrak{D} = T(G)$, $\Gamma = \mathrm{Mod}(G)$, $Y = M_g$ and $f = J$. So we have to see that three conditions (1), (2) and (3) in the theorem are satisfied in this case.

If $p, q \in R_g$ and $\tilde{p}, \tilde{q} \in T(G)$ such that $\pi(\tilde{p}) = p$ and $\pi(\tilde{q}) = q$, then

$$d'_{R_g}(p, q) = \inf_{\tilde{p}} d_{T(G)}(\tilde{p}, \tilde{q}) ,$$

where the infimum is taken over all $\tilde{p} \in T(G)$ satisfying $\pi(\tilde{p}) = p$. Since $T(G)$ is a bounded domain in C^{3g-3} and since $\mathrm{Mod}(G)$ acts properly discontinuously on $T(G)$, the pseudo-distance $d_M{}'$ is a distance.

Since M_g is compact, the condition (2) is satisfied obviously.

Given a point $[S_0]$ on ∂R_g and a neighborhood U of $[S_0]$ in M_g, there is a neighborhood N of $\langle \mathrm{id} \rangle$ in $X(S_0)$ such that N is stable under $\Gamma_0(S_0)$ and such that $N/\Gamma_0(S_0)$ is a neighborhood of $[S_0]$ in M_g.

(See Bers [9].) We may assume that $\gamma(N) \cap N = \phi$ for every $\gamma \in \Gamma(S_0) - \Gamma_0(S_0)$ and $U = N/\Gamma_0(S_0)$. Take a neighborhood $N_0(\Subset N)$ of $\langle id \rangle$ in $X(S_0)$ such that N_0 is stable under $\Gamma_0(S_0)$ and such that $V = N_0/\Gamma_0(S_0)$ is a neighborhood of $[S_0]$ in M_g. Let $a : S \to S_0$ be a deformation and let $a_* : T(G) \to X(S_0)$ be an allowable mapping which sends $[S, f, S']$ into $\langle S', a \circ f^{-1}, S_0 \rangle$. Then

$$d'_{R_g}(R_g \cap (M_g - U), R_g \cap V)$$

$$= \inf_{p,q} \{ \inf_{\tilde{p}} d_{T(G)}(\tilde{p}, \tilde{q}) \}$$

$$\geq \inf_{p,q} \{ \inf_{\tilde{p}} d_{X_0(S_0)}(a_*(\tilde{p}), a_*(\tilde{q})) \} \,,$$

where the first infimum is taken over all $p \in R_g \cap (M_g - U)$ and all $q \in R_g \cap V$, and the second infimum is taken over all $\tilde{p} \in T(G)$ with $\pi(\tilde{p}) = p$ and a point $\tilde{q} \in T(G)$ with $\pi(\tilde{q}) = q$ and $a_*(\tilde{q}) \in N_0$. Since $N_0 \Subset N$ and $d_{X_0(S_0)} \geq d_X(S_0)$ and since $d_{X(S_0)}$ is a distance, we have

$$d'_{R_g}(R_g \cap (M_g - U), R_g \cap V)$$

$$\geq \inf_{\hat{p},\hat{q}} d_{X(S_0)}(\hat{p}, \hat{q}) > 0 \,,$$

where the infimum is taken over all $\hat{p} \in X(S_0) - N$ and all $\tilde{q} \in N_0$.

Therefore, the conditions (1), (2), (3) in the above theorem are satisfied and we see that the locally liftable holomorphic mapping $J : D^* \to R_g$ has a holomorphic extension $\hat{J} : D \to M_g$. Thus we have Lemma 1.

Now we set $\hat{J}(0) = [S_0]$. Then we can take a deformation $a : S \to S_0$ such that $a_*([S, f_r, S_r]) = \langle S_r, a \circ f_r^{-1}, S_0 \rangle$ converges to $\langle id \rangle$ in $X(S_0)$ as $r \to 0$ through $r \in (0, 1)$. We shall prove the following lemma.

LEMMA 2. *For a convergent subsequence* $\{\phi_{r_n}\}$ *of* $\{\phi_r\} = [S, f_r, S_r]$ *having its limit* $\phi_0 \in \partial T(G)$ *as* $r_n \to 0$, *the boundary group* $G_0 = \chi_{\phi_0}(G)$ *is a regular b-group and*

$$D_0 \cup \{\text{parabolic fixed points of } G_0 \text{ on } \overline{D}_0\}/G_0$$

is isomorphic to S_0, *where* $D_0 = \Omega(G_0) - \Delta(G_0)$, $\Omega(G_0)$ *is the region of discontinuity of* G_0 *and* $\Delta(G_0)$ *is the invariant component of* G_0.

To prove this lemma, we use a parametrization of all Riemann surfaces with nodes which can be deformed into a given surface S_0, which will be explained briefly as follows (see Bers [9], §3):

Let S_0 have r parts $\Sigma_1, \cdots, \Sigma_r$ and k nodes $P_1, \cdots, P_k$. Let Σ_j have genus g_j and n_j punctures. We choose r Fuchsian groups $H_1, \cdots, H_r$ acting on discs $\Delta_1, \cdots, \Delta_r$ with disjoint closures, such that H_j has n_j non-conjugate maximal elliptic subgroups with the same fixed order $\nu > 3$, the Riemann surface Δ_j/H_j, with the images of all elliptic vertices removed, is conformally equivalent to Σ_j and such that $H_1, \cdots, H_r$ generate a Kleinian group H being their free product and having an invariant component Δ_0. Clearly Δ_0/H is a compact Riemann surface Σ of genus $g_1 + \cdots + g_r$. We assign to each node P_i of S_0 two non-conjugate maximal elliptic subgroups Γ_i', Γ_i'' of H so that, if P_i joins Σ_j to Σ_ℓ, then $\Gamma_i' \subset H_j$ and $\Gamma_i'' \subset H_\ell$. Two elliptic vertices not in Δ_0 is called related if they are fixed under elliptic subgroups conjugate to either Γ_i' or Γ_i''. The Γ_i are chosen such that the union of the Δ_j/H_j, with the images of any two related elliptic vertices identified, is isomorphic to S_0.

If $s_i \in C$, $|s_i| > 0$ and small, then there exists a unique loxodromic Möbius transformation h_{i,s_i}, which conjugates the Γ_i' into Γ_i'', has the multiplier s_i and has fixed points in Δ_j and Δ_ℓ (j and ℓ being as before). Set $s = (s_1, \cdots, s_k)$. If $|s| = \max |s_i|$ is small, the group $H_{0,s}$ generated by H and by all elements h_{i,s_i} with $s_i \neq 0$ is a Kleinian group.

Let s be as before and let W be a quasi-conformal automorphism of $\hat{C}$ such that W leaves $0, 1, \infty$ fixed and $W \circ H_{0,s} \circ W^{-1}$ is a Kleinian group and such that $W_{|\Delta_0}$ is conformal. Then $W_{|\Delta_j}$, $j = 1, \cdots, r$, defines an element τ_j of the Teichmüller space $T(H_j)$, which we represent as a bounded domain in $C^{3g_j-3+n_j}$. If $s_i \neq 0$, set $t_i = a_i - \hat{a}_i$ where a_i is the repelling fixed point of $W \circ h_{i,s_i} \circ W^{-1}$ and $\hat{a}_i$ is the fixed point of $W \circ \Gamma_i' \circ W^{-1}$ in $W(\Delta_j)$. If $s_i = 0$, set $t_i = 0$. Then the point

$$(\tau, t) = (\tau_1, \cdots, \tau_r, t_1, \cdots, t_k) \in C^{3g-3}$$

determines the group $H_{\tau,t} = W \circ H_{0,s} \circ W^{-1}$ completely. We denote by $X_\alpha(S_0)$ the set of all (τ, t) for which a group $H_{\tau,t}$ is defined. Here α represents the choices made; the group H_j and the subgroups Γ_i', Γ_i''. Then $X_\alpha(S_0)$ is a bounded domain in C^{3g-3}.

We denote by $\Omega(\tau, t)$ the part of the region of discontinuity of $H_{\tau,t}$ which does not map onto Σ, and by $\Omega^0(\tau, t)$ the complement in $\Omega(\tau, t)$ of the set of elliptic vertices. Then $\Omega(\tau, t)/H(\tau, t)$, with images of related elliptic vertices identified, is a Riemann surface with node $S_{\tau,t}$. The genus of $S_{\tau,t}$ is g and $S_{\tau,t}$ has as many nodes as there are zeros in $(t_1, \cdots, t_k)$. Furthermore, each $S_{\tau,t}$ is equipped with a deformation up to equivalence. Then $X_\alpha(S_0)$ is in a natural one-to-one correspondence with $X(S_0)$.

Proof of Lemma 2. Let Σ_j be one of parts of S_0. Take a point O on Σ_j. Set $S_j' = a^{-1}(\Sigma_j)$, $S_{j,r}' = f_r(S_j')$, $O' = a^{-1}(O)$ and $O_r' = f_r(O')$ for $r \in (0, 1)$. Let G_j' be a subgroup of G corresponding to the subgroup $\pi_1(S_j', O')$ of the fundamental group $\pi_1(S, O')$. Set $G_{j,r}' = \chi_{\phi_r}(G_j')$.

If $a_*([S, f_r, S_r])$ is represented by the point (τ^r, t^r) in $X_\alpha(S_0)$, then we denote by $<S_{\tau^r,t^r}, a_{\tau^r,t^r}, S_0>$ the point in $X(S_0)$ induced by (τ^r, t^r) and denote by $[\Sigma_j, F_{j,r}, \Sigma_{j,r}]$ the point in $T(\Sigma_j)$ corresponding to the point τ_j^r in $T(H_j)$. Let $\tilde{H}_j$ be the Fuchsian group determined by the universal converging of Σ_j, that is, $\tilde{H}_j$ corresponds to $\pi_1(\Sigma_j, O)$. Then $\tilde{H}_j$ is isomorphic to G_j' by a. Regard $[\Sigma_j, F_{j,r}, \Sigma_{j,r}]$ in $T(\Sigma_j)$ as a point $\phi_{j,r}$ in $T(\tilde{H}_j)$. Since $a_*([S, f_r, S_r]) \to <id>$, we have $\phi_{j,r} \to 0$ in $T(\tilde{H}_j)$ as $r \to 0$.

Since $<S_r, a \circ f_r^{-1}, S_0> = <S_{\tau^r, t^r}, a_{\tau^r, t^r}, S_0>$, by definition, there

exists a homeomorphism $g_r : S_{\tau^r, t^r} \to S_r$ such that $a_{\tau^r, t^r} = a \circ f_r^{-1} \circ g_r$ and

g_r is homotopic to a product of ν-th powers of Dehn twists about Jordan

curves mapped by a_{τ^r, t^r} into nodes, followed by an isomorphism. Hence,

if this product of Dehn twists is denoted by d_r, then $[S, f_r, S_r] =$

$[S, d_r \circ g_r^{-1} \circ f_r, S_{\tau^r, t^r}]$ in $T(G)$. We can assume that $a : S \to S_0$ is locally

quasiconformal and for any sufficiently small neighborhood δ_i, $i = 1, \cdots, k$,

of P_i in S_0, we can also assume that there exists a quasiconformal

mapping $h_r : S \to S_{\tau^r, t^r}$ such that $[S, f_r, S_r] = [S, h_r, S_{\tau^r, t^r}]$ and $h_r =$

$d_r \circ g_r^{-1} \circ f_r$ on $S - a^{-1}(\delta)$, where $\delta = \delta_1 \cup \cdots \cup \delta_k$. (See Bers [5].)

Let $\pi : U \to U/G = S$ and $\tilde{\pi} : \tilde{U} \to \tilde{U}/\tilde{H}_j = \Sigma_j$ be the natural projec-

tions, where $\tilde{U}$ is the upper-half plane in the $\tilde{z}$-plane, let Δ_j be a con-

nected component of $\pi^{-1} \circ a^{-1}(\Sigma_j - \delta)$ and let $\tilde{\Delta}_j = \tilde{\pi}^{-1}(\Sigma_j - \delta)$. Since

$\pi_1(S_j', O')$ and $\pi_1(\Sigma_j, O)$ are conjugate by a, we can lift a to a quasi-

conformal mapping $A : \Delta_j \to \tilde{\Delta}_j$ such that G_j' and $\tilde{H}_j$ are conjugate by A.

Let $W_r = W_{\phi_r}$ be the quasiconformal automorphism of $\hat{C}$ defined by

$[S, h_r, S_{\tau^r, t^r}]$ and let $W_{j,r} = W_{j, \phi_{j,r}}$ be the quasiconformal automorphism

of $\hat{C}$ defined by $[\Sigma_j, F_{j,r}, \Sigma_{j,r}]$. Then

$$V_{j,r} = W_r \circ A^{-1} \circ W_{j,r}^{-1} : W_{j,r}(\tilde{\Delta}_j) \to W_r(\Delta_j)$$

is a conformal mapping, because W_r and $W_{j,r} \circ A$ have the same

Beltrami coefficient on Δ_j. We can assume that $W_{j,r}$ converges to the

Mobius transformation $W_{j,0} : z \mapsto \dfrac{1}{\tilde{z} + i}$ uniformly on any compact subset

of $\tilde{U}$ as $r \to 0$, because $F_{j,r}$ converges to identity as $r \to 0$. If we set

$W_{j,0}(\tilde{\Delta}_j) = \tilde{\Delta}_{j,0}$, then $\{V_{j,r}\}$ for $r \in (0, 1)$ forms a normal family on $\tilde{\Delta}_{j,0}$

and hence $\{W_r\}$ is also a normal family on Δ_j.

Now we can prove that $G_{j,0}' = \chi_{\phi_0}(G_j')$ is not degenerate: We may

assume that V_{j,r_n} converges to a holomorphic mapping $V_{j,0}$ uniformly on any compact subset of $\tilde{\Delta}_{j,0}$ as $r_n \to 0$. Suppose that $V_{j,0}$ is a constant mapping with a value c. There exist two loxodromic elements $\tilde{\eta}$ and $\tilde{\gamma}$ in $\tilde{H}_j$ and a point ζ in $\tilde{\Delta}_j$ such that $\tilde{\eta}(\zeta)$, $\tilde{\gamma}(\zeta)$ are contained in $\tilde{\Delta}_j$ and $\tilde{\eta}\tilde{\gamma} \neq \tilde{\gamma}\tilde{\eta}$. Since $\tilde{H}_j$ is conjugate to G'_j by A, we see that, if $\eta = A^{-1}\tilde{\eta}A$, $\gamma = A^{-1}\tilde{\gamma}A$ and $\eta_0 = \chi_{\phi_0}(\eta)$, $\gamma_0 = \chi_{\phi_0}(\gamma)$, then η_0 and γ_0 are of infinite order and $\eta_0\gamma_0 \neq \gamma_0\eta_0$. If we set $\tilde{\eta}_n = \chi_{\phi_{j,r_n}}(\eta)$, $\tilde{\gamma}_n = \chi_{\phi_{j,r_n}}(\gamma)$, $\eta_n = \chi_{\phi_{r_n}}(\eta)$, $\gamma_n = \chi_{\phi_{r_n}}(\gamma)$ and $\zeta_n = W_{j,r_n}(\zeta)$, then we have $\eta_n \circ V_{j,r_n}(\zeta_n) = V_{j,r_n} \circ \tilde{\eta}_n(\zeta_n)$. By taking limit of this equality, we obtain $\eta_0 \circ V_{j,0}(\zeta) = V_{j,0} \circ \eta(\zeta)$, which implies $\eta_0(c) = c$. Similarly, we have $\gamma_0(c) = c$. Three properties obtained above that η_0, γ_0 are of infinite order, $\eta_0 \circ \gamma_0 \neq \gamma_0 \circ \eta_0$ and $\eta_0(c) = \gamma_0(c) = c$ contradict the discreteness of G_0. Therefore, $V_{j,0}$ is a univalent holomorphic mapping.

Using above functions $V_{j,0}$ and contracting each δ_j to the point P_j, we can construct a conformal mapping $\tilde{V}_{j,0}$ of $\tilde{U}$ onto $D_{j,0} = \tilde{V}_{j,0}(\tilde{U})$ such that $\tilde{V}_{j,0}$ conjugates $\tilde{H}_j$ into $G'_{j,0}$. Then $D_{j,0} \subset \Omega(G'_{j,0})$ and, for any $g_0 \in G_0 - G'_{j,0}$, we see $g_0(D_{j,0}) \cap D_{j,0} = \phi$ and $\tilde{V}_{j,0}$ induces an isomorphism of $\Sigma_j = \tilde{U}/\tilde{H}_j$ onto $D_{j,0}/G'_{j,0}$. Hence $G'_{j,0}$ is quasi-Fuchsian and $D_{j,0}$ is an invariant component of $G'_{j,0}$. Therefore, $D_{j,0}$ is a component of $\Omega(G_0)$ and the stability group of $D_{j,0}$ is $G'_{j,0}$.

Therefore, $G_0 = \chi_{\phi_0}(G)$ is a regular b-group and $D_0 \cup \{$parabolic fixed points of G_0 on $\bar{D}_0\}/G_0$ is isomorphic to S_0. This establishes Lemma 2.

§6. *The case where* $\tilde{\mathfrak{M}}$ *is of infinite order*

First we shall prove the following.

THEOREM 3. *Suppose that the homotopical monodromy* $\tilde{\mathfrak{M}}$ *is of infinite order. If* $D_1 = \{t \in D | \theta_1 < \arg t < \theta_2\}$ *is an angular domain in* D, *then* $\Phi(t) = \phi_t$ *converges to* ϕ_0 *uniformly, as* t *tends to zero through* D_1, *where* ϕ_0 *is a point on* $\partial T(G)$ *and does not depend on* D_1.

The boundary group $\chi_{\phi_0}(G)$ *corresponding to* ϕ_0 *is a regular b-group and* $\chi_{\phi_t}(G)$ *converges to* $\chi_{\phi_0}(G)$ *uniformly as* t *tends to zero through* D_1 .

If $\tilde{\mathfrak{M}}$ *is regarded as a mapping from* $\hat{T}(G)$ *to* $\hat{T}(G)$, *then* ϕ_0 *is a fixed point of* $\tilde{\mathfrak{M}}$ *in* $\hat{T}(G)$, *where* $\hat{T}(G)$ *is the augment space of* $T(G)$ *in the sense of Abikoff* [1].

Proof. By Lindelöf's theorem, it is sufficient to prove that ϕ_t converges to ϕ_0 for a fixed argument of t as $r = |t| \to 0$. We may assume arg t = 0 and will prove that ϕ_r converges to an element ϕ_0 in $\partial T(G)$ as $r \to 0$, where r is a positive real number.

For a convergent subsequence $\{\phi_{r_n}\}$ of $\{\phi_r\}$ with its limit $\phi_0 \in \overline{T(G)}$ as $r_n \to 0$, we have $\phi_0 \in \partial T(G)$. In fact, assume $\phi_0 \in T(G)$, then the image $\hat{J}(0) = [S_0]$ of $t = 0$ by the holomorphic mapping $\hat{J} : D \to M_g$ in §5 is contained in R_g, that is, S_0 is a non-singular Riemann surface. Hence there exists a sufficiently small neighborhood U_0 of ϕ_0 in $T(G)$ such that $\tilde{\mathfrak{M}}^{\ell}(\phi_t) \in U_0$ for a sufficiently small $t \in D$ and for every integer ℓ. Since $\tilde{\mathfrak{M}}$ is of infinite order, this contradicts the discreteness of $\mathrm{Mod}(G)$. This implies $\phi_0 \in \partial T(G)$.

The facts stated in §5 show that, for any convergent subsequence $\{\phi_{r_n}\}$ of $\{\phi_r\}$ with its limit $\phi_0 \in \partial T(G)$ as $r \to 0$, $\chi_{\phi_{r_n}}(G)$ converges to a regular b-group $G_0 = \chi_{\phi_0}(G)$ and that $D_0 \cup \{$parabolic fixed points of G_0 on $\overline{D}_0\}/G_0$ is isomorphic to S_0. For another convergent subsequence $\{\phi_{r_n'}\}$ of $\{\phi_r\}$ with its limit $\phi_0' \in \partial T(G)$ as $r_n' \to 0$, $G_0' = \chi_{\phi_0'}(G)$ is also a regular b-group and $D_0' \cup \{$parabolic fixed points of G_0' on $\overline{D}_0'\}/G_0'$ is isomorphic to S_0, where $D_0' = \Omega(G_0') - \Delta(G_0')$, $\Omega(G_0')$ is the region of discontinuity of G_0' and $\Delta(G_0')$ is the invariant component of G_0'. From the univalent holomorphic functions $\tilde{V}_{j,0}$, $j = 1, \cdots, r$, constructed in the proof of Lemma 2, we can construct a conformal mapping $h : \Omega(G_0) \to \Omega(G_0')$ such that h conjugates G_0 into G_0'. Hence, by a theorem due to Abikoff [2] and Marden [14], the mapping h is a complex

Möbius transformation. Hence G_0 is conjugate to $G_0{}'$ in $SL'(2; C)$. On
the other hand, by Bers' theorem [6], every boundary group of G is conjugate, in $SL'(2; C)$, to precisely $m-1$ other boundary group of G, where
m is the index of G in its normalizer in $SL'(2; R)$. Since m is finite
and since the mapping $\Phi_{|(0,1)} : (0, 1) \to T(G)$ sending r into ϕ_r is con-
tinuous, we have $\phi_0 = \phi_0{}'$. Therefore, ϕ_r converges to an element
$\phi_0 \in \partial T(G)$ as $r \to 0$.

By Lindelof's theorem, it is obvious that ϕ_0 is a fixed point of $\tilde{\mathfrak{M}}$.
This completes the proof.

Further we have the following.

THEOREM 4. *If the homotopical monodromy $\tilde{\mathfrak{M}}$ is of infinite order, then
a two-dimensional normal complex space $\hat{\tilde{\mathfrak{S}}}$ can be canonically constructed in the following way:*

*(i) $(\hat{\tilde{\mathfrak{S}}}, \hat{\pi}, D)$ is a holomorphic family of compact Riemann surfaces
of genus g with a singular fiber over $t = 0$,*

*(ii) if S_0 is the Riemann surface with nodes corresponding to
$\phi_0 \in \partial T(G)$ in Theorem 3 and if Γ_0 is an automorphism group of S_0 in-
duced by $\tilde{\mathfrak{M}}$, then the fiber $\hat{S}_0 = \hat{\pi}^{-1}(0)$ over $t = 0$ of $\hat{\tilde{\mathfrak{S}}}$ is isomorphic
to S_0/Γ_0, and*

*(iii) $\mathfrak{S}$ is naturally isomorphic to $\hat{\tilde{\mathfrak{S}}} - \hat{S}_0$ so that the following
diagram is commutative:*

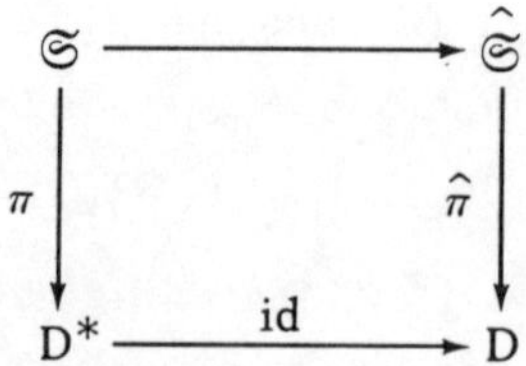

Proof. We use the notations in §5. Let f_* be an element of $\mathrm{Mod}(S)$
corresponding to $\tilde{\mathfrak{M}}$ of $\mathrm{Mod}(G)$. Since the mapping $J : D \to R_g$ sending
t into $[S_t]$ has a holomorphic extension $\hat{J} : D \to M_g$ with $\hat{J}(0) = [S_0]$
and since there is a neighborhood N of $<\mathrm{id}>$ in $X(S_0)$ such that N is
stable under the finite group $\Gamma_0(S_0)$ and $N/\Gamma_0(S_0)$ is a neighborhood of

$[S_0]$ in M_g, there exists a positive integer ρ such that f^ρ is homotopic to a product of ν-th powers of Dehn twists about Jordan curves mapped by α into nodes. Set $\Delta = (|\zeta| < 1)$ and $\Delta^* = \Delta - \{0\}$ in the ζ-plane and let $\kappa : \Delta \to D$ be the mapping sending ζ into ζ^ρ. Consider the holomorphic family (δ, Π, Δ^*) constructed by $(\mathfrak{S}, \pi, D^*)$ from the relation $t = \zeta^\rho$. Then the analytic mapping $K : \Delta^* \to X(S_0)$ sending ζ into $<S_\zeta, \alpha \circ f_\zeta^{-1}, S_0>$ is single-valued. Hence K has a holomorphic extension $\hat{K} : \Delta \to X(S_0)$ with $\hat{K}(0) = <\mathrm{id}>$.

We denote by $H(\zeta)$ and $\Omega(\zeta)$ the Kleinian group $H(\tau, t)$ and its components $\Omega(\tau, t)$, respectively, determined by the point $K(\zeta) = (\tau, t)$ in $X_\alpha(S_0)$.

Now we can canonically construct a completion $\hat{\delta}$ of δ as follows: We set

$$\hat{\delta}_1 = \{(\zeta, [z]) | \zeta \in \Delta^*, [z] \in \Omega(\zeta)/H(\zeta)\} .$$

Then, by its construction, $\hat{\delta}_1$ becomes a two-dimensional complex manifold. If $\hat{\Pi}_1 : \hat{\delta}_1 \to \Delta^*$ is the mapping sending $(\zeta, [z])$ into ζ, then $(\hat{\delta}_1, \hat{\Pi}_1, \Delta^*)$ is a holomorphic family of compact Riemann surfaces of genus g. Let Φ and Ψ be the analytic mappings (many-valued) of (δ, Π, Δ^*) and $(\hat{\delta}_1, \hat{\Pi}_1, \Delta^*)$ into $T(G)$, respectively. We may assume that $\Phi = \Psi$ on Δ^* and that (δ, Π, Δ^*) and $(\hat{\delta}_1, \hat{\Pi}_1, \Delta^*)$ have the same homotopical monodromy for a certain positive integer ρ. We set

$$\hat{\delta}_0 = \hat{\delta}_1 \cup \{\Omega(0)/H(0) \text{ with the images of all elliptic vertices removed}\}$$

and

$$\hat{\delta} = \hat{\delta}_1 \cup \{\Omega(0)/H(0) \text{ with the images of related elliptic vertices identified}\}.$$

Then, by its construction, $\hat{\delta}_0$ becomes a two-dimensional complex manifold. We can naturally define a locally compact Hausdorff topology on $\hat{\delta}$ such that $\hat{\delta}_0$ is an open dense subset of $\hat{\delta}$, By Cartan's theorem on the continuation of normal complex spaces, we can induce a normal complex structure $\mathcal{R}$ on $\hat{\delta}$ such that the restricted structure $\mathcal{R}|\hat{\delta}_0$ to $\hat{\delta}_0$ is the same one given on $\hat{\delta}_0$ and $\hat{\delta} - \hat{\delta}_0$ is a proper analytic subset of $\hat{\delta}$. (The functions which separate the points of $\hat{\mathsf{S}}_0$ are obtained from the automorphic forms constructed by Bers [9].) Then the projection $\hat{\Pi} : \hat{\mathsf{S}} \to \Delta$ sending $(\zeta, [z])$ into ζ is holomorphic and $(\hat{\mathsf{S}}, \hat{\Pi}, \Delta)$ is a completion of $(\mathsf{S}, \Pi, \Delta^*)$.

Finally we construct a completion $\widehat{\mathfrak{S}}$ of $\mathfrak{S}$ as follows: For an element $\omega \in N(G)$ with $<\omega> = \tilde{\mathfrak{M}}$, the automorphism $[\omega]_*$ of $F(G)$ induces an automorphism $\hat{\gamma}_1$ of $\hat{\mathsf{S}}_1$. By the similar reasoning to that in the proof of Lemma 1, $\hat{\gamma}_1$ has a holomorphic extension $\hat{\gamma}_0 : \hat{\mathsf{S}}_0 \to \hat{\mathsf{S}}$. Similarly, if $\hat{\delta}_1$ is the inverse mapping of $\hat{\gamma}_1$, then $\hat{\delta}_1$ has a holomorphic extension $\hat{\delta}_0 : \hat{\mathsf{S}}_0 \to \hat{\mathsf{S}}$. We can prove that the mappings $\hat{\gamma}_0$ and $\hat{\delta}_0$ are not constant on each part of S_0 as follows: Let $P_1, \cdots, P_k$ be the nodes of S_0. Denote by $\hat{\mathsf{s}}_\zeta$ the fiber of $\hat{\mathsf{S}}_1$ over $\zeta \in \Delta^*$. If $\hat{\gamma}_0$ has a constant value $q_0 \in S_0$ on a part Σ_ℓ of S_0, then there exist a small neighborhood A of q_0 in $\hat{\mathsf{S}}$ and a small neighborhood b_j of P_j in S_0, $j = 1, \cdots, k$, such that $\hat{\gamma}_0\left(\Sigma_\ell - \bigcup\limits_{j=1}^{k} b_j\right)$ is contained in A and such that $\hat{\mathsf{s}}_\zeta \cap A$ is

homeomorphic to a disc or an annular domain for every small $\zeta \in \Delta^*$. We can take a sufficiently small neighborhood B of $\Sigma_\ell - \bigcup\limits_{j=1}^{k} b_j$ in $\hat{\mathsf{S}}$ such that $\hat{\gamma}_0(\hat{\mathsf{s}}_\zeta \cap B)$ is contained in A and $\hat{\mathsf{s}}_\zeta \cap B$ is homeomorphic to $\Sigma_\ell - \bigcup\limits_{j=1}^{k} b_j$ for every small $\zeta \in \Delta^*$. Therefore, $\hat{\mathsf{s}}_\zeta \cap B$ must be schlicht for every small $\zeta \in \Delta^*$. Then $\hat{\mathsf{s}}_\zeta \cap B$ has at least three boundary curves c_1, c_2, c_3 such that each of them is not contractible to a point in $\hat{\mathsf{s}}_\zeta$. On the other hand, the images $\hat{\gamma}_0(c_1)$, $\hat{\gamma}_0(c_2)$, $\hat{\gamma}_0(c_3)$ are contained in $\hat{\mathsf{s}}_{\zeta_1} \cap A$ with $\hat{\mathsf{s}}_{\zeta_1} = \hat{\gamma}_0(\hat{\mathsf{s}}_\zeta)$. Since $\hat{\mathsf{s}}_{\zeta_1} \cap A$ is homeomorphic to a disc or an annular domain, at least one of the curves $\hat{\gamma}_0(c_1)$, $\hat{\gamma}_0(c_2)$, $\hat{\gamma}_0(c_3)$ is contractible to a point in $\hat{\mathsf{s}}_{\zeta_1}$. This contradicts the fact such that $\hat{\gamma}_0$ induces an isomorphism of $\hat{\mathsf{s}}_\zeta$ onto $\hat{\mathsf{s}}_{\zeta_1}$. Similarly, $\hat{\delta}_0$ is not constant on each part of S_0.

If $\hat{\gamma}_0$ is not holomorphic at some node of S_0, then $\hat{\delta}_0$ is constant on a certain part of S_0, which implies $\hat{\gamma}_0$ has a holomorphic extension $\hat{\gamma} : \hat{\mathsf{S}} \to \hat{\mathsf{S}}$. Similarly, $\hat{\delta}_0$ has a holomorphic extension $\hat{\delta} : \hat{\mathsf{S}} \to \hat{\mathsf{S}}$. Since $\hat{\gamma}\hat{\delta} = \hat{\delta}\hat{\gamma} = \mathrm{id}$ on $\hat{\mathsf{S}}$, $\hat{\gamma}$ is an automorphism of $\hat{\mathsf{S}}$.

Therefore, the automorphism $[\omega]_*$ of $F(G)$ induces a finite group $\hat{\Gamma}$ of automorphisms of $\hat{\delta}$. Then the quotient space $\hat{\tilde{\mathfrak{S}}} = \hat{\delta}/\hat{\Gamma}$ becomes a normal complex space by a Cartan's theorem. If $\hat{\pi}$ is the projection of $\hat{\tilde{\mathfrak{S}}}$ onto D, then by its construction, $(\hat{\tilde{\mathfrak{S}}}, \hat{\pi}, D)$ is a completion of $(\mathfrak{S}, \pi, D^*)$ and $\hat{S}_0 = \hat{\pi}^{-1}(0)$ is isomorphic to S_0/Γ_0, where Γ_0 is a finite automorphism group of S_0 induced by $\hat{\Gamma}$. Thus we have Theorem 4.

§7. *An extension theorem*

THEOREM 5. *Let* $(\mathfrak{S}, \pi, D^*)$ *be a holomorphic family of compact Riemann surfaces of genus* g *and* $(\hat{\tilde{\mathfrak{S}}}, \hat{\pi}, D)$ *the completion of* $(\mathfrak{S}, \pi, D^*)$ *constructed canonically as above and let* $\mathfrak{S}'$ *be the image of* $\mathfrak{S}$ *by the inclusion map of* $\mathfrak{S}$ *into* $\hat{\tilde{\mathfrak{S}}}$. *Then, for a complex manifold* X *and for a locally closed complex submanifold* A *of* X, *every locally liftable holomorphic mapping* $f : X - A \to \mathfrak{S}'$ *extends to a holomorphic mapping* $\hat{f} : X \to \hat{\tilde{\mathfrak{S}}}$.

Proof. From the construction of $\hat{\tilde{\mathfrak{S}}}$, this theorem can be proved by the same reasoning as that in the proof of Lemma 1.

THEOREM 6. *Let* $(\mathfrak{S}, \pi, D)$ *be a holomorphic family of compact Riemann surfaces of genus* g *with a fiber* S_0 *over* $t = 0$ *and* $(\hat{\tilde{\mathfrak{S}}}, \hat{\pi}, D)$ *the completion of* $(\mathfrak{S} - S_0, \pi, D^*)$ *constructed canonically as above. Then the inclusion mapping* $j : \mathfrak{S} - S_0 \to \hat{\tilde{\mathfrak{S}}}$ *induces a bimeromorphic equivalence* $\hat{j} : \mathfrak{S} \to \hat{\tilde{\mathfrak{S}}}$.

Proof. The set $\mathrm{Sing}(S_0)$ of singular points of S_0 has at least codimension 2, that is, $\mathrm{Sing}(S_0)$ is the set of finite points. By Theorem 5, $j : \mathfrak{S} - S_0 \to \hat{\tilde{\mathfrak{S}}} - \hat{S}_0$ has a holomorphic extension $\tilde{j} : \mathfrak{S} - \mathrm{Sing}(S_0) \to \hat{\tilde{\mathfrak{S}}}$.

The graph $G_{\tilde{j}}$ of $\tilde{j}$ is the set $\{(p, \tilde{j}(p)) | p \in \mathfrak{S} - \mathrm{Sing}(S_0)\}$ and $G_{\tilde{j}}$ is an analytic subset of $\mathfrak{S} \times \hat{\tilde{\mathfrak{S}}} - \mathrm{Sing}(S_0) \times \hat{S}_0$. Since $\dim \mathrm{Sing}(S_0) \times \hat{S}_0 = 1$ and $\dim_x G_{\tilde{j}} = 2$ for every point $x \in G_{\tilde{j}}$, Remmert-Stein's theorem shows that the closure $\bar{G}_{\tilde{j}}$ of $G_{\tilde{j}}$ in $\mathfrak{S} \times \hat{\tilde{\mathfrak{S}}}$ is an analytic subset of $\mathfrak{S} \times \hat{\tilde{\mathfrak{S}}}$. Further, for projections $\omega : \mathfrak{S} \times \hat{\tilde{\mathfrak{S}}} \to \mathfrak{S}$ and $\hat{\omega} : \mathfrak{S} \times \hat{\tilde{\mathfrak{S}}} \to \hat{\tilde{\mathfrak{S}}}$, the mappings

$\omega|_{\bar{G}_{\tilde{j}}}$ and $\hat{\omega}|_{\bar{G}_{\tilde{j}}}$ are proper. Hence j extends to a bimeromorphic mapping $\hat{j} : \hat{\mathfrak{S}} \to \hat{\mathfrak{S}}$. This completes the proof of Theorem 6.

DEPARTMENT OF MATHEMATICS
COLLEGE OF GENERAL EDUCATION
OSAKA UNIVERSITY
TOYANAKA, OSAKA, 560 JAPAN

REFERENCES

[1] W. Abikoff, Moduli of Riemann surfaces, in "A crash course on Kleinian groups," Springer Lecture notes, No. 400 (1974), 79-93.

[2] ________, On boundaries of Teichmüller spaces and on Kleinian groups: III, Acta Math. 134 (1975), 211-237.

[3] L. Ahlfors, The complex analytic structure of the space of closed Riemann surfaces, in "Analytic functions," Princeton Univ. Press (1960), 45-66.

[4] ________, Lectures on quasiconformal mappings, Van Nostrand Math. Studies #10 (1966).

[5] L. Bers, Uniformization by Beltrami equations, Comm. Pure Appl. Math. 14 (1961), 215-228.

[6] ________, On boundary of Teichmüller spaces and Kleinian groups: I, Ann. of Math. 91 (1970), 570-600.

[7] ________, Uniformization, Moduli and Kleinian groups, Bull. of London Math. Soc. 4 (1972), 257-300.

[8] ________, Fibre spaces over Teichmüller spaces, Acta Math. 130 (1973), 89-126.

[9] ________, Spaces of degenerating Riemann surfaces, in "Discontinuous groups and Riemann surfaces," Ann. of Math. Studies 79, Princeton Univ. Press, (1974), 43-55.

[10] C. J. Earle, On the Carathéodory metric in Teichmüller spaces in "Discontinuous groups and Riemann surfaces," Ann. of Math. Studies 79, Princeton Univ. Press, (1974), 99-103.

[11] P. A. Griffiths, Complex analytic properties of certain Zariski-open sets on algebraic varieties, Ann. of Math. 94 (1971), 21-51.

[12] S. Kobayashi, Hyperbolic manifolds and holomorphic mappings, Marcel Dekker Inc., New York, 1970.

[13] S. Kobayashi and T. Ochiai, Satake compactification and the great Picard theorem, J. Math. Soc. Japan 23 (1971), 340-350.

[14] A. Marden, The geometry of finitely generated Kleinian groups, Ann. of Math. 99 (1974), 383-462.

[15] B. Maskit, On boundary of Teichmüller spaces and on Kleinian groups: II, Ann. of Math. 91 (1970), 607-639.

[16] T. Nishino, Nouvelles recherches sur les fonctions entières de plusieurs variables complexes: [V], Fonctions qui se réduisent au polynômes, J. Math. Kyoto Univ., 15 (1975), 527-553.

COMMUTATORS IN SL(2, C)

Troels Jørgensen[*]

As usual, we denote by τ the trace function on SL(2, C). With use of the two well-known identities

1)
$$\tau(xy) + \tau(xy^{-1}) = \tau(x)\tau(y)$$

and

2)
$$\tau(xyx^{-1}y^{-1}) - 2 = [\tau(x) - \tau(y)]^2 - [\tau(xy) - 2][\tau(xy^{-1}) - 2],$$

we will prove the following:

PROPOSITION. *If two elements* x *and* y *with equal traces generate a non-elementary discrete subgroup of* SL(2, C), *then*

$$|\tau(xyx^{-1}y^{-1}) - 2| > \frac{1}{8}.$$

Under the assumptions of the proposition, one can show that $|\tau(xyx^{-1}y^{-1}) - 2|$ attains a global minimum, and that this minimum is less than $\frac{1}{4}$. I do not know its exact value. The method used below would allow us to obtain a constant slightly larger than $\frac{1}{8}$, but not the best possible constant.

____

[*]Supported by the National Science Foundation MCS 78-00949.

If no hypothesis is made about the traces of x and y , then the trace of the commutator may be arbitrarily close to 2 . This was shown in [2].

Let us put

$$\alpha = |\tau(xy) - 2|$$
$$\beta = |\tau(xy^{-1}) - 2|$$
$$\gamma = |\tau(xyx^{-1}y^{-1}) - 2|$$
$$\delta = |\tau^2(x) - 4| .$$

Assuming that $\tau(x) = \tau(y)$, we obtain from 2)

3) $$\alpha \cdot \beta = \gamma$$

and from 1), with use of the triangle inequality,

4) $$\alpha + \beta \geq \delta .$$

Now recall the following result:

LEMMA. *If* A *and* B *generate a discrete subgroup of* SL(2, C), *then*

$$|\tau^2(A) - 4| + |\tau(ABA^{-1}B^{-1}) - 2| \geq 1 ,$$

unless $BAB^{-1} \epsilon \{A, A^{-1}\}$, *in which case the subgroup is elementary.*

This was proved in [1]. When applied to x and y , one obtains with use of 4) the inequality

5) $$\alpha + \beta + \gamma \geq 1 .$$

If instead the lemma is applied with xy or xy^{-1} as A and y as B , then one sees almost immediately that

6) $$\alpha^2 + 4\alpha + \gamma \geq 1$$

and

7) $$\beta^2 + 4\beta + \gamma \geq 1 .$$

With this information available, we are ready to prove the proposition.

By symmetry, we may assume that

8)
$$\alpha \le \sqrt{\gamma} \le \beta .$$

If $\alpha \le \frac{1}{5}$, then 6) shows that $\gamma > \frac{1}{8}$.

If $\alpha > \frac{1}{5}$ and $\gamma \le \frac{1}{8}$, then we deduce from 3) and 5) and 3) and 5) that

$$\beta < \frac{5}{8} \quad \text{and} \quad \alpha > \frac{1}{4} \quad \text{and} \quad \beta < \frac{1}{2}$$

and

$$\alpha > \frac{3}{8} > \sqrt{\frac{1}{8}} > \sqrt{\gamma} ,$$

contrary to 8). This finishes the proof.

UNIVERSITY OF MINNESOTA

REFERENCES

[1] T. Jørgensen, On discrete groups of Möbius transformations, Amer. J. Math. 98 (1976), 739-749.

[2] ________, Comments on a discreteness condition for subgroups of SL(2, C), Can. J. Math. 31 (1979), 87-92.

TWO EXAMPLES OF COVERING SURFACES

T. Jørgensen, A. Marden, C. Pommerenke[*]

1. *Introduction*

We will present two classes of Riemann surfaces which are, in a sense, opposite extremes:

1. Surfaces S which cover nothing,

2. Surfaces S which cover themselves.

By a surface S we mean a *maximal* Riemann surface. That is, one for which there is no embedding $S < S'$ as a proper subset such that the inclusion of the fundamental group $\pi_1(S) \to \pi_1(S')$ is injective.

By a cover we mean an analytic covering $\pi: S \to S_0$ such that (i) every point $q \in S_0$ is the center of a closed disk Δ such that each component of $\{\pi^{-1}(\Delta)\}$ is compact (in short, closed arcs can be lifted except at branch points), and (ii) the branching, if any, is constant on each fiber $\{\pi^{-1}(q)\}$, $q \in S_0$. We do not admit the trivial covering $S_0 = S$ and $\pi = \mathrm{id}$.

Example (2) requires and example (1) becomes more difficult to find if we stipulate that $\pi_1(S)$ must have infinite rank.

The examples will be constructed in the context of fuchsian groups acting on hyperbolic 2-space **H**, which will be realized as the upper half plane or unit disk D depending on which is more convenient under

---

[*]Authors supported in part by the National Science Foundation.

the circumstances. The connection with maximal surfaces arises through the fact that a surface is maximal if and only if the fuchsian covering group in its universal covering surface is of the first kind.

2. *Maximal fuchsian groups*

Let Γ be a fuchsian group acting on a particular realization of $\mathbf{H}$. Γ is called *maximal* if the augmented group $\langle \Gamma, t \rangle$ is not discrete for each Möbius transformation t acting on $\mathbf{H}$ with $t \notin \Gamma$. Maximal groups are necessarily of the first kind. Finitely generated ones were investigated principally by L. Greenberg [2]. He showed that for most signatures, the groups which are maximal form a dense open set, and the others lie on subvarieties representing lower dimensional Teichmüller spaces. In particular for finite $n \geq 4$, most free groups of rank n are maximal. In contrast to Greenberg's analysis however, we will exhibit fairly explicitly maximal groups of rank n. At the same time our analysis works for the infinite rank case, whereas dimensional arguments break down.

THEOREM 1. *For every* n, $4 \leq n \leq \aleph_0$, *there exists a maximal fuchsian group* Γ_n *which is free of rank* n, *and* $\mathbf{H}/\Gamma_n$ *is an* (n+1)-*punctured sphere.*

The proof depends on a property of groups close to the group representing the triply punctured sphere.

PROPOSITION. *Suppose* $G = \langle a, \beta \rangle$ *is a fuchsian group generated by parabolic transformations* a, β. *There exists a universal constant* $\varepsilon > 0$ *such that if*

$$(1) \qquad\qquad 0 < \mathrm{tr}^2 a\beta - 4 < \varepsilon$$

then no group of the form $\langle a^{1/n}, \beta \rangle$, $n \geq 2$, *is discrete.*

The hypothesis of the proposition is equivalent to the condition that $\mathbf{H}/\langle a, \beta \rangle$ be a twice punctured disk with $a\beta$ determined by a simple loop surrounding the two punctures (see §4). Let $\sigma(a\beta)$ denote the length of the geodesic in the Poincaré metric in the free homotopy class of this loop. Condition (1) is equivalent to the relation,

$$(2) \qquad\qquad \sigma(\alpha\beta) < \epsilon^* ,$$

where

$$\epsilon^* = 2 \log(\epsilon^{\frac{1}{2}} + (\epsilon+4)^{\frac{1}{2}})/2 .$$

As is customary, the notation $\text{tr}\,\gamma$ will denote the trace $(a+d)$ of the normalized Möbius transformation $\gamma(z) = (az+b)/(cz+d)$, $ad-bc = 1$. In general this is determined only up to sign. For our work here we will always choose the matrices representing parabolic transformations to have trace $+2$ and then we will not encounter ambiguities of sign. The symbol $a^{1/n}$ designates that parabolic transformation with the same fixed point as a such that $(a^{1/n})^n = a$.

We will prove the proposition after proving the theorem.

Proof of Theorem 1. We will first construct a maximal group Γ_∞ of rank $\aleph_0$. Set $x_0 = 0$ and choose an infinite sequence of disjoint closed intervals $[x_{2k}, x_{2k+1}]$ on the positive real axis, $k = 0, 1, \cdots$, with $x_{2k+1} - x_{2k} < 1/2$. Space the intervals far enough apart so that the disks $D_k = \{z : |z-(x_{2k}+x_{2k+1})/2| < 1\}$ are mutually disjoint. Set $\Omega = C\backslash\cup\{x_i\}$. Now push the endpoints of each interval $[x_{2k}, x_{2k+1}]$ close enough together so that the length of the simple geodesic in Ω (in the Poincaré metric of Ω) separating x_{2k} and x_{2k+1} from all other x_j has length $< \epsilon^*$. This can be fulfilled by estimation in D_k: the Poincaré metric of $D_k\backslash\{x_{2k}\}\cup\{x_{2k+1}\}$ is larger than that of Ω restricted to D_k. Finally, we require that the length of these geodesics approach zero, as $k \to \infty$.

Let Γ_∞ denote a universal covering group for Ω, acting in the unit disk D.

Choose a point $O \epsilon \Omega$ on the negative real axis, the point $z = -1$, say, and fix a point $O^* \epsilon D$ over O. We may assume O^* is the origin. Let $\mathcal{P}$ denote the Poincaré fundamental polygon for Γ with center at O^*.

We claim that the interior of $\mathcal{P}$ projects one-to-one on the region $\Omega_0 \subset \Omega$ obtained by slitting the complex plane along the positive real axis $[0, +\infty)$.

This is true because Ω is invariant under the reflection $z \mapsto \bar{z}$ which is therefore an isometry in the Poincare metric. Consequently, the geodesic from O in Ω to any point in the upper or lower half plane does not cross the real axis. It follows that the lift of Ω_0 from O^* is exactly the interior of $\mathcal{P}$.

We can describe $\partial \mathcal{P}$ by travelling around $\partial \Omega_0$. We see that $\partial \mathcal{P} \cap \partial D$ is the union of a countable number of parabolic fixed points of Γ_∞ and one point ζ which is the limit point of these fixed points. $\mathcal{P}$ is symmetric about the diameter of D through ζ.

REMARK. We could have imposed the additional requirement that the segments $[x_{2k}, x_{2k+1}]$ approach $+\infty$ so fast that the length of the geodesics in Ω, belonging to the free homotopy classes of the circles $\{z : |z| = (x_{2k-1} + x_{2k})/2\}$, are uniformly bounded above. Then the sides of $\mathcal{P}$ would approach ζ non-tangentially.

Now we are ready to show that Γ_∞ is maximal. Start by fixing a free set of generators $\{y_{2k}, y_{2k+1}\}$, $k = 0, 1, \cdots$, with the following property: y_{2k}, y_{2k+1} are induced by simple loops in Ω from O, contractable in Ω to x_{2k}, x_{2k+1} respectively, and so oriented that $y_{2k+1} y_{2k}$ is homotopic in Ω to a simple loop surrounding the segment $[x_{2k}, x_{2k+1}]$, separating it from all other x_j. Thus the group $G_k = \langle y_{2k}, y_{2k+1} \rangle$ satisfies the hypothesis of the proposition.

Suppose that Γ_∞ is not maximal. Then there is a Möbius transformation $t \notin \Gamma$ such that the extended group $\Gamma^* = \langle \Gamma_\infty, t \rangle$ is fuchsian. We distinguish three cases.

Case 1. For each i, the parabolic transformation $t y_i t^{-1}$ is conjugate in Γ_∞ to a power of some y_j. In this case $t \Gamma_\infty t^{-1} \subset \Gamma_\infty$. Consequently, t induces a holomorphic map $f : \Omega \to \Omega$. It follows that f has removable singularities at all the punctures $\{x_j\}$, and ∞ is a pole because f is an isometry in the Poincaré metric on Ω. Therefore f is a polynomial. Because $\{f(x_j)\} \subset \{x_j\}$ and $\{f^{-1}(x_j)\} \subset \{x_j\}$, f must be linear. Finally, the fact that the geodesic lengths around $[x_{2k}, x_{2k+1}]$ tend to zero ensures that f is the identity. This can happen only if $t \in \Gamma_\infty$, a contradiction.

Case 2. For some i, the fixed point of $t\gamma_i t^{-1}$ is also the fixed point of (say) $\sigma\gamma_{2k}\sigma^{-1}$ for some $\sigma \in \Gamma_\infty$, but is not a power of this element. But then, by the proposition, the group $<t\gamma_i t^{-1}, \sigma\gamma_{2k+1}\sigma^{-1}> \subset \Gamma^*$ is not discrete, a contradiction.

Case 3. The only remaining possibility is that for some i, $t\gamma_i t^{-1}$ does not share its fixed point ξ with any element of Γ_∞. Let Δ be a horocycle at ξ with respect to Γ^*: $\gamma^*(\Delta) \cap \Delta = \emptyset$ for each $\gamma^* \in \Gamma^*$, $\gamma^* \neq$ id., which does not have ξ as its fixed point. The Poincaré fundamental polygon $\mathcal{P}^*$ for Γ^* with center at O^* is a proper subset of $\mathcal{P}$. For some $\gamma^* \in \Gamma^*$, the horocycle $\gamma^*(\Delta)$ has its fixed point $\gamma^*(\xi) \in \partial\mathcal{P}^*$. Since $\partial\mathcal{P}^* \cap \partial D \subset \partial\mathcal{P} \cap \partial D$, the only possible location for $\gamma^*(\xi)$ is at ζ. Consequently ζ itself is a parabolic fixed point for Γ^* (but not for Γ_∞).

Note that $\gamma^*(\Delta) \subset \mathcal{P} \cup \{\zeta\}$ since $\gamma^*(\Delta)$ is symmetric with respect to the diameter through ζ and contains no two points equivalent under Γ_∞.

Let δ be a generator of the cyclic subgroup of parabolic transformations in Γ^* fixing ζ. In forming $\mathcal{P}^*$ from $\mathcal{P}$, in particular we remove from $\mathcal{P}$ those two portions separated from O^* by the perpendicular bisectors of the two segments $[O^*, \delta^{\pm 1}(O^*)]$. These bisectors meet and form a cusp at ζ. When the two portions are removed from $\mathcal{P}$ we are left with a polygon of finitely many sides. Therefore $\mathcal{P}^*$ itself has finitely many sides and Γ^* is finitely generated of the first kind. But again we have arrived at a contradiction. For Γ_∞ and hence Γ^* contain infinitely many elements with uniformly bounded but distinct traces, namely $\gamma_{2k+1}\gamma_{2k}$ as $k \to \infty$. This cannot occur in a finitely generated group.

REMARK. If we had imposed the additional condition on Ω suggested in the remark above, then $\mathcal{P}$ could not contain the internally tangent disk $\gamma^*(\Delta)$ at ζ for metric reasons. This fact alone would prevent Case 3 from arising.

Finally, to complete the proof of Theorem 1, we will show how the argument above can be modified to yield the existence of maximal free groups of any rank $n \geq 4$.

The case of odd n (an even number of punctures) is easiest. Choose the pairs $[x_{2k}, x_{2k+1}]$ close together as above so as to obtain the two generator groups G_k satisfying the hypothesis of the proposition. In addition, position the $(n+1)$-punctures so that there is no conformal automorphism of Ω (this cannot be done if there are only four punctures). Only Cases 1 or 2 can arise for the corresponding group Γ_n and both of these are excluded for the reasons cited above.

For the case of even n we proceed in the same way except that an additional observation is necessary. Let γ_{n+1} be a parabolic transformation corresponding to the unpaired puncture. Assume that for one of the paired transformations γ_{2k}, say, it is true that $t\gamma_{2k}t^{-1}$ has the same fixed point ξ as $\sigma\gamma_{n+1}\sigma^{-1}$, some $\sigma \epsilon \Gamma_n$, but is not a power of it. Let δ be a generator of the parabolic subgroup of $\Gamma^* = <\Gamma_n, t>$ fixing ξ. Then $\beta = \sigma^{-1}\delta\sigma$ has the same fixed point as γ_{n+1} and $\gamma_{n+1} = \beta^m$ for some integer m.

We claim that $\beta\Gamma_n\beta^{-1} \subset \Gamma_n$. Consider for example γ_{2j}. If $\beta\gamma_{2j}\beta^{-1} \notin \Gamma_n$ then $\beta\gamma_{2j}\beta^{-1}$ has the same fixed point as $\sigma_1\gamma_{n+1}\sigma_1^{-1}$, some $\sigma_1 \epsilon \Gamma_n$, but is not a power of it. Therefore, for some integer d,

$$\beta\gamma_{2j}\beta^{-1} = \sigma_1\beta^d\sigma_1^{-1},$$

or

$$\gamma_{2j} = (\beta^{-1}\sigma_1)\beta^d(\beta^{-1}\sigma_1)^{-1} = \beta_0^d$$

and γ_{2j} is a power of $\beta_0 \epsilon \Gamma^*$. Since $<\beta_0, \gamma_{2j+1}>$ is not discrete, this is impossible. We conclude that $\beta\Gamma_n\beta^{-1} \subset \Gamma_n$ but this is not possible either.

3. The group for the triply punctured sphere

In this section we will prove the proposition. The proof is based on the following results.

LEMMA 1. *Given $M > 0$ there exists $N > 0$ such that for any non-cyclic fuchsian group $G = {<}x, y{>}$ with two parabolic generators x, y satisfying $|\mathrm{tr}\ xy| < M$, the group ${<}x^{1/n}, y{>}$ is not discrete for any $n \geq N$.*

LEMMA 2. *Suppose $G = {<}a, \beta{>}$ is the group of the triply punctured sphere with the parabolic generators $a(z) = z + 2$, $\beta(z) = z/(-2z + 1)$. Then*

(i) *$G_1 = {<}a^{1/2}, \beta{>}$ contains G as a normal subgroup of index two, and $a^{1/2}\beta$ is elliptic or order two,*

(ii) *$G_2 = {<}a^{1/4}, \beta{>}$ contains G as a normal subgroup of index six, $a^{1/4}\beta$ is elliptic of order three, and G_2 is conjugate to the modular group. The cases (i) and (ii) are the only discrete groups of the form ${<}a^{1/n}, \beta{>}$, $n \geq 2$.*

Proof of Lemma 1. By conjugating G we may assume $x : z \mapsto z + a$, $y : z \mapsto z/(-2z + 1)$. Replacing x by x^{-1} if necessary we may assume $a > 0$. According to Siegel [5] or to [4], when G is a non-elementary discrete group, $a \geq 1/2$. Thus if $a = (2 - \mathrm{tr}\ xy)/2 \leq M_1$, for $n > 2M_1$ the group ${<}x^{1/n}, y{>}$ cannot be discrete.

Proof of Lemma 2. When $a = 2$, the only possibilities for adjoining roots are $n = 2, 3$ or 4 since $a \geq n/2$. The case $n = 3$ cannot occur because then $a^{1/3}\beta$ would be an elliptic element of trace $2/3$. This is not the trace of an element of finite order. The other cases are as indicated.

Proof of the proposition. It is known that for ε sufficiently small, say $\varepsilon < \varepsilon_0$, (1) implies that $a\beta$ is a simple hyperbolic element. This in turn implies that $H/{<}a, \beta{>}$ is a twice punctured disk.

If the conclusion is false then for each $k > 1/\varepsilon_0$ there is a group ${<}x_k, y_k{>}$ with $0 < \mathrm{tr}\ x_k y_k - 4 < 1/k$. Normalize so that $x_k : z \mapsto z + a_k$ and $y_k = \beta : z \mapsto z/(-2z + 1)$. Because $x_k y_k$ is a simple hyperbolic element, $a_k > 2$. In the notation of Lemma 2, $\lim x_k = a$.

There are only a finite number of possible n for which ${<}x_k^{1/n}, y_k{>}$ is discrete (Lemma 1). Since ${<}x_k^{1/n}, y_k{>} \to {<}a^{1/n}, \beta{>}$, for large k

only the cases $n = 2$ and $n = 4$ have any chance of yielding discrete groups. For otherwise $\langle a^{1/n}, \beta \rangle$ would not be discrete (Lemma 2) and $\langle x_k^{1/n}, y_k \rangle$ would contain elements arbitrarily close to the identity as $k \to \infty$. This is impossible by [5] or [4]. But the same line of reasoning shows that the cases $n = 2$, $n = 4$ cannot occur either, for large k. For if, for example, the group $\langle x_k^{1/2}, y_k \rangle$ were discrete for infinitely many k , it would contain elliptic elements of arbitrarily high order and hence elements arbitrarily close to the identity as $k \to \infty$, since $\operatorname{tr} x_k^{1/2} y_k = 2 - a_k$ and $a_k \to 2$. We conclude that for sufficiently small ε , the proposition is true.

REMARK. Instead of the proposition, another approach is to observe that, with the normalization of the generators x, y of $\langle x, y \rangle$ as in the proof of Lemma 1, if $2 < a < 4$ then $x^{1/n}y$ is elliptic of trace $2 - 2a/n$, $n \geq 2$. Consequently, if a is known to be transcendental, no group $\langle x^{1/n}, y \rangle$, $n \geq 2$ is discrete. In the context of our example Γ_∞ , this can be arranged for each G_k as follows. Set up Γ_∞ as above but introduce the parameter $\lambda = x_1 - x_0$, keeping the other points x_j fixed. As λ decreases, the geodesic length σ_k of the simple loop in Ω surrounding $[x_{2k}, x_{2k+1}]$ also strictly decreases. The values of λ for which $\exp(\sigma_k)$ is an algebraic number for some $k \geq 1$ are countable. Hence for all except a countable number of values of λ , the traces of all $y_{2k+1} y_{2k}$ are transcendental.

4.   *Conformal automorphisms of surfaces*

The following is a rather striking application of Theorem 1. Without the condition that the surfaces involved are maximal, it was obtained by Greenberg [1].

COROLLARY. *Let* Φ *be an abstract n-generator group,* $4 \leq n \leq \aleph_0$. *There exists a maximal Riemann surface S whose group of conformal automorphisms is isomorphic to Φ . The surface S can be taken to be a covering surface of the* $(n+1)$ *or* $(n+2)$ *-punctured sphere.*

Proof. Working with our maximal free group Γ_n of rank n, there is an isomorphism

$$\Phi \cong \Gamma_n/N$$

where N is the normal subgroup of Γ_n generated by the ''relations'' of Φ. We must, however, ensure that $N \neq \{id.\}$. Therefore if Φ is itself free of rank $n < \aleph_0$, use Γ_{n+1} instead of Γ_n, or if Φ is free of rank $\aleph_0$, choose the isomorphism to send Φ into Γ_∞, not onto. Then N will be a non-elementary group (one having more than two limit points) and therefore, having the same limit set as Γ_n, will be a group of the first kind. Let S be the maximal surface

$$S = H/N \ .$$

The group of conformal automorphism Aut S of S is determined by the normalizer of N in the group of all Möbius transformations acting on H. This group is discrete, hence because of maximality it is precisely Γ_n or Γ_{n+1}. That is, Aut $S \cong \Gamma_n/N$ or Γ_{n+1}/N (if Φ is free of rank n).

REMARK. Aut S is exactly the group of cover transformations of S over the $(n+1)$ or $(n+2)$ -punctured sphere. Thus Φ is an n-generator finite group if and only if S is a finite sheeted cover of the $(n+1)$-punctured sphere.

5. *Groups conjugate to a proper subgroup*

In this section we will investigate fuchsian groups F for which $tFt^{-1} < F$ for some Mobius transformation t (the symbol $<$ denotes proper inclusion). This phenomenon does not appear for finitely generated groups nor, according to Heins [3], for groups without elliptic elements which are of divergence type (i.e., the quotient surface does not support a Green's function). Here we will construct large families of such groups. Since we allow groups with elliptic elements, the examples presented are more general than those suggested in the introduction. According to Heins [3], the question for covering surfaces was originally raised by H. Hopf.

Without the requirement that F be a group of the first kind, trivial examples abound. For instance let C_1 denote the semi-circle $\{z : |z-1/2| = 1/4,\ \mathrm{Im}\ z \geq 0\}$ and C_2 the semi-circle $\{z : |z-3/2| = 1/4,\ \mathrm{Im}\ z \geq 0\}$. Let A_0 denote a Möbius transformation sending the upper half plane to itself and the exterior of C_1 in it onto the interior of C_2. Let t denote the translation $z \mapsto z+2$ and set $A_k = t^k A_0 t^{-k}$, $k \geq 0$. Then the group F generated by all A_k, $k \geq 0$, has the property that $tFt^{-1} < F$.

When the conjugation t is parabolic, we can, in a sense, describe all the possibilities.

THEOREM 2. *If for some fuchsian group F of the first kind and parabolic transformation t it is true that $tFt^{-1} < F$, then $G = <F,t>$ is a fuchsian group that splits into a free product of the form $G = G_0 * <t>$. Conversely, if G is a fuchsian group of the first kind with a decomposition $G = G_0 * <t>$ where $t \in G$ is parabolic, then there exist groups F of the first kind with $G = <F,t>$ and $tFt^{-1} < F$.*

Proof. If $tFt^{-1} < F$, the groups $t^k F t^{-k}$, $-\infty < k < \infty$, are nested; the larger the k, the smaller the group. Set $H = \bigcup_{k=-\infty}^{\infty} t^k F t^{-k}$. Then H satisfies $tHt^{-1} = H$ while $t^n \notin H$ for all $n \neq 0$.

Set $G = <H,t> = <F,t>$. Then G is also a fuchsian group of the first kind (because H is a normal subgroup) and satisfies,

(i) $G = \{t^n h : h \in H\}$, $G/H \cong <t>$,

(ii)   the commutator subgroup $[G,G]$ lies in H,

(iii) for $n \neq 0$ and $h \in H$, $t^n h$ is not elliptic or the identity.

These are true because in any word in G, the letter t can be moved to the left end by virtue of the fact $tHt^{-1} = H$.

If G is finitely generated it has a standard generator system of hyperbolic elements $\{a_i, b_i\}$ elliptic $\{e_i\}$, and parabolic $\{p_i\}$ satisfying the relation $\Pi[a_i, b_i]\Pi e_i \Pi p_i = 1$. By (ii) the commutator $[a_i, b_i] \in H$ and by (iii) $e_i \in H$. We can assume one p_i is t. Therefore if this relation is to hold, there must be at least two parabolic transformations p_i. This

means the surface H/G has at least two punctures, one of which determines t. Draw a simple arc on this surface from one of these punctures
to the other. This arc determines a splitting of G as required.

If G is not finitely generated, then the surface H/G has a puncture
determined by t and at least one other ideal boundary component. Con-
nect these two by a simple arc. In the same way as above, this determines
a splitting of G.

To prove the converse we begin by constructing the fuchsian group
$H \subset G$ which is the free product of the groups $t^k G_0 t^{-k}$, $-\infty < k < \infty$,

$$H = \cdots * t^{-1} G_0 t * G_0 * t G_0 t^{-1} * \cdots .$$

Because H is a (non-elementary) normal subgroup of G, it has the same
limit set and therefore is also of the first kind.

Define the projection $p_k : H \to t^k G_0 t^{-k}$ which is a homomorphism as
follows. For $x \in t^j G_0 t^{-j}$, $j \neq k$, set $p_k(x) = \mathrm{id.}$; for $x \in t^k G_0 t^{-k}$, set
$p_k(x) = x$, and extend to a homomorphism of all H.

Let N be any proper normal subgroup of G_0, for example, $N = \{\mathrm{id.}\}$.
Define
$$F = \{h \in H : p_k(h) \in t^k N t^{-k} \quad \text{for all} \quad k < 0\} .$$

The group F is not elementary and it is a normal subgroup of H. Consequently F is of the first kind. Furthermore, $tFt^{-1} < F$ since $tFt^{-1} \cap G_0 = N$
while $F \cap G_0 = G_0$.

EXAMPLE. Perhaps the simplest example of the construction above is the
following. Let G be the $(2, \infty, \infty)$ triangle group. It splits as
$G = \langle e_0 \rangle * \langle t \rangle$ where e_0 is elliptic of order two and t is parabolic. H
is the free product of groups of order two,

$$H = \cdots * \langle e_{-1} \rangle * \langle e_0 \rangle * \langle e_1 \rangle * \cdots ,$$

where $e_k = t^k e_0 t^{-k}$. If we take t as $z \mapsto z + 1$ the geometric picture in
the upper half plane is quite nice: Each e_k has its fixed point at the

north pole of the circle of radius $1/2$ centered at $z = k$. The group F can be constructed as above taking $N = \{id.\}$, or as follows:

$$F = \{h \in H : \text{the total number of } e_k\text{'s with } k < 0 \text{ appearing in } h \text{ is even}\}.$$

Then F is a subgroup of index two in H and $e_0 \notin tFt^{-1}$ while $e_0 \in F$.

REMARK. Suppose $tFt^{-1} < F$ for a hyperbolic transformation t. We apply the methods above and use the notation introduced there. It is true that for some n and $h \in H$, $t^n h$ is a simple hyperbolic transformation, that is, represents a simple loop δ on H/G. Either δ is non-dividing or it divides H/G into two parts, neither of which is relatively compact.

Conversely, if there is a splitting $G = G_0 * <t>$, t hyperbolic but not necessarily simple, we can construct as above a group F with $tFt^{-1} < F$.

On the other hand, suppose G is a surface group of genus $g + 1 \geq 3$. Cutting the surface along a simple, non-dividing loop gives rise to a representation of G as an HNN extension of a subgroup G_0 by a hyperbolic transformation t. The presentation of G_0 looks like

$$G_0 = <a_i, b_i, x, t^{-1}xt : \Pi[a_i, b_i]xt^{-1}xt = 1> ,$$

where $1 \leq i \leq g$. A homomorphism $\phi : G_0 \to \Sigma_g$ onto the surface group of genus g, $\Sigma_g = <a_i', b_i' : \Pi[a_i', b_i'] = 1>$, is determined by setting $\phi(a_i) = a_i'$, $\phi(b_i) = b_i'$ and $\phi(x) = \phi(t^{-1}xt) = 1$. Let ϕ_k denote the corresponding homomorphism $t^k G_0 t^{-k} \to \Sigma_g$.

Consider the free product with amalgamation,

$$H = \cdots \underset{<\cdot>}{*} t^{-1}G_0 t \underset{<t^{-1}xt>}{*} G_0 \underset{<x>}{*} tG_0 t^{-1} \underset{<\cdot>}{*} \cdots .$$

We can extend each ϕ_k to a homomorphism $\phi_k : H \to \Sigma_g$ after requiring $\phi_k(x) = id.$ for $x \in t^j G_0 t^{-j}$, $j \neq k$. Now form the normal subgroup,

$$F = \{h \in H : \phi_k(x) = id. \text{ for all } k < 0\}.$$

Again, $tFt^{-1} < F$ and F is of the first kind.

The point of this last example is that $H/\langle F, t \rangle$ can even be a happy, closed surface.

UNIVERSITY OF MINNESOTA

REFERENCES

[1] L. Greenberg, Conformal transformations of Riemann surface, Amer. J. Math. *82*(1960), 749-760.

[2] _________, Maximal groups and signatures, in Discontinuous Groups and Riemann Surfaces, L. Greenberg, ed., Annals of Math. Studies 79, Princeton Univ. Press (1974).

[3] M. Heins, On a problem of H. Hopf, Jour. de Math. *37*(1958), 153-160.

[4] T. Jørgensen, On discrete groups of Möbius transformations, Amer. J. Math. *98*(1976), 739-749.

[5] C. L. Siegel, Über einige Ungleichungen bei Bewegungsgruppen in der nichteuklidischen Ebene, Math. Ann. *133*(1957), 127-138.

DEFORMATIONS OF SYMMETRIC PRODUCTS

George R. Kempf[*]

Let C be a smooth complete algebraic curve of genus g over an algebraically closed field k. If $n > 2g-2$, any universal abelian integral $\int_n : C^{(n)} \to J$ from the n-th symmetric product $C^{(n)}$ to the Jacobian J of C is a locally trivial P^r-bundle, where, $r = n-g$.

If one deforms J as an abelian variety, one may ask whether the projective bundle $\int_n$ may also be simultaneously deformed along with J. The main result of this paper is that such a deformation is impossible even in first order unless the deformation of J comes from a deformation of C, if C is not hyperelliptic.

We first show that any deformation of the variety $C^{(n)}$ must come from a deformation of C, when C is not hyperelliptic. From this, the main result easily follows.

My proof uses heavily the deformation theory of Kodaira-Spencer-Grothendieck. These methods allow effective linearization of the deformation problems to accessible problems in global linear algebra.

The problem of understanding explicitly how these projective bundles $\int_n : C^{(n)} \to J$ are twisted up remains the unsolved inversion problem. One would like to know explicitly how they become unraveled as one pulls

[*] Partly supported by NSF Grant No. MPS 75-05578

them back to the universal covering space of the Jacobian variety. Further, background information about these bundles may be found in [5].

One may view this problem from the point of view of the curve and its symmetric products or invert the problem and look at it from the point of view of the Jacobian. Both points of view have their own merits. In this paper, I am proceeding from the curve's viewpoint. In the previous paper [6], I was looking at the essentially same situation from the Jacobian point of view.

I will use without particular references standard facts from deformation theory. General references may be found in [1, 3, 4, 10, 11 and 12]. Further material related to curves may be found in [2 and 8]. Good references for the basic facts about symmetric products are [7 and 13].

§1. *Symmetric products*

Let n be a non-negative integer. Let C^n denote the product of C with itself n-times. Let $C^{(n)}$ denote the symmetric product of C with itself n-times. Then $C^{(n)}$ is the quotient of $C^n/Sym(n)$, where $Sym(n)$ is the symmetric group which permutes the factors of C^n. Let $\rho_n : C^n \to C^{(n)}$ be the quotient morphism, which sends $(c_1, \cdots, c_n)$ to $c_1 + \cdots + c_n$.

Recall that C^n and $C^{(n)}$ are smooth varieties of dimension n. Furthermore, ρ_n is a faithfully flat finite morphism of degree $n!$ If $n > 0$, we have another morphism $\sigma_n : C \times C^{(n-1)} \to C^{(n)}$, which sends $(c, c_1 + \cdots + c_{n-1})$ to $c + c_1 + \cdots + c_{n-1}$. It is known that $(\pi_C, \sigma_n) : C \times C^{(n-1)} \to C \times C^{(n)}$ maps $C \times C^{(n-1)}$ isomorphically onto a divisor D_n on $C \times C^{(n)}$. The induced isomorphism $C \times C^{(n-1)} \xrightarrow{\approx} D_n$ will be denoted by γ_n.

For any smooth variety X of dimension r, let ω_X denote the invertible $\mathcal{O}_X$-module of r-forms, $\Lambda^r \Omega^1_X$, on X. Also, Θ_X will denote the sheaf of regular vector fields, $\mathcal{H}om_{\mathcal{O}_X}(\Omega^1_X, \mathcal{O}_X)$, on X.

We will later need to know the following facts from differential calculus.

LEMMA 1.1. *We have natural isomorphisms,*

a) $\omega_{C \times C^{(n)}} \approx \pi_C^* \omega_C \otimes \pi_{C^{(n)}}^* \omega_{C^{(n)}}$,

b) $\sigma_n^* \omega_{C^{(n)}}(D_{n-1}) \approx \omega_{C \times C^{(n-1)}}$ *if* $n > 0$,

c) $\omega_{D_n} \approx \omega_{C \times C^{(n)}}(D_n)|_{D_n}$ *and*

d) $\gamma_n^*(\mathcal{O}_{C \times C^{(n)}}(D_n)|_{D_n}) \approx \pi_C^* \Theta_C(D_{n-1})$ *if* $n > 0$.

Proof. By advanced calculus, $\Omega^1_{C \times C^{(n)}} \approx \pi_C^* \Omega_C^1 \oplus \pi_C^* \Omega_{C^{(n)}}^1$. By exterior multiplication, we have an isomorphism $\Lambda^{n+1} \Omega^1_{C \times C^{(n)}} \approx \pi_C^* \Omega_C^1 \otimes \pi_C^* \Lambda^n \Omega_{C^{(n)}}^1$. This gives the isomorphism a). To prove b), notice that we have a natural $\mathcal{O}_{C \times C^{(n-1)}}$-homomorphism, $\tau : \sigma_n^* \omega_{C^{(n)}} \to \omega_{C \times C^{(n-1)}}$, between invertible sheaves. Thus, we have an induced isomorphism $\sigma_n^* \omega_{C^{(n)}}(E) \to \omega_{C \times C^{(n-1)}}$, where E is the divisor of ramification of σ_n. Thus, b) means that σ_n ramifies once along D_{n-1}, which may be easily checked. For c), we use the usual identification of ω_D with $\omega_{C \times C^{(n)}}|_{D_n}$ tensored with the normal sheaf $\mathcal{O}_{C \times C^{(n)}}(D_n)|_{D_n}$ of the smooth divisor D_n on $C \times C^{(n)}$.

The isomorphism of d) will be deduced from the others. By c), we have an isomorphism $\mathcal{O}_{C \times C^{(n)}}(D_n)|_{D_n} \approx \omega_D \otimes (\omega_{C \times C^{(n)}}|_{D_n})^{\otimes -1}$. Thus, $\gamma_n^*(\mathcal{O}_{C \times C^{(n)}}(D_n)|_{D_n}) \approx \omega_{C \times C^{(n-1)}} \otimes \gamma_n^*(\omega_{C \times C^{(n)}}|_{D_n})^{\otimes -1}$. On the other hand, $\gamma_n^*(\omega_{C \times C^{(n)}}|_{D_n}) \approx (\pi_C, \sigma_n)^*(\omega_{C \times C^{(n)}}) \approx (\pi_C, \sigma_n)^*(\pi_C^* \omega_C \otimes \pi_{C^{(n)}}^* \omega_{C^{(n)}})$ $\approx \pi_C^* \omega_C \otimes \sigma_n^* \omega_{C^{(n)}}$ by a). Using b), we get an isomorphism $\gamma_n^*(\omega_{C \times C^{(n)}}|_{D_n})$ $\approx \pi_C^* \omega_C \otimes \omega_{C \times C^{(n-1)}}(-D_{n-1})$. Summing up, we have $\gamma_n^*(\mathcal{O}_{C \times C^{(n)}}(D_n)|_{D_n})$ $\approx \omega_{C \times C^{(n-1)}} \otimes (\pi_C^* \omega_C \otimes \omega_{C \times C^{(n-1)}}(-D_{n-1}))^{\otimes -1} \approx \pi_C^* \Theta_C(D_{n-1})$. Q.E.D.

Let c be a point of C. We have the morphism $\sigma(c): C^{(n-1)} \to C^{(n)}$ which sends $c_1 + \cdots + c_{n-1}$ to $c + c_1 + \cdots + c_{n-1}$. This morphism induces an isomorphism $\gamma_n(c): C^{(n-1)} \to c + C^{(n-1)}$, where $c + C^{(n-1)}$ is a divisor on $C^{(n)}$. We will also need to know

LEMMA 1.2. *If* $g > 0$ *and* $n > 0$, $k = \Gamma(C^{(n)}, \mathcal{O}_{C^{(n)}}) \xrightarrow{\approx} \Gamma(C^{(n)}, \mathcal{O}_{C^{(n)}}(c + C^{(n-1)}))$.

Proof. As $g > 0$, $k = \Gamma(C, \mathcal{O}_C) \approx \Gamma(C, \mathcal{O}_C(c))$. Let $\mathfrak{M} = \bigotimes_{1 \leq i \leq n} \pi_i^* \mathcal{O}_C(c)$ be the product sheaf on C^n. Thus, as $\Gamma(C^n, \mathfrak{M}) \approx \otimes_i \Gamma(C, \mathcal{O}_C(c)) \approx k$, we know that any everywhere regular function on C^n with at most simple poles along the divisor $S = \sum_{1 \leq i \leq n} C \times \cdots \times C \times c \times C \times \cdots \times C$ must be a constant. One may easily check that $\rho_n^{-1}(c + C^{(n-1)}) = S$ and, hence, we have an injection $\Gamma(C^{(n)}, \mathcal{O}_{C^{(n)}}(c + C^{(n-1)})) \hookrightarrow \Gamma(C^n, \mathcal{O}_{C^n}(S)) = k$. Therefore, $\Gamma(C^{(n)}, \mathcal{O}_{C^{(n)}}(c + C^{(n-1)}))$ consists only of constants. Q.E.D.

We will later need to know about the first cohomology group $H^1(C^{(n)}, \mathcal{O}_{C^{(n)}})$ of the symmetric product. Before stating what we will need to know, we will be required to understand a more general fact.

Let G be an algebraic group. A principal G-bundle over a variety X is a smooth morphism $\pi : Y \to X$ together with a group action $G \times Y \to Y$ such that

    1) $\pi(g \cdot y) = \pi(y)$ for all g in G and y in Y and

    2) the induced morphism $G \times Y \to Y \times_X Y$ sending (g, y) to (gy, y) is an isomorphism.

A trivial principal G-bundle over X is the product $\pi_X : G \times X \to X$, where G acts on $G \times X$ by $g_1 \cdot (g_2, x) = (g_1 \cdot g_2, x)$. If $f : Z \to X$ is a morphism and $\pi : Y \to X$ is a principal G-bundle, then we have an induced principal G-bundle $f^* \pi \equiv \pi_Z : Y \times_X Z \to Z$, where G acts on $Y \times_X Z$ via the factor Y.

With all this fancy language, we may state the next result.

LEMMA 1.3. *Let* G *be an affine group. Let* $\pi : Y \to C^{(n)}$ *be a principal* G-*bundle. Then* $\rho_n^* \pi$ *is a trivial bundle over the product* C^n *if and only if* π *is a trivial bundle over* $C^{(n)}$.

Proof. Clearly, if π is trivial, then $\rho_n^* \pi$ is trivial. Conversely, assume that $\rho_n^* \pi : Y' \equiv Y \times_{C^{(n)}} C^n \to C^n$ is trivial. Let $\tau : C^n \to Y'$ be a section of $\rho_n^* \pi$. Let τ' be another section of $\rho_n^* \pi$. Then, $\tau' = \sigma \cdot \tau$, where $\sigma : C^{(n)} \to G$ is a morphism. As C^n is connected and complete and G is affine, σ must be a constant. Therefore, we have proven that any section of $\rho_n^* \pi$ must have the form $g \cdot \tau$ for some point g of G.

The symmetric group $\mathrm{Sym}(n)$ acts on Y' in a natural way such that $\rho_n^* \pi : Y' \to C^n$ is equivariant. If we can show that the image $\tau(C^n)$ is invariant under the action of $\mathrm{Sym}(n)$ on Y', then $\tau(C^n)/\mathrm{Sym}(n) \subseteq Y'/\mathrm{Sym}(n) = Y$ will be isomorphic to $C^{(n)}$ via $\pi : Y \to C^{(n)}$ and, hence, π will be trivial bundle.

Let h be a permutation in $\mathrm{Sym}(n)$. Then, $\tau(C^n) * h$ must have the form $g \cdot \tau(C^n)$ for some g in G by the first paragraph. If a point a in C^n is a fixed point of h, then h acts trivially on the fiber of $\rho^* \pi$ over a. Thus, $\tau(C^n) * h = \tau(C^n)$ if h has a fixed point; e.g. h is a transposition. As $\mathrm{Sym}(n)$ is generated by transpositions, $\tau(C^n)$ is fixed by the action of $\mathrm{Sym}(n)$. Therefore, π is trivial. Q.E.D.

Recalling that principal G_a-bundles over a variety X are classified by their cohomology classes in $H^1(X, \mathcal{O}_X)$, we see that we have proven

COROLLARY 1.4. *The homomorphism* $H^1(C^{(n)}, \mathcal{O}_{C^{(n)}}) \to H^1(C^n, \mathcal{O}_{C^n})$ *induced by* ρ_n *is injective. In fact, the image is contained in the* $\mathrm{Sym}(n)$-*invariant part of* $H^1(C^n, \mathcal{O}_{C^n})$.

To use this result, recall that $H^1(C^n, \mathcal{O}_{C^n}) \approx \bigotimes_{1 \leq i \leq n} H^1(C, \mathcal{O}_C)$ by the Kunneth formula. Hence, $H^1(C^n, \mathcal{O}_{C^{(n)}})^{\mathrm{Sym}(n)} \approx \{w \oplus w \cdots \oplus w \mid w \in H^1(C, \mathcal{O}_C)\} \approx H^1(C, \mathcal{O}_C)$. Therefore, we may conclude the truth of

COROLLARY 1.5. *For any point* d *of* $C^{(n-1)}$, *let* $C \to C^{(n)}$ *be the morphism sending* c *to* $c + d$. *This morphism induces an injection* $H^1(C^{(n)}, \mathcal{O}_{C^{(n)}}) \to H^1(C, \mathcal{O}_C)$, *which does not depend on the choice of* d.

REMARK. It is well known that this injection is actually an isomorphism.

§2. *The divisor* D_n *regarded as a family of divisors*

Let D be an effective divisor on the product $X \times Y$ of two varieties. Let y be a point of Y. Set $\psi_y : X \to X \times Y$ equal to the morphism sending x to (x,y). If $\psi_y^{-1}D$ is defined (i.e., $X \times \{y\} \not\subset D$), let $D|_y = \psi_y^{-1}D$. We will assume that $\psi_y^{-1}D$ is defined for all points y of Y. Thus, we have an algebraic family $\{D|_y\}$ of divisors on X parametrized by Y.

Deformation theory [1, 3, 4, 10, 11 and 12] may be used to study how the divisor $D|_y$ moves when y changes. The first order infinitesimal changes of the divisor $D|_y$ are the easiest to understand.

Basically, as $D|_y$ moves away from $D_0 \equiv D|_{y_0}$, the most significant part of the movement will be movement normally to D_0 at each of its points. Thus, it should not be too surprising that the family D give rise to a characteristic mapping,

$$\mathcal{X}_{y_0} : \text{Tangent space of } Y \text{ at } y_0 \to \text{Sections of the normal bundle}$$
$$\text{to } D_0 \text{ in } X$$
$$\|$$
$$\Gamma(X, \mathcal{O}_X(D_0)|_{D_0}) .$$

This characteristic mapping is the linearization at y_0 of the mapping $Y \to \{\text{Divisors on } X\}$ given by $y \mapsto D|_y$.

It is also interesting to see how the invertible sheaf $\mathcal{O}_X(D|_y)$ on X changes as y moves. As the additive group G_a is the tangent space of the multiplicative group G_m at 1, one expects a first order infinitesimal change of an invertible sheaf on X (principal G_m-bundle) to give rise to a cohomology class in $H^1(X, \mathcal{O}_X)$, which classifies principal G_a-bundles. Thus, the commutative diagram,

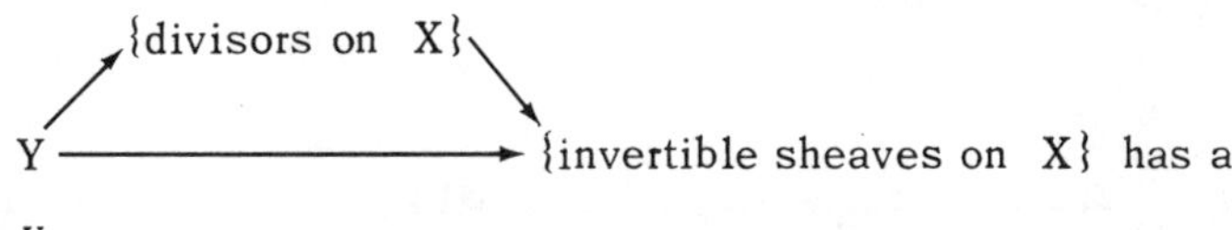

linearization at y_0

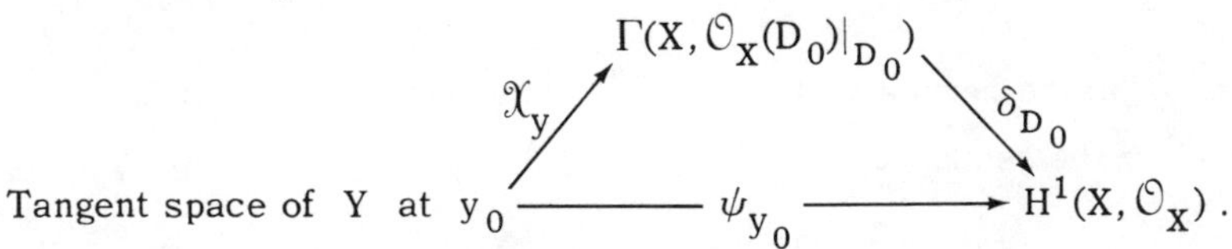

Here, ψ_{y_0} measures the linear part of the change of $\mathcal{O}_X(D|_y)$ with y

at $y = y_0$. The homomorphism δ_{D_0} is the boundary mapping

$\Gamma(X, \mathcal{O}_X(D_0)|_{D_0}) \to H^1(X, \mathcal{O}_X)$ in the long exact sequence of cohomology

of the sequence $0 \to \mathcal{O}_X \to \mathcal{O}_X(D_0) \to \mathcal{O}_X(D_0)|_{D_0} \to 0$ of sheaves on X.

Next, I will explain what these abstract ideas mean in this particular

case of interest. Consider the divisor D_n on $C \times C^{(n)}$ from section 1.

For each point $d = c_1 + \cdots + c_n$ of $C^{(n)}$, $D_n|_d$ is the divisor $D \equiv$

$[c_1] + \cdots + [c_n]$ on C. Thus, one may confuse points of $C^{(n)}$ with effec-

tive divisors of degree n of C. This confusion does not lead to any

infinitesimal ambiguities because

A. The characteristic mapping, $\mathcal{X}_d :$ Tangent space of $C^{(n)}$ at $d \to$

   $\Gamma(C, \mathcal{O}_C(D)|_D)$, is an isomorphism of n-dimensional vector spaces.

   One may even globalize the characteristic mappings to get

A'. The characteristic mapping $\mathcal{X} : \Theta_{C^{(n)}} \to \pi_{C*}(\mathcal{O}_{C \times C^{(n)}}(D_n)|_{D_n})$ gives

   an isomorphism of locally free $\mathcal{O}_{C^{(n)}}$-modules of rank n.

   In this case,

B. the mapping $\psi_d : \Theta_{C^{(n)}, c} \to H^1(C, \mathcal{O}_C)$ is the linearization of the

   abelian integral $\int_n : C^{(n)} \to P_n = \{\eta \epsilon H^1(C, \mathcal{O}_{C*}) | \deg \eta = n\}$.

   More globally,

B'. we have a commutative diagram of homomorphisms of $\mathcal{O}_{C^{(n)}}$-modules,

$$\Theta_{C^{(n)}} \xrightarrow[\approx]{\mathfrak{X}} \pi_{C*}(\mathcal{O}_{C\times C^{(n)}}(D_n)|_{D_n}) \xrightarrow{\delta} R^1\pi_{C*}\mathcal{O}_{C\times C^{(n)}} \approx H^1(C,\mathcal{O}_C)\otimes_k \mathcal{O}_{C^{(n)}},$$

$$\Theta_{C^{(n)}} \xrightarrow{\psi} R^1\pi_{C*}\mathcal{O}_{C\times C^{(n)}} \approx H^1(C,\mathcal{O}_C)\otimes_k \mathcal{O}_{C^{(n)}},$$

where δ is the boundary in the long exact sequence of direct images $R^*\pi_{C*}$ of the sequence

$$0 \to \mathcal{O}_{C\times C^{(n)}} \to \mathcal{O}_{C\times C^{(n)}}(D_n) \to \mathcal{O}_{C\times C^{(n)}}(D_n)|_{D_n} \to 0 .$$

The following special case will be very useful.

C'. If $n = 1$, we have a commutative diagram

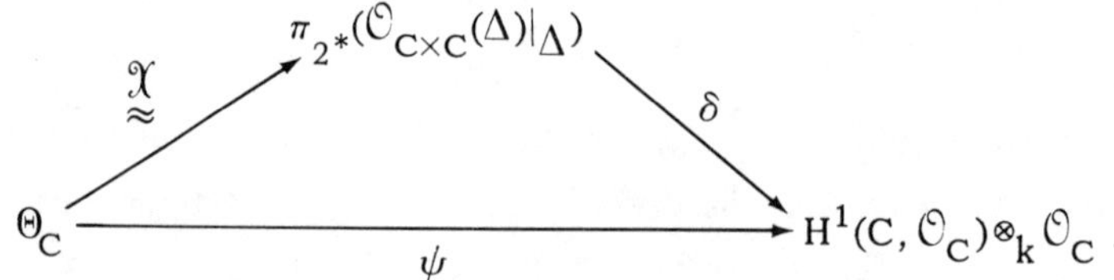

$$\Theta_C \xrightarrow[\approx]{\mathfrak{X}} \pi_{2*}(\mathcal{O}_{C\times C}(\Delta)|_\Delta) \xrightarrow{\delta} H^1(C,\mathcal{O}_C)\otimes_k \mathcal{O}_C .$$

$$\Theta_C \xrightarrow{\psi} H^1(C,\mathcal{O}_C)\otimes_k \mathcal{O}_C .$$

The dual $\psi^\wedge : H^1(C,\mathcal{O}_C)^\wedge \otimes_k \mathcal{O}_C \to \Theta_C^{\otimes -1} = \omega_C$ induces the usual isomorphism $H^1(C,\mathcal{O}_C)^\wedge \xrightarrow{\approx} \Gamma(C,\omega_C)$ on the global sections of these sheaves. If $g > 0$, $\delta|_c$ and $\psi|_c$ are injective at each point c of C.

Next we will reverse the roles of C and $C^{(n)}$ and regard the divisor D_n as defining a family of divisors on $C^{(n)}$ parameterized by C. Let c be a point of C. The divisor $D_n|_c$ on $C^{(n)}$ is in fact $c + C^{(n-1)}$ from section 1.

The general infinitesimal theory asks us to look at the commutative diagram

$$\mathfrak{X}_c : \Gamma(C^{(n)}, \mathcal{O}_{C^{(n)}}(c+C^{(n-1)})|_{c+C^{(n-1)}})$$

$$\Psi_c \searrow \qquad \downarrow \delta_c$$

$$H^1(C^{(n)}, \mathcal{O}_{C^{(n)}}) .$$

* Tangent space of C at c

An easy point at which to begin our study is

LEMMA 2.1. *Let* d *be any point of* $C^{(n-1)}$. *Let* $\phi: H^1(C^{(n)}, \mathcal{O}_{C^{(n)}}) \hookrightarrow$
$H^1(C, \mathcal{O}_C)$ *be the injection from Lemma 1.6. Then, we have a commutative
diagram*:

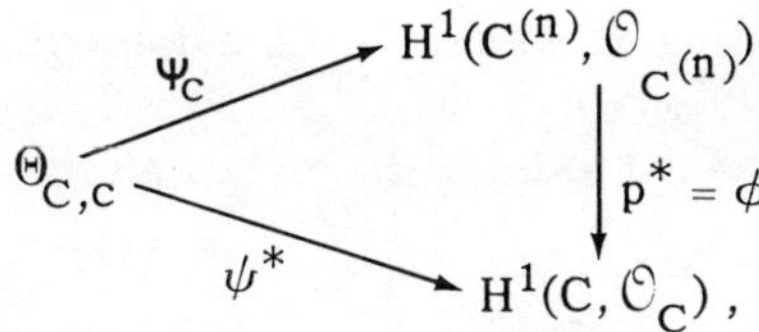

where ψ_c *is the value at* c *of the homomorphism* ψ *of part* C.

Proof. Let $p: C \to C^{(n)}$ be the morphism sending c to $c + d$. Then,

$$(1_C, p)^{-1}(D_n) = \{(c_1, c_2)| \text{ either } c_1 = c_2 \text{ or } c_1 + d' = d \text{ for some } d' \text{ in}$$

$$= \Delta + \pi_1^{-1} D$$

$C^{(n-2)}\}$, where D is the divisor on C corresponding to d. Therefore,
we have a commutative diagram,

where ψ^* is the infinitesimal classifying mapping of the family
$\mathcal{O}_C(c_1 + D)$ of invertible sheaves on C parameterized by c_1 in C. As
D does not depend on c_1, $\psi^* = \psi_c$, where ψ_c is the classifying mapping for the family $\mathcal{O}_C(c_1)$. This proves the lemma. Q.E.D.

   This last fact neatly globalizes to give

COROLLARY 2.2. *We have a commutative diagram of* $\mathcal{O}_C$-*homomorphisms*,

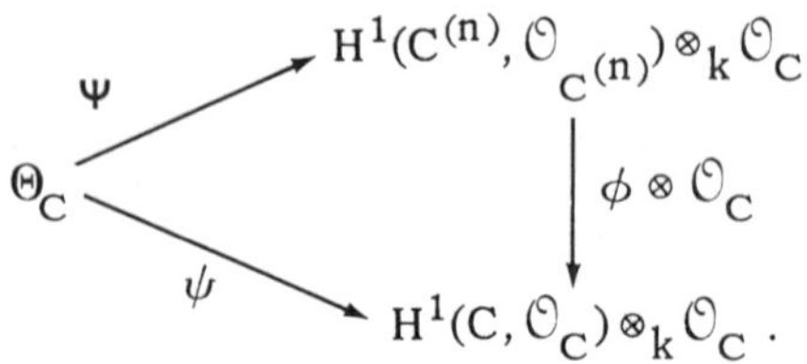

In fact, ϕ is an isomorphism.

Proof. The existence follows from the basic continuity of the diagram of Lemma 2.1 with respect to the variable c . We need only check that the injection ϕ is surjective. Dualizing the diagram and taking global sections, we have a commutative diagram,

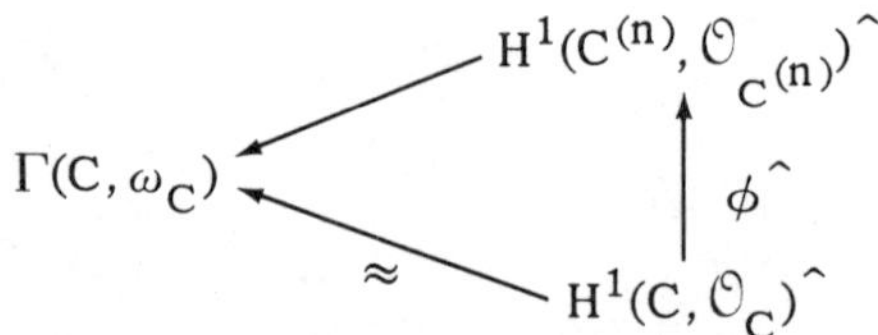

where we have seen that the bottom arrow is an isomorphism in point C .
Formally, $\hat{\phi}$ is injective or, rather, ϕ is surjective. Q.E.D.

We are now ready to compute the characteristic homomorphism $\mathfrak{X}_c$:
$$\Theta_{C,c} \to \Gamma(C^{(n)}, \mathcal{O}_{C^{(n)}}(c + C^{(n-1)})|_{c+C^{(n-1)}}).$$ The result is

LEMMA 2.4. *For all points c of C , $\mathfrak{X}_c$ is an isomorphism if $g > 0$.*

Proof. By the diagram $*$ and the last lemma, $\mathfrak{X}_c$ must be injective if ψ_c is injective. In point C , we have seen that ψ_c is in fact injective. Thus, to prove the lemma, we need only check that $\Gamma(C^{(n)}, \mathcal{O}_{C^{(n)}}(c + C^{(n-1)})|_{c+C^{(n-1)}})$ is one-dimensional.

By Lemma 1.1.d, $\gamma_n^*(\mathcal{O}_{C \times C^{(n)}}(D_n)|_{D_n}) \approx \pi_C^* \Theta_C(D_{n-1})$. If we evaluate this isomorphism over c in C , we get an isomorphism,

$$\mathcal{O}_{C^{(n)}}(c + C^{(n-1)})|_{c+C^{(n-1)}} \approx \Theta_{C,c} \otimes_k \mathcal{O}_{C^{(n-1)}}(c + C^{(n-1)}),$$

where $c + C^{(-1)}$ is the empty set. By Lemma 1.2, $\Gamma(C^{(n-1)}, \mathcal{O}_{C^{(n-1)}}(c + C^{(n-2)}))$ is one-dimensional if $g > 0$. By the isomorphism, we may conclude that $\Gamma(C^{(n)}, \mathcal{O}_{C^{(n)}}(c + C^{(n-1)}|_{c+C^{(n-1)}}))$ must be one-dimensional and we are done. Q.E.D.

We may easily globalize this result.

D. We have a global characteristic homomorphism $\mathfrak{X}$, which is the $\mathcal{O}_C$-homomorphism that fits into the commutative diagram,

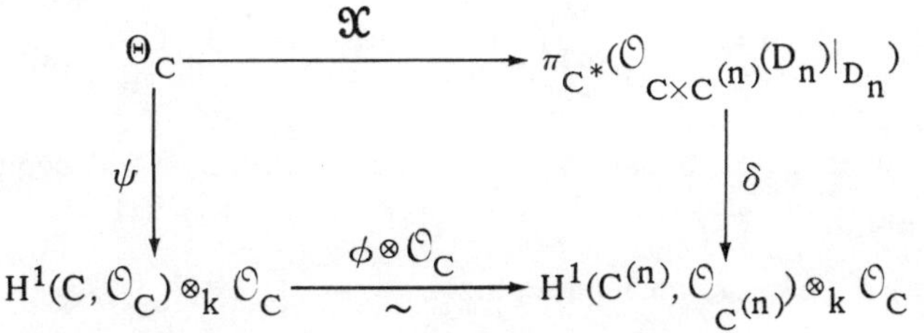

where δ is the differential in the long exact sequence of direct images of

$$0 \to \mathcal{O}_{C \times C^{(n)}} \to \mathcal{O}_{C \times C^{(n)}}(D_n) \to \mathcal{O}_{C \times C^{(n)}}(D_n)|_{D_n} \to 0 \, .$$

§3. *The infinitesimal deformations of symmetric products*

In this section, we will apply the previous work to compute $H^1(C^{(n)}, \Theta_{C^{(n)}})$, which classifies the first order infinitesimal deformations of the symmetric product. For good measure, we may easily see that $C^{(n)}$ possesses no global vector fields if $g > 1$.

By the point A' of section 2, we have the characteristic isomorphism, $\psi : \Theta_{C^{(n)}} \to \pi_{C*}(\mathcal{O}_{C \times C^{(n)}}(D_n)|_{D_n})$. As $\mathcal{O}_{C \times C^{(n)}}(D_n)|_{D_n}$ is a coherent sheaf on D_n and the projection $D_n \to C^{(n)}$ is finite (hence, affine), we have a natural isomorphism,

$$H^i(D_n, \mathcal{O}_{C \times C^{(n)}}(D_n)|_{D_n}) \xrightarrow{\sim} H^i(C^{(n)}, \pi_{C*}(\mathcal{O}_{C \times C^{(n)}}(D_n)|_{D_n})) \, ,$$

for all i. Putting these two isomorphisms together, we have proven

LEMMA 3.1. *There is a natural isomorphism,*

$$H^i(C^{(n)}, \Theta_{C^{(n)}}) \approx H^i(D_n, \mathcal{O}_{C \times C^{(n)}}(D_n)|_{D_n}) .$$

If $g > 0$, we have another characteristic isomorphism, $\psi: \Theta_C \xrightarrow{\approx} \pi_{C*}(\mathcal{O}_{C \times C^{(n)}}(D_n)|_{D_n})$, (see the end of section 2). We want to use this isomorphism to help us understand the cohomology of $\mathcal{O}_{C \times C^{(n)}}(D_n)|_{D_n}$. We will use the Leray spectral sequence,

$$E^{i,j} = H^i(C, R^j \pi_{C*}(\mathcal{O}_{C \times C^{(n)}}(D_n)|_{D_n})) \implies H^{i+j}(D_n, \mathcal{O}_{C \times C^{(n)}}(D_n)|_{D_n}) .$$

As $E^{i,j} = 0$ if $i > 0$ because C is a curve, this spectral sequence is very degenerate.

In fact, it gives as an isomorphism,

$$(\dagger) \qquad H^0(C, \pi_{C*}(\mathcal{O}_{C \times C^{(n)}}(D_n)|_{D_n})) \approx H^0(D_n, \mathcal{O}_{C \times C^{(n)}}(D_n)|_{D_n}) ,$$

and a short exact sequence,

$$(\dagger\dagger) \quad 0 \to H^1(C, R^{j-1} \pi_{C*}(\mathcal{O}_{C \times C^{(n)}}(D_n)|_{D_n})) \to H^j(D_n, \mathcal{O}_{C \times C^{(n)}}(D_n)|_{D_n}) \to$$

$$H^0(C, R^j \pi_{C*}(\mathcal{O}_{C \times C^{(n)}}(D_n)|_{D_n})) \to 0$$

for all $j > 0$.

From the isomorphism ψ and Lemma 3.1, we may conclude

LEMMA 3.2. *If $g > 0$, we have*

  a) *a natural isomorphism* $\Gamma(C^{(n)}, \Theta_{C^{(n)}}) \approx \Gamma(C, \Theta_C)$, *and*

  b) *a natural short exact sequence,*

$$0 \to H^1(C, \Theta_C) \to H^1(C^{(n)}, \Theta_{C^{(n)}}) \to \Gamma(C, R^1 \pi_{C*}(\mathcal{O}_{C \times C^{(n)}}(D_n)|_{D_n})) \to 0 .$$

An immediate consequence of a) is

COROLLARY 3.3. *If $g > 1$, $\Gamma(C^{(n)}, \Theta_{C^{(n)}})$ is zero.*

Proof. As $\deg(\Theta_C) = -2g+2$ is negative, $\Gamma(C, \Theta_C)$ is zero. Thus, $\Gamma(C^{(n)}, \Theta_{C^{(n)}})$ is zero for all positive n by Lemma 3.2.a. Q.E.D.

Our main objective is to prove that the homomorphism $H^1(C, \Theta_C) \to H^1(C^{(n)}, \Theta_{C^{(n)}})$ is an isomorphism at least for most curves. By Lemma 3.2.b, we will need to see when $\Gamma(C, R^1\pi_{C*}(\mathcal{O}_{C\times C^{(n)}}(D_n)|_{D_n}))$ is zero. Recall from Lemma 1.1.d, that, under the C-isomorphism $\gamma_n : C \times C^{(n-1)} \to D_n$, we have an isomorphism $\gamma_n^*(\mathcal{O}_{C\times C^{(n)}}(D_n)|_{D_n}) \approx \pi_C^* \Theta_C(D_{n-1})$. Therefore, we have isomorphisms,

$$R^1\pi_{C*}(\mathcal{O}_{C\times C^{(n)}}(D_n)) \approx R^1\pi_{C*}(\pi_C^* \Theta_C(D_{n-1})) \approx \Theta_C \otimes_{\mathcal{O}_C} R^1\pi_{C*}(\mathcal{O}_{C\times C^{(n-1)}}(D_{n-1})).$$

For the record, we will list the global effect of this isomorphism.

LEMMA 3.4. *There is a natural isomorphism,*

$$\Gamma(C, R^1\pi_{C*}(\mathcal{O}_{C\times C^{(n)}}(D_n)|_{D_n})) \approx \Gamma(C, \Theta_C \otimes_{\mathcal{O}_C} R^1\pi_{C*}(\mathcal{O}_{C\times C^{(n-1)}}(D_{n-1})).$$

The last technical result that we need, is

LEMMA 3.5. *For all* $p > 0$ *and all non-hyperelliptic curves* C,

$$\Gamma(C, \Theta_C^{\otimes p} \otimes_{\mathcal{O}_C} R^1\pi_{C*}(\mathcal{O}_{C\times C^{(n)}}(D_n)))$$

is zero.

Before I prove this, I will explain how it implies the main result of this section.

THEOREM 3.6. *If* C *is a non-hyperelliptic curve, then we have a natural isomorphism,*

$$H^1(C, \Theta_C) \xrightarrow[\approx]{} H^1(C^{(n)}, \Theta_{C^{(n)}}).$$

Proof. By Lemma 3.5, $\Gamma(C, \Theta_C \otimes_{\mathcal{O}_C} R^1\pi_{C*}(\mathcal{O}_{C\times C^{(n-1)}}(D_{n-1})))$ is zero. By Lemma 3.4, $\Gamma(C, R^1\pi_{C*}(\mathcal{O}_{C\times C^{(n)}}(D_n)|_{D_n}))$ is zero. Therefore, the exact sequence of Lemma 3.2.b gives our theorem. Q.E.D.

We now begin the proof of Lemma 3.5. Consider the exact sequence,
$$0 \to \mathcal{O}_{C\times C^{(n)}} \to \mathcal{O}_{C\times C^{(n)}}(D_n) \to \mathcal{O}_{C\times C^{(n)}}(D_n)|_{D_n} \to 0,$$ of sheaves on $C \times C^{(n)}$. Its long exact sequence of direct images via π_{C*} is

$$\Theta_C \xrightarrow{\ \Psi\ } H^1(C, \mathcal{O}_C) \otimes_k \mathcal{O}_C$$
$$\wr\wr \qquad\qquad\qquad \wr\wr$$
$$0 \longrightarrow \mathcal{O}_C \approx \pi_{C*}\mathcal{O}_{C\times C^{(n)}}(D_n) \xrightarrow{\ 0\ } \pi_{C*}(\mathcal{O}_{C\times C^{(n)}}(D_n)|_{D_n}) \xrightarrow{\ \delta\ } H^1(C^{(n)}, \mathcal{O}_{C^{(n)}})$$

$$\Theta_C \otimes_{\mathcal{O}_C} R^1\pi_C(\mathcal{O}_{C\times C^{(n-1)}}(D_{n-1}))$$
$$\wr\wr$$
$$\otimes_k \mathcal{O}_C \longrightarrow R^1\pi_{C*}(\mathcal{O}_{C\times C^{(n)}}(D_n)) \longrightarrow R^1\pi_{C*}(\mathcal{O}_{C\times C^{(n)}}(D_n)|_{D_n}) \cdots .$$

The important thing to see here is that we have a short exact sequence,

$$\#) \quad 0 \to \mathrm{Cok}(\Psi) \to R^1\pi_{C*}(\mathcal{O}_{C\times C^{(n)}}(D_n)) \to \Theta_C \otimes_{\mathcal{O}_C} R^1\pi_{C*}(\mathcal{O}_{C\times C^{(n-1)}}(D_{n-1})),$$

of sheaves on C. As the sequence of its global sections of $\# \otimes_{\mathcal{O}_C} \Theta_C^{\otimes p}$ is exact, if Lemma 3.5 is true for $p+1$ and $n-1$, we have an isomorphism,

$$\Gamma(C, \mathrm{Cok}(\Psi) \otimes_{\mathcal{O}_C} \Theta_C^{\otimes p}) \approx \Gamma(C, \Theta_C^{(p)} \otimes_{\mathcal{O}_C} R^1\pi_{C*}(\mathcal{O}_{C\times C^{(n)}}(D_n))) .$$

Hence, Lemma 3.5 will be proven by induction on n, as soon as we prove

SUBLEMMA 3.7. *If C is not hyperelliptic and $p > 0$,*

$$\Gamma(C, \mathrm{Cok}(\Psi) \otimes_{\mathcal{O}_C} \Theta_C^{\otimes p})$$

is zero.

Proof. We have the short exact sequence,

$$0 \longrightarrow \Theta_C^{\otimes p+1} \approx \Theta_C \otimes_{\mathcal{O}_C} \Theta_C^{\otimes p} \xrightarrow{\psi \otimes \Theta_C^{\otimes p}} H^1(C, \mathcal{O}_C) \otimes_k \mathcal{O}_C \otimes_{\mathcal{O}_C} \Theta_C^{\otimes p} \longrightarrow$$

$$\mathrm{cok}(\psi) \otimes \Theta_C^{\otimes p} \longrightarrow 0 \; ,$$

of locally free sheaves on C. As $\Gamma(C, *)$ of the first two terms is zero, we need to see that $H^1(C, \psi \otimes \Theta_C^{\otimes p})$ is injective. By duality, it will be enough to check that $\Gamma(C, \Omega_C \otimes (\psi \otimes \Theta_C^{\otimes p})\hat{\;})$ is surjective.

Now, $\Omega_C \otimes (\psi \otimes \Theta_C^{\otimes p})\hat{\;} \approx \Omega_C \otimes \Omega_C^{\otimes p} \otimes \psi\hat{\;} : \Omega_C^{\otimes p+1} \otimes_k H^1(C, \mathcal{O}_C)\hat{\;} \to \Omega_C^{\otimes p+2}$ is isomorphic to $\beta : \Omega_C^{\otimes p+1} \otimes_k \Gamma(C, \Omega_C) \to \Omega_C^{\otimes p+2}$ where β is given by multiplying a rational p+1-differential by a regular differential. Thus,

$$\Gamma(C, \beta) : \Gamma(C, \Omega_C^{\otimes p+1}) \otimes \Gamma(C, \Omega_C) \to \Gamma(C, \Omega_C^{\otimes p+2})$$

is just multiplication in the canonical ring of C. For non-hyperelliptic curves, the surjectivity of $\Gamma(C, \beta)$ is slightly weaker than Max Noether's famous theorem. Therefore, $\Gamma(C, \Omega_C \otimes (\psi \otimes \Theta_C^{\otimes p})\hat{\;})$ is in fact surjective. Q.E.D.

§4. *The actual deformations of symmetric products*

In this section, we will extend the first order infinitesimal results of the last two sections to results about actual or, at least, arbitrarily high order infinitesimal deformations. We will begin by recalling general facts about the deformations of divisors on a fixed variety.

Let D be an effective divisor on the product $X \times Y$ of two complete varieties such that, for each point y of Y, the divisor $D|_y$ on X is defined. One would like to know when the family $\{D|_y\}$ gives all possible deformations of its members $D|_y$ in a unique way. More precisely, given any scheme S (of finite type over k) and an effective (Cartier) divisor E on $X \times S$ such that $E|_S$ is defined and equal to some $D|_y$ for each point s of S, does there exist a unique morphism $a_E : S \to Y$ such that $E = (\pi_X, a_E)^{-1}D$?

We will say that D gives a full family of deformations of its members $\{D|_y\}$ if this question always has an affirmative answer. At first sight, it may seem to be difficult to verify that a given divisor D gives a full family, but, in simple cases, one may use the following

CRITERION. The divisor D on $X \times Y$ gives a full family if
  a) the divisors $D|_y$ are distinct as y runs through Y,
  b) Y is smooth, and
  c) the characteristic homomorphism, $\mathfrak{X}_y : \Theta_{Y,y} \to \Gamma(X, \mathcal{O}_X(D))_D)$, is
     an isomorphism at each point y of Y.

We may apply this criterion to verify

PROPOSITION 4.1. *Consider the divisor* D_n *on* $C \times C^{(n)}$. *Then,*
  a) D_n *gives a full family of divisors on* C *parameterized by* $C^{(n)}$.
  b) *If* $g > 0$, D_n *gives a full family of divisors on* $C^{(n)}$ *parameterized by* C.

Proof. Both C and $C^{(n)}$ are smooth complete varieties. We have noted in A (section 2) that the characteristic homomorphism for the family in a) is an isomorphism. By Lemma 2.4, we know that the characteristic homomorphism of the family in b) is also an isomorphism. It remains to check the univalence condition a) of the criterion. In case a), the point $c_1 + \cdots + c_n$ is clearly determined by the divisor $[c_1] + \cdots + [c_n]$. In case b), the point c is determined by the divisor $c + C^{(n-1)}$ (for instance, because $c + C^{(n-1)} \cap c' + C^{(n-1)} = c + c' + C^{(n-2)}$ if $c \ne c'$). Thus, the criterion implies the truth of the proposition at hand. Q.E.D.

We will now begin studying the deformations of a variety. Let X be a smooth complete variety. A (formal) deformation of X is a formal scheme $\mathcal{J}$ with one point t, a smooth proper morphism $s : \mathcal{X} \to \mathcal{J}$ and an isomorphism $X \approx \mathcal{X}_t \equiv s^{-1}(t)$.

Deformation theory provides a method of measuring the first order deviation of a deformation $s : \mathcal{X} \to \mathcal{J}$ from the trivial deformation (i.e., $\pi_{\mathcal{J}} : \mathcal{J} \times X \to \mathcal{J}$). This deviation is measured by a linear mapping,

$$\mu_s : \text{Tangent space of } \mathcal{J} \text{ at } t \to H^1(X, \Theta_X) .$$

A universal deformation of X is a formal deformation $s : \mathcal{X} \to \mathcal{J}$ such that any formal deformation $s' : \mathcal{X}' \to \mathcal{J}'$ is induced from s by a unique morphism $\mathcal{J}' \to \mathcal{J}$. Clearly, if s is a universal deformation of X, then μ_s must be an isomorphism. We have another criterion for the universality of a given deformation.

CRITERION. If $s : \mathcal{X} \to \mathcal{J}$ is a deformation of X such that

    a) the linear mapping μ_0 is an isomorphism,

    b) $\Gamma(X, \Theta_X)$ is zero, and

    c) $\mathcal{J}$ is a smooth formal scheme, then $s : \mathcal{X} \to \mathcal{J}$ is a universal deformation of X.

The most well-known case of a universal deformation is when X is a smooth complete curve C of genus $g > 1$. Then it is well known that there exists a universal deformation $s : \mathcal{C} \to \mathcal{J}$ of C, where $\mathcal{J}$ is a smooth formal scheme. In fact, if C is a compact Riemann surface, $\mathcal{J}$ is the formal completion of Teichmüller space at a point corresponding to C.

With our universal deformation $s : \mathcal{C} \to \mathcal{J}$ of the curve C, we may construct a deformation $s^n : \mathcal{C}^n \to \mathcal{J}$ of the product C^n by taking $\mathcal{C}^n = \mathcal{C} \times_{\mathcal{J}} \mathcal{C} \cdots \times_{\mathcal{J}} \mathcal{C}$. Furthermore, we have a deformation $s^{(n)} : \mathcal{C}^{(n)} \to \mathcal{J}$ of $C^{(n)}$ where $\mathcal{C}^{(n)} = \mathcal{C}^n/\mathrm{Sym}(n)$. A key result of this paper is

THEOREM 4.2. *If C is a non-hyperelliptic curve of genus > 2, then $s^{(n)} : C^{(n)} \to \mathcal{J}$ is a universal deformation of the symmetric product $C^{(n)}$.*

Proof. We will use the above criterion to check the universality of $C^{(n)}$. The condition b) that $\Gamma(C^{(n)}, \Theta_{C^{(n)}}) = 0$ has been verified in Corollary 3.3. The condition c) is verified because $\mathcal{J}$ is smooth. Furthermore, we know that the linear mapping $\mu_s : \Theta_{\mathcal{J},t} \to H^1(C, \Theta_C)$ is an isomorphism.

It remains to verify condition c); i.e. $\mu_{s^{(n)}} : \Theta_{\mathcal{J},t} \to H^1(C^{(n)}, \Theta_{C^{(n)}})$ is an isomorphism. Thus, we need to show that $\eta \equiv \mu_{s^{(n)}} \circ \mu_s^{-1} : H^1(C, \Theta_C) \to H^1(C^{(n)}, \Theta_{C^{(n)}})$ is an isomorphism. By Theorem 3.6, we know that $H^1(C, \Theta_C)$ and $H^1(C^{(n)}, \Theta_{(n)})$ are isomorphic. Hence, at least, they have the same finite dimension $3g - 3$. Therefore, we will have verified condition c) and proved our theorem once we prove

LEMMA 4.3. *If* $g > 0$, *then* $\eta : H^1(C, \Theta_C) \to H^1(C^{(n)}, \Theta_{C^{(n)}})$ *is injective.*

Proof. The statement means that any non-trivial first order infinitesimal deformation of C induces a non-trivial first order infinitesimal deformation of $C^{(n)}$. Let $\kappa : \mathcal{C} \to S = \mathrm{Spec}\,(k[\delta]/(\delta^2))$ be a deformation of C. Assume that $\kappa^{(n)} : \mathcal{C}^{(n)} \to S$ is the trivial deformation of $C^{(n)}$. We need to see that κ must be trivial.

In the usual way, one constructs an effective Cartier divisor $\mathcal{D}_n$ in $\mathcal{C} \underset{S}{\times} \mathcal{C}^{(n)}$ which is a deformation of D_n. As $\kappa^{(n)}$ is trivial, we have an S-isomorphism, $\mathcal{C}^{(n)} \approx S \times C^{(n)}$. Then, $\mathcal{C} \underset{S}{\times} \mathcal{C}^{(n)}$ is $\mathcal{C}$-isomorphic to $\mathcal{C} \times C^{(n)}$ and we may regard $\mathcal{D}_n$ as a divisor on $\mathcal{C} \times C^{(n)}$. By Proposition 4.1, we have a morphism $f : \mathcal{C} \to C$, which gives the isomorphism $\mathcal{C}|_s \approx C$, where s is the unique point of S. Therefore, $(\kappa, f) : \mathcal{C} \to S \times C$ is an isomorphism, which shows that κ is a trivial deformation of C. Q.E.D.

§5. *The deformations of the bundles over the Jacobian*

Let $s : \mathcal{C} \to \mathcal{M}$ be a deformation of the smooth complete curve C. In the last section, we have discussed the deformations $s^n : \mathcal{C}^n \to \mathcal{M}$ and $s^{(n)} : \mathcal{C}^{(n)} \to \mathcal{M}$ of the ordinary and symmetric products C^n and $C^{(n)}$.

We will next discuss the deformations of the Picard varieties P_d of C. Recall that the d-th Picard variety P_d classifies linear equivalence classes of divisors on C of degree P_d. Furthermore, we have a natural morphism $\int_d : C^{(d)} \to P_d$, which sends a point in $C^{(d)}$ to the class of the corresponding divisor. For each pair of integers d and d', we have a morphism $P_d \times P_{d'} \to P_{d+d'}$ which corresponds to addition of divisor classes. This addition makes P_d a principal homogeneous space under the abelian variety P_0.

Grothendieck has constructed relative Picard schemes, $s_d : \mathcal{P}_d \to \mathcal{M}$, for the family $s : \mathcal{C} \to \mathcal{M}$ such that s_d is a deformation of P_d. Also, we have $\mathcal{M}$-morphisms $\int_d : \mathcal{C}^{(d)} \to \mathcal{P}_d$ and $\mathcal{P}_d \times_{\mathcal{M}} \mathcal{P}_{d'} \to \mathcal{P}_{d+d'}$, which generalize the above concepts. Furthermore, $\mathcal{P}_0 \to \mathcal{M}$ is an abelian $\mathcal{M}$-scheme and each $\mathcal{P}_d$ is a principal $\mathcal{P}_0$-homogeneous $\mathcal{M}$-scheme. More details may be found in [4].

As I can't find it in the literature, I will prove a generalization of the classical universal mapping property of abelian integrals.

PROPOSITION 5.1. *Let* $\rho : \mathcal{C}^{(d)} \to \mathcal{G}$ *be any* $\mathcal{M}$-*morphism, where* $\mathcal{G}$ *is a group scheme over* $\mathcal{M}$. *Then,* ρ *factors as the composition* $\psi \circ \int_d$, *where* ψ *is an* $\mathcal{M}$-*morphism* $\mathcal{P}_d \to \mathcal{G}$.

Proof. Assume that $d > 2g-2$. Then the integral $\int_d : \mathcal{C}^{(d)} \to \mathcal{P}_d$ is a locally trivial family of projective spaces. By descent theory, the morphism $\rho : \mathcal{C}^{(d)} \to \mathcal{G}$ factors through $\int_d$ if and only if $\rho \circ \pi_1 = \rho \circ \pi_2$ as morphism from $\mathcal{C}^{(d)} \times_{\mathcal{P}_d} \mathcal{C}^{(d)}$ to $\mathcal{G}$. By Corollary 6.2 of [9], we know that $\rho \circ \pi_1 = \rho \circ \pi_2 * g$, where $*g$ denotes multiplication in $\mathcal{G}$ by a morphism $g : \mathcal{C}^{(d)} \times_{\mathcal{P}_d} \mathcal{C}^{(d)} \to \mathcal{G}$, which factors through the structure morphism $\mathcal{C}^{(d)} \times_{\mathcal{P}_d} \mathcal{C}^{(d)} \to \mathcal{M}$. On the other hand, g must be the identity of $\mathcal{G}$, when it is restricted to the diagonal Δ of $\mathcal{C}^{(d)} \times_{\mathcal{P}_d} \mathcal{C}^{(d)}$. Therefore, g

itself is the identity because Δ dominates $\mathfrak{M}$. This proves the proposition when $d > 2g - 2$.

Let ρ be any $\mathfrak{M}$-morphism $\mathcal{C} \to \mathcal{G}$. Consider the two morphisms, $\rho \circ \pi_1 * \rho \circ \pi_2$ and $\rho \circ \pi_2 * \rho \circ \pi_1$ from $\mathcal{C} \times_{\mathfrak{M}} \mathcal{C} \to \mathcal{G}$. As before, $\rho \circ \pi_1 * \rho \circ \pi_2 = \rho \circ \pi_2 * \rho \circ \pi_1 * g$, where g is an $\mathfrak{M}$-morphism $\mathcal{C} \times_{\mathfrak{M}} \mathcal{C} \to \mathcal{G}$ which only depends on $\mathfrak{M}$. Thus, evaluating g on the diagonal of $\mathcal{C} \times_{\mathfrak{M}} \mathcal{C}$, we again see that g must be the identity. Hence, $\rho \circ \pi_1 * \rho \circ \pi_2 = \rho \circ \pi_2 * \rho \circ \pi_1$ or, rather, $\rho \circ \pi_1$ commutes with $\rho \circ \pi_2$. Therefore, the $\mathfrak{M}$-morphism,

$$\rho_n \equiv \rho \circ \pi_1 * \cdots * \rho \circ \pi_n : \mathcal{C} \times_{\mathfrak{M}} \cdots \times_{\mathfrak{M}} \mathcal{C} = \mathcal{C}^n \to \mathcal{G},$$

is invariant under the action of $\mathrm{Sym}(n)$ on $\mathcal{C}^n$. Hence, ρ_n comes from a $\mathfrak{M}$-morphism $\rho_{(n)} : \mathcal{C}^{(n)} \to \mathcal{G}$. By the first paragraph, if $n > 2g - 2$, $\rho_{(n)}$ factors through a morphism $\psi : \mathcal{P}_n \to \mathcal{G}$.

Let σ be a section of $s : \mathcal{C} \to \mathfrak{M}$ and $1 \leq d \leq n$. We have a commutative diagram,

$$
\begin{array}{ccc}
\mathcal{C}^{(d)} & \xrightarrow{\int_d} & \mathcal{P}_d \\
{\scriptstyle + (n-d)\sigma} \downarrow & & \downarrow {\scriptstyle + (n-d)\int_1 \circ \sigma} \\
\mathcal{C}^{(n)} & \xrightarrow{\int_n} & \mathcal{P}_n
\end{array}
$$

Therefore, $\mu \equiv \psi \circ (+ (n-d)\int_1 \circ \sigma) \circ \int_d : \mathcal{C}^{(d)} \to \mathcal{G}$ is a $\mathfrak{M}$-morphism which factors through $\int_d$. Hence, by Corollary 6.2 of [9], any $\mathfrak{M}$-morphism $\mathcal{C}^{(d)} \to \mathcal{G}$ factors through $\int_d$, if one does. Thus, we have proved the proposition if there is a $\mathfrak{M}$-morphism $\mathcal{C} \to \mathcal{G}$.

Now, in the proposition, we are given a $\mathfrak{M}$-morphism $\rho : \mathcal{C}^{(d)} \to \mathcal{G}$. Therefore, we have an $\mathfrak{M}$-morphism $\rho \circ (+ (d-1)\sigma) : \mathcal{C} \to \mathcal{G}$, where σ is a section of s. Hence, the assumption of the last paragraph holds and we are done. Q.E.D.

We will begin the final part with an abstract definition. Let $f : X \to Y$ be a morphism between two smooth complete varieties. A deformation of the morphism f is a $\mathfrak{M}$-morphism $F : \mathcal{X} \to \mathcal{Y}$ between deformations $\mathcal{X} \to \mathfrak{M}$ and $\mathcal{Y} \to \mathfrak{M}$ of X and Y such that the diagram

$$F_m : \mathcal{X}|_m \longrightarrow \mathcal{Y}|_m$$
$$\wr\wr \qquad\qquad \wr\wr \qquad \text{commutes}$$
$$f : X \longrightarrow Y$$

where m is the unique point of $\mathfrak{M}$.

We want to determine explicitly the deformations of an abelian integral $\int_n : C^{(n)} \to P_n$ (at least for most curves) where the genus of $C > 1$. We will first give the obvious deformations of $\int_n$. Let $s : \mathcal{C} \to \mathcal{T}$ be a universal deformation of C. Let $\mathfrak{M}$ be a formal scheme with one point m. Let $\alpha : \mathfrak{M} \to \mathcal{T}$ be a morphism and let $\beta : \mathfrak{M} \to \mathcal{P}_0$ be a morphism such that $\beta(m) = $ zero in P_0 and $s_0 \circ \beta = \alpha$.

Thus, we have an $\mathfrak{M}$-morphism $\mathfrak{M} \times \int_n : \mathfrak{M} \times_{\mathcal{T}} \mathcal{C}^{(n)} \to \mathfrak{M} \times_{\mathcal{T}} \mathcal{P}_n$ using α. Furthermore, using the addition $\mathcal{P}_0 \times_{\mathcal{T}} \mathcal{P}_n \to \mathcal{P}_n$ we have a $\mathfrak{M}$-isomorphism, $+\beta : \mathfrak{M} \times_{\mathcal{T}} \mathcal{P}_n \to \mathfrak{M} \times_{\mathcal{T}} \mathcal{P}_n$, given by addition by β. Denote the composition $(+\beta) \circ (\mathfrak{M} \times \int_n)$ by $D(\alpha, \beta)$. Trivial, one may check that $D(\alpha, \beta) : \mathfrak{M} \times_{\mathcal{T}} \mathcal{C}^{(n)} \to \mathfrak{M} \times_{\mathcal{T}} \mathcal{P}_n$ is a deformation of $\int_n : C^{(n)} \to P_n$ as we have assumed that $\beta(m) = $ zero.

The converse is usually true.

THEOREM 5.2. *If C is a non-hyperelliptic curve of genus > 3, then any deformation of the integral $\int_n : C^{(n)} \to P_n$ has the form $D(\alpha, \beta)$, where these morphisms α and β are uniquely determined by the deformation.*

Proof. Let $F : \mathcal{R} \to \mathcal{Q}$ be the deformation of $\int_n$. In Theorem 4.2, we have

determined that any deformation of the symmetric product $C^{(n)}$ has the

form $\mathfrak{M} \times_{\mathcal{J}} C^{(n)} \to \mathfrak{M}$ where $a : \mathfrak{M} \to \mathcal{J}$ is a unique morphism. If we replace

$s : \mathcal{C} \to \mathcal{J}$ by $\mathfrak{M} \times s : \mathfrak{M} \times_{\mathcal{J}} \mathcal{C} \to \mathfrak{M}$, we may assume that the deformation

$\mathcal{R} \to \mathfrak{M}$ of $C^{(n)}$ is the $s^{(n)} : \mathcal{C}^{(n)} \to \mathfrak{M}$ of the first part of this section.

Next, consider the deformation $\mathcal{Q} \to \mathfrak{M}$ of P_n. As we may take a sec-

tion τ of $\mathcal{Q} \to \mathfrak{M}$ and P_n is a principal homogeneous space under an

abelian variety, the Theorem 6.14 of [9] implies that $\mathcal{Q} \to \mathfrak{M}$ possesses a

unique structure of an abelian $\mathfrak{M}$-scheme with identity τ. Thus, $\mathcal{Q}$ is a

group scheme over $\mathfrak{M}$ and we may now apply Proposition 5.1 to the

$\mathfrak{M}$-morphism $F : \mathcal{C}^{(n)} \to \mathcal{Q}$. Therefore, F factors through an $\mathfrak{M}$-morphism

$\psi : \mathcal{P}_n \to \mathcal{Q}$. As ψ must be a deformation of the identity of P_n, ψ is an

$\mathfrak{M}$-isomorphism.

Thus far, we have shown that our deformation must have the form

$\kappa \circ D(a, \text{zero}) : \mathfrak{M} \times_{\mathcal{J}} \mathcal{C}^{(n)} \to \mathfrak{M} \times_{\mathcal{J}} \mathcal{P}_n$, where κ is an $\mathfrak{M}$-isomorphism

$\mathfrak{M} \times_{\mathcal{J}} \mathcal{P}_n$, which is a deformation of the identity of P_n. By Corollary 6.2

of [9], κ must be translation by a unique $\mathcal{J}$-morphism $\beta : \mathfrak{M} \to \mathcal{P}_0$, which

vanishes at the point m of $\mathfrak{M}$. Therefore, our deformation is in fact

$D(a, \beta)$. Q.E.D.

Now, I may explain why I wrote this paper. Let $n > 2g - 2$. Then,

$\int_n : C^{(n)} \to P_n$ is a locally trivial projective bundle. If one chooses a

point in P_n, one may identify P_n with the Jacobian of C. I was trying

to answer the question: Given a deformation $\mathcal{A} \to \mathcal{N}$ of J as an abelian

variety, when can the projective bundle $\int_n$ be extended to a projective

bundle $\mathcal{O} \to \mathcal{A}$?

As the bundle $\mathcal{O} \to \mathcal{A}$ would be a deformation of the integral

$\int_n : C^{(n)} \to P_n$, the Theorem 5.2 implies that we must be deforming J as

a Jacobian variety, when C is not hyperelliptic.

PRINCETON UNIVERSITY
AND THE JOHNS HOPKINS UNIVERSITY

REFERENCES

[1]   H. Cartan, Seminare: Families d'espaces complexes et fondements de la géométrie analytique, Secretariat mathematique, Paris, 1960-1961.

[2]   P. Griffiths, Some remarks and examples on continuous systems and moduli. J. Math. Mech. 16 (1967), 789-802.

[3]   A. Grothendieck, Seminare de géométrie algébraique, I.H.E.S., Bures-sur-Yvette, 1960-1961.

[4]   ________, Fondements de la géométries algébraique (Extracts du Seminare Bourbaki) Secrétariat mathématique, Paris, 1962.

[5]   R. Gunning, Riemann surfaces and generalized theta functions, Springer-Verlag, Berlin, 1976.

[6]   G. Kempf, Toward the inversion of abelian integrals I, to appear.

[7]   A. Mattuck, Symmetric products and Jacobians, Amer. J. Math. 83 (1961), 189-206.

[8]   A. Mayer, Rauch's variational formula and the heat equation, Math. Ann. 181 (1969), 53-59.

[9]   D. Mumford, Geometric Invariant Theory, Springer-Verlag, Berlin, 1965.

[10] ________, Lectures on curves on an algebraic surface, Princeton University Press, Princeton, 1966.

[11] K. Kodaira and D. C. Spencer, On deformations of complex analytic structures. Annals of Math. 67 (1958), 328-466.

[12] ________, A theorem of completeness of characteristic systems of complete continuous systems. Amer. of Math. 81 (1959), 477-500.

[13] J.-P. Serre, Groupes algébraiques et corps de classes. Hermann, Paris, 1959.

REMARKS ON PROJECTIVE STRUCTURES

Irwin Kra and Bernard Maskit[*]

In this note the authors continue their previous investigations of deformations of Fuchsian groups [10, 14]. Throughout this paper Γ denotes a finitely generated, purely loxodromic (including hyperbolic) Fuchsian group of the first kind operating on the upper half plane U. Then Γ operates discontinuously and freely on U, and $S = U/\Gamma$ is a closed Riemann surface of genus $g \geq 2$. We denote the space of holomorphic quadratic differentials on S by $B_2(\Gamma)$. We do not differentiate between differentials on S and automorphic forms for Γ, so that we regard $\phi \, \epsilon \, B_2(\Gamma)$ as a holomorphic function on U, with

$$\phi(Az)A'(z)^2 = \phi(z), \quad \text{for all} \quad A \, \epsilon \, \Gamma, \quad \text{and for all} \quad z \, \epsilon \, U .$$

We use the Nehari norm [20]

$$\|\phi\| = \sup_{z \, \epsilon \, U} \{|\phi(z)| \, (2 \, \text{Im} \, z)^2\} .$$

(We will on occasion need to regard U as the unit disc, in which case

$$\|\phi\| = \sup_{z \, \epsilon \, U} \{|\phi(z)| \, (1 - |z|^2)^2\} .)$$

---

[*]Research supported in part by NSF grant MCS 7801248.

Given $\phi \in B_2(\Gamma)$, we can find a locally schlicht meromorphic function f on U with $\{f, z\} = \phi(z)$ [9, pp. 376-377]; here $\{f, \cdot\}$ is the Schwarzian differential operator

$$\{f, \cdot\} = \left(\frac{f''}{f'}\right)' - \frac{1}{2}\left(\frac{f''}{f'}\right)^2 .$$

The function f will be unique once we normalize by requiring

$$f(z) = (z-i)^{-1} + O(|z-i|), \quad \text{near } i .$$

In general, f may be replaced by $a \circ f$, with a a Möbius transformation $(a \in \mathrm{PSL}(2; C))$. It will, on occasion, be useful to choose other normalizations.

The function f determines a homomorphism

$$\theta : \Gamma \to \mathrm{PSL}(2; C)$$

defined by

$$f \circ A = \theta(A) \circ f, \quad A \in \Gamma .$$

When f is replaced by $a \circ f$ (with $a \in \mathrm{PSL}(2; C)$), θ is replaced by $a\theta a^{-1}$. With these identifications, there exist canonical correspondences among the three maps ϕ, f and θ: any one determines the other two. The only non-trivial part of this assertion is Poincaré's theorem (see, for example, Kra [11]) that $\phi \mapsto \theta$ is an injective mapping. The maps f (or ϕ or θ) are the deformations of Γ.

We set $G = \theta(\Gamma)$. It was remarked by Kra [10] that generically G is purely loxodromic and isomorphic to Γ; Maskit [14] showed that if this generic G was also discontinuous (that is, Kleinian), then G has a simply connected invariant component (of its set of discontinuity). Maskit [14] further showed that this invariant component need not be f(U).

If $\|\phi\|$ is sufficiently small (less than 2), then G is quasi-Fuchsian (by Ahlfors-Weill [1], for example); f is schlicht on U; and f maps U onto one (of the at most two) invariant components of G. In this case, we say that ϕ belongs to the Bers embedded Teichmüller space $T(\Gamma)$ (see

[4]). The Teichmüller space $T(\Gamma) \subset B_2(\Gamma)$ is bounded (it is contained in the ball of radius 6 about the origin); the points of its boundary $\partial T(\Gamma)$ are isomorphisms of Γ onto groups with one simply-connected invariant component (here f maps U onto the invariant component).

What happens outside of $T(\Gamma)$ is largely unknown. It was shown by Gunning [7] that either f maps U onto $C \cup \{\infty\}$ or that f is a cover map of a domain Δ. Kra [10] added the equivalent condition that G acts discontinuously on $\Delta = f(U)$ if (and only if) f is a covering map of Δ. It was shown by Maskit [14] that for appropriate choice of Γ, there are isomorphisms θ for which G is a Fuchsian group and f is not a cover map.

In this paper we shall be mostly concerned with the case where f is a cover map; we then set $\tilde{S} = \Delta/G$, and let $\tilde{f} : S \to \tilde{S}$ be the induced map. We denote the natural projections by $p : U \to S$, and $\tilde{p} : \Delta \to \tilde{S}$, so that we have a commutative diagram

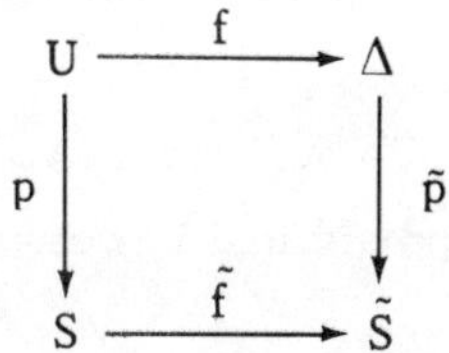

Since f and p are cover maps, and $\tilde{p}$ is a possibly branched cover map, $\tilde{f}$ is also a possibly branched cover map. In our first example we show that $\tilde{f}$ can be a non-trivial covering, and in our second example we show that $\tilde{f}$ can be branched.

Our first result describes the set of "nice" deformations of Γ.

THEOREM 1. *The set of $\phi \in B_2(\Gamma)$ for which f is a cover map is compact.*

That this set is closed was shown by Kra [10]; we show here that it is bounded, and we explicitly exhibit a bound.

In the generic case, where S covers no other surface, the set of cover maps can be identified with the set of uniformizations of S: A uniformization of S is a Kleinian group G (called the uniformizing group) with an invariant component Δ where G acts freely on Δ and $\Delta/G = S$. We regard two uniformizations of S as being the same if the uniformizing groups are conjugate in the group of Möbius transformations. Two uniformizations (Δ, G) and (Δ^*, G^*) of S are called topologically equivalent if there is a homeomorphism w of Δ onto Δ^* that conjugates G onto G^*. This homeomorphism establishes a correspondence (isomorphism) between the elements of G and G^*.

A parabolic element A in a uniformizing group G is called accidental if there is a topologically equivalent uniformization of S with uniformizing group G^* where the element $A^* \epsilon G^*$ corresponding to A is loxodromic (or hyperbolic).

A Kleinian group G is geometrically finite if it has a finite sided fundamental polyhedron for its action on hyperbolic 3-space. Equivalent conditions for a Kleinian group to be geometrically finite can be found in Beardon-Maskit [2].

Our next result shows how a deformation that yields an accidental parabolic element can be approximated by deformations that give rise to elliptic elements.

THEOREM 2. *Let $\phi \epsilon B_2(\Gamma)$ be such that* f *is a cover map onto a region* Δ, $\Delta/G = S$, G *is geometrically finite, and* G *contains an accidental parabolic element* $\theta(A_0)$, $A_0 \epsilon \Gamma$. *Then there exists a sequence of points* $\phi_n \epsilon B_2(\Gamma)$, $\phi_n \to \phi$, *so that the corresponding maps* f_n *are cover maps with* $f_n(U) = \Delta_n$, $\Delta_n/G_n = S$, G_n *is geometrically finite, and* $\theta_n(A_0)$ *is elliptic of order* n.

We remark that the regular b-groups (see Bers [5] and Maskit [15]) on $\partial T(\Gamma)$ satisfy the hypotheses of this theorem.

§1. *Example one*

In this section we construct an example where $\tilde{f} : S \to \tilde{S}$ is a regular (unbranched) covering of index 2.

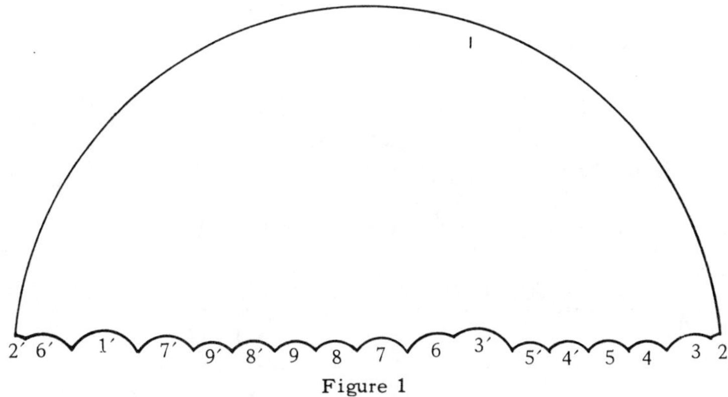

Figure 1

We start with a Fuchsian group Γ^* defined by a fundamental polygon P^* with 18 sides (see Figure 1). The sides are labelled with the following identification scheme: 1, 2, 3, 4, 5, 4′, 5′, 3′, 6, 7, 8, 9, 8′, 9′, 7′, 1′, 6′, 2′; Γ^* is to be generated by the Möbius transformations identifying side j to side j'. One easily sees that there are four cycles of vertices; we choose P^* so that the sum of the angles of each cycle is 2π. A simple calculation (using the Euler-Poincaré index) shows that U/Γ^* has genus 3.

We next construct a Kleinian group G of Schottky type (see Chuckrow [3]) representing a closed Riemann surface of genus 2; G is the uniformization obtained by requiring that some non-dividing loop lift to a loop. The group G is then a free product in the sense of Klein's combination theorem (see Maskit [16]) of a torus group (that is, a purely parabolic group isomorphic to $Z \oplus Z$) and a cyclic loxodromic group. A fundamental domain D for G is shown in Figure 2.

In this figure, we have drawn in two extra lines and have labelled the sides. Note that side "a" joins two equivalent points on the boundary of D. One can find a reasonably smooth local homeomorphism from P^* to D which covers D twice and preserves boundary identifications. One starts with a map on the boundary that is given as follows: $1 \to a$, $2 \to b$, $3 \to c$, $4 \to d$, $5 \to e'$, $4' \to d'$, $5' \to e$, $3' \to c$, $6 \to b$, $7 \to c$, $8 \to d$, $9 \to e'$, $8' \to d'$, $9' \to e$, $7' \to c$, $1' \to a$, $6' \to b'$, $2' \to b'$. One easily sees that this boundary map extends to a local homeomorphism of P^* to D. The actions of Γ^*

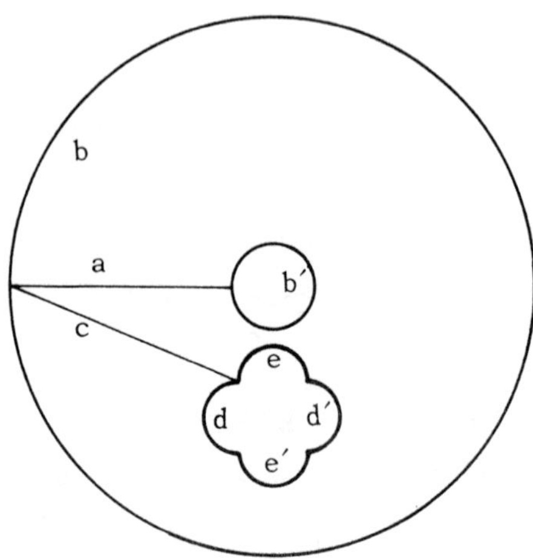

Figure 2

on U and G on $\Omega(G)$, the ordinary set of G, allows one to extend
the constructed map to a local homeomorphism of U onto $\Omega(G)$ that conjugates Γ^* onto G. Finally, we use standard techniques in quasiconformal mappings (see, for example, Maskit [14]) to replace Γ^* by Γ, a
quasiconformal deformation of Γ^*, so that there is an analytic local
homeomorphism conjugating Γ onto G.

§2. *Example two*

Our second example is closely related to the first. Here we start with
a Fuchsian group Γ^* representing a surface of genus 4 ; Γ^* has a fundamental polygon P^* whose 18 sides are labelled in order: 1, 2, 3, 2′, 3′,
4, 5, 4′, 5′, 1′, 6, 7, 6′, 7′, 8, 9, 8′, 9′ (Figure 3).

For our group G we choose a Kleinian group with an invariant component Δ, where Δ/G is a surface of genus 2, and the covering is
branched to order two over two points in Δ/G. The covering is defined
by the requirement that the dividing loop in Figure 4 lift to a loop.

The group G can be realized (Maskit [16]) as a free product of two
Fuchsian groups, where each Fuchsian group represents a surface of
genus 1 and the covering is branched to order two over one point. (The
group G will then represent the three surfaces discussed here.) A funda-
domain for G is shown in Figure 5.

We remark that each of the vertices of each of the polygons in Figure
5 has angle $\pi/4$.

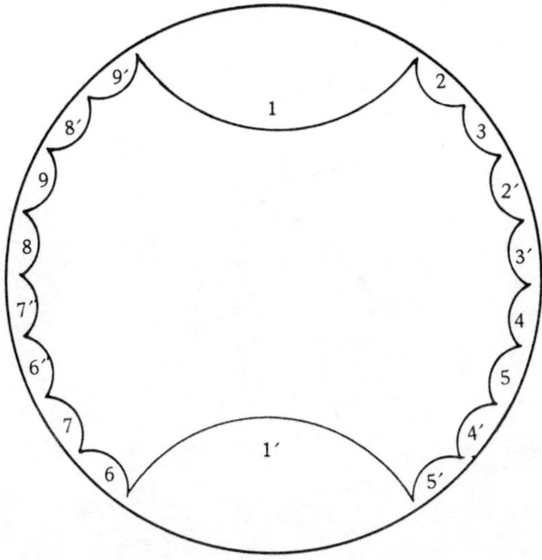

Figure 3

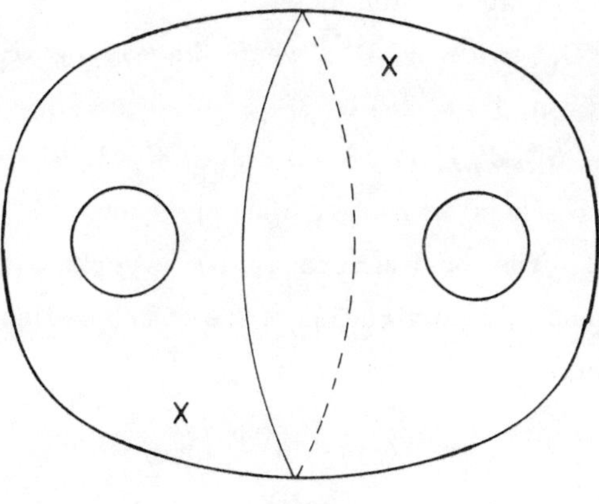

Figure 4

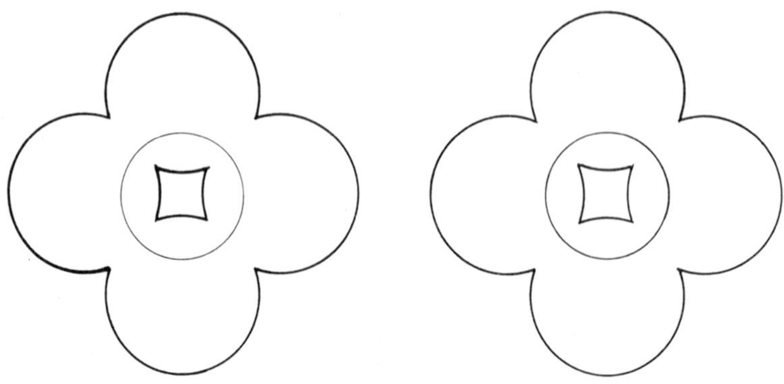

Figure 5

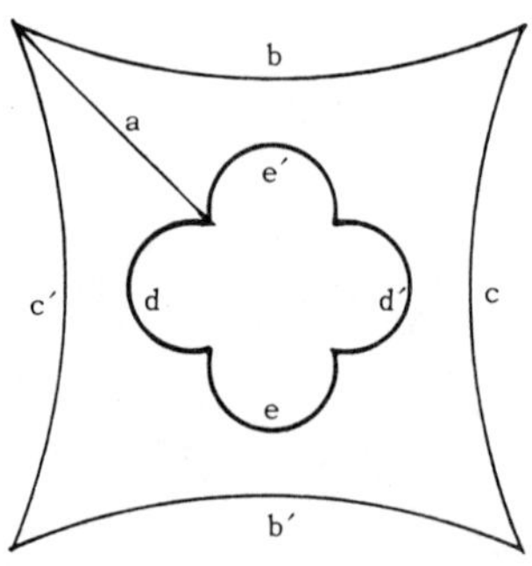

Figure 6

For simplicity, we redraw, in Figure 6, that part D_0 of D which represents the action of G on Δ, where we have changed the normalization, added a line and labelled the sides.

We now cover D_0 twice by P^*, where the mapping on the sides is given by: $1 \to a$, $2 \to b$, $3 \to c$, $2' \to b'$, $3' \to c'$, $4 \to b$, $5 \to c$, $4' \to b'$, $5' \to c'$, $1' \to a$, $6 \to d$, $7 \to e$, $6' \to d'$, $7' \to e'$, $8 \to d$, $9 \to e$, $8' \to d'$, $9' \to e'$. As before, this map extends to a local homeomorphism of U onto Δ which conjugates Γ^* onto G. The local homeomorphism is replaced by an analytic local homeomorphism using a variation of the complex structure through quasiconformal mappings.

§3. *A problem*

We can regard our second example from a slightly different point of view. We start with the Fuchsian group $\tilde{\Gamma}$ representing the closed surface $\tilde{S}$ of genus 2, and we observe that there is a subgroup Γ, of finite index in $\tilde{\Gamma}$, and there is a uniformization (Δ, G) of $\tilde{S}$, so that the covering map $\tilde{f} : U \to \Delta$ conjugates the *proper subgroup* Γ onto all of G. One observes that this can occur only if $\tilde{f}$ is not injective.

PROBLEM. *For which uniformizations* (Δ, G) *of a Riemann surface* $\tilde{S}$ *can one choose a proper subgroup* Γ *of the Fuchsian group* $\tilde{\Gamma}$ *representing* $\tilde{S}$, *so that the cover map* $f : U \to \Delta$ *conjugates* Γ *onto all of* G?

§4. *Generalizations of the examples*

Our two examples involve low genera and low index $[\tilde{\Gamma} : \Gamma]$; these examples can easily be generalized to higher genera and higher index. In fact it was only because of the low genus that we included the first example. If we start with a deformation ϕ, f, $\theta : \Gamma \to G$, where $S = U/\Gamma$ is a branched covering of $\tilde{S}(= f(U)/G = \Delta/G)$, then we let G_0 be a torsion-free normal subgroup of finite index in G (Selberg [21]). One easily sees that $\Gamma/\theta^{-1}(G_0) = \Gamma/\Gamma_0$ is isomorphic to G/G_0, and so the coverings $U/\Gamma_0 \to U/\Gamma$ and $\Delta/G_0 \to \Delta/G$ have the same number of sheets. It follows that f projects to a covering $U/\Gamma_0 \to \Delta/G_0$ which has the same number of sheets as the covering $U/\Gamma \to \Delta/G$. Of course, since G_0 is torsion free, the covering $U/\Gamma_0 \to \Delta/G_0$ is unbranched.

Except for genus 2, the generic Fuchsian group is not contained in any other Fuchsian group (Greenberg [6]), and so our examples of non-trivial coverings are exceptional. Of course, once we have such an example ϕ, f, $\theta : \Gamma \to G$, then we can obtain a family of such examples by varying $\tilde{\Gamma}$, the extension of Γ. In fact, we can regard $T(\tilde{\Gamma})$, the Teichmüller space of $\tilde{\Gamma}$, as a submanifold of $T(\Gamma)$, and every point of this submanifold admits a deformation which yields a non-trivial covering.

§5. *Theorem 1*

In this section we prove Theorem 1; as we have already remarked, it suffices to show that the set of ϕ for which f is a cover map is bounded. Our proof makes use of a generalization of the Kraus [13] -Nehari [20] theorem.

LEMMA 5.1. *Let* f *be a locally schlicht meromorphic function on the unit disc* U, *and let* $\phi = \{f, \cdot\}$.

(a) *If* f *is schlicht on every non-Euclidean disc of radius* δ, *then* $\|\phi\| \le 6 \tanh^{-2} \delta$.

(b) *If* $\|\phi\| \le k$, *and* $k > 2$, *then* f *is schlicht on every non-Euclidean disc of radius* $\frac{1}{2} \log \frac{\sqrt{k} + \sqrt{2}}{\sqrt{k} - \sqrt{2}}$. (*If* $k \le 2$ *then, by Nehari's theorem* [20], f *is schlicht in* U.)

We actually only need part (a) of this lemma. To prove this part, we set $A(z) = \delta^* z$, where $\delta^* = \tanh \delta$. Observe that $f \circ A$ is schlicht in the unit disc, and hence

$$(5.1) \qquad |\phi(0)| = |\{f, 0\}| = |\{f, A(0)\}| = |\{f \circ A, 0\} A'(0)^{-2}|$$

$$\le \|\{f \circ A, \cdot\}\| \, |\delta^*|^{-2} \le 6(\delta^*)^{-2} \, ;$$

where we have used the Cayley identity, and Kraus' theorem [13] which asserts that if ϕ is the Schwarzian of a function that is schlicht in the unit disc, then $\|\phi\| \le 6$.

Now for any $z \ne 0$, in U, we let T_z be some non-Euclidean motion mapping 0 to z. Then, using the Cayley identity, the invariance of the Poincaré metric and (5.1), we obtain

$$(5.2) \qquad |\{f, z\}| (1 - |z|^2)^2 = |\{f \circ T_z, 0\}| \, |T_z'(0)^{-2}| (1 - |z|^2)^2$$

$$= |\{f \circ T_z, 0\}| \le 6(\delta^*)^{-2} \, .$$

Combining (5.1) and (5.2) we obtain

$$\|\phi\| \le 6(\delta^*)^{-2} = 6 \tanh^{-2} \delta \, .$$

The proof (although not the present formulation) of part (b) appears in Kra [10]. It is based on Nehari's theorem [20].

We turn now to the proof of Theorem 1.

We have already observed that $\tilde{f}$ is a possibly ramified cover map and that hence $\tilde{f} \circ p = \tilde{p} \circ f$ is a possibly ramified covering of $\tilde{S}$. Let $\tilde{\Gamma}$ be the corresponding cover group (the Fuchsian model of G). Since $\Gamma \subset \tilde{\Gamma}$, $\tilde{\Gamma}$ must be a finite extension of Γ. We cover $\tilde{S}$ by a finite number of open sets $\{U_j\}$, so that for each j, $(\tilde{p} \circ f)^{-1}(U_j)$ is a disjoint union $\{U_{ja}\}$ of open circular discs, each of which is precisely invariant under a (possibly trivial) finite cyclic subgroup of $\tilde{\Gamma}$. It is easy to see that f is schlicht on each U_{ja}. Since we can cover a fundamental domain for $\tilde{\Gamma}$ by a finite number of the U_{ja}, we have shown that there is a $\delta > 0$ so that for every $z \in U$, f restricted to the non-Euclidean disc of radius δ about z is schlicht.

The above remarks hold for any given covering map f, and, of course, δ depends, not on f, but on $\tilde{S}$. Since Γ has at most finitely many extensions, we can in fact find a δ independent of $\tilde{S}$. (This δ still depends on S.)

§6. *The radius of univalence*

In this section we obtain another (more geometric) lower bound on the radius of univalence of cover maps f (producing an alternate proof of Theorem 1).

We may assume that f is not univalent. We let

$$\delta = \delta(\Gamma) = \inf_{\substack{z \in U \\ A \in \Gamma \setminus \{1\}}} \rho(z, Az) > 0 ,$$

where ρ is the non-Euclidean distance function on U. The quantity $\delta(\Gamma)$ is the length of the shortest closed geodesic on S.

(1) If $\tilde{f}$ is one-to-one, then f is clearly univalent in every non-Euclidean disc of radius $\leq \delta(\Gamma)$.

(2) If $\tilde{f}$ is an unramified n-sheeted cover, then f is univalent in every disc of radius $\leq \delta(\tilde{\Gamma})$, where $\tilde{\Gamma}$ is the cover group of $\tilde{S}$. But it is easy to see that

$$\delta(\tilde{\Gamma}) \geq \frac{\delta(\Gamma)}{n} ,$$

and from Riemann-Hurwitz

$$n \leq g - 1 \quad (g = \text{genus of } S) .$$

Thus f is univalent in every disc of radius $\leq \delta(\Gamma)/(g-1)$.

(3) Finally, if $\tilde{f}$ is ramified we pass to a torsion-free finite index sub-group G_0 of G. Let $\Gamma_0 = \theta^{-1}(G_0)$. Then Γ_0 is a subgroup of Γ and the index n of Γ_0 in Γ equals the index of G_0 in G. Clearly

$$\delta(\Gamma_0) \geq \delta(\Gamma) .$$

As before, we need only estimate the number of sheets k in the covering $\tilde{f} : \Delta/\Gamma_0 \to D/G_0$. Since the following diagram commutes

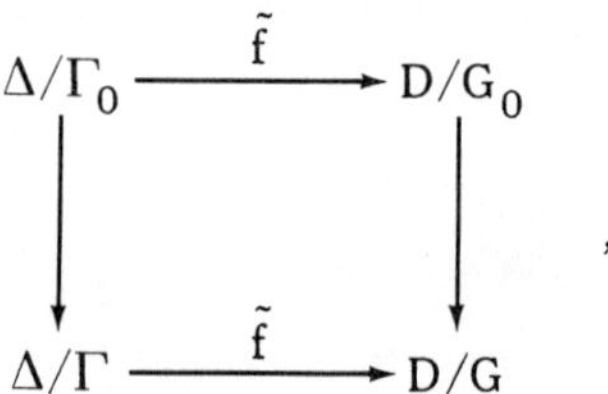

and each of the vertical maps is n-sheeted, we see that k is the number of sheets in $\tilde{f} : \Delta/\Gamma \to D/G$. But it is well known that

$$k = \frac{\text{Area}(\Delta/\Gamma)}{\text{Area}(\Delta/\tilde{\Gamma})} ,$$

where $\tilde{\Gamma}$ is the branched universal cover group of $\tilde{S}$ ($=$ full pre-image of G via the map f). Now Area $\Delta/\Gamma = 2\pi(2g-2)$, and Area$(\Delta/\tilde{\Gamma}) \geq \pi/3$ (see, for example, Kra [12, pp. 76-78]). This last estimate follows from the fact that G cannot be a triangle group (D would have to be simply connected for G to be a triangle group). We conclude that in this case f is univalent in every disc of radius $\leq \delta(\Gamma)/12(g-1)$.

§7. *Convergence of cover maps*

In this section, we discuss one example of convergence of cover maps in $B_2(\Gamma)$ and prove Theorem 2; our discussion is based on the classification of geometrically finite Kleinian groups with an invariant component (Maskit [18], [19]).

We start with the situation described in the hypotheses of Theorem 2.

We can find a simple loop w_0 on S, so that $\theta(A_0) \in G$ corresponds to the lifting of w_0 (Maskit [17]); there is, in fact, a set of simple disjoint loops $w_0, w_1, \cdots, w_k$ on S, and a set of "integers" $\infty, a_1, \cdots, a_k$, $1 \leq a_j \leq \infty$, so that the following hold (Maskit [17]):

(1) If $a_j < \infty$, then the element of G corresponding to w_j is elliptic (or the identity) of order a_j.

(2) All the relations in G follow from the one relation in the fundamental group $\pi_1(S) \cong \Gamma$ and the relations implicit in (1).

(3) If $a_j = \infty$, then the element of G corresponding to w_j is accidental parabolic.

(4) Every accidental parabolic element of G is a power of a conjugate of one of those elements given in (3).

It was also shown by Maskit [17] that these loops and integers determine the topological type of the other surfaces that appear in $\Omega(G)/G$. These other surfaces are obtained as follows. If $a_j = 1$, we cut S along w_j and fill in the two holes with discs. If $1 < a_j < \infty$, we cut S along w_j and fill in the two holes with discs, choose a point in each of these discs, and require that the projection map from $\Omega(G)$ be branched to order a_j over these two points. If $a_j = \infty$, we cut S along w_j and fill in these two holes with punctured discs. We obtain this way a collection of surfaces with signatures. We retain all the surfaces so produced except for those corresponding to elementary groups.

It was also shown in [17], that with few very special exceptions (that are eliminated by the hypotheses of this theorem), a set of loops and "integers" as above defines a uniformization of S, and if we prescribe the conformal structure on the other components of $\Omega(G)/G$, then the group is unique.

We now define G_n to be that Kleinian group—defined up to conjugation—which uniformizes S where the loops $w_0, w_1, \cdots, w_k$ are given above and the associated integers are $n, a_1, \cdots, a_k$, and $\Omega(G_n)/G_n = \Omega(G)/G$; the last equality is taken in the sense that the two omitted points in $\Omega(G)/G$ corresponding to w_0 are taken onto the corresponding new branch points of order n.

We define $\Omega'(G_n)$ to be $\Omega(G_n)$ with the points lying over the two branch points corresponding to w_0 removed. Then there is an analytic cover map

$$g_n : \Omega(G) \to \Omega'(G_n)$$

that conjugates G onto G_n. Note that on each component, except possibly on Δ and on those components of $\Omega(G)$ lying over the components that contain points in $\Omega(G_n)\backslash\Omega'(G_n)$, g is schlicht (each such component is simply connected). We normalize G, G_n, and g_n so that ∞ belongs to the invariant component Δ of G and Δ_n of G_n and so that

$$g_n(z) = z + O(|z|^{-1}), \quad \text{near} \quad \infty .$$

We choose a fundamental domain D for the action of G on Δ, where D is bounded by finitely many circular arcs, and $\theta(A_0)$ identifies two of these boundary arcs. It follows easily from the way G_n has been defined that if W is an open arc in Δ which projects to a loop in Δ_n, then W passes through at least n translates of D. We conclude that there is an increasing sequence U_n of open sets with $\bigcup_n U_n = \Delta$ and g_n schlicht in U_n. Hence we can choose a subsequence which we again call $\{g_n\}$, so that g_n converges to a schlicht function g on Δ.

For each $B \in G$, and each n, there is a Möbius transformation $\chi_n(B)$ so that $g_n \circ B = \chi_n(B) \circ g_n$. For n sufficiently large, D and $B(D)$ are contained in U_n, hence $\chi_n(B) \to g \circ B \circ g^{-1}$, a Möbius transformation, and the mapping $B \mapsto g \circ B \circ g^{-1}$ is an isomorphism.

We choose a finite number of non-conjugate other components of G; call them $\Delta_1, \cdots, \Delta_s$. Either $g_n|\Delta_j$ is already schlicht, or as above, we can exhaust Δ_j by $\{U_n\}$ with g_n schlicht in U_n. In either case, after passing to an appropriate subsequence, we have that $g_n \to g$, where g is either schlicht or constant on each component of G. But g cannot be constant on any component Δ_j; for if it were, then $\chi_n(B)$ would converge to a constant for every B stabilizing Δ_j, and we already know that χ_n converges to an isomorphism. We conclude that g is schlicht on all of $\Omega(G)$; by the uniqueness theorem of Maskit [17], g is the identity.

We now let $f_n = g_n \circ f$. Since g_n is converging to the identity uniformly on compact subsets, and since Γ has a compact fundamental domain, $\{g_n, \cdot\}$ is converging to O uniformly on $\Omega(G)$. Hence by the Cayley identity, $\{f_n, \cdot\}$ is converging to $\{f, \cdot\}$ in $B_2(\Gamma)$.

REMARKS. (1) We have shown convergence only on a subsequence of $\{\phi_n\}$. However, Theorem 1 implies convergence of the entire sequence. (2) The above proof holds in the more general case of a cover map f with $\Delta/G = \tilde{S}$ (which may not necessarily equal S). Here the conclusion $\Delta_n/G_n = \tilde{S}$ is to be understood as equality of coverings with ramifications. We note that once we have established that there do exist non-trivial coverings $\tilde{f} : S \to \tilde{S}$ among the deformations, it is easy to see that there are such deformations where the image group G contains accidental parabolic elements.

§8. *Uniformizations of surfaces of genus 2*

We conclude this paper with the following simple

THEOREM 3. *Let (Δ, G) be a uniformization of the closed (hyperelliptic) surface S of genus 2. Let the hyperelliptic involution be denoted by j. Then there is an elliptic Mobius transformation J of order 2, where $J\Delta = \Delta$, and $\tilde{p} \circ J = j \circ \tilde{p}$ ($\tilde{p}: \Delta \to \Delta/G = S$ is the natural projection).*

Proof. Let Γ be the universal covering group of S. Then (Δ, G) corresponds to a deformation of Γ: $\phi_0 = \{f_0, \cdot\}$, $f_0 : U \to \Delta$, $\theta_0 : \Gamma \to G$. We can lift the hyperelliptic involution to U so as to obtain an extension Γ_1

of Γ, where $[\Gamma_1 : \Gamma] = 2$, and where there is an elliptic element A of order 2 in Γ_1. It is well known that $B_2(\Gamma) = B_2(\Gamma_1)$; hence θ_0 extends to a homomorphism of Γ_1. Since the trace function $\phi \mapsto \mathrm{trace}^2\theta(A)$ is holomorphic, and $\mathrm{trace}^2\theta(A) = 0$ for all ϕ with $\|\phi\| < 2$; we conclude that $\theta_0(A)$ is elliptic of order 2.

BIBLIOGRAPHICAL REMARK. See Gunning's paper [8] in this volume for a more complete bibliography on deformations of Fuchsian groups.

STATE UNIVERSITY OF NEW YORK
STONY BROOK, NEW YORK 11794

REFERENCES

[1] L. V. Ahlfors and G. Weill, A uniqueness theorem for Beltrami equations, *Proc. Amer. Math. Soc.*, *13*(1962), 975-978.

[2] A. F. Beardon and B. Maskit, Limit points of Kleinian groups and finite sided fundamental polyhedra, *Acta Math.*, *132*(1974), 1-12.

[3] V. Chuckrow, Subgroups and automorphisms of extended Schottky type groups, *Trans. Amer. Math. Soc.*, *150*(1970), 121-129.

[4] L. Bers, A non-standard integral equation with applications to quasiconformal mappings, *Acta Math.*, *116*(1966), 113-134.

[5] _________, On boundaries of Teichmüller spaces and on Kleinian groups: I, *Ann. of Math.*, *91*(1970), 570-600.

[6] L. Greenberg, Maximal Fuchsian groups, *Bull. Amer. Math. Soc.*, *69* (1963), 569-573.

[7] R. C. Gunning, Special coordinate coverings of Riemann surfaces, *Math. Ann.*, *170*(1967), 67-86.

[8] _________, Affine and projective structures on Riemann surfaces, this volume.

[9] E. Hille, Analytic Function Theory, Volume II, Blaisdell, Waltham, Mass., 1962, 496 pp.

[10] I. Kra, Deformations of Fuchsian groups, *Duke Math. J.*, *36*(1969), 537-546.

[11] _________, A generalization of a theorem of Poincaré, *Proc. Amer. Math. Soc.*, *27*(1971), 299-302.

[12] _________, Automorphic Forms and Kleinian Groups, Benjamin, Reading, Mass., 1972, 464 pp.

[13] W. Kraus, Uber den Zusammenhang einiger Charakteristiken eines einfach zusammenhangenden Bereiches mit der Kreisabbildung, *Mitt. Math. Sem. Giessen*, *21*(1932), 1-28.

[14] B. Maskit, On a class of Kleinian groups, *Ann. Acad. Sci. Fenn.*, Ser. A. *442* (1969), 8 pp.

[15] ————, On boundaries of Teichmüller spaces and on Kleinian groups: II, *Ann. of Math.*, *91* (1970), 607-639.

[16] ————, On Klein's combination theorem III, *Advances in the theory of Riemann surfaces*, Ann. of Math. Studies, No. 66 (1971), 297-316.

[17] ————, Decomposition of certain Kleinian groups, *Acta. Math.*, *130* (1973), 243-263.

[18] ————, On the classification of Kleinian groups: I - Koebe groups, *Acta Math.*, *135* (1975), 249-270.

[19] ————, On the classification of Kleinian groups II - signatures, *Acta Math.*, *138* (1977), 17-42.

[20] Z. Nehari, Schwarzian derivatives and schlicht functions, *Bull. Amer. Math. Soc.*, *55* (1949), 545-551.

[21] A. Selberg, On discontinuous groups in higher-dimensional symmetric spaces, *Contributions to Function Theory*, Bombay (1960), 147-164.

SOME REMARKS ON KLEINIAN GROUPS

S. L. Krushkal'

Kleinian groups acting in the plane have been extensively studied; considerable progress has also been made in the investigation of higher dimensional Kleinian groups. Here, I will consider some "non-standard" questions in both the planar and space theories of Kleinian groups and discuss a number of results in these theories. The proofs have appeared elsewhere.

§1. *Some properties of fundamental domains*

Let G be (non-elementary) Kleinian group acting on the extended complex plane $\overline{C}$. By $\Omega(G)$ and $\Lambda(G)$ we denote respectively the discontinuity set and the limit set of the group G. Let $\Omega_1, \Omega_2, \cdots$ be a complete list of non-equivalent components of the group G, and $G_{\Omega_1}, G_{\Omega_2}, \cdots$ be their respective stabilizers. Then Ω_j/G_{Ω_j} are Riemann surfaces, and

$$\Omega(G)/G = (\Omega_1/G_{\Omega_1}) \cup (\Omega_2/G_{\Omega_2}) \cup \cdots$$

is their finite or countable union.

It is also well known that every Kleinian group has a fundamental domain P_G, that is, a set of points $z \in \Omega(G)$ containing one point from each orbit Gz, $z \in \Omega(G)$, and such that all $P_j = P \cap \Omega_j$ are connected.

© 1980 Princeton University Press
Riemann Surfaces and Related Topics
Proceedings of the 1978 Stony Brook Conference
0-691-08264-2/80/000361-06 $00.50/1 (cloth)
0-691-08267-7/80/000361-06 $00.50/1 (paperback)
For copying information, see copyright page

This domain can always be chosen in such a way that the boundaries of the components P_j in Ω_j consist of smooth arcs which are pairwise identified by the generators of G_{Ω_j}. A fundamental domain corresponding to canonical dissections of the surfaces Ω_j/G_{Ω_j} is called *standard*.

It is often necessary to determine whether a given set P is a fundamental domain for a Kleinian group G. The question reduces immediately to the study of the action of G_{Ω_i} on Ω_i. When Ω_i is simply connected we have the following useful result.

PROPOSITION 1. *Let Ω_0 be a simply-connected component of a finitely generated Kleinian group without torsion and D be a domain in Ω_0. We assume (i) ∂D consists of a finite number of Jordan arcs, pairwise identified by the generators of G_{Ω_0}, and (ii) the domain D, after an obvious identification of its boundary points, gives a Riemann surface which is homeomorphic to Ω_0/G_{Ω_0}. Then D (after adjoining to it the part of the boundary) is a fundamental domain for the action of G_{Ω_0} on Ω_0.*

The proof of the proposition is obtained by using the Riemann-Hurwitz relation for the coverings of compact Riemann surfaces.

Partial result of such type has been obtained earlier in [2].

One can go further and set up the analogous question for the domains which are not known *a priori* to lie in $\Omega(G)$. For such domains we have the following theorem

THEOREM 1. *Let G_0 be a finitely generated torsion-free subgroup of Kleinian group G which is the stabilizer of some simply-connected component $\Omega_0 \subset \overline{\Omega}(G)$. Further let D be a domain in $\overline{C}$ whose boundary consists of a finite number of Jordan arcs, pairwise identified by some set of generators of G_0. Assume (i) the domain D and its images under the action of these generators give a Riemann surface, homeomorphic to Ω_0/G_0 and (ii) the domain D and its adjacent translates $A(D)$ (for corresponding $A \in G_0$) lie together univalently in $\overline{C}$. Then D is a fundamental domain of the group G_0 in Ω_0.*

These statements show that (under the conditions mentioned above) every domain in the plane which determines the Riemann surface, homeomorphic to the initial one, can be taken as a fundamental one. In the proofs available, especially in the one of Theorem 1, simple connectivity plays an essential role. The question of the validity of such type statements for an arbitrary finitely generated Kleinian group remains open.

§2. *Functions concentrated on limit set*

Usually one considers the functions and deformations associated with a Kleinian group G and having support in the discontinuity set $\Omega(G)$. The investigation of the corresponding functions concentrated on the limit set $\Lambda(G)$ is also of great interest. These complementary situations are strikingly different.

The considerations which follow are suitable for Kleinian groups acting both on a plane and in a multi-dimensional Euclidean space. Therefore, we directly consider the general case.

To the Euclidean space R^n, $n \geq 1$, we adjoin one infinitely distant point and obtain the *Möbius space* $\bar{R}^n = R^n \cup \{\infty\}$; by $m_n E$ we denote the n-dimensional Lebesgue measure of the sets $E \subset \bar{R}^n$.

Let $M(n+1)$ be the group of orientation preserving isometries $x \mapsto Ax$ of the half-space $R_+^{n+1} = \{x = (x_1, \cdots, x_{n+1}) \in R^{n+1} : x_{n+1} > 0\}$ which models hyperbolic n-space. Let G be a discrete subgroup of $M(n+1)$ acting discontinuously in R_+^{n+1}, and, perhaps, on some (open) set $\Omega(G) \subset \bar{R}^n = \partial R_+^{n+1}$. Then the limit set $\Lambda(G)$ of the group G lies in $\bar{R}^n$ and either coincides with $\bar{R}^n$ or is nowhere dense in $\bar{R}^n$. However, for $n \geq 2$ we may start from discrete groups acting in $\bar{R}^n$; their continuations in R_+^{n+1} (obtained in the well-known way) will be discontinuous there.

For the elements $A \in M(n+1)$ there is well defined the value

$$|A'(x)| = |dAx|/|dx| \, ,$$

which is the linear expansion at the point x; it is independent of the direction. We assume that $m_n \Lambda(G) > 0$.

In $\Lambda(G)$ one may form (using the axiom of choice) a fundamental set e_Λ which contains one point of each orbit Gx for almost all $x \in \Lambda(G)$, or rather, for all $x \in \Lambda(G)$, different from the fixed points of the elements of G (see [3]). As it has been shown in [3], there is a condition which guarantees that this set is non-measurable. If we pass from R_+^{n+1} to the unit ball $|x| < 1$, then the above-mentioned condition is (*) the intersection of isometric fundamental polyhedron with $\Lambda(G)$ has zero measure. More generally, we have

PROPOSITION 2. *If G satisfies condition (*), every subset $e \subset e_\Lambda$ with positive exterior measure is non-measurable.*

It is not difficult to construct the Kleinian groups which have measurable e_Λ. We shall not dwell upon them here.

Henceforth *we consider only these groups for which Proposition 2 is valid.*

For an arbitrary set $E \subset \Lambda(G)$ we may define the multiplicity function

$$k(x; E) = \text{card } E \cap Gx, \quad x \in \Lambda(G).$$

It is clear that if $E_1 \subset E$, then $k(x; E_1) \leq k(x; E)$. For the measurable set E the function $k(x; E)$ is also measurable, and we have the decomposition

$$E = \left(\bigcup_{p=1}^{\infty} E_p \right) \cup E_\infty,$$

where

$$E_p = \{x \in E : k(x; E) = p\}, \quad E_\infty = E \setminus \bigcup_{p=1}^{\infty} E_p.$$

We then have

THEOREM 2. *If E is a measurable subset of $\Lambda(G)$ having positive measure, then almost everywhere on E the function $k(x; E) = \infty$, that is, the intersection $E \cap Gx$ is either empty or consists of an infinite set of points.*

This is in sharp contrast with the fact that any compact set $F \subset R_+^{n+1} \cup \Omega(G)$ can intersect only a finite number of images $A(P_G)$ where P_G is a fundamental domain for the group G in $R_+^{n+1} \cup \Omega(G)$.

Such rigidity must, of course, raise the possibility of the existence of non-trivial automorphic forms and deformations with supports on $\Lambda(G)$. For many forms this question is completely solved by the following theorem.

THEOREM 3. *For* $p \in Z \setminus \{0\}$ *the equation*

$$\phi(Ax)|A'(x)|^P = \phi(x)(A \in G, x \in \Lambda(G))$$

has, in the class of measurable a.e. finite functions on $\Lambda(G)$, *only the zero solution.*

In particular, if $p + q \neq 0$ *every measurable G-form* $\phi(z)dz^P d\bar{z}^q(z \in C)$, *concentrated on* $\Lambda(G)$, *assumes only the values* 0 *or* ∞.

§3. *The rigidity of deformations*

Theorem 3 does not embrace the automorphic functions $\phi(x)(p = 0)$ and, for $n = 2$ (in $\bar{C}$), also the measurable forms $\phi(z)dz^P d\bar{z}^{-P}$, $p \in Z$, with supports on $\Lambda(G)$. These include the Beltrami differentials $\phi(z)d\bar{z}/dz$ which are of special interest and connected with planar quasi-conformal mappings.

The question of the existence of non-trivial differentials of such type presents a special case of the general rigidity problem of quasiconformal deformations of Kleinian groups in $\bar{R}^n$, $n \geq 3$. This is connected to the rigidity problem for Riemannian manifolds with negative curvature. Precisely, one questions whether there can exist quasiconformal automorphisms $f : \bar{R}^n \to \bar{R}^n (n \geq 2)$ deforming G into an isomorphic Kleinian group which is not conformal on $\Lambda(G)$. In particular, whether any quasiconformal automorphism of $\bar{R}^n$ compatible with the discrete nondiscontinuous group G, must be reduced to a Möbius one (that is, the composition of a finite number of inversions in spheres). Such deformations are, for instance, obtained as the boundary values of quasiconformal automorphisms of the half-space.

The affirmative answer to the latter question would mean that every two discrete subgroups G_1, G_2 from $M(n+1)$, $n \geq 2$, which are conjugate by a quasiconformal automorphism of R_+^{n+1} are also conjugate in $M(n+1)$. For groups whose quotients have finite volume, this result has been obtained by Mostow [4]. However, B. Apanasov [1] has recently shown that the rigidity theorem fails in the general case.

INSTITUTE OF MATHEMATICS
SIBERIAN BRANCH OF THE USSR ACADEMY OF SCIENCES
NOVOSIBIRSK

REFERENCES

[1] Apanasov, B. N., On Mostow's rigidity theorem, Dokl. Akad. Nauk SSSR, t. 243, No. 4 (1978), 829-832.

[2] Knapp, A. W., Doubly generated Fuchsian groups, Mich. Math. J., v. 15, No. 3 (1968), 289-304.

[3] Krushkal', S. L., On a property of limit sets of Kleinian groups, Dokl. Akad. Nauk SSSR, t. 225, No. 3 (1975), 500-502; t. 237, No. 2 (1977), 258.

[4] Mostov, G. D., Strong rigidity of locally symmetric spaces. Princeton, Princeton Univ. Press, 1973, 195 p. (Ann. of Math. Stud., No. 78).

REMARKS ON WEB GROUPS

Tadashi Kuroda, Seiki Mori and Hidenori Takahashi

1. Let G be a finitely generated non-elementary Kleinian group and let $\Omega(G)$ be the region of discontinuity of G. The limit set $\Lambda(G)$ of G is the complementary set of $\Omega(G)$ with respect to the extended complex plane $\hat{C}$. For a (connected) component Ω_* of $\Omega(G)$, we denote by G_{Ω_*} the stabilizer subgroup of G for Ω_*, that is, $G_{\Omega_*} = \{\gamma \in G : \gamma(\Omega_*) = \Omega_*\}$. This subgroup G_{Ω_*} of G is called the component subgroup of G for Ω_*.

In the following, we call G a web group, if $\Omega(G)$ has neither an invariant component under G nor just two components and if every component subgroup of G is quasi-Fuchsian. Since a web group is, of course, finitely generated by the definition, we see from Ahlfors' finiteness theorem that every component subgroup of a web group is a quasi-Fuchsian group of the first kind.

In this note, we shall deal with characterization of web groups in connection with properties of the so-called residual limit sets.

2. In his paper [1], Abikoff investigated the residual limit set of a finitely generated non-elementary Kleinian group G in detail. Here we summerize Abikoff's results briefly.

Assume that, for the closure $\overline{\Omega_*}$ of a component Ω_* of $\Omega(G)$, the set $\hat{C} - \overline{\Omega_*}$ is not empty. The set $\hat{C} - \overline{\Omega_*}$ consists of at most a countable number of components $\{\Omega_{*_j}\}_{j=1}^n$, $n \leq \infty$. The boundary $\partial\Omega_{*_j}$ of

every Ω_{*_j} is called a separator of G and the set of all the separators

of G is denoted by $S(G)$. Ahlfors' finiteness theorem and a well-known

Maskit's theorem imply that every separator of G is a quasi-circle.

Abikoff proved the fact that the residual limit set $\Lambda_0(G) = \Lambda(G) - \bigcup_{\Omega_*} \partial\Omega_*$

is empty if and only if $\Omega(G)$ has a component invariant under G or has

just two components. So, a finitely generated Kleinian group G is a web

group (in our sense) if and only if not only every component subgroup of

G is quasi-Fuchsian but also $\Lambda_0(G) \neq \emptyset$. He also showed that there

exists a finitely generated Kleinian group G with the property $\Lambda_0(G) \neq \emptyset$.

This last fact was also proved by Sasaki [3] from another point of view.

From now on, we assume $\infty \in \Omega(G)$. About separators of G, the

following properties are known:

a) $S(G)$ consists of at most a countable number of separators of G.

b) For every $\sigma \in S(G)$, the image $\gamma(\sigma)$ of σ under every $\gamma \in G$ is
   also a separator of G.

c) If, for a separator σ of G, (σ) denotes the component of $\hat{C} - \sigma$
   not containing ∞, then, for any two separators σ, σ' of G,
   either $(\sigma) \cap (\sigma') = \phi$ or one of (σ) and (σ') contains the other.

d) For every infinite sequence $\{\sigma_n\}_{n=1}^\infty$ of distinct separators of G,
   the diameter of σ_n tends to zero as n tends to infinity.

e) If there is an element $\gamma \in G$ for σ, $\sigma' \in S(G)$ such that $\sigma' = \gamma(\sigma)$,
   then we write $\sigma \sim \sigma'$. This relation is an equivalence relation
   and by this equivalence relation, $S(G)$ can be classified into a
   finite number of equivalence classes.

Abikoff classified points of $\Lambda_0(G)$ into two kinds. We denote by

$L_1(G)$ the set of all the points $\lambda \in \Lambda_0(G)$ such that there exists a

sequence $\{\sigma_n\}_{n=1}^\infty$ of distinct separators of G with $\bigcap_{n=1}^\infty \overline{(\sigma_n)} = \{\lambda\}$ and

we put $L_2(G) = \Lambda_0(G) - L_1(G)$. Assume $\lambda \in L_2(G)$ and, for a point

$z \in \Omega(G)$, denote by $S(G; z, \lambda)$ the set of all the separators of G which

separate λ from z. It is observed that there exists a separator $\sigma(z,\lambda)$ in $S(G; z,\lambda)$ such that there is no separator in $S(G; z,\lambda)$ which lies in a component of $\hat{C} - \sigma(z,\lambda)$ containing λ. This separator $\sigma(z,\lambda)$ is called a maximal separator in $S(G; z,\lambda)$ for λ and $M(\lambda)$ denotes the set of all the maximal separators for λ, that is, $M(\lambda) = \{\sigma(z,\lambda): z \in \Omega(G)\}$. The set $\Phi(\lambda) = \overline{\underset{\sigma \in M(\lambda)}{U} \sigma}$ is called a web of G with center $\lambda \in L_2(G)$.
Abikoff also proved that $\Phi(\lambda)$ is connected and that a web $\Phi(\lambda)$ of G is identical with a web $\Phi(\lambda')$ of G, if there is a separator σ of G satisfying $\sigma \subset \Phi(\lambda) \cap \Phi(\lambda')$.

3. To characterize a web group G by means of its residual limit set, we prepare two lemmas.

LEMMA 1. *Let σ and σ' be two distinct separators of a finitely generated Kleinian group G. If $\gamma \in G$ is loxodromic and if one of fixed points of γ lies on $\sigma - \sigma'$, then the other fixed point of γ lies also on $\sigma - \sigma'$.*

Proof. Sasaki [3] proved that, if one of fixed point ζ of γ lies on $\sigma \in S(G)$, then the other fixed point ζ' of γ also lies on σ. Hence $\zeta \in \sigma \cap \sigma'$ is equivalent to $\zeta' \in \sigma \cap \sigma'$. Therefore $\zeta \in \sigma - \sigma'$ yields $\zeta' \in \sigma - \sigma'$.

LEMMA 2. *Let σ_1, σ_1', σ_1'' be three distinct separators of a finitely generated Kleinian group G. If $\sigma_1' \subset \overline{(\sigma_1)}$ and if $(\sigma_1) \cap \sigma_1'' = \phi$, then $L_1(G) \cap (\sigma_1) \neq \phi$ and $L_1(G) \cap (\hat{C} - \overline{(\sigma_1)}) \neq \phi$.*

Proof. Since the set of fixed points of loxodromic elements of G is dense in $\Lambda(G)$, we can find the attractive fixed point $\zeta \in \sigma_1' - \sigma_1$ of a loxodromic element $\gamma \in G$ and take a neighborhood U of ζ with $U \subset (\sigma_1)$. Lemma 1 implies that the repelling fixed point of γ lies also on $\sigma_1' - \sigma_1$. Hence, for a sufficiently large integer m, a separator $\sigma_2 = \gamma^m(\sigma_1)$ is contained in U. Clearly we can see that the one σ_2' of $\gamma^m(\sigma_1')$ and $\gamma^m(\sigma_1'')$ lies in $\overline{(\sigma_2)}$ and the other σ_2'' satisfies $(\sigma_2) \cap \sigma_2'' = \phi$. Hence σ_2, σ_2', σ_2'' are distinct separators of G and $\sigma_2' \subset \overline{(\sigma_2)}$,

$(\sigma_2) \cap \sigma''_2 = \phi$ and $(\sigma_2) \subset (\sigma_1)$. Repeating the above procedure by using σ_2, σ'_2 and σ''_2 instead of σ_1, σ'_1, σ''_1, we have distinct separators σ_3, σ'_3, σ''_3 of G, which satisfy $\sigma'_3 \subset \overline{(\sigma_3)}$, $(\sigma_3) \cap \sigma''_3 = \phi$ and $(\sigma_3) \subset (\sigma_2)$. Thus we have a sequence of separators $\{\sigma_n\}_{n=1}^{\infty}$ with $(\sigma_{n+1}) \subset (\sigma_n)$, $n = 1, 2, \cdots$. From the property d) of separators of G, we see that $\bigcap_{n=1}^{\infty} \overline{(\sigma_n)}$ reduces to a point which belongs to $L_1(G)$. Therefore $L_1(G) \cap (\sigma_1) \neq \phi$. In the above discussion, by replacing σ'_1 with σ''_1, we can see $L_1(G) \cap (\hat{C} - \overline{(\sigma_1)}) \neq \phi$.

4. By using Lemma 2, we can prove the following characterization of web groups.

THEOREM 1. *Given a Kleinian group* G, *the following four propositions are equivalent to each other*:

   i) G *is a web group.*

   ii) G *is finitely generated and* $\Lambda_0(G) = L_2(G) \neq \phi$.

   iii) G *is finitely generated,* $L_2(G) \neq \phi$ *and* $M(\lambda) = S(G)$ *for every* $\lambda \in L_2(G)$.

   iv) G *is finitely generated,* $L_2(G) \neq \phi$ *and* $M(\lambda) = M(\lambda')$ *for arbitrary two* λ, $\lambda' \in L_2(G)$.

Proof. First we suppose i). We note $\Lambda_0(G) \neq \phi$. If $L_1(G)$ is not empty, then we can find a sequence $\{\sigma_n\}_{n=1}^{\infty}$ of distinct separators of G such that $\bigcap_{n=1}^{\infty} (\sigma_n) = \{\lambda\} \in L_1(G)$. The properties c) and d) of separators of G imply $\sigma_n \subset \overline{(\sigma_1)}$ for every n. By the definition of separators of G, there is a component Ω_n of $\Omega(G)$ such that $\sigma_n = \partial\Omega_{nj}$ for some component Ω_{nj} of $\hat{C} - \overline{\Omega}_n$. Furthermore, the definition of web groups implies that $\partial\Omega_n$ is a quasi-circle. Hence $\sigma_n = \partial\Omega_n$. Since $\lambda \notin \Omega_n$, we see that Ω_n is the component of $\hat{C} - \sigma_n$ not containing λ. This fact holds for every n and we have $\Omega_1 = \Omega_n$ for every n. This contradicts the fact $\bigcap_{n=1}^{\infty} \overline{(\sigma_n)} = \{\lambda\}$. Hence i) implies $L_1(G) = \phi$. Therefore we have $\Lambda_0(G) = L_2(G) \neq \phi$. So i) implies ii).

Next we prove that ii) implies iii). It suffices to show that $S(G) \subset M(\lambda)$ for every $\lambda \in L_2(G)$ under the condition ii). Take an arbitrary $\sigma \in S(G)$ and choose a point $z \in \Omega(G)$ which lies in a component of $\hat{C} - \sigma$ not containing λ. Obviously $\sigma \in S(G; z, \lambda)$. Suppose that σ is not the maximal separator in $S(G; z, \lambda)$ for λ, that is, suppose $\sigma \neq \sigma(z, \lambda)$. Take a point $z' \in \Omega(G)$ in a component of $\hat{C} - \sigma(z, \lambda)$ containing λ and choose a separator $\sigma' \in S(G; z', \lambda)$ of G. Then $\sigma(z, \lambda)$, σ and σ' satisfy the condition for σ_1, σ'_1, σ''_1 in Lemma 2 and we have $L_1(G) \neq \phi$. Hence ii) implies $\sigma = \sigma(z, \lambda)$ and $\sigma \in M(\lambda)$, which proves $S(G) \subset M(\lambda)$.

The proposition iv) is immediately obtained from iii).

Now we prove that iv) yields iii). For the purpose we assume that under the condition iv) there exists a point $\lambda \in L_2(G)$ with $M(\lambda) \neq S(G)$. Let σ be a separator of G belonging to $S(G) - M(\lambda)$ and let $z \in \Omega(G)$ be a point which lies in a component of $\hat{C} - \sigma$ not containing λ. Then $\sigma \neq \sigma(z, \lambda)$. Clearly, in a component of $\hat{C} - \sigma(z, \lambda)$ not containing σ, there is a separator $\sigma' \in M(\lambda)$. By Lemma 2, there exists a point $\lambda' \in L_1(G)$ inside a component Δ of $\hat{C} - \sigma(z, \lambda)$ not containing the point λ. Hence there is an infinite sequence $\{\sigma_n\}_{n=1}^{\infty}$ in $S(G)$ such that $\bigcap_{n=1}^{\infty} \overline{(\sigma_n)} = \{\lambda'\}$, $\sigma_{n+1} \subset \overline{(\sigma_n)} \subset \Delta$ $(n = 1, 2, \cdots)$ and $\sigma_i \neq \sigma_j$ $(i \neq j)$. Choose the attractive fixed point $\zeta \in \sigma_2 - (\sigma_1 \cup \sigma_3)$ of a loxodromic element γ of G and let U be a neighborhood of ζ in Δ. Note that the repelling fixed point of γ lies on σ_2 by Lemma 1. By virtue of the property b) of separators of G, we can easily see that the set $\gamma^m(\Phi(\lambda))$ for every integer m is also a web of G with center $\gamma^m(\lambda)$. Since, in Δ, there is no separator of G belonging to $M(\lambda)$, a web $\gamma^m(\Phi(\lambda))$ of G with center $\gamma^m(\lambda)$ lies in $U(\subset \Delta)$ for a sufficiently large integer m and $\gamma^m(\Phi(\lambda))$ is different from $\Phi(\lambda)$. Hence we obtain $M(\gamma^m(\lambda)) \cap M(\lambda) = \phi$, which contradicts iv). Therefore, $M(\lambda) = S(G)$ for every $\lambda \in L_2(G)$. In other words, iv) implies iii).

It remains to prove that iii) implies i). It suffices to show that, for an arbitrary component Ω_* of $\Omega(G)$, the component subgroup G_{Ω_*} is quasi-Fuchsian. Denoting by Ω_*^0 the component of $\hat{C} - \overline{\Omega}_*$ containing λ, we see that $\sigma_0 = \partial\Omega_{*_0}$ is a separator of G. Let Δ_0 be the component of $\hat{C} - \sigma_0$ not containing λ. Since $M(\lambda) = S(G)$, we see that, in $\overline{\Delta_0}$, there is no separator of G different from σ_0. Hence $\Delta_0 = \Omega_*$. The component subgroup G_{Ω_*} of G for Ω_* makes $\Omega_* = \hat{C} - \overline{\Omega_{*_0}}$ invariant and is a quasi-Fuchsian group. Since $L_2(G) \neq \phi$, we see that G is a web group, which shows that iii) implies i).

Thus we have our Theorem.

5. Here we shall prove a result on the number of webs of a finitely generated Kleinian group.

THEOREM 2. *Let G be a finitely generated Kleinian group with the property $L_2(G) \neq \phi$. Then the number of webs of G different from each other is one or countably infinite.*

Proof. If G is a web group, then from Theorem 1, the number of webs of G is equal to one. So we assume that G is not a web group. By virtue of the property a) of separators of G, it is evident that the number of webs of G is at most countable. Since G is not a web group and since $L_2(G) \neq \phi$, we see from Theorem 1 that $L_1(G) \neq \phi$ and that G has at least two webs $\Phi(\lambda_1)$ and $\Phi(\lambda_2)$. Note that $M(\lambda_1)$ and $M(\lambda_2)$ have no common separator of G. For a point $\lambda_0 \in L_1(G)$, we can find an infinite sequence $\{\sigma_n\}_{n=1}^\infty$ in $S(G)$ such that $\bigcap_{n=1}^\infty \overline{(\sigma_n)} = \{\lambda_0\}$ and such that (σ_1) contains no separator in $M(\lambda_1)$ and $M(\lambda_2)$. Choose the attractive fixed point $\zeta_k \in \sigma_{2k+1} - (\sigma_{2k} \cup \sigma_{2k+2})$ of some loxodromic $\gamma_k \in G$ $(k=1,2,\cdots)$ and take a neighborhood U_k of ζ_k inside a domain bounded by σ_{2k} and σ_{2k+2}. By Lemma 1, the repelling fixed point of γ_k must lie on σ_{2k+1}. So two fixed points of γ_k do not lie on $\Phi(\lambda_j)$, $j = 1, 2$. Hence, by the same reasoning as in the proof of Theorem 1, we see that, for a sufficiently large integer m_k, $\gamma_m^{m_k}(\Phi(\lambda_j))(\subset U_k)$ is a web of $G\,(j = 1, 2)$ and that $\gamma_k^{m_k}(\Phi(\lambda_j))$ $(j = 1, 2; k = 1, 2, \cdots)$ are different from each other. Therefore, we have an infinite number of webs of G. Thus we have our Theorem.

6. In this section, we prove the following theorem.

THEOREM 3. *Let* G *be a web group with* $\infty \in \Omega(G)$. *If* $\{\sigma_n\}_{n=1}^{\infty}$ *is the set of all the separators of* G *and if* δ_n *is the diameter of* σ_n, *then the series* $\sum_{n=1}^{\infty} \delta_n^2$ *is convergent.*

To prove the theorem, we note the following property of a quasi-circle which was shown by Sasaki [2]. Let σ be a quasi-circle which is the image of a circle by a quasi-conformal mapping with the maximal dilatation K_σ. Then, for arbitrary four points z_1, z_3, z_2, z_4 lying on σ in this order, it holds that

$$(*) \qquad \left| \frac{z_3 - z_1}{z_2 - z_1} \right| \leq c_0(K_\sigma) \left| \frac{z_4 - z_3}{z_4 - z_2} \right| \, ,$$

where $c_0(K_\sigma)$ is a constant depending only on K_σ. By using this fact, we first show the following lemma.

LEMMA 3. *Let* σ *be a quasi-circle not passing through the point at infinity. If* δ *denotes the diameter of* σ *and if* A *is the area of the interior* (σ) *of* σ, *then there exists a constant* $c(K_\sigma)$ *depending only on* K_σ *such that*

$$\delta^2 \leq c(K_\sigma)A \, .$$

Proof. Let σ be represented by $z = z(t)$, $t \in [0,1](z(0)=z(1))$. We may assume that $z(0) = z_1 = \frac{\delta}{2}$ and $z\left(\frac{1}{2}\right) = z_2 = -\frac{\delta}{2}$. Consider two subarcs $\sigma': z = z(t)$, $t \in \left[0, \frac{1}{2}\right]$ and $\sigma'': z = z(t)$, $t \in \left[\frac{1}{2}, 1\right]$ of σ. Put

$$\sigma_0' = \sigma' \cap \left\{ z : |\mathrm{Re}\, z| \leq \frac{\delta}{4} \right\} \quad \text{and} \quad \sigma_0'' = \sigma'' \cap \left\{ z : |\mathrm{Re}\, z| \leq \frac{\delta}{4} \right\} \, .$$

The distance γ between σ_0' and σ_0'' is positive and there exist points $\zeta' \in \sigma_0'$ and $\zeta'' \in \sigma_0''$ satisfying $\gamma = |\zeta' - \zeta''|$. The set $\theta_x = (\sigma) \cap \{z : \mathrm{Re}\, z = x\}$ for every $x \in \left[-\frac{\delta}{4}, \frac{\delta}{4}\right]$ consists of a countable number of open segments and among them there is a segment $\tilde{\theta}_x$, one of whose end points lies on σ'_0 and the other on σ''_0. Denote by $\theta(x)$ and by $\tilde{\theta}(x)$ the length of θ_x and of $\tilde{\theta}_x$, respectively. Obviously

$$\gamma \leq \tilde{\theta}(x) \leq \theta(x) .$$

On the other hand, four points z_1, ζ', z_2, ζ'' lie on σ in this order and (*) yields

$$(**) \qquad \left|\frac{\zeta'-z_1}{z_2-z_1}\right| \leq c_0(K_\sigma) \left|\frac{\zeta'-\zeta''}{z_2-\zeta''}\right| .$$

Evidently, $\frac{\delta}{4}$ is not greater than $|\zeta'-z_1|$ and $|z_2-\zeta''|$. So we have

$$\left(\frac{\delta}{4}\right)^2 \leq |\zeta'-z_1||z_2-\zeta''| \leq c_0(K_\sigma)|z_2-z_1||\zeta'-\zeta''| = c_0(K_\sigma)\delta\gamma ,$$

or

$$\delta \leq 16c_0(K_\sigma)\gamma \leq 16c_0(K_\sigma)\theta(x) .$$

This holds for every $x \in \left[-\frac{\delta}{4},\frac{\delta}{4}\right]$ and, by integrating both sides, we have

$$\frac{1}{2}\delta^2 = \int_{-\frac{\delta}{4}}^{\frac{\delta}{4}} \delta \, dx \leq 16c_0(K_\sigma)\int_{-\frac{\delta}{4}}^{\frac{\delta}{4}} \theta(x)\,dx \leq 16c_0(K_\sigma)A .$$

Thus, by taking $c(K_\sigma) = 32\,c_0(K_\sigma)$, we obtain the required.

REMARK. We see easily that $c(K_{a(\sigma)}) = c(K_\sigma)$ for an arbitrary bilinear transformation a.

Now we can prove Theorem 3. Theorem 1 shows that the set $\{\sigma_n\}_{n=1}^\infty$ of all the separators of G coincides with $M(\lambda)$ for an arbitrary taken $\lambda \in L_2(G)$. We may assume $\sigma_1 = \sigma(\infty, \lambda)$. The definition of $\sigma(\infty, \lambda)$ yields $\sigma_n \subset \overline{(\sigma_1)}$ for every n. Furthermore, Theorem 1 and Lemma 2 show that $(\sigma_n) \cap (\sigma_m) = \phi$ for arbitrary n, $m(\neq 1, n)$. By the property e) of separators of G, $\{\sigma_n\}_{n=1}^\infty$ are classified into a finite number of equiva- lence classes. We may assume that $\{\sigma_n\}_{n=1}^N$ $(1 \leq N < \infty)$ is the set of representatives of those equivalence classes. Consider the constant $c(K_{\sigma_n})$ $(1 \leq n \leq N)$ which appeared in Lemma 3 corresponding to the quasi-circle σ_n. From the remark of Lemma 3, we have $c(K_{\gamma(\sigma_n)}) = c(K_{\sigma_n})$ for every $\gamma \in G$. Therefore, if we put $c(G) = \underset{1 \leq n \leq N}{\text{Max}} \, c(K_{\sigma_n})$ and denote by A_n the area of the interior (σ_n) of σ_n, then Lemma 3 implies

$$\delta_n^2 \leq c(G) A_n$$

for every n. Since $\sum_{n=2}^{\infty} A_n$ does not exceed A_1, we have our theorem.

The following is an immediate consequence of Theorem 3.

COROLLARY. *Under the conditions in Theorem 3, the sum of squares of diameters of all the components of* G *not containing* ∞ *is convergent.*

REFERENCES

[1] W. Abikoff, The residual limit sets of Kleinian groups, Acta Math., 130 (1973), 127-144.

[2] T. Sasaki, Boundaries of components of Kleinian groups, Tohoku Math. J., 28 (1976), 267-276.

[3] _________, The residual limit sets and the generators of finitely generated Kleinian groups, to appear in Osaka J. Math.

REMARKS ON FUCHSIAN GROUPS ASSOCIATED WITH
OPEN RIEMANN SURFACES

Yukio Kusunoki and Masahiko Taniguchi[*]

Introduction

Every Riemann surface R whose universal covering surface is hyperbolic can be represented by a Fuchsian group G_R without elliptic elements acting on the unit disk $U = \{|z| < 1\}$. The properties of R naturally should reflect on those of G_R, and vice versa. In the case of Riemann surfaces of finite type, such informations have been much obtained newly in connection with the recent development of the theory of Kleinian groups, but few in case of general open Riemann surfaces. As for the studies to the latter case, we refer for example P. J. Myrberg [8], M. Tsuji [13], L. Bers [2], Ch. Pommerenke [10], and so forth.

A Riemann surface R is called of the type I or II, respectively, according as the Fuchsian group G_R is of the first kind or second kind (Rodlitz [11]). A surface of the type II is just one whose boundary contains some "border" arc, but generally it is not so easy to see whether a given surface satisfying some abstract properties has actually a "border" arc or not. On the other hand, a surface R of finite type is of the type I if and only if R does not have the Green's function (namely, $R \epsilon O_G$), so one might expect

[*]The authors are grateful to the referee for valuable comments.

some characterizations of general Riemann surfaces of the type I from the viewpoint of classification theory. This was our motivation to the present article.

Now in case of open Riemann surfaces, there exists a surface of the type II where there are no non-constant bounded (or Dirichlet finite) analytic functions, that is, the existence of a "border" arc does not necessarily guarantee the existence of such analytic functions (cf. [8] and Example 3 below), and so the situation is quite complicated. While, it seems intuitively true that a surface R is of the type I if and only if R has a small boundary in some sense. In this respect we can prove that surfaces with a small boundary in the harmonic sense are of the type I (cf. Theorems 1 and 4), and also we note that a surface of the type I has not always a small boundary in the analytic sense (cf. Example 2 and Theorem 6). These tell us that a characterization of the class of surfaces of the type I from the classification theoretic point of view is very difficult.

In this paper we shall give a study, including above-mentioned facts, on the type of open Riemann surfaces belonging to several familiar or new classes in the classification theory. The last section §3, II) contains some supplements to the results of Pommerenke [10] from our point of view.

§1. *Riemann surfaces of arbitrary genus*

First we shall recall some of the definitions. By HD (resp. HB) we denote the space of harmonic functions u such that the Dirichlet integral $\|du\|^2$ of u is finite (resp. u is bounded). Replacing harmonic functions above by analytic functions we have the spaces AD and AB . A function $f \in X$ (X = HD , etc.) is called often an X-function. The class O_{KD} (resp. O_{KB}) is by definition the set of Riemann surfaces on which there are no nonconstant HD- (resp. HB-) functions u with the property that

(*)   *du is semiexact, i.e. $\int_\gamma$ *du $= 0$ for every dividing cycle γ,

where *du denotes the conjugate (harmonic) differential of du .

Then the following inclusion relations are known:

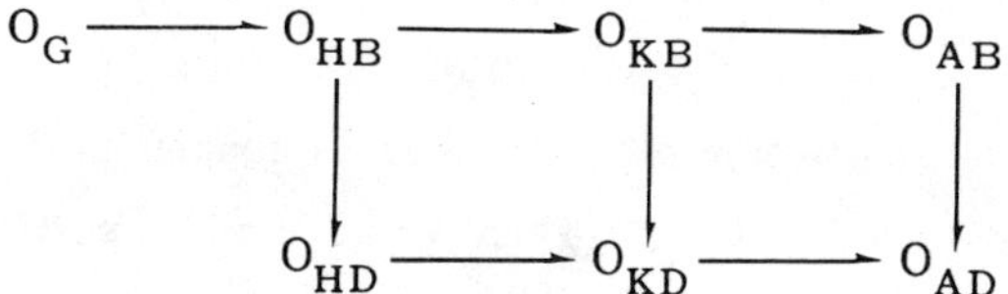

where $X \to Y$ means that X is a proper subset of Y (cf. [1], [3], [12]).

Now we shall show that if a Riemann surface R belongs to one of the classes except O_{AB} and O_{AD} in the above table, then R must be of the type I. The well-known example due to Myrberg [9] gives obviously a Riemann surface of class O_{AB} ($\subset O_{AD}$) which is of the type II.

THEOREM 1. *Every surface in class O_{KD} is of the type I.*

Proof. Suppose that $R \in O_{KD}$ is of the type II. Then there is a free boundary arc b ($\subset \partial U$) for the Fuchsian group G_R of R such that $g(b) \cap b = \emptyset$ for all $g \in G_R$. Let D be the disk in the z-plane such that ∂D is orthogonal to $\partial U = \{|z| = 1\}$ and passes through the endpoints of b. Then the (open) Riemann surface

$$\tilde{R} = (U \cup D)/G_R$$

contains R and $\tilde{R} - \overline{R}$ has interior points, $\overline{R}$ being the closure of R in $\tilde{R}$. Let p be an interior point of $\tilde{R} - \overline{R}$ and $O_p = \{|z_p| < 1\}$ be a local parameter disk at p completely contained in $\tilde{R} - \overline{R}$. Now given a singularity $\theta = d(1/z_p)$ at p, there exists an abelian differential ω on $\tilde{R}$ such that

(1) $\omega - \theta$ is regular in O_p,

(2) ω is semiexact outside of $\{p\}$,

(3) Re ω is exact on $\tilde{R} - \{p\}$, that is, Re $\omega = du$ with a single-valued harmonic function u on $\tilde{R} - \{p\}$, and

(4) $$\|du\|_{(\tilde{R}-O_p)} < +\infty .$$

Such a differential ω (satisfying further certain boundary condition) does exist on arbitrary open Riemann surfaces (Kusunoki [5], also cf. Ahlfors-Sario [1]). Thus $u - \mathrm{Re}(1/z_p)$ is regular harmonic at p, hence u is non-constant. The restriction of u to R is a nonconstant KD-function on R, which implies $R \notin O_{KD}$. Thus we have our assertion. q.e.d.

Next we consider another sequence of classes for which the following inclusion system is known:

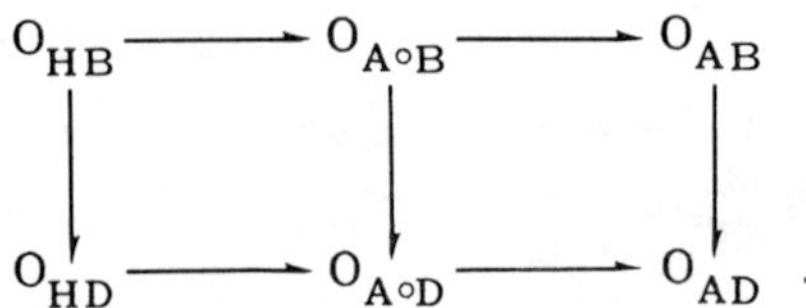

The class $O_{A \circ D}$ (resp. $O_{A \circ B}$) denotes the set of Riemann surfaces R such that on every bordered subregion R' of R with the compact or non-compact border $\partial R'$ there are no nonconstant AD- (resp. AB-) functions f on $R' \cup \partial R'$ with $\mathrm{Re}\, f = 0$ on $\partial R'$. The following theorem shows also that Riemann surfaces of class O_{HD} ($\supset O_{HB} \supset O_G$) are of the type I.

THEOREM 2. *Every surface of class* $O_{A \circ D}$ *must be of the type* I.

Proof. Suppose that $R \in O_{A \circ D}$ is of the type II. We take a free boundary arc b for G_R and the disk D as in the proof of Theorem 1. Consider the conformal mapping g of $D \cap U$ to a half disk such that $\partial D \cap U$ is mapped on the imaginary axis. Let $\pi: U \to R = U/G_R$ be the projection, then the pull-back $g \circ \pi^{-1}$ is a nonconstant AD-function on $R' = \pi(D \cap U)$ whose real part vanishes on $\partial R'$. Thus $R \notin O_{A \circ D}$, which proves the theorem. q.e.d.

§2. *The case of finite genus*

For surfaces of finite genus, it is known that

$$O_G = O_{HB} = O_{HD} \to O_{AB} \to O_{KD} = O_{AD} .$$

Hence Theorem 1 implies the following

THEOREM 1′. *Suppose* R *is of finite genus. If* R *belongs to class* O_{AD}, *then* R *is of the type* I.

This theorem can be improved as follows. Now we shall introduce new classes of Riemann surfaces. Let $O_{AD,n}$ $(1 \leq n \leq \infty)$ denotes the class of Riemann surfaces on which there are no nonconstant AD-functions which are at most n-valent, i.e. take every complex numbers at most n times. We may consider $O_{AD,\infty} \equiv O_{AD}$. Every surface of positive genus belongs to $O_{AD,1}$, while for planar surfaces $O_{AD,1}$ is identical with the class O_{SD} introduced by Ahlfors-Beurling, and $O_{AD} \to O_{AD,1}$ (cf. [1], [12]). Generally $O_{AD} \subset O_{AD,n+1} \subset O_{AD,n}$ for every n. One can show a surface belonging to

$$\left(\bigcap_{n=1}^{\infty} O_{AD,n} \right) - O_{AD}$$

as follows.

EXAMPLE 1. Let R^* be a two-sheeted covering surface of genus g over the whole Riemann sphere $\overline{C} = C \cup \{\infty\}$, and $\pi: R^* \to \overline{C}$ be the projection. Take a compact set E on C which does not contain the branch points of R^* and belongs to the class $N_{SB} - N_D$. Put $\tilde{E} = \pi^{-1}(E)$ and $R = R^* - \tilde{E}$. Then $R \notin O_{AD}$, for there exists a nonconstant AD-function f on $\overline{C} - E$ and $f \circ \pi$ belongs to AD(R). Next we show that $R \in O_{AD,n}$ for every n. Otherwise there is a nonconstant m-valent function $F \in AD(R)$, and F can be extended to a homeomorphism of R^* onto another compact Riemann surface S^* of genus g, because $\tilde{E}$ has a planar neighborhood on R^* and of class N_{SB}. S^* is an m-sheeted covering surface of $\overline{C}$ and $S^* - F(R)$ belongs to the class N_{SB}. On the other hand we can find that $S^* - F(R)$ is not in the class N_{SB}, as the area of F(R) is finite. This is a contradiction.

THEOREM 3. *Suppose* R *is a Riemann surface of genus* $g(0 \leq g < \infty)$.

Then,

(1) *if* R *belongs to* $O_{AD,g+1}$, *then* R *is of the type* I . *In particular, every planar surface of* O_{SD} *is of the type* I .

(2) *There exists a planar surface of the type* I *which does not belong to the class* $O_{SD} \supset O_{AD}$.

Proof. Suppose that R is of the type II. We consider, as in the proof of Theorem 1, the open Riemann surface $\tilde{R}$ with the same genus g as R. Let p be an interior point of $\tilde{R} - \bar{R}$ and consider the divisor $\delta = p^{g+1}$. Then by Riemann-Roch theorem on open Riemann surface $\tilde{R}$ (Kusunoki [5]) there is a nonconstant meromorphic function f such that f is the multiple of $1/\delta$ and df is (exact) canonical. It is known that f is at most $(g+1)$-valent on R and $\|df\|$ is finite outside of a neighborhood of $\{p\}$. Thus the restriction of f on R is a non-constant AD-function on R, hence $R \notin O_{AD,g+1}$, which proves the first statement.

The second statement comes from the following example.

EXAMPLE 2. Let $\Delta = \{|z| < 1\}$ and E be its countable subset such that E does not cluster in Δ and the closure $\bar{E}$ contains $\partial\Delta$, (for example, $E = \{z_{n,k}\}_{n,k=1}^{\infty}$, where $z_{n,k} = \left(1 - \frac{1}{n}\right) \cdot \exp\left(\sqrt{-1}\,\frac{2\pi k}{n}\right)$). Then $R = \Delta - E$ is the required for the statement (2).

For the sake of the completeness we include the proof. First $R \notin O_{SD}$, because R admits a non-constant schlicht AD-function z . Now suppose that R is of the type II, and consider an abstract Riemann surface $\tilde{R} = (U \cup D)/G_R$ as in the proof of Theorem 1. Since $\tilde{R}$ is planar, by the classical uniformization theorem we find that there exists a conformal mapping Φ of $\tilde{R}$ onto a bounded domain in the complex plane. We denote by R' the image of U/G_R (which we may regard as R) by Φ. Then $\partial R'$ contains a compact analytic boundary arc b', the image of a free boundary arc b for G_R . Each point of E is clearly removable for Φ, thus Φ can be extended to a conformal mapping of Δ to a bounded domain Δ' which contains R' densely. Moreover, by Carathéodory theorem the conformal mapping Φ^{-1} can be extended to a continuous mapping from $\Delta' \cup b'$ into $\overline{\Delta}$. Let B' be a simply connected subregion in R' corresponding to $D' \cap U$, D' being a smaller disk in D such that $\partial D'$ is orthogonal to b. Let $B = \Phi^{-1}(B') \subset \Delta$. We note that B does not contain any point of E and that ∂B terminates at some points p_1 and p_2 on $\partial\Delta$. If $p_1 \neq p_2$, B would contain an infinite number of points of E, which contradicts with the above. Hence $p_1 = p_2$, but this is also impossible by Riesz-Lusin-Privaloff theorem. Consequently R must be of the type I. q.e.d.

§3. *Other classes of Riemann surfaces*

I) The classes of Riemann surfaces we shall be concerned here are those related implicitly with the compactification theory.

For any positive integer n we denote by O_{HD}^n the class of Riemann surfaces R such that $\dim HD(R) \leq n$, where $HD(R)$ is the real vector space of HD-functions on R. Similarly $O_{HB}^n = \{R; \dim HB(R) \leq n\}$. Evidently $O_{HD}^1 = O_{HD}$ and $O_{HB}^1 = O_{HB}$. Every surface of $O_{HD}^n - O_G$ is of infinite genus (cf. [6], [12]).

Next by U_{HB} (resp. $U_{\widetilde{HD}}$) we denote the class of hyperbolic Riemann surfaces on which there exists a nonconstant HB- (resp. $\widetilde{HD}$-) minimal function. The following inclusion relations are known:

$$
\begin{array}{ccccccc}
O_{HB} & \longrightarrow & O_{HB}^n & \longrightarrow & U_{HB} \cup O_G & \longrightarrow & O_{AB} \\
\downarrow & & \downarrow & & \downarrow & & \downarrow \\
O_{HD} & \longrightarrow & O_{HD}^n & \longrightarrow & U_{\widetilde{HD}} \cup O_G & \longrightarrow & O_{AD}
\end{array}
$$

THEOREM 4. *Every Riemann surface of class* O_{HD}^n *is of the type* I. *There exists a surface of the class* U_{HB} *which is of the type* II.

Proof. The first statement is almost a direct consequence from the following theorems (Kusunoki-Mori [6], and S. Mori [7]);

1) $R \in O_{HD}^n - O_{HD}^{n-1}$ (resp. $O_{HB}^n - O_{HB}^{n-1}$) if and only if the Royden's (resp. Wiener's) harmonic boundary Δ_R (resp. Δ_W) consists of exactly n points.

2) Let G be a noncompact subregion on any open Riemann surface R. Then $G \notin SO_{HD}$ (i.e. there exists a nonconstant HD-function on G which vanishes on ∂G), if and only if $\overline{G} - \overline{\partial G}$ contains some Royden's harmonic boundary points, where the bar stands for the closure on the Royden's compactification of R.

In fact, if $R \in O_{HD}^n$ is of the type II, there exists a free boundary arc b for the Fuchsian group G_R. Take $n+1$ disjoint disks D_j such that $D_j \cap U$ are not of type SO_{HD} (cf. the proof of Theorem 2), and we find $n+1$ disjoint noncompact subregions G_j of R corresponding to $D_j \cap U$ for which $G_j \notin SO_{HD}$ $(j=1, \cdots, n+1)$. Hence $R \notin O_{HD}^n$ by 1) and 2), which is absurd.

The second statement is due to the following example.

EXAMPLE 3. Take a hyperbolic Riemann surface $S \in U_{HB}$. Let S^* be the Wiener's compactification of S and $\Delta_W(S)$ be the harmonic boundary of S^*. Then $\Delta_W(S)$ contains an isolated point q (cf. [12] IV 3K corollary 1). Now take a closed disk D in S and let $R = S - D$. Then we claim that R belongs to U_{HB} and is of the type II.

In fact by the localization theorem (cf. [3] Satz 9.11, [12] IV 5C) we see that the harmonic boundary of the Wiener's compactification of R also contains an isolated point corresponding to q, which implies that $R \in U_{HB}$. It is clear that R is of the type II. q.e.d.

REMARK. 1) Using the localization theorem for the Royden's compactification and the characterization of $O_{\widetilde{HD}}^\infty$, we can prove further that every surface of class $O_{\widetilde{HD}}^\infty$ is of the type I. The class $O_{\widetilde{HD}}^\infty$ was introduced by Constantinescu-Cornea and it is known that

$$\bigcup_n O_{HD}^n \to O_{\widetilde{HD}}^\infty \to U_{\widetilde{HD}} \cup O_G .$$

2) The class O_G, O_{HD}, O_{KD}, and O_{HD}^n are quasiconformally invariant, but not O_{KB}, O_{AB}, and O_{AD}. We don't know whether O_{HB} and $O_{A \circ D}$ are quasiconformally invariant or not.

II) Let G be a Fuchsian group on $U = \{|z| < 1\}$ of convergence type, that is,

$$\sum_{\gamma \epsilon G} (1 - |\gamma(z)|^2) = (1 - |z|^2) \cdot \sum_{\gamma \epsilon G} |\gamma'(z)| < \infty$$

for every $z \epsilon U$. Then the product (the Green's function for G)

$$g(z) = \prod_{\gamma \epsilon G} [e^{-\sqrt{-1} \cdot \arg \gamma(0)} \cdot \gamma(z)]$$

converges and represents a bounded analytic function on U. Considering the angular derivative $\tilde{g}'(\theta)$ of $g(z)$, we say (Pommerenke [10]) that

1) G is of accessible type if $\tilde{g}'(\theta)$ exists on a set of positive measure on ∂U,

2) G is of fully accessible type if $\tilde{g}'(\theta)$ exists almost everywhere on ∂U, and

3) G is of Widom type if $g'(z)$ is of bounded characteristic on U. Now for each j $(j = 1, 2, 3)$, let P_j be the class of hyperbolic Riemann surfaces R such that the Fuchsian group G_R satisfies the condition j) above. We denote

$$O_j = \{R : R \not\in P_j\} \quad \text{for} \quad j = 1, 2, 3 .$$

It is known that $R \not\in O_G$ if and only if the Fuchsian group G_R is of convergence type (P. J. Myrberg [8], Tsuji [13]). Pommerenke [10] proved that $R \not\in O_3$ if and only if R satisfies the Widom condition for some (hence every) $q \epsilon R$, that is, letting $g(p, q)$ be the Green's function on R with pole at q and $B(t, q)$ the first Betti number of $\{p \epsilon R; g(p, q) > t\}$, it holds that

$$\int_0^\infty B(t, q) dt < \infty .$$

THEOREM 5. *The following system of strict inclusion relations holds*:

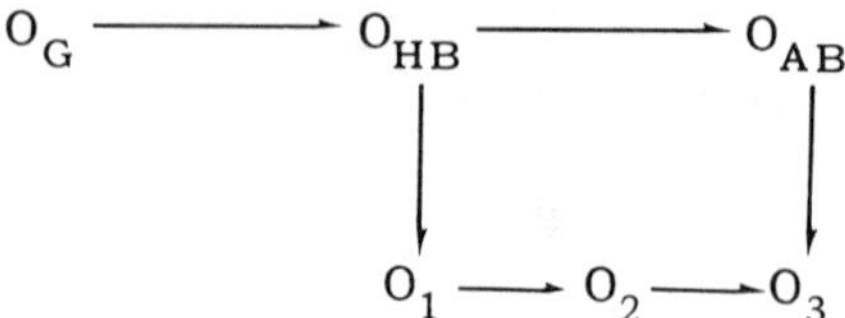

Proof. The inclusion $O_1 \subset O_2 \subset O_3$ follows at once from the definitions, and the inclusion $O_{AB} \subset O_3$ is proved by Widom [14]. Next the examples 1 and 2 in [10] give Riemann surfaces which belong to $O_1 - O_{AB}$ and $O_2 - O_1$ respectively. Hence we find that $O_1 - O_{HB}$, $O_2 - O_1$, and $O_3 - O_{AB}$ are not empty. Thus we need only to show that $O_{HB} \subset O_1$ and $O_3 - O_2$ is not empty.

First we show that $O_{HB} \subset O_1$. Let $R \notin O_1$, and $G = G_R$ be the Fuchsian group associated with R. Then it is known ([10] Theorem 1) that there exists a set B of positive measure on ∂U which contains no two G-equivalent points. Taking a subset if necessary, we may assume that Σ meas. $\gamma(B) < 2\pi$. Let $E = \underset{\gamma \in G}{\cup} \gamma(B)$, and f_E be the characteristic function of E on ∂U. Then there exists a unique bounded harmonic function $u(z)$ with the boundary function f_E, which is obviously nonconstant and positive. For every $\gamma \in G$, $u(\gamma(z))$ also has the boundary function f_E, for $f_E \circ \gamma \equiv f_{\gamma(E)} \equiv f_E$. So we have $u(\gamma(z)) \equiv u(z)$ for every $\gamma \in G$. Hence, $u(z)$ can be projected to a nonconstant bounded harmonic function on R, which implies $R \notin O_{HB}$. This proves $O_{HB} \subset O_1$.

Next we give a Riemann surface in $O_3 - O_2$. Let $R = \Delta - \left\{ \exp\left(-\frac{1}{n}\right) \right\}_{n=1}^{\infty}$ where $\Delta = \{|z| < 1\}$, then it is clear that $G = G_R$ is of the second kind, and $\partial R = \partial \Delta - \{1\}$ is lifted univalently onto a component of the complement $\partial U - \Lambda(G)$ of the limit set $\Lambda(G)$ of G. Now construct a unique bounded G-invariant harmonic function u on U with the boundary function $f_{\Lambda(G)}$ as above, and we have a bounded harmonic function u on R such that $u = 0$ on ∂R. Hence we can see that $u \equiv 0$ on R, that is, the linear measure of $\Lambda(G)$ vanishes (cf. [4]). Thus we conclude that G is of fully accessible type, i.e. $R \notin O_2$ (cf. [10] Theorem 1).

Now noting that $g(z, 0) = -\log |z|$, and

$$B(t, 0) = n \quad \text{if} \quad \frac{1}{n+1} \leq t < \frac{1}{n} ,$$

one sees that

$$\int_0^\infty B(t, 0)\,dt = \sum_{n=1}^\infty \frac{1}{n+1} = +\infty ,$$

hence $R \in O_3$. Thus R is a desired surface. q.e.d.

THEOREM 6. 1) *If* $R \in O_1$, *then* R *is of the type* I.

2) *There exists an* R *in* O_2 *which is of the type* II.

3) *There exists an* R *not in* O_3 *which is of the type* I.

Proof. 1) is just a restatement of the obvious fact that any Fuchsian group of the second kind is of accessible type, and the example 2 in [10] shows the assertion 2). Hence it remains to show a surface R not in O_3 which is of the type I. Let $p_{n,k} = \exp\left(-\frac{1}{n^3} + \sqrt{-1}\,\frac{2\pi k}{n}\right)$ for every n (≥ 1) and k, and $R = U - \{p_{n,k}\}_{n,k=1}^\infty$. Then as we have seen before (Example 2), the Fuchsian group G_R is of the first kind. And noting that $g(z, 0) = -\log |z|$ and

$$B(t, 0) = \sum_{k=1}^n k = \frac{n(n+1)}{2} \quad \text{if} \quad \frac{1}{(n+1)^3} \leq t < \frac{1}{n^3} ,$$

it follows that

$$\int_0^\infty B(t, 0)\,dt = \frac{1}{2} \sum_{n=1}^\infty \frac{3n^2 + 3n + 1}{n^2(n+1)^2} < +\infty ,$$

hence R satisfies the Widom condition, that is, $R \notin O_3$. q.e.d.

Finally we shall construct explicitly a Fuchsian group of the first kind and of accessible type.

EXAMPLE 4. Let $\{p_i\}_{i=1}^{\infty}$ be a sequence of real numbers such that each $p_i > 1$ and the generalized Cantor set E on $[0,1]$ with respect to $\{p_i\}_{i=1}^{\infty}$ has a positive linear measure (e.g. take $\{p_i\}$ so that $\sum_i (1/p_i) < \infty$). For every k, let $\{L_{k,j} : j = 1, \cdots, 2^{k-1}\}$ be the set of components of $[0,1] - E$ such that meas. $(L_{k,j}) = 2^{1-k} \cdot \dfrac{1}{p_k} \cdot \prod_{i=1}^{k-1} \left(1 - \dfrac{1}{p_i}\right)$, and $a_{k,j}$, $b_{k,j}$ be the endpoints of $L_{k,j}$ with $a_{k,j} < b_{k,j}$. Now set

$$g_{0,0}(z) = z + 1 ,$$

and

$$g_{k,j}(z) = \frac{(3a_{k,j} + b_{k,j}) \cdot z - (a_{k,j} + b_{k,j})^2}{4z - (a_{k,j} + 3b_{k,j})} .$$

Note that $g_{k,j}$ maps the exterior of the circle

$$C_{k,j} = \{|z - (a_{k,j} + 3b_{k,j})/4| = (b_{k,j} - a_{k,j})/4\}$$

onto the interior of the circle

$$C'_{k,j} = \{|z - (3a_{k,j} + b_{k,j})/4| = (b_{k,j} - a_{k,j})/4\} .$$

Let G be the group generated by $g_{0,0}$ and all $g_{k,j}$, then it is easily seen that G is a Fuchsian group on the upper half plane U, and that $E - \{a_{k,j}, b_{k,j}\}_{k,j=1}^{\infty} - \{0,1\}$ is a set of positive measure which contains no two G-invariant points. Hence by Theorem 1 of [10], we conclude that G is of accessible type.

Next we show that G is of the first kind. Let $\Lambda(G)$ be the limit set of G, then $E \subset \Lambda(G)$. In fact $\overline{\bigcup_{k,j} \{a_{k,j}, b_{k,j}\}} = E$, that is, for any $p \in E$ we can find a sequence $\{c_k\}_{k=1}^{\infty}$ converging to p, where $c_k = a_{k,j}$ or

$b_{k,j}$ with some j. Then there exists a sequence $\{d_k\}_{k=1}^{\infty}$ of parabolic

fixed points of G such that $|c_k - d_k| \leq 2^{-k} \cdot \dfrac{1}{p_k} \cdot \prod_{i=1}^{k-1} \left(1 - \dfrac{1}{p_i}\right)$ for every

k. Hence d_k converges to p and we conclude that $E \subset \Lambda(G)$. Now let

a real p be arbitrarily given, and W be as any disk with the center p.

Then noting that $\bigcup\limits_{g \in G} g(D)$ is dense in U, where D is the region in U

surrounded by all $C_{k,j}$, $C'_{k,j}$, and lines $\{\operatorname{Re} z = 0\}$, $\{\operatorname{Re} z = 1\}$, we can

find a $g \in G$ such that $g(D) \cap W \neq \emptyset$, which implies that $\Lambda(G) \cap W \neq \emptyset$.

Thus we conclude that $p \in \Lambda(G)$, hence G is of the first kind.

KYOTO UNIVERSITY

REFERENCES

[1] Ahlfors, L. and Sario, L., Riemann surfaces. Princeton Univ. Press (1960).

[2] Bers, L., Automorphic forms and Poincaré series for infinitely generated fuchsian groups. Amer. J. Math. 87 (1965), 196-214.

[3] Constantinescu, C. and Cornea, A., Ideale Ränder Riemannscher Flächen. Springer-Verlag (1963).

[4] Earle, C. and Marden, A., On Poincaré series with applications to H_p spaces on bordered Riemann surfaces. Ill. J. Math. 13 (1969), 202-219.

[5] Kusunoki, Y., Theory of Abelian integrals and its applications to conformal mappings. Mem. Col. Sci. Univ. Kyoto, Ser. A. Math. 32 (1959), 235-258.

[6] Kusunoki, Y. and Mori, S., On the harmonic boundary of an open Riemann surface, I. Jap. J. Math. 29 (1959), 52-56.

[7] Mori, S., On a compactification of an open Riemann surface and its application. J. Math. Kyoto Univ. 1 (1961), 21-42.

[8] Myrberg, P. J., Über Existenz der Greenschen Funktionen auf einer gegebenen Riemannschen Fläche. Acta Math. 61 (1933), 39-79.

[9] ________, Über die analytische Fortsetzung von beschränkten Funktionen. Ann. Acad. Sci. Fenn. Ser. A. I. 58 (1947), 7 pp.

[10] Pommerenke, Ch., On the Green's function of Fuchsian groups. *Ibid.*, vol. 2 (1976), 409-427.

[11] Rodlitz, E., Locally trivial deformations of open Riemann surfaces. Comm. Pure Appl. Math. 14 (1961), 157-168.

[12] Sario, L. and Nakai, M., Classification theory of Riemann surfaces. Springer-Verlag (1970).

[13] Tsuji, M., Potential theory in modern function theory. Maruzen (1959).

[14] Widom, H., H_p sections of vector bundles over Riemann surfaces. Ann. of Math. (2) 94 (1971), 304-324.

ON GENERALIZED WEIERSTRASS POINTS AND RINGS
WITH NO PRIME ELEMENTS

Henry B. Laufer*

I. Introduction

Let X be a Riemann surface of genus $g \geq 2$. Let $x \in X$. Let
$0 < r_1 < r_2 < \cdots < r_g < 2g$ be the Weierstrass gaps at x [10, pp. 120-128].
Necessarily $r_1 = 1$. Then r_i is a gap if and only if there is a section h
of the canonical bundle κ on X such that h has a zero of order $r_i - 1$
at x. x is a Weierstrass point on X if the following two equivalent conditions are satisfied: (i) for some integer w, $1 \leq w \leq g-1$, there does
not exist $h \in \Gamma(X, \mathcal{O}(\kappa))$ with a zero of order w at x and (ii) for some
integer d, $g \leq d \leq 2g-2$, there is an $h \in \Gamma(X, \mathcal{O}(\kappa))$ with a zero of order
d at x.

Let $D_n = \dim \Gamma(X, \mathcal{O}(\kappa^n))$. Observe that $g = D_1$. For $n \geq 2$, $D_n =$
$(2n-1)(g-1)$. Then x is a generalized Weierstrass point of order n [4]
if (i) there does not exist $h \in \Gamma(X, \mathcal{O}(\kappa^n))$ with a zero at x of some given
order at most $D_n - 1$ or equivalently (ii) there exists $h \in \Gamma(X, \mathcal{O}(\kappa^n))$
with a zero at x of order at least D_n.

Rauch [18], see also [6], [15], [2], has studied the distribution in
Teichmüller space of Riemann surfaces having Weierstrass points with a

*The author is an Alfred P. Sloan Fellow. This research was also partially
supported by NSF Grant MCS7604969A01.

given lowest non-gap, i.e., using condition (i) above. We generalize this result to arbitrary n (Theorem 2.2). We also prove for arbitrary n the corresponding statement about the distribution in Teichmüller space of Riemann surfaces having Weierstrass points with a given highest gap, i.e. using condition (ii) above (Theorem 2.1). A special case of Theorem 2.1 has been shown by Pinkham [17, Theorem 14.7, pp. 114-115].

Let R be the local ring of germs of holomorphic functions at a point p on the variety V. Suppose that V is irreducible at p and that dim $V \geq 3$. Recall that an element in a ring is called prime if the principal ideal which it generates is a prime ideal. Then Flenner [7, Satz 4.10, p. 108] has shown that R contains prime elements. Using Theorem 2.1 for $g = 3$, we show that the corresponding result is not true for dimension two. Let $C<x,y,z>$ be the ring of convergent power series at $0 = (0,0,0)$. Let $R = C<x,y,z>/(f)$, where f is a homogeneous polynomial of degree 4. Then (Theorem 3.1) for a generic choice of coefficients of f, R is normal and has no prime elements. More strongly, let I be a principal ideal in R. Then the radical of I is not prime.

Some of the work for this paper was done while the author was visiting Princeton University. The author thanks Princeton for its gracious hospitality.

II. *Generalized Weierstrass points*

A point, call it o, in Teichmüller space T_g represents a Riemann surface X_o of genus g plus additional structure. Above T_g one can construct a fiber space V, $\pi : V \to T_g$ such that the fiber above each point in T_g is the Riemann surface represented by that point [9]. In this paper, however, we shall have no need for the additional structure on X_o nor for the global construction of V. We shall only be interested in any small neighborhood T of o in T_g and in $\mathfrak{X} = \pi^{-1}(T)$. For our purposes, $\pi : \mathfrak{X} \to T$ is more conveniently constructed as the complete, effectively parameterized deformation of X_o in the Kodaira-Spencer sense [13], [12], [14]. Since the tangent space to T_g at o may be identified with $H^1(X_o, \Theta)$, where Θ is the tangent sheaf to X_o [5, p. 131], $\pi : V \to T_g$ is locally complete and effectively parameterized. So the constructions from [9] and [12] coincide locally. One can also prove this coincidence from the universal property in [9, Theorem 3.1, pp. 7-8].

For $x_o \in X_o$, let $d_n(x_o)$ be the maximal order of a zero at x_o of an $h \in \Gamma(X_o, \mathcal{O}(\kappa^n))$. Recall that $D_n = \dim \Gamma(X_o, \mathcal{O}(\kappa^n))$.

Observe that since the singular points of a subvariety are nowhere dense and since the number of generalized Weierstrass points of order n is bounded in terms of g [4] the first condition on h in Theorem 2.1 below is the generic case.

THEOREM 2.1. *Let* X_o *be a Riemann surface of genus* $g \geq 2$. *Let* $\pi : \mathfrak{X} \to T$ *be the complete effectively parameterized deformation of* X_o. *Let* $x_o \in X_o$. *Fix* $n \geq 1$. *Suppose that* $d_n(x_o) \geq D_n$. *Let* $G = \{x \in \mathfrak{X},$ x *near* $x_o | d_n(x) = d_n(x_o)\}$. *In case* $n \geq 2$, $d_n(x_o) = 2n(g-1)$ *and* $d_1(x_o) = 2g-2$, G *coincides with* $G_1 = \{x \in \mathfrak{X}, x$ *near* $x_o | d_1(x) = d_1(x_o)\}$. *In all other cases,* G *is a submanifold of* $\mathfrak{X}$ *of dimension* $3g - 3 + D_n - d_n(x_o)$. $\pi : G \to T$ *is a finite proper map. If there does not exist an* $h \in \Gamma(X_o, \mathcal{O}(\kappa^n))$ *with a zero of order* $d_n(x_o) - 1$ *at* x_o, *then* G *is transverse to* X_o, $\pi(G)$ *is a submanifold of* T *and* $\pi : G \to \pi(G)$ *is a biholomorphic map. If there exists* $h \in \Gamma(X_o, \mathcal{O}(\kappa^n))$ *with a zero of order* $d_n(x_o) - 1$ *at* x_o, *then* G *is not transverse to* X_o *and* $\pi(G)$ *is singular or some points in* $\pi(G)$ *parameterize Riemann surfaces with more than one point* x *near* x_o *with* $d_n(x) = d_n(x_o)$.

Proof. We shall first consider the case $n \geq 2$, $d_n(x_o) = 2n(g-1)$ and $d_1(x_o) = 2g-2$. By taking tensor products, we see that $G_1 \subset G$. Let ζ_x denote the point bundle at x on X. Let $\sim$ denote equivalence of line bundles. Then $x \in G$ if and only if $\kappa^n \sim \zeta_x^{2n(g-1)}$. Then $(\kappa \zeta_x^{2-2g})^n \sim 0$. Since $d_1(x_o) = 2g-2$, $\kappa \zeta_{x_o}^{2-2g} \sim 0$. Let K be the line bundle over $\mathfrak{X}$ which restricts to the canonical bundle on each fiber. By [3, Theorem II, p. 208] or [19, (4.7), p. 52], for each $h \in \Gamma(X_o, \mathcal{O}(\kappa))$, there exists

$H \in \Gamma(\mathfrak{X}, \mathcal{O}(K))$ which restricts to h on X_0. Looking at the zeroes of one such H gives a continuous (on T) choice of divisors which represent κ on each fiber. Choosing a basis of such h and corresponding H gives a continuous (on T) choice of the abelian differentials needed to apply Abel's Theorem [10, Corollary 2, pp. 161-162]. Then, since x is near to x_0, $\kappa \zeta_x^{2-2g} \sim 0$ for $x \in G$. Then $G \subset G_1$.

We now consider the other values of n, $d_n(x_0)$ and $d_1(x_0)$. Let $d = d_n(x_0)$ and $D = D_n$. Let $h_1, h_2, \cdots, h_D \in \Gamma(X_0, \mathcal{O}(\kappa^n))$ have zeroes of order $0 = s_1 < s_2 < \cdots < s_D = d$ at x_0. We shall choose L so that the s_i include all non-negative integers up to and including L. Thus let

$$
\begin{aligned}
&\text{for } n = 1, \ L = 0\\
(2.1) \qquad &\text{for } n \ge 2, d < 2n(g-1), \ L = (2n-2)(g-1) - 2\\
&\text{for } n \ge 2, d = 2n(g-1), \ L = (2n-2)(g-1) - 1 \ .
\end{aligned}
$$

Let K, as above, be the line bundle over $\mathfrak{X}$ which restricts to the canonical bundle on each fiber. By [3, Theorem II, p. 208] or [19, (4.7), p. 52], for each i, $1 \le i \le D$, there exists $H_i \in \Gamma(\mathfrak{X}, \mathcal{O}(K^n))$ which restricts to h_i on X_0. Since $\{h_i\}$, $1 \le i \le D$, is a basis of $\Gamma(X_0, \mathcal{O}(\kappa^n))$, $\{H_i\}$, $1 \le i \le D$, restricts to a basis of $\Gamma(X, \mathcal{O}(\kappa^n))$ for all X sufficiently near to X_0. Let

$$
\begin{aligned}
V_s = \{(x, h), x \in X \text{ near } & X_0, h \in \Gamma(X, \mathcal{O}(\kappa^n)) |\\
& h \text{ has a zero of order at least } s \text{ at } x\}/C^* \ ,
\end{aligned}
$$

where the action of C^* is multiplication on h. Then V_s can be readily seen, as follows, to be a subvariety of $\mathfrak{X} \times P^{D-1}$. Let z be a local coordinate on X_0 and $t = (t_1, \cdots, t_{3g-3})$ local coordinates on T. (z, t) provides local coordinates for $\mathfrak{X}$. Let $(u_1, \cdots, u_D)$ be homogeneous coordinates on P^{D-1}. Let $H = u_1 H_1(z, t) + \cdots + u_D H_D(z, t)$. Then

$$
(2.2) \qquad\qquad V_s = \{H = \cdots = \partial^{s-1} H / \partial z^{s-1} = 0\} \ .
$$

Let $x_0 \in X_0$ correspond to $(z, t) = (0, 0)$. Near $z = 0$, $V_d \cap \{t = 0\} = (0, 0, (0, \cdots, 0, 1))$ and $V_{d+1} = \emptyset$. Let $\tau : \mathfrak{X} \times P^{D-1} \to \mathfrak{X}$ be projection.

Let V be V_d intersected in $\mathfrak{X} \times \mathbf{P}^{D-1}$ with a suitably small neighborhood of $(0, 0, (0, \cdots, 0, 1))$. Then $G = \tau(V)$. Let $u_D = 1$ and $u = (u_1, \cdots, u_{D-1})$ be local inhomogeneous coordinates for $\mathbf{P}^{D-1}$. Since $V \cap \{t = 0\} = (0, 0, 0)$, by replacing T by a suitably small neighborhood of 0 in T and then correspondingly replacing V, we may make $\pi \circ \tau : V \to T$ proper. Then also $\pi : G \to T$ is proper.

We shall now show that the defining equations in (2.2) make V a submanifold. Expand the H_i in power series in t. We now describe one possible choice of first order terms for H_D; any other choice differs only by linear combinations of the h_i. Recall that $H_D(z, o) = h_D(z)$. For ease of computation, we choose local z-coordinates so that $h_D(z) = z^d dz^n$. Choose a Stein cover (U_1, U_2) [11, Theorem VI. D. 4, p. 189 and Theorem IX. B. 10, p. 220] for X_o with U_1 a local coordinate disc $\{|z| < \delta\}$ about $z = 0$ and $U_2 = X_o - \{z \in U_1 \mid |z| \le \delta/2\}$. Then $A = U_1 \cap U_2 = \{\delta/2 < |z| < \delta\}$. By [1] the deformation $\pi : \mathfrak{X} \to T$ can be realized by joining the cartesian products $U_1 \times T$ and $U_2 \times T$. On A, the following represent linearly independent cohomology classes in $H^1(X_o, \Theta)$:

$$z^{-1} \partial/\partial z, \cdots, z^{3-2g} \partial/\partial z .$$

These classes are in fact dual to quadratic differentials having zeroes of order $0, \cdots, 2g - 4$ at $z = 0$.

Let $z_1 = z$ be a local coordinate on U_1 and $z_2 = z$ be a local coordinate on A as a subset of U_2. Let the t_i-axis in T have tangent direction $z^{-i} \partial/\partial z$, $1 \le i \le 2g-3$. Then along the t_i-axis, the change of coordinates on $\mathfrak{X}$ is given to first order by $z_2 = z_1 + t_i z_1^{-i}$. Comparing $h_D(z) = z^d dz^n$ in the z_1- and z_2-coordinate systems gives

$$\eta = t_i(d - in) z^{d-1-i} dz^n$$

on A, with $\mathrm{cls}[\eta] \in H^1(X_o, \mathcal{O}(\kappa^n))$ the first obstruction to extending h_D to H_D along the t_i-axis. Of course, $\mathrm{cls}[\eta] = 0$. But for $d - 1 - i \ge 0$, η is trivially a coboundary since it has a holomorphic extension to U_1. Recall the numerical values given for L in (2.1) above. Then we may choose the power series expansion for H_D in $U_2 \times T$ to begin $\{z_2^d + 0 \cdot t_1 + \cdots + 0 \cdot t_{d-2-L} + \cdots\} dz_2^n$ and in $U_1 \times T$ to begin, $z = z_1$,

$$H_D = \{z^d + t_1(d-n)z^{d-2} + t_2(d-2n)z^{d-3} + \cdots +$$

(2.3)

$$t_{d-2-L}(d-n(d-2-L))z^{L+1} + \cdots\} dz^n .$$

Observe that the indicated coefficients in (2.3) are not 0. Recall that the h_i, $1 \le i \le L+1$, have zeroes at $z = 0$ of all orders less than or equal to L. Then (2.4) below gives the Jacobian for (2.2) at $(z, t, u) = (0, 0, 0)$. In (2.4), we have reordered the variables, as indicated, to clarify the results. We have also replaced each variable by a suitable (not indicated) non-zero scalar multiple of itself so as to make the important non-zero entries equal to 1.

(2.4)

$$
\begin{array}{l|ccccccccc}
 & H & \partial H/\partial z & \cdots & \partial^L H/\partial z^L & \cdots & & \partial^{d-1}H/\partial z^{d-1} \\
\hline
\partial/\partial u_1 & 1 \\
 & 0 & 1 \\
\vdots & 0 & 0 & 1 \\
 & \vdots & \vdots & & \ddots \\
\partial/\partial u_{L+1} & 0 & 0 & 0 & 1 \\
\partial/\partial t_{d-2-L} & 0 & 0 & 0 \cdots 0 & 1 \\
 & 0 & 0 & 0 \cdots 0 & 0 & 1 \\
\vdots & \vdots & \vdots & \vdots & \vdots & \vdots & & \ddots \\
\partial/\partial t_1 & 0 & 0 & 0 \cdots 0 & 0 & & 1 \\
\partial/\partial z & 0 & 0 & 0 \cdots 0 & 0 \cdots & & 0 & 1 \\
\partial/\partial u_{L+2} & 0 & 0 & 0 \cdots 0 \\
\vdots & \vdots & \vdots & \vdots & \vdots \\
\partial/\partial u_{D-1} & 0 & 0 & 0 & 0 \\
\partial/\partial t_{d-1-L} \\
\vdots \\
\partial/\partial t_{3g-3}
\end{array}
$$

Observe that the first d rows of (2.4) are an upper-diagonal matrix. Hence V is a submanifold of dimension $3g-3+D-d$.

The columns in (2.4) give defining equations for the tangent space to V at $(z,t,u) = (0,0,0)$ in $\mathfrak{X} \times P^{D-1}$. To exhibit $\tau : V \to G$ as an isomorphism, we must solve these defining equations for the u_i, $1 \leq i \leq D-1$, in terms of (z,t). This solution is immediate from the form of the $\partial/\partial u_i$ rows. So G is a submanifold of $\mathfrak{X}$ as claimed. Also, defining equations for the tangent space to G at $(z,t) = (0,0)$ in $\mathfrak{X}$ may be obtained as follows. In (2.4), in each $\partial/\partial u_i$ row, there is a first non-zero entry. Delete these $D-1$ columns from (2.4). Then delete the $\partial/\partial u_i$ rows. The columns of the remaining matrix give defining equations for the tangent space to G.

We now distinguish between the cases h_{D-1} has a zero of order less than $d-1$ and h_{D-1} has a zero of order $d-1$. In the first case, the $\partial/\partial u_{D-1}$ row in (2.4) has its first non-zero entry before the last column. So in the matrix for the tangent space to G, the $\partial/\partial z$ row is non-zero. So we can solve for z in terms of t. Thus G is transverse to X_o and $\pi : G \to \pi(G)$ is an isomorphism. In the second case, the $\partial/\partial u_{D-1}$ row in (2.4) has its first (and only) non-zero entry in the last column. So in the matrix for the tangent space to G, the $\partial/\partial z$ row is zero. Hence the tangent space to X_o at $(z,t) = (0,0)$ is contained in the tangent space to G at $(z,t) = (0,0)$. So the tangent map $\pi : G \to T$ does not have maximal rank. Since π is proper, $\pi(G)$ is a subvariety of T [11, Theorem V. C. 5, p. 162]. Then if $\pi(G)$ is a submanifold, π cannot be one-to-one. This is one of the possibilities under Theorem 2.1. $\pi(G)$ being singular is the other possibility. This concludes the proof of Theorem 2.1.

Let $x \in X$, a Riemann surface. Let $w_n(x)$ denote the smallest non-negative integer such that there does not exist $h \in \Gamma(X, \mathcal{O}(\kappa^n))$ with a zero at x of order $w_n(x)$.

THEOREM 2.2. *Let X_o be a Riemann surface of genus $g \geq 2$. Let $\pi : \mathfrak{X} \to T$ be the complete, effectively parameterized deformation of X_o. Let $x_o \in X_o$. Fix $n \geq 1$. Suppose that $w_n(x_o) \leq D_n - 1$. Let $W = \{x \in \mathfrak{X}, x$ near $x_o | w_n(x) = w_n(x_o)\}$. In case $n \geq 2$, $w_n(x_o) = (2n-2)(g-1)-1$ and*

$d_1(x_0) = 2g - 2$, W *coincides with* $G_1 = \{x \in \mathfrak{X}, x$ *near* $x_0 | d_1(x) = d_1(x_0)\}$
from Theorem 2.1. In all other cases, W *is a submanifold of* $\mathfrak{X}$ *of dimen-*
sion $3g - 2 - D_n + w_n(x_0)$. $\pi : W \to T$ *is a finite proper map. If there*
exists an $h \in \Gamma(X_0, \mathcal{O}(\kappa^n))$ *with a zero of order* $w_n(x_0) + 1$ *at* x_0, *then*
W *is transverse to* X_0, $\pi(W)$ *is a submanifold of* T *and* $\pi : W \to \pi(W)$
is a biholomorphic map. If there does not exist $h \in \Gamma(X_0, \mathcal{O}(\kappa^n))$ *with a*
zero of order $w_n(x_0) + 1$ *at* x_0, W *is not transverse to* X_0 *and* $\pi(W)$
is singular or some points in $\pi(W)$ *parameterize Riemann surfaces with*
more than one point x *near* x_0 *with* $w_n(x) = w_n(x_0)$.

Proof. We retain the notation of the proof of Theorem 2.1. Let
$w = w_n(x_0)$.

By hypothesis, the h_i, $1 \leq i \leq w$, have zeroes of order $i - 1$ at x_0.
For $w + 1 \leq j \leq D$, let

$$
(2.5) \qquad \det_j = \det
\begin{pmatrix}
H_1 & \cdots & H_w & \cdots & H_j \\[4pt]
\dfrac{\partial H_1}{\partial z} & & \dfrac{\partial H_w}{\partial z} & & \dfrac{\partial H_j}{\partial z} \\[8pt]
\vdots & & & & \\[8pt]
\dfrac{\partial^w H_1}{\partial z^w} & & \dfrac{\partial^w H_w}{\partial z^w} & & \dfrac{\partial^w H_j}{\partial z^w}
\end{pmatrix}
$$

Then

$$
(2.6) \qquad W = \{(z, t) | \det_j(z, t) = 0, \; w + 1 \leq j \leq D\}.
$$

Observe that at $z = t = 0$, the $(w{+}1)$-st column in (2.5) is zero. For
$j \geq w + 1$, let $h_j(z, t) = z^k \varepsilon(z) dz^n$ in local coordinates, with $\varepsilon(z)$ holomorphic and $\varepsilon(0) = 1$. $k \geq j$. As in (2.1), $(2n-2)(g-1) \leq k \leq 2n(g-1)$. Let
$I = k - w - 1$. Observe that $0 \leq I \leq 2g - 2$. $I = 0$ or $I = 2g - 2$ will lead to
the various cases of this theorem. For $1 \leq I \leq 2g - 3$, compute, as in the
proof of Theorem 2.1, a power series expansion for H_j in $U_1 \times T$:

$$H_j(z,t) = \{z^k \varepsilon(z) + t_1(k-n)z^{k-2}\varepsilon_1(z)$$

$$+ t_2(k-2n)z^{k-3}\varepsilon_2(z) + \cdots + t_I(k-In)z^W \varepsilon_I(z) + \cdots \}dz^n$$

with $k - in \neq 0$, $\varepsilon_i(0) = 1$ and $\varepsilon_i(z)$ holomorphic for $1 \leq i \leq I$.

Then from (2.5), $\partial(\det_j(z,t))/\partial t_i = 0$ for $1 \leq i \leq I-1$ and $\partial(\det_j(z,t))/\partial t_I \neq 0$.

Now consider $I = 0$. This occurs precisely when h_{w+1} has a zero of order $w + 1$ at $z = 0$. Then $\partial(\det_{w+1}(z,t))/\partial z \neq 0$. If h_j, any $j \geq w+1$, has a zero of order greater than $w + 1$ at $z = 0$, $\partial(\det_j(z,t))/\partial z = 0$.

Unless the case $I = 2g - 2$ occurs, we now can compute the Jacobian for (2.6) and see that it has maximal rank. Then W is a submanifold of $\mathcal{X}$ of dimension $3g - 2 - D + w$, as claimed. $I = 2g - 2$ only occurs when $w = (2n-2)(g-1) - 1$ and $d_n(x_0) = 2n(g-1)$. The above computation shows that W is a subvariety of a submanifold W' of dimension $2g - 1$. $\kappa^n \sim \zeta_{x_0}^{2n(g-1)}$. If κ is not equivalent to $\zeta_{x_0}^{2g-2}$, then there exists $h \in \Gamma(X_0, \mathcal{O}(\kappa^n))$ with a zero of order $(2n-2)(g-1) - 1$ at x_0. Since in fact $w = (2n-2)(g-1) - 1$, $\kappa \sim \zeta_{x_0}^{2g-2}$. Hence by the Riemann-Roch theorem, there does not exist $h \in \Gamma(X, \mathcal{O}(\kappa^n))$ with a zero at x of order $(2n-2)(g-1) - 1$. Since this is the minimum possible value for $w(x)$, $w(x) = (2n-2)(g-1) - 1$, i.e. $G_1 \subset W$. But by Theorem 2.1, $\dim G_1 = 2g - 1$. Hence $W = G_1$, as claimed.

The rest of Theorem 2.2 follows as in the proof of Theorem 2.1.

III. *Application to rings*

The following is standard material [11]. Let $f = (x, y, z)$ be a holomorphic function in a neighborhood U of $0 = (0,0,0)$. Assume that $f(0,0,0) = 0$. Let $V = V(f) = \{(x,y,z) \in U \mid f(x,y,z) = 0\}$. Let R be the ring of germs at 0 of holomorphic functions on V, i.e. consider all functions given by convergent power series at $(0,0,0)$ and then restrict these

functions to V . Let I be an ideal in R . I is necessarily finitely generated. Choosing representatives $g_1, \cdots, g_r$ for generators gives loc I =
$\{(x, y, z) \epsilon V | g_i(x, y, z) = 0, 1 \leq i \leq r\}$ as a subvariety of some neighborhood
V_1 of 0 in V . The Nullstellensatz then says that Id (loc I), i.e. the
ideal of germs of holomorphic functions on R which vanish on loc I ,
satisfies Id (loc I) = Rad (I), the radical of I . We will be interested in
prime ideals. A prime ideal is necessarily its own radical. If $I_1 = $ Rad (I),
then I_1 is prime if and only if loc I_1 is irreducible, i.e. cannot be written
as the union of two proper subvarieties. Of course, loc I_1 = loc I . In
particular, if g is an element of R , I_1 = Rad (g) is prime if and only if
loc (g) is irreducible.

Note [16, Lemma 1, p. 261] that R is normal if and only if 0 is an
isolated singular point of V .

THEOREM 3.1. *Let f = f(x, y, z) be a homogeneous polynomial of degree
4 with complex coefficients. Let $C < x, y, z >$ be the ring of convergent
power series at 0 = (0, 0, 0). Let $R = C < s, y, z >/(f)$. Let W be the
manifold of possible coefficients for f which yield normal rings R . Then
for coefficients of f in W but off a countable union of nowhere dense
subvarieties of W , R has the following property. Let I be a principal
ideal in R . Then the radical of I is not prime. In particular, I is not
prime.*

Proof. 0 is an isolated singularity of V if and only if the projective
variety X in P^2 given by f is non-singular. Thus W is indeed a
manifold. For R normal, X is necessarily a Riemann surface of genus 3.
V is a cone over X .

Now let X be a Riemann surface of genus 3. Let M be the total
space of the dual to the canonical bundle over X . Let $\pi : M \to V$ be the
blow-down [8] of M . $\pi(X) = p$. For X other than a hyperelliptic surface,
V is a hypersurface in C^3 with a defining equation f which is a homogeneous polynomial of degree 4. Moreover, by [3, Theorem II, p. 208] or
[19, (4, 7), p. 52], the blow-down can be done simultaneously over Teich-

müller space near X non-hyperelliptic. Suppose that for X there does not exist any pair (x, n), $x \in X$, $n \geq 1$, such that $d_n(x) = 2n(g-1)$. By Theorem 2.1, such X exist off the countable union of proper submanifolds of Teichmüller space.

We see as follows that R has the indicated property. Let $I = (g)$. g is holomorphic on V near $p = 0$, $g(0) = 0$. $\pi^*(g)$ is holomorphic on M and has a zero of order $n \geq 1$ on X. Let $\mathcal{I}$ be the ideal sheaf of X in M. Then $\mathcal{I}^n/\mathcal{I}^{n+1} = \mathcal{O}(\kappa^n)$. Then $\pi^*(g)$ projects to a non-zero element $h \in \Gamma(X, \mathcal{O}(\kappa^n))$. By our choice of X, h has more than one distinct zero. Hence near X, $V(\pi^*(g))$ has at least two irreducible components in addition to X. Hence $V(g)$ on V has more than one irreducible component near p. Thus R is indeed the desired ring.

Finally, we must show that the coefficients of f may be chosen off a nowhere dense subvariety of W. Blowing up at $(0, 0, 0)$ simultaneously resolves all the $V(f)$ with isolated singularities at 0 and provides a holomorphic family $\mathcal{F}$ of Riemann surfaces of genus 3 parameterized by W. For each n, the set of $X \in W$ such that there exist $x \in X$ with $d_n(x) = 2n(g-1)$ is a subvariety N of W. By Theorem 2.1 and the fact that all non-hyperelliptic surfaces appear in $\mathcal{F}$, N is a nowhere dense subvariety of W. This concludes the proof of Theorem 3.1.

Note that if one does not care about a hypersurface example for Theorem 3.1, a similar argument works in the following case. X is a fixed Riemann surface of genus $g \geq 2$. L is a negative line bundle over X such that no power of L is equivalent to a multiple of a point bundle. Such an L exists since the Picard variety has dimension $g > 1$, while bundles of fixed Chern class which are multiples of points lie on a one-dimensional subvariety of the Picard variety. Blowing down X in L provides the desired ring R such that the radical of no principal ideal is prime.

STATE UNIVERSITY OF NEW YORK
STONY BROOK, NEW YORK 11794

REFERENCES

[1] Andreotti, A., and Vesentini, E., On the pseudo-rigidity of Stein manifolds, Annali d. Scuola Norm. di Pisa, *16*(1962), 213-223.

[2] Arbarello, E., Weierstrass points and moduli of curves, Comp. Math. *29*(1974), 325-342.

[3] Bers, L., Holomorphic differentials as functions of moduli, Bull. A.M.S. *67*(1961), 206-210.

[4] Duma, A., Weierstrass — Punkte and kanonische Familien von Riemannschen Metriken auf regulären Familien kompakter Riemannscher Flachen (II), Math. Ann. *216*(1975), 265-268.

[5] Earle, C., and Eells, J., Deformations of Riemann Surfaces, Lectures in Modern Analysis and Applications I, C. T. Taam editor, Lecture Notes in Mathematics, vol. 103, 1969, Springer, pp. 122-149.

[6] Farkas, H., Special divisors and analytic subloci of Teichmüller space, Amer. J. Math. *88*(1966), 881-914.

[7] Flenner, H., Die Sätze von Bertini für lokale Ringe, Math. Ann. *229* (1977), 97-111.

[8] Grauert, H., Über Modifikationen und exzeptionelle analytische Mengen, Math. Ann. *146*(1962), 331-368.

[9] Grothendieck, A., Techniques de construction en géométrie analytique, I: Description axiomatique de l'espace de Teichmüller et de ses variantes, Sem. H. Cartan *13*(1960-61) Numbers 7 and 8.

[10] Gunning, R., *Lectures on Riemann surfaces*, Princeton Math. Notes, Princeton Univ. Press 1966.

[11] Gunning, R., and Rossi, H., Analytic functions of several complex variables, Prentice-Hall, Englewood Cliffs, N. J., 1965.

[12] Kodaira, K., Nirenberg, L., and Spencer, D., On the existence of deformations of complex analytic structures. Ann. Math. *68*(1958), 450-459.

[13] Kodaira, K., and Spencer D., On deformations of complex analytic structures, I, II, Ann. of Math. *67*(1958), 328-466.

[14] ________, A theorem of completeness for complex analytic fibre spaces, Acta Math. *100*(1958b), 281-294.

[15] Lax, R., Weierstrass points of the universal curve, Math. Ann. *216* (1975), 35-42.

[16] Oka, K., Sur les functions analytique de plusieurs variables VIII, J. Math. Soc. Japan *3*(1951), 204-214, 259-278.

[17] Pinkham, H., Deformations of algebraic varieties with G_m action, Asterisque *20*(1974).

[18] Rauch, H., Weierstrass points, branch points, and moduli of Riemann surfaces, Comm. Pure and Applied Math. *12*(1959), 543-560.

[19] Riemenschneider, O., Über die Anwendung algebraischer Methoden in der Deformationstheorie komplexen Räume, Math. Ann. *187*(1970), 40-55.

THE TOPOLOGY OF ANALYTIC SURFACES:
DECOMPOSITIONS OF ELLIPTIC SURFACES

Richard Mandelbaum[*]

§1. As is well known the topology of a compact Riemann surface R is completely determined by its genus g. The genus can, of course be described either, analytically, as the complex dimension of the space of holomorphic forms on R or, purely topologically, as the number of handles in a topological model for R. In any event the topology of the Riemann surface is relatively uninteresting and thus its analytic properties are the topic of prime importance.

Furthermore the question of when a given C^∞ manifold admits a complex structure is also not of any interest in two dimensions since it is easily shown that all orientable 2-manifolds have Riemann surface structures.

The picture changes drastically when we go up one complex dimension. Our basic object of inquiry, the complex analytic surface, is now a 4-dimensional real manifold and 4-dimensional topology is a largely unexplored domain. (See [Man 3].)

If we wish to characterize 4-dimensional manifolds we must first note that a complete set of homeomorphism invariants can't exist. This is due

—————————
[*]Author partially supported by NSF Grant MCS77-04165.

to the undecidability of the word problem for finitely presented groups [BHP] and the fact that any finitely presented group can be realized as the fundamental group of a 4-manifold.

We might thus begin our inquiry by restricting our attention to 4-manifolds with trivial fundamental group. Now results begin to be more encouraging. Firstly results of Whitehead and Milnor [Wh, Mil] show that the cohomology ring of a compact simply-connected 4-manifold completely determines its homotopy type. Since the isomorphism class of the cohomology ring of such a 4-manifold is completely determined by the intersection form on that manifold we in turn have:

THEOREM 1. *Let* V_1 *and* V_2 *be compact simply-connected four-dimensional manifolds. Let* L_{V_i} *be the symmetric bilinear form on* $H^2(V_i, Z)$ *induced by the cup-product pairing.*

Then V_1 *is homotopy equivalent to* V_2 *if and only if* L_{V_1} *is congruent to* L_{V_2}.

Unfortunately, homotopy equivalent does not, in general, imply homeomorphism. Rather than jump directly from homotopy equivalence to the finer equivalence relation of homeomorphism, we proceed by stages. We recall [Mil 2] that two compact oriented n-manifolds M_1^n, M_2^n are said to be (oriented) cobordant if there exists an oriented $n+1$ manifold W whose oriented boundary $\partial W = M_1 - M_2$. If the inclusion map $1 : M_1 \hookrightarrow W$ is a homotopy equivalence then M_1 and M_2 are said to be h-cobordant. Clearly homeomorphic implies h-cobordant implies homotopy equivalent. The reverse implications are not, in general, true. However if $n \geq 5$ and the M_i are simply-connected the h-cobordism theorem [KS, RS, Sm] says that h-cobordant implies homeomorphic. Thus for simply-connected manifolds of dimension greater than 4 the two equivalence relations of h-cobordism and homeomorphism are equivalent. With this in mind we recall the following theorem of Novikov and Wall.

THEOREM 2 [N, W]. *Suppose that* M_1, M_2 *are compact simply connected 4-manifolds. Then* M_1 *is h-cobordant to* M_2 *if and only if they are homotopy equivalent.*

If the h-cobordism theorem were true for 4-dimensional manifolds our problem would be solved. However it is presently not known whether the theorem extends and there are some indications (see [Man 3]) that it doesn't. Thus for our 4-dimensional problem it appears that we can't progress beyond our h-cobordism classification! If we weaken our demands some progress is possible though.

One possible approach is to ask how many different homeomorphism classes exist in one homotopy type. For higher dimensional simply-connected manifolds it is known [Br, N, Sul] that fixing the homotopy type and characteristic classes determines the homeomorphism type up to a finite ambiguity. This result is also not available for 4-dimensional manifolds. If we restrict ourselves to those simply-connected compact 4-dimensional manifolds which admit a complex structure we do obtain the following theorem of Enriques-Kodaira. ([Kod], see also [Msh 2, Man 3].)

THEOREM 3. *Let* M *be a compact simply-connected complex manifold of real dimension* 4.

Then M *is diffeomorphic to an algebraic submanifold of* CP^5 *which is either 1) rational or 2) elliptic or 3) of general type.*

We shall henceforth call a non-singular algebraic manifold of pure complex dimension 2 *an algebraic surface and refer to complex manifolds of that dimension as analytic surfaces.*

Theorem 3 shows us that we can now study the topology of analytic surfaces by analyzing the above 3 categories of algebraic surfaces. Before doing this we simplify the topological nature of our classification process by introducing the notion of a minimal model.

We know that in the case of Riemann surfaces, birational equivalence implies analytic equivalence. This however is not true for algebraic surfaces. That is, if V_1 and V_2 are algebraic surfaces, they may have isomorphic fields of meromorphic functions without being either analytically or even topologically the same. They will however be related by a series of blowing up's and blowing down's [Zar, Saf].

We recall [Kod] that if V is an analytic surface and $p \in V$, then there exists a new analytic surface $\tilde{V}$ and a map $\sigma : \tilde{V} \to V$ such that if $\sigma^{-1}(p) = L$ then L is an embedded CP^1 in $\tilde{V}$ with self-intersection -1 and $\sigma | \tilde{V} - L$ is a biholomorphic map between $\tilde{V} - L$ and $V - \{p\}$. $\tilde{V}$ is said to be obtained by blowing V up at the point p or by a σ-process on V at p. If we let $Q = \overline{CP}^2$ be the complex projective plane with orientation opposite to the usual then it is not difficult to see that $\tilde{V}$ is just topologically the connected sum $V \# Q$. Conversely if V is a complex manifold admitting an analytically embedded 2-sphere L with $L \cdot L = -1$ then the criterion of Castelnuevo-Kodaira guarantees that there exists an analytic manifold W with $V = W \# Q$ being W blown up at some point $X \in W$. L is called an exceptional curve (of the first kind) on V and we say W is obtained by blowing down L in V. It is then clear that by successive 'blowing down's' we obtain that any analytic manifold V is biholomorphic to $V' \# nQ$, for some $n \geq 0$ and some analytic surface V' possessing no exceptional curves. We call such a V' a minimal surface and note that for purposes of topological classification it suffices to consider only minimal surfaces. (We note however that minimal models are not necessarily unique. For example the non-singular cubic hypersurface V_3 in CP^3 has both $P = CP^2$ and the non-singular quadric $V_2 \subseteq CP^2$ as its minimal models. The basic theorem on birational equivalence and minimal models says that if X_1 and X_2 are birationally equivalent then there exists an analytic manifold X_3 and birational morphisms $\sigma_1 : X_3 \to X_1$ and $\sigma_2 : X_3 \to X_2$, each of which can be factored into a succession of blowing-down's. For example V_2 and $P = CP^2$ are birationally equivalent. Setting $X_3 = V_2 \# Q$ we can show that $X_3 \approx P \# 2Q$ can be obtained by either blowing V_2 up at one point or blowing P up at 2 points.)

We now return to define the various categories of surfaces. We shall say S is a rational surface if and only if its field of meromorphic functions $\mathfrak{M}_S$ is isomorphic to $C(x, y) =$ the field of rational function in two variables. A more numerical criterion due to Castelnuevo states that S

is rational if and only if its first betti-number is zero and it possesses no holomorphic quadratic 2-forms.

Minimal rational surfaces have been completely classified and if V is a minimal rational surface it is diffeomorphic to either P or $S^2 \times S^2$. We thus have:

THEOREM 4. *Let* V *be a rational surface. Then* V *is diffeomorphic to either* $S^2 \times S^2$ *or* $P \# nQ$, *for some* $n \geq 0$.

The category of elliptic surfaces is defined as follows: S is an elliptic surface if and only if for some Riemann surface R, there exists a holomorphic map $f : S \to R$ such that $f^{-1}(x)$ is a complex torus for all but finitely many $x \in R$.

Lastly a simply-connected surface is of general type if it is neither rational nor elliptic.

As a consequence of work of [Msh 1] one can prove:

THEOREM 5. *Let* V *be a minimal simply-connected surface of general type. Then there exist at most a finite number of diffeomorphism classes of analytic surface which are homotopy equivalent to* V *but not diffeomorphic to it.*

There is however no counterpart to Theorem 5 for elliptic surfaces. In fact we have the following:

Let $\Lambda = \{(p, m, n) \in Z^3 \,|\, p \geq 0, m \geq 1, n \geq 1 \text{ and } \text{g.c.d.}(m,n)=1\}$ and for $p \geq 0$ let $\Lambda_p = \{(m, n)\,|\,(p, m, n) \in \Lambda\}$ and let $\sim_h$ denote "is h-cobordant to" and $=$ denote "is diffeomorphic to." Now let V_4 be a non-singular quartic in CP^3 and for every integral $p \geq 0$ let $W_p^I = (2p+1)P \#(10p+q)Q$ and for odd p let $W_p^{II} = \left(\frac{p+1}{2}\right) V_4 \# \left(\frac{p-1}{2}\right) (S^2 \times S^2)$.

Then we have: (See [Man 2, 3, 4])

THEOREM 6. *Let* p *be a non-negative integer. Then for every* $(m, n) \in \Lambda_p$ *there exists a simply-connected compact elliptic surface* $V_{(p,m,n)}$ *such that if either* p *is even or* p *is odd and* $m+n$ *is odd*

then $V_{(p,m,n)} \sim_h W_p^I$; *while if* p *is odd and* $m+n$ *is even then*
$V_{(p,m,n)} \sim_h W_p^{II}$.

NOTE. (1) If $p > 0$ then no pair $V_{(p,m,n)}$, $V_{(p,m',n')}$ are known to be diffeomorphic to one another except in the trivial case when the set $\{m,n\} = \{m',n'\}$. In particular for every non-negative integer p there exists an infinite family of h-cobordant simply-connected elliptic surfaces, no two of which are known to be diffeomorphic to one another.

(2) If $p = 0$, then it can be shown that $V_{(0,1,m)} = V_{(0,1,1)} = V_{(0,n,1)} = W_0^I$ for all $m,n \geq 1$. However if $(m,n) \in \Lambda_0$ with $m \geq 2$ and $n \geq 2$ then the conclusions of (1) above apply unchanged.

The elliptic surface $V_0 = V_{(0,1,1)}$ is called the rational elliptic surface and a model for it can be obtained as follows:

Let C_1, C_2 be non-singular elliptic curves in CP^2 with transversal intersection, (for example let C_1 be given by $Z_2^3 + Z_1^3 + Z_0^3 = 0$ and C_2 by $Z_2^3 + 2Z_1^3 + 3Z_0^3 = 0$). Let $V_0 \subset CP^1 \times CP^2$ be given by $\lambda C_1 + \mu C_2 = 0$; (in our case by $(\lambda + \mu)Z_2^3 + (\lambda + 2\mu)Z_1^3 + (\lambda + 3\mu)Z_0^3 = 0$) and let $\Phi: V_0 \to CP^1$ be the restriction to V_0 of the projection map $\pi_1 \ CP^1 \times CP^2 \to CP^1$. Then it can be verified that V_0 is a minimal elliptic surface. (V_0 is not a minimal surface. It has exceptional curves of the first kind, however, each of them are transverse to all the fibers and so do not lie in a fiber. Thus V_0 is nevertheless minimal elliptic. It is quite easy to see that V_0 is analytically just CP^2 blown up at the 9 distinct point of intersection in $C_1 \cap C_2$. In particular if $\pi: V_0 \to CP^2$ is the restriction of the projection map $\pi_2: CP^1 \times CP^2 \to CP^2$, then π is the blowing down map.)

We are now again at an impasse. There may indeed exist infinitely many inequivalent diffeomorphism classes of simply-connected elliptic surfaces all homotopy equivalent to one another. We are thus led to another approach, inspired by the idea of birational equivalence in the algebraic case. We first define a $\bar{\sigma}$-process, or anti-holomorphic blowing up in exactly the same way as a σ-process with the exception, that we now demand $L^2 = +1$. Then if V' is obtained by a single $\bar{\sigma}$-process from

V we have that topologically $V' = V \# P$. We then state the following
consequence of a result of Wall [W].

THEOREM 7. *Let* M_1, M_2 *be homotopy equivalent simply-connected compact manifolds. Then there exist non-negative integers* k_1 , k_2 *such that*

$$M_1 \# k_1 P \# k_2 Q = M_2 \# k_1 P \# k_2 Q .$$

Furthermore for any simply-connected such M *there exist integers* ℓ_1 *and*
ℓ_2 *such that*

(∗) $\qquad\qquad\qquad M \# \ell_1 P \# \ell_2 Q = aP \# bQ .$

What Wall's result says is that homotopy type does determine diffeomorphism type up to some 'stability' requirement expressed by the neces-
sity of doing additional 'blowing-up's' to get true diffeomorphic
equivalence.

Unfortunately Wall's result is a pure existence theorem and offers no
clue on how to estimate k_1 , k_2 . Furthermore if M_i is an infinite
sequence of such homotopy equivalent simply-connected 4-manifolds there
is no guarantee that a uniform pair of (ℓ_1, ℓ_2) can be found such that
$M_i \# \ell_1 P \# \ell_2 Q$ decomposes as above for all i . It is therefore of some
interest to find intrinsic estimates of ℓ_1 and ℓ_2 in terms of topological
invariants of M .

If we can take $\ell_1 = \ell_2 = 0$ in (∗) above we say that M is completely
decomposable. In the papers [Man 1, 2 MM 1, 2] we have shown that for a
wide variety of complex surfaces, V , V # P is completely decomposable
(ACD). In particular in [Man 2] we show that all simply-connected elliptic
surfaces are almost completely decomposable. Thus for example
$V_{(p,m,n)} \# P$ is diffeomorphic to $V_{(p,\varnothing,\varnothing)} \# P$ for any $(m, n) \in \Lambda_p$.

What can be done in classifying non-simply connected 4-manifolds?
Clearly things will in general be more complicated. For example it is no
longer true that the homotopy type of a manifold is determined by its cohomology ring nor that homotopy equivalent manifolds are h-corbordant. For

example in [CS] a manifold Q is exhibited which is homotopy equivalent
to RP^4 but not h-cobordant to it! Despite all this in many significant
cases $M \# P$ still decomposes nicely. In particular we can prove the
following. [Man 4]

THEOREM 8. *Let* V *be a compact complex elliptic surface of standard
type* (see §2 for definition). *Let* $g = \frac{1}{2} \dim_C H_1(V, C)$ *and let*
$n = \dim_C H^2(V, \mathcal{O}_V) + 1$, *where* $\mathcal{O}_V$ *is the sheaf of germs of holomorphic
functions on* V . *Then*

$$V \# P = S^2 \times F_g \# aP \# bQ$$

where F_g *is a compact Riemann surface of genus* g *and* $a = 2(g+n) - 1$,
$b = 2(g+5n-1)$.

In the remainder of this article we shall discuss the proof of Theorem 8
as well as the general procedure underlying the proofs of almost complete decomposability results. We also will discuss some open questions.
Before doing this we note that in a private communication Kas has pointed
out that any elliptic surface with no multiple fibers and at least one degenerate fiber is of standard type if $4n \geq 5g$.

§2. The basic procedure used in obtaining the decomposability results
quoted above is a topological version of a degeneracy argument long known
in the study of Riemann surfaces and deformations of algebraic or analytic
manifolds or varieties. That is, suppose we are interested in the topology
of the surface V . We proceed as follows:

1) We find an analytic family V_t of surfaces, $t \in D^2$, such that for
$t \neq 0$, $V_t \approx V$ and V_0 is degenerate (in an appropriate sense). 2) We
relate the topology of V_t to the hopefully simpler topology of V_0 by
means of an appropriate deformation theorem.

For example, suppose we were interested in analyzing the topology of
non-singular algebraic plane curves, (i.e. non-singular hypersurfaces of
CP^2). It is not difficult to show that for any $n > 1$, there exists a family

$_n V_t$ of plane curves, such that for $t \neq 0$, $_n V_t$ is diffeomorphic to a non-singular curve, C_n, of degree n in CP^2, (recall that all non-singular hypersurfaces of CP^N of fixed degree k are diffeomorphic; see [Man 3] for a proof), and V_0 is a transverse intersection of a C_{n-1} and a C_1. We then make use of the following Deformation theorem of [MM 2].

THEOREM 9 (Deformation Theorem [MM 2]). *Let* V_t, $t \in D^2$ *be a complex analytic family with* V_{t_1} *diffeomorphic to* V_{t_2} *whenever* $t_1, t_2 \in D^2$ $-\{0\}$. *Suppose* V_0 *is a transversal intersection of compact-connected complex manifolds* S_1, X_2 *(of multiplicity 1) and let* $C = X_1 \pitchfork X_2$ *denote their transversal intersection.*

Then there exists a diffeomorphism $h : \partial T_1 C \to \partial T_2 C$, *where the* $T_i C$ *are (small) tubular neighborhoods of* C *in* X_i, *such that for all*

$t \in D^2 - \{0\}$ V_t *is diffeomorphic to* $\overline{X_1 - T_1 C} \cup_h \overline{X_2 - T_2 C}$.

In particular in the case of the degenerating family of plane curves we see that $C_1 \pitchfork C_{n-1}$ is just $n-1$ distinct points. We thus get that

$$C_n = C_1 - \left\{ \begin{matrix} \text{disjoint} \\ \text{neighborhoods} \end{matrix} \text{ of } n-1 \text{ points} \right\} \cup_h C_{n-1} - \left\{ \begin{matrix} \text{neighborhood of} \\ n-1 \text{ points} \end{matrix} \right\}.$$

Noting that C_1 is just a 2-sphere we see that our result is that C_n is just C_{n-1} with $(n-2)$ handles attached. That is, $g_n = \text{genus}(C_n) = \text{genus}(C_n) + (n-2)$, which by induction gives us the well-known fact that $g_n = 1/2(n-1)(n-2)$.

If we now attempt to use this procedure for studying analytic surfaces we immediately meet an obstacle. That is, the deformation theorem above, will tell us that the object we are interested in V, is diffeomorphic to a 'sum' of the form $\overline{X_1 - T_1} \cup_h \overline{X_2 - T_2}$ where the T_i are tubular neighborhood of the Riemann surface C in X_i. Unfortunately the topology of the 'sum' $\overline{X_1 - T_1} \cup \overline{X_2 - T_2}$ is not much clearer than that of our original V, even if X_1, X_2 have much simpler structure.

We are thus led to a compromise. We can't currently read off much in-
formation about V from our 'sum' $\overline{X_1-T_1} \cup_h \overline{X_2-T_2}$. However, if we
allow additional blowing ups a great deal of information can be obtained.

In particular we have:

THEOREM 10 (Irrational Connected Sum [Man 2, 3]). *Suppose* X_1, X_2 *are
oriented compact 4-manifolds and* S_1, S_2 *are oriented genus* g *compact
2-submanifolds of* X_1, X_2 *respectively. Let* $X_2' \xrightarrow{\sigma} X_2$ *be* X_2 *blown
up by a* σ-*process at some point* $p \in S_2$ *and let* $S_2' = \overline{\sigma^{-1}(S_2-p)}$.

Let T_1, T_2' *be tubular neighborhood of* S_1, *resp.* S_2', *in* X_1, *resp.*
X_2'. *Let* $\eta: \partial T_1 \to \partial T_2'$ *be an orientation-reversing bundle isomorphism.*

Let Y_2 *be* X_2 *surgered along disjointly embedded circles* e_α,
$\alpha = 1, \cdots, 2g$ *in* X_2 *such that for some wedge* $K = VS_\alpha^1$ *of 1-spheres in*
S_2, e_α *is isotopic to* S_α^1 *in* X_2 *for* $\alpha = 1, \cdots, 2g$, *and* K *is a deformation retract of* $S_2 - \{p\}$.

Set
$$V = \overline{X_1-T_1} \cup_\eta \overline{X_2'-T_2}$$

Then if X_1 *is simply connected* $V \# P = X_1 \# Y_2$ *while if both* X_1 *and
X_2 *are simply connected then* $V \# P = X_1 \# X_2 \# 2g(P \# Q)$.

Now if V_n is a non-singular hypersurface in CP^3 then one can show
that V_n degenerates to $V_{n-1} \cup \sigma V_1$, where $\sigma V_1 = V_1 \# n(n-1)Q$ and
$V_{n-1} \pitchfork \sigma V_1 = \mathcal{C}_{n-1}$. Thus using Theorems 9 and 10 we obtain that
[MM 1, 2] $V_n \# P = V_{n-1} \# V_1 \# [n(n-1)-1]Q \#(n-2)(n-3)(P \# Q)$ and so by
induction $V_n \# P = k_n P \# \ell_n Q$ with $k_n = 1/3n(n^2-6n+1))$ and $\ell_n =
1/3(n-1)(2n^2-4n+3)$.

Similarly if X_p is any simply-connected minimal elliptic surface of
geometric genus p then it can be shown that $X_p \# P = X_{p-1} \# 3P \# 10Q$
and so by induction we obtain $X_p \# P = (2p+2)P \#(10p+9)Q$. Thus for
example the classes of non-singular hypersurfaces of CP^3 as well as the
class of all simply-connected elliptic surfaces are almost completely
decomposable.

To discuss Theorem 8 we need to recall some additional facts about elliptic surfaces. Thus suppose V is an elliptic surface. There thus exists [Kod] a proper holomorphic map $\phi : V \to R$ of V onto a non-singular compact Riemann surface R and R is analytically unique. We consider V to be an analytic fiber space over R and call R the base curve and Φ the projection map of V. We define the fiber $\Phi^*(x)$ over x in R as the divisor of the holomorphic function $\mathcal{J} \circ \Phi - \mathcal{J}(x)$, where $\mathcal{J}$ is any local uniformization variable on a neighborhood of x in R with value $\mathcal{J}(x)$ at x. We shall assume that any fiber of V is free from exceptional curves. (Such a surface is called a minimal elliptic surface.)

We shall say $\Phi^*(x)$ is a regular fiber if $d(\mathcal{J} \circ \psi)$ is nowhere zero on $\Phi^{-1}(x)$. Having assumed that V was minimally elliptic we see that if $\Phi^*(x)$ is a regular fiber then $\Phi^*(x) = \Phi^{-1}(x)$ is a non-singular elliptic curve.

Clearly an elliptic surface can have at most a finite number of singular fibers $C_{a_i} = \Phi^*(a_i)$, $a_1, \cdots, a_n \in R$. We write each such singular fiber in the form $C_{a_i} = \Sigma_j n_{ij} \Gamma_{ij}$ where the Γ_{ij} are irreducible curves and the n_{ij} are positive integers. We let $m_i = \text{g.c.d.}(n_{i1}, n_{i2} \cdots)$ and call m_i the multiplicity of the fiber C_{a_i} at a_i. We note that even if $\Phi^{-1}(a_i)$ is non-singular, C_{a_i} may still be a singular-fiber by virtue of having multiplicity > 1.

If $\Phi^{-1}(a_i)$ has singular points we shall call the fiber C_{a_i} a degenerate singular fiber or sometimes simply a degenerate fiber.

Now suppose $V \xrightarrow{\Phi} R$ is a compact elliptic surface and $f : S \to R$ is a holomorphic map of the non-singular Riemann surface S onto R. Then there exists a unique elliptic surface $f^*V \xrightarrow{f^*\Phi} S$ making the following diagram commute:

$$
\begin{array}{ccc}
f^*V & \xrightarrow{\tilde{f}} & V \\
\downarrow {\scriptstyle f^*\Phi} & & \downarrow {\scriptstyle \Phi} \\
S & \xrightarrow{f} & R
\end{array}
$$

We call f^*V the pullback of $V \xrightarrow{\phi} R$ by f. If the elliptic surface X is diffeomorphic to the pullback of $V_0 \xrightarrow{\pi} S^2$ (V_0 as defined in the addendum to Theorem 6), by some map $f : R \to S^2$, where R is a compact Riemann surface and the Branch locus $B(f) \subset S^2$ of f is disjoint from the critical values $\{a_i\}$ of π, we shall say that it is an elliptic surface of standard type.

What is then of crucial importance about Elliptic Surfaces of Standard Type? The key point is that if V is of standard type with base curve F_g, then it can be shown [Man 5] that there exists a 2-disc D on F_g such that V is diffeomorphic to $\overline{F_g \times T^2 - D^2 \times T^2} \cup_h \overline{X - TN}$ where $N \approx D^2 \times T^2$, T^2 a 2-torus and h is an autodiffeomorphism of T^3. We can then apply Theorem 10 in an inductive fashion to obtain Theorem 8. We refer the reader to [Man 4] for details. We note that if X is a simply-connected surface without multiple fibers then a theorem of Moishezon [Msh] guarantees that it is of standard type. We note also that if $X \to R$ is of standard type then it must have no multiple fibers and at least one singular fiber. Restricting attention to only such elliptic surfaces we may then ask whether they are all of standard type in the non-simply-connected case as well. The answer to this is as yet unknown and we thus are left with the important question of characterizing surfaces of standard type!

Actually a partial generalization of Theorem 8 to surfaces which are not necessarily of standard type as well as to non-necessarily complex 4-manifolds is known. We refer the interested reader to [Man 4] for details. In addition more information on related questions and more general problems of 4-manifolds can be found in [Man 3] and [Man 5].

THE UNIVERSITY OF ROCHESTER
ROCHESTER, N. Y. 14627

REFERENCES

[BHP]   Boone, W. W., Haken, W. and Poenarv, V., On recursively unsolv-
         able problems in topology and their classification. Cont. to Math.
         Logic (Hanover Coll. 1966), 37-74, N. Holland Publ. Amsterdam
         1968.

[BR]   Browder, W., Surgery on Simply-Connected Manifolds Ergebnisse
         der Math. Band 65, Springer-Verlag, N. Y. 1972.

[CS]   Cappel, S. and Shaneson, J., Some New Four-Manifolds, Annals of
         Math 104(1976), 61-72.

[KS]   Kirby, R. C. and Siebenmann, L. C., Foundational Essays on
         Topological Manifolds, Smoothing and Triangulation, Princeton
         Univ. Press (1977).

[Kod]   Kodaira, K., On Compact Analytic Surfaces I, II, III, Annals of
         Math 71(1960), 111-152; 77(1963), 563-626; 78(1963), 1-40.

[Man 1] Mandelbaum, R., On Irrational Connected Sums, Trans. Am. Math.
         Soc. (1979), 247, 137-156.

[Man 2] _________, On the Topology of Elliptic Surfaces, Advances in
         Math (1979). Supplem Series. Vol 5; 143-166.

[Man 3] _________, Four-Dimensional Topology, Bull. Am. Math. Soc.
         (1980). Vol. 2, #1, 1-159.

[Man 4] _________, Lefschetz Fibrations and Decompositions of Complex
         Elliptic Surfaces. (to appear)

[Man 5] _________, Decomposing Analytic Surfaces, Proc. Georgia Topology
         Conference of 1977. Academic Press 1979, 147-218.

[MM 1]   Mandelbaum, R. and Moishezon, B., On the Topological Structure
         of Non-Singular Algebraic Surfaces in CP^3. Topology, Vol. 15
         (1976), 23-40.

[MM 2]   _________, On the Topology of Algebraic Surfaces, Trans. Am.
         Math. Soc. (to appear)

[Mil 1] Milnor, J., On Simply-Connected 4-Manifolds. "Symp. Internat.
         de Top. Alg. (Mexico 1976) Mexico 1958, 122-128.

[Mil 2] _________, Lectures on the h-cobordism theorem, Princeton Math.
         Notes, 1965.

[Msh 1] Moishezon, B., Surfaces of Fundamental Type, in Algebraic Sur-
         faces, "Proceedings of the Steklov Instituet" No. 75(1965). Am.
         Math. Soc., Providence (1967), 113-161.

[Msh 2] _________, Complex Surfaces. Lecture Notes in Math #603,
         Springer-Verlag (1977).

[N]   Novikov, S. P., Homotopically Equivalent Smooth Manifolds I.
         Translations Am. Math. Soc. 48(1965), 271-396.

[RS]   Rourke, C. P. and Sanderson, B. J., Introduction to Piecewise
         Linear Topology, Springer-Verlag, N. Y. 1972 Ergebnisse Math.
         Band 69.

[Shaf] Shafarevitch, I. R., Lectures on Minimal Models. Tata Institute Lecture Notes, Bombay (1966).

[Sm]   Smale, S., Generalized Poincaré's Conjecture in Dimensions Greater than 4. Ann. of Math. 74(1961), 391-406.

[Sul]   Sullivan, D., Thesis, Princeton 1965.

[W]   Wall, C.T.C., On Simply-Connected 4-Manifolds, J. London Math. Soc. 39(1964), 141-149.

[Wh]   Whitehead, J.H.C., On Simply-Connected 4-Dimensional Polyhedra, Comm. Math. Helv. 22(1949), 48-92.

[Zar]   Zariski, O., Introduction to the Problem of Minimal Models, Publ. Math. Soc. Japan 4(1958).

DENSE GEODESICS IN MODULI SPACE

Howard Masur

Introduction and statement of the theorem

Let T_g be the Teichmüller space of genus $g > 1$, M_g the Teichmüller modular group and $R_g = T_g/M_g$ the moduli space. Let Q be the vector bundle of quadratic differentials over T_g and $Q_0 \subset Q$ the unit subbundle with respect to the norm $\int |q|$. For each $q \in Q_0$ lying over $x_0 \in T_g$ consider the *directed* Teichmüller line ℓ_q determined by q. For $t \in R$ let $x_t \in \ell_q$ be of directed Teichmüller distance t from x_0 and let $q_t \in Q_0$ over x_t be the quadratic differential determining the same directed line. In the language of extremal maps, if $t = 0$, $q_t = q$. If $t > 0$, q_t is the terminal quadratic differential given by the Teichmüller map with complex dilatation $\tanh t\,\bar{q}/|q|$. If $t < 0$, q is the terminal quadratic differential for the Teichmüller map with dilatation $\tanh(-t)\bar{q}_t/|q_t|$. The map $T_t : Q_0 \to Q_0$ given by $q \to q_t$ defines a flow on Q_0.

It is easy to check that M_g commutes with the flow. The purpose of this paper is to prove

THEOREM. *There is an orbit q_t in Q_0 such that for either $t > 0$ or $t < 0$ the translates of q_t under M_g are dense in Q_0. Equivalently, both the projection of q_t, $t > 0$ and $t < 0$ are dense in Q_0/M_g.*

COROLLARY. *There is a geodesic in T_g whose projection is dense in both directions in R_g.*

© 1980 Princeton University Press
Riemann Surfaces and Related Topics
Proceedings of the 1978 Stony Brook Conference
0-691-08264-2/80/000417-22 $01.10/1 (cloth)
0-691-08267-7/80/000417-22 $01.10/1 (paperback)
For copying information, see copyright page

There exist finite coverings of R_g whose universal cover is T_g. These are then smooth, hyperbolic quasi-projective varieties. The theorem provides an example of such an object whose universal cover admits extremal discs which do not cover algebraic curves.

Let α and β be simple closed curves on a surface X in minimal position such that the components of $X - \alpha \cup \beta$ are all contractible. We will show that products of Dehn twists about α and inverse of twists about β are elements of M_g which preserve orbits in Q_0. The existence of such closed orbits determined by "hyperbolic" elements in M_g was discovered by Thurston [14] and Bers [3]. The existence of the particular examples mentioned above was stated by Thurston. The idea of the proof of the theorem is to show that these special closed orbits in Q_0/M_g are dense and can be used to approximate a dense orbit.

The techniques involve heavy use of the interplay between measured foliations and quadratic differentials, much of which was developed in [8]. We will often refer to that paper. Along the way in §2 we will prove that quadratic differentials with closed horizontal and closed vertical trajectories are dense in Q_0. §1 reviews the basis results on measured foliations, §2 contains results on approximation of foliations, §3 discusses hyperbolic homeomorphisms and §4 contains the proof of the theorem.

We assume the reader is familiar with basic results of Teichmüller theory and quadratic differentials; in particular, Teichmüller's theorem on extremal quasiconformal mappings. References are [1], [2], [6], and [8]. Our approach to Teichmüller theory is the one found by Earle and Eells [6].

The author would like to thank the referee for several helpful suggestions.

§1. *Quadratic differentials, measured foliations and the space* R^S

The ideas and definitions in this section were introduced in [14] and expanded on in [8]. We retain the notation of [8] and refer to it for details. Let M be a compact C^∞ surface of genus $g > 1$. By F we mean a measured foliation on M; S refers to the set of homotopy classes of simple closed curves on M; R^S refers to the linear space of functions from S to R given the product topology. A measured foliation F defines a nonzero element in R^S by sending $[\gamma]$ to $\inf_{\gamma \epsilon [\gamma]} V_F(\gamma)$ where $V_F(\gamma)$ is the transverse length or measure of γ. Two measured foliations are *equivalent* if their images in R^S coincide. We also consider the projective space PR^S corresponding to the linear space R^S and the natural projection $\Pi : R^S - \{0\} \to PR^S$. The set $\mathcal{F}$ of projective classes of foliations in PR^S is homeomorphic to the sphere S^{6g-7}. This is Theorem 2 [14] and also Theorem 4 [8]. It is easy to see that M_g acts naturally on R^S restricting to an action on the projective foliations; there also defined by pull-back.

A quadratic differential q determines a pair of transverse measured foliations, a *horizontal* foliation F_q defined by $\operatorname{Im} q^{1/2} dz$ and a *vertical* foliation F_{-q} defined by $\operatorname{Re} q^{1/2} dz$. Conversely, transverse measured foliations F_1 and F_2 define a complex structure X on M together with a quadratic differential q denoted (F_1, F_2) on X; that is, a point in Q. Away from a zero we can find local coordinates (x, y) such that F_1 is given by dy, F_2 by dx. Then $z = x + iy$ are complex coordinates and $q = dz^2$. In a neighborhood of a singularity of F_1 and F_2 of order k, we can find local coordinates z so that F_1 is given by $\operatorname{Im} z^{k/2} dz$ and F_2 by $\operatorname{Re} z^{k/2} dz$. Then $q = z^k dz^2$.

The family $(\lambda^{1/2} F_1, \lambda^{-1/2} F_2)$, $0 < \lambda < \infty$ of Riemann surfaces X_λ and quadratic differentials q_t, $t = -1/2 \log \lambda$ is the Teichmüller orbit $T_t q$. Indeed $(\lambda^{1/2}, \lambda^{-1/2})$ is precisely the result of stretching q along the horizontal trajectories by $\lambda^{-1/2}$ and the vertical trajectories by $\lambda^{1/2}$ which is a Teichmüller map with dilatation $\frac{1-\lambda}{1+\lambda} \bar{q}/|q|$.

Notice here $\tanh t = \frac{1-\lambda}{1+\lambda}$. If f is a diffeomorphism and F_1 and F_2 are transverse, then f is a conformal map from the Riemann surface of (F_1, F_2) to the surface given by (fF_1, fF_2) and $[f] \epsilon M_g$ maps the quadratic differentials to each other as a mapping of Q. The Teichmüller orbits correspond under $[f]$ as well.

§2. *Approximation techniques in* Q

In this section we will prove several results on approximating differentials and foliations.

Let X_0 be a Riemann surface, q_0 a quadratic differential on X_0 with *simple* zeroes. Let $\tilde{X}_0$ be the Riemann surface of $q_0^{1/2}$, a double cover of X_0 ramified over the zeroes. There is an involution τ on $\tilde{X}_0$ commuting with the natural projection $\Pi : \tilde{X}_0 \to X_0$ and an odd abelian differential ω_{q_0} on $\tilde{X}_0$, "the square root" of q_0 satisfying $\omega_{q_0} = \Pi^* q_0$. We will consider the homology group $H_1(\tilde{X}_0, C)^-$ which is odd with respect to τ and has complex dimension $6g-6$. Let U be a simply-connected open neighborhood of q_0 in Q such that all $q \,\epsilon\, U$ have only simple zeroes. The local system on U defined by the groups $H_1(\tilde{X}_q, C)^-$ is trivial so we can identify $H_1(\tilde{X}_q, C)^-$ and $H_1(\tilde{X}_0, C)^-$.

For q near q_0 pick ω_q near ω_{q_0} and for each $q \,\epsilon\, U$ let $\theta_q \,\epsilon\, \mathrm{Hom}\,(H_1(\tilde{X}_0, C)^-$ be the homomorphism $\tilde{\gamma} \mapsto \int_{\tilde{\gamma}} \omega_q$. Denote by $\mathrm{Re}\,\theta_q$ and $\mathrm{Im}\,\theta_q$ the real and imaginary parts of θ_q.

PROPOSITION 2.1. *The map* $q \to \theta_q$ *is differentiable in* U; *its derivative at* q_0 *is an isomorphism on the tangent space to* Q *at* q_0.

Proof. That the map is smooth follows from the fact that the surfaces $\tilde{X}_q$ form a proper and smooth family over U.

We first restrict the map to $U \cap H^0(X_0, \Omega^{\otimes 2})$. By differentiating under the integral sign (see also Prop. 1 of [5]) the derivative at q_0 in the direction $\nu \,\epsilon\, H^0(X_0, \Omega^{\otimes 2})$ evaluated at $\tilde{\gamma}$ is $\frac{1}{2} \int_{\tilde{\gamma}} \frac{\Pi^* \nu}{\omega_q}$. Now the map $\nu \to \frac{\Pi^* \nu}{\omega_q}$ is an isomorphism of $H^0(X, \Omega^{\otimes 2})$ and $H^0(\tilde{X}_0, \Omega)^-$. Furthermore, the map $H^0(\tilde{X}_0, \Omega)^- \to \mathrm{Hom}\,(H_1(\tilde{X}_0, C)^-, R)$ given by taking the imaginary part of the integral is an isomorphism as it is the restriction to the odd parts of an isomorphism. We have shown that the derivative of $q \to \mathrm{Im}\,\theta$ is an isomorphism on the vertical tangent space $H^0(X_0, \Omega^{\otimes 2})$ to Q at q_0.

Now let E_0 be the fiber containing q_0 of the map $q \to \mathrm{Im}\,\theta_q$ defined in U.

By the above result, the derivative is an isomorphism on $U \cap H^0(X_0, \Omega^{\otimes 2})$ so E_0 is a submanifold of U and its tangent space at q maps isomorphically onto the tangent space to T_g. Now $\mathrm{Hom}(H_1(\tilde{X}_0, C)^-, R)$ has real dimension $6g - 6$ so to prove the proposition, it is enough to show that if $\mathrm{Re}\,\theta_q$ is restricted to E_0, its derivative is injective at q_0. If there is an element in the kernel there is a nontrivial curve $q_s \in E_0$ lying over $X_s \in T_g$ such that both

$$\mathrm{Re} \int_{\tilde{\gamma}} \omega_{q_s} = \mathrm{Re} \int_{\tilde{\gamma}} \omega_0 + o(s) \quad \text{and} \quad \mathrm{Im} \int_{\tilde{\gamma}} \omega_{q_s} = \mathrm{Im} \int_{\tilde{\gamma}} \omega_0 + o(s)$$

for all $\tilde{\gamma} \in H_1(\tilde{X}_0, C)^-$. Now $H_1(\tilde{X}_0, C)^-$ is generated by the lifts of arcs γ in X_0 that join the zeroes. We take a polygon decomposition of X_0 such that the vertices are zeroes and the sides are geodesic arcs in the metric $|q_0|^{1/2}$. The nearby q_s has a similar decomposition and the lengths of the sides vary from those of q_0 by a quantity which is $o(s)$. Then it is easy to see there are quasiconformal maps $f_s : X_0 \to X_s$ which map polygons to polygons such that the dilatation of f_s is $o(s)$. This is a contradiction.

REMARK. The manifold E_0 consists of all q near q_0 such that the horizontal foliation F_q is equivalent to the horizontal foliation F_{q_0}. This will be useful in the next two results. The first consequence is the following.

PROPOSITION 2.2. *The quadratic differentials with closed horizontal and closed vertical trajectories each determining one cylinder are dense.*

Proof. We need only approximate q_0 with simple zeroes. By Theorem 1 of [12] there are sequences $\phi_n \to q_0$ and $\psi_n \to -q_0$ such that ϕ_n and ψ_n have closed horizontal trajectories and one cylinder. Clearly $-\psi_n$ has closed vertical trajectories. Let $\theta_n \in \mathrm{Hom}(H_1(\tilde{X}_0, C)^-, C)$ be the homomorphism whose imaginary part is $\mathrm{Im}\,\theta_{\phi_n}$ and real part is $\mathrm{Re}\,\theta_{-\psi_n}$.

Since

$$\text{Im} \int_{\tilde{\gamma}} \omega_{\phi_n} \to \text{Im} \int_{\tilde{\gamma}} \omega_{q_0} \quad \text{and} \quad \text{Re} \int_{\tilde{\gamma}} \omega_{-\psi_n} \to \text{Re} \int_{\tilde{\gamma}} \omega_{q_0}, \theta_n \to \theta_{q_0}.$$

By the inverse function theorem and Proposition 2.1, $\theta_n = \theta_{q_n}$ for some $q_n \to q_0$.

Now $\text{Im }\theta_{\phi_n} = \text{Im }\theta_{q_n}$ so q_n and ϕ_n determine equivalent foliations and by Lemma 2.10 of [8], q_n has closed horizontal trajectories homotopic to those of ϕ_n. Similarly q_n has closed vertical trajectories homotopic to those of ψ_n. q.e.d.

PROPOSITION 2.3. *Suppose* $q \in H^0(X, \Omega^{\otimes 2})$ *has closed horizontal and vertical trajectories with only one cylinder each. Let* α *and* β *be a closed horizontal and vertical trajectory in each of the two homotopy classes. Then each component of* $X - \alpha \cup \beta$ *is contractible.*

Proof. First one sees that β must intersect every segment in the critical graph Γ of q. If not consider a closed vertical leaf β' homotopic to β intersecting a critical segment γ that β does not intersect. There is a homotopy of β' and β through closed regular vertical leaves. But such a homotopy must pass through an endpoint of γ which is a zero of q.

Now suppose some component of $X - \alpha \cup \beta$ is not contractible hence contains a nontrivial Jordan curve δ. Since $\delta \cap \alpha = \phi$ we may assume $\delta \subset \Gamma$. Hence $\delta \cap \beta \neq \phi$ which is a contradiction.

The converse is also true.

PROPOSITION 2.4. *Suppose* α *and* β *are two simple closed curves on a surface* M *in minimal position and such that each component of* $M - \alpha \cup \beta$ *is contractible. Then there is a Riemann surface* X *and* $q \in H^0(X, \Omega^{\otimes 2})$ *such that all regular horizontal and vertical leaves of* q *are closed and homotopic to* α *and* β *resp.*

Proof. Each component of $X - \alpha \cup \beta$ is a polygon with $2n$ sides, $n \geq 2$, which are alternately segments of α and β. Given a component U with $2n$ sides find transverse measured foliations of U with a singularity of order $n-2$ in the interior of U such that for one foliation the segments of α are leaves, for the transverse foliation, the segments of β are leaves.

The situation for $n = 3$ is drawn below. The measures are chosen in each U so that all segments of β have the same length measured with respect to the horizontal foliation and all segments of α have the same length measured with respect to the vertical. This means the measures agree on common boundaries and critical trajectories always cut each segment of α and β in half. This latter forces all critical trajectories to be compact. (See the figure.) It is clear from the construction now that all trajectories are homotopic to either α or β.

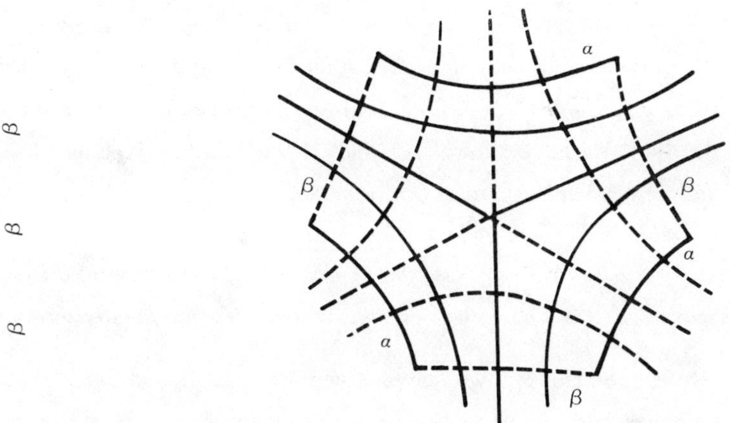

We now need to prove a result which says that a pair of foliations which are close in R^S to a transverse pair are still transverse.

PROPOSITION 2.5. *Let* F_{q_0} *and* F_{-q_0} *be the horizontal and vertical foliations of* $q_0 \in Q$ *on* X_0 *and assume* q_0 *has simple zeroes. Any two equivalence classes of foliations* H_1 *and* H_2 *sufficiently near* F_{q_0} *and* F_{-q_0} *in* R^S *have transverse representatives still denoted* H_1 *and* H_2 *which are the horizontal and vertical foliations of a* q *near* q_0.

Proof. By the main theorem of [8], we may find unique quadratic differentials ψ_1 and ψ_2 on X_0 whose horizontal foliations are in the classes H_1 and H_2 resp. The map $p: Q \to R^S$ given by $q \to ([\gamma] \mapsto \inf_{\gamma \in [\gamma]} \int_\gamma |\operatorname{Im} q^{1/2} dz|)$ is continuous by a routine application of Ascolis' theorem. By Proposition 4.16 of [8] if two quadratic differentials on X have the same image in R^S they coincide. Together the last two statements mean that ψ_1 and ψ_2 vary continuously with H_1 and H_2 and thus ψ_1 is near q_0, ψ_2 near $-q_0$. In particular, $\operatorname{Im} \theta_{\psi_2}$ is near $\operatorname{Re} \theta_{q_0}$ and $\operatorname{Im} \theta_{\psi_1}$ is near $\operatorname{Im} \theta_{q_0} = -\operatorname{Re} \theta_{-q_0}$. By the inverse function theorem and Proposition 2.1, choose ϕ near q_0 so that $\operatorname{Im} \theta_\phi = \operatorname{Im} \theta_{\psi_1}$ and $\operatorname{Re} \theta_\phi = \operatorname{Im} \theta_{\psi_2}$. Then ϕ is equivalent to ψ_1. Now choose ψ near $-q_0$ so that $\operatorname{Im} \theta_\psi = \operatorname{Im} \theta_{\psi_2}$, so ψ is equivalent to ψ_2, and also such that $-\operatorname{Re} \theta_\psi = \operatorname{Im} \theta_\phi$. The two choices mean $\theta_\psi = \theta_{-\phi}$ which again by the inverse function theorem gives $\psi = -\phi$. We have shown H_1 and H_2 can be realized as the horizontal and vertical foliations of ϕ near q_0.

COROLLARY 2.6. *Any* q_0 *may be approximated by quadratic differentials as in Prop. 2.2 where the vertical trajectories are nondividing curves.*

Proof. By Proposition 2.2 we may assume q_0 has closed horizontal trajectories homotopic to some curve α and closed vertical trajectories homotopic to some β. Let F_{q_0} be the horizontal foliation of q_0 and F_{-q_0} the horizontal foliation of $-q_0$. Assume now β is dividing and let τ_β be the Dehn twist about β. Let γ be a simple closed curve intersecting β twice. Then $\tau_\beta^n(\gamma)$ is nondividing. Let $i(\rho, \cdot)$ be the element in R^S induced by a simple closed curve ρ where $i(\cdot, \cdot)$ is the geometric intersection number. Then $\tau_\beta^n(\gamma)$ converges to β in PR^S as

$n \to \infty$. Therefore we may take a foliation F_n all of whose leaves are closed and homotopic to $\tau_{\beta}^n(\gamma)$ and such that $F_n \to F_{-q_0}$ in R^S. The foliations F_{q_0} and F_n satisfy the conditions of Proposition 2.5.

The main theorem of [8] states that every equivalence class of measured foliations is realized uniquely on each Riemann surface as the horizontal foliation of a quadratic differential. We indicated above that a pair of transverse foliations define an element in Q. We now show this procedure is defined on *equivalence* classes as well.

PROPOSITION 2.7. *Let* F_1 *and* F_2 *be transverse foliations,* G_1 *and* G_2 *transverse and* F_i *equivalent to* G_i $i = 1, 2$. *Then the quadratic differentials* $q_1 = (F_1, F_2)$ *and* $q_2 = (G_1, G_2)$ *coincide in* Q.

REMARK. We need to show there is a conformal map of the underlying Riemann surfaces homotopic to the identity, mapping one quadratic differential to the other. We note that for arbitrary equivalent foliations there need not be a homeomorphism homotopic to the identity sending one to the other. The notion of equivalence includes other deformations as well (see [8]). The point here is that the equivalent foliations have transverse foliations which are also equivalent.

We will prove the result for F_2 and G_2 with closed leaves and $3g{-}3$ cylinders and deduce the general case by approximation. In the specific case we consider for each of F_2 and G_2 the complement of a closed leaf in each homotopy class. The complement is $2g{-}2$ spheres each with three holes, each of which is a leaf. The union of the critical leaves is a graph Γ whose complement in each sphere is three annuli. There are, up to homotopy, three possibilities.

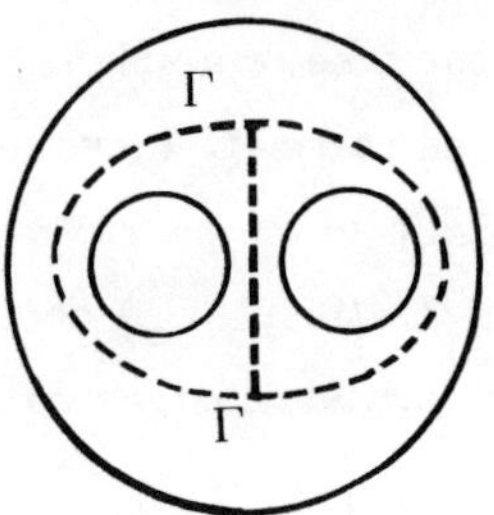
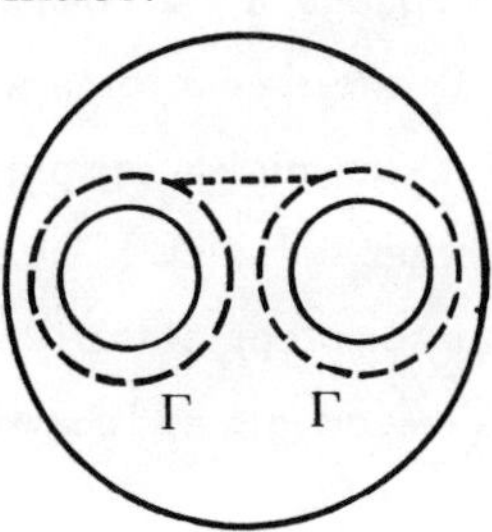
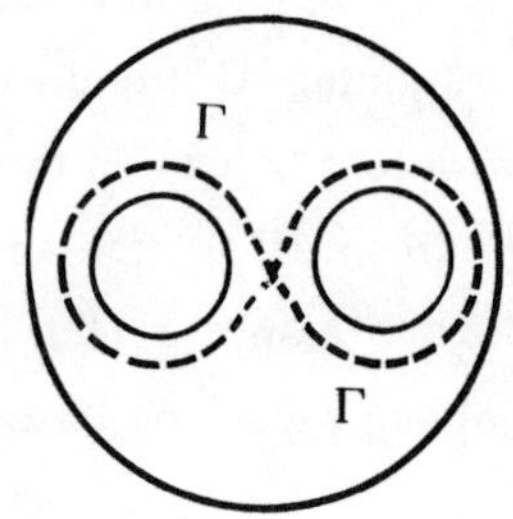

In the first case, the length of any leaf (measured with respect to the transverse foliation), is less than the sum of the other two, in the second, the length of one is greater than the sum of the other two, and in the third case, there is equality. Since F_2 and G_2 are equivalent, their leaves are homotopic and the corresponding cylinders have the same heights (Lemma 2.10 of [8]).

Since F_1 is equivalent to G_1, the horizontal lengths of closed leaves are the same. By the above characterization of the possible graphs, this means that for each three-holed sphere, the graphs of F_2 and G_2 are the same type and corresponding edges have the same length. There is a homeomorphism f homotopic to the identity taking F_2 to G_2.

Now each cylinder C of F_2 has two boundary components consisting of compact leaves joining zeroes.

Let u and v be zeroes on opposite components of C (they may coincide on the surface) and u′ and v′ the corresponding zeroes under f for the cylinder C′ with leaves homotopic to those of C. Measured in the cylindrical coordinates defined by q_1 and q_2, the lines in C and C′ from u to v and u′ to v′ make angles α and α', resp. with the closed trajectories. Suppose $\alpha = \alpha'$. Since the lengths of edges on each component of C are equal to the corresponding lengths on the boundary of C′ there must be equality for all other possible angles and we can map C to C′ by a homeomorphism extending to the boundary, which not only sends $F_2 \cap C$ to $G_2 \cap C'$, but also preserves horizontal distances, and therefore maps $F_1 \cap C$ to $G_1 \cap C'$ as well. If this is true for each cylinder, we would have the necessary homeomorphism of the surface. To prove $\alpha = \alpha'$ cut the surface along one leaf in every cylinder except C (resp. C′) and take the component of the component of the complement containing C (resp. C′). It is either a torus with one hole or a sphere with four holes. In the first case, pick a simple closed curve β crossing C (resp. C′) once and no other cylinder. The geodesic in the metric $|q_1|^{1/2}$ (resp. $|q_2|^{1/2}$) in that homotopy class crosses C (resp. C′) once joining zeroes on opposite components and may also consist of edges on the boundary. For each edge of the boundary of C traversed by the geodesic, the corresponding edge on the boundary of C' is traversed. Since the lengths of these edges are equal and since the total horizontal lengths of the geodesics are equal, the lengths of the pieces crossing the cylinders must also be equal, forcing a and a' to either be equal or to have their sum equal to zero (mod 2π). The latter possibility cannot occur, as a twist of β about the closed leaf produces geodesics of unequal length.

In the case of a sphere with three holes, find a geodesic β which crosses the cylinder twice, with other pieces on the boundary. If the crossing pieces coincide, the analysis is the same as above, otherwise, set one boundary component of each cylinder to correspond to $h = 0$ in the cylindrical coordinates and set x_1, x_2 to be the horizontal lengths of the pieces crossing C , y_1, y_2 the lengths of pieces crossing C' with the added stipulation that a number is negative if the slope of the segment is negative; i.e., it leaves $h = 0$ heading to the left.

Then we have $|x_1| + |x_2| = |y_1| + |y_2|$ and $x_1 - y_1 = x_2 - y_2$. The first equation holds because the horizontal lengths of the geodesics are equal, the second because C and C' differ only by the rotation angle $a - a'$ and this difference is cancelled by subtraction. Moreover, again twisting β about a closed leaf shows the first equation must hold if the x_i and y_i are replaced by $x_i + c$ and $y_i + c$ where c is the circumference of the cylinder. This forces $x_1 = x_2$, $y_1 = y_2$ and again the angles are equal. This finishes the proof in the special case and (F_1, F_2) and (G_1, G_2) define the same point in Q .

Finally, for arbitrary equivalent G_1 and G_2 , approximate by equivalence classes G_n with $3g-3$ cylinders of closed leaves. That this can be done follows from the density theorem in [5]. (See also [13].) Then by Proposition 2.5, there exists G_{in} in the equivalence class of G_n such that $(F_i, G_{in}) \to (F_i, G_i)$ i $= 1, 2$. But $(F_1, G_{1n}) = (F_2, G_{2n})$ by the special case argument proving $(F_1, G_1) = (F_2, G_2)$.

§3. *Hyperbolic elements of* M_g

Recall that M_g has a natural action on the space $\mathcal{F}$. We are interested in the case where $[f] \in M_g$ has two transverse fixed projective foliations. This was considered first by Thurston in his classification theorems 5 in [14].

PROPOSITION 3.1 (Thurston [14]). *Suppose* $[f]$ *is an isotopy class of diffeomorphisms fixing two transverse projective classes of foliations* $[F_1]$ *and* $[F_2]$. *Then there is a* $\lambda > 0$, *foliations* F_1 *and* F_2 *in the respective classes,* $f \in [f]$ *such that* $f(F_1) = \lambda F_1$, $f(F_2) = 1/\lambda F_2$. *If* $\lambda = 1$, f *is a conformal map fixing* q. *In either case,* $[f] \in M_g$ *fixes the Teichmüller orbit* $T_t q$ *in the first case acting as a translation.*

Proof. Let $g \in [f]$ be any diffeomorphism and $F_1 \in [F_1]$ $F_2 \in [F_2]$ transverse foliations. Then $g(F_1)$ is equivalent to $\mu_1 F_1$ and $g(F_2)$ is equivalent to $\mu_2 F_2$ for some $\mu_1, \mu_2 > 0$. By Proposition 2.7, $(g(F_1), g(F_2))$ determines the same element of Q as $(\mu_1 F_1, \mu_2 F_2)$. Therefore, there is a map h homotopic to the identity such that $\text{hog}(F_1) = \mu_1 F_1 = (\mu_1^{1/2}/\mu_2^{1/2})\mu_1^{1/2}\mu_2^{1/2} F_1$ and $\text{hog}(F_2) = \mu_2 F_2 = (\mu_2^{1/2}/\mu_1^{1/2})\mu_1^{1/2}\mu_2^{1/2} F_2$. Now suppose $\mu_1 \neq \mu_2$. Then $f = \text{hog}$ is a Teichmüller map with terminal quadratic differential $\mu_1^{1/2}\mu_2^{1/2} q$ and maximal dilatation μ_1/μ_2 or μ_2/μ_1 depending on which is larger than one. But Teichmüller maps are area preserving with respect to the area elements of the initial and terminal quadratic differentials. This proves $\mu_1^{1/2}\mu_2^{1/2} = 1$ and we let $\lambda = \mu_1 \neq 1$. If $\mu_1 = \mu_2$, then f is a conformal map, hence area preserving forcing $\mu_1^{1/2}\mu_2^{1/2} = 1$.

REMARK. We note that in the first case, f is a conformal map from the Riemann surface given by (F_1, F_2) to the Riemann surface given by $\left(\lambda F, \frac{1}{\lambda} F_2\right)$. As a self mapping of (F_1, F_2) it is a Teichmüller map and hence not even differentiable at the singularities.

We are interested primarily in the case $\lambda \neq 1$. Then $[f]$ preserves the orbit determined by q and the corresponding orbit in R_g is closed.

These f were discovered by Thurston [14] and he calls them "pseudo-Anosov" diffeomorphisms. He gives necessary and sufficient topological conditions on an element in M_g to guarantee it has a pseudo-Anosov representative. In [3], Bers gave a "Teichmuller theory" proof of the same result and because of the invariant orbit calls these hyperbolic elements in M_g. We will not need to use these topological conditions. Instead we explicitly construct the class of hyperbolic homeomorphisms mentioned in the introduction and use them to generate others.

Let α and β be simple closed curves determining a quadratic differential q as in Proposition 2.4. As we saw in Proposition 2.2 these are dense. Now let τ_α and τ_β be the Dehn twists about α and β.

PROPOSITION 3.2. *For each positive integer* n , $g = \tau_\alpha^n \tau_\beta^{-n}$ *is hyperbolic with fixed transverse projective classes* $[F_{1n}]$ *and* $[F_{2n}]$. *There are representatives* F_{1n} *and* F_{2n} *such that* $q_n = (F_{1n}, F_{2n}) \to q$ *as* $n \to \infty$. *Furthermore, for* K , *a compact set of projective foliations not containing* $[F_{2n}]$ *and* n *fixed,* $g^k(K) \to [F_{1n}]$ *as* $k \to \infty$. *For* K *not containing* $[F_{1n}]$, $g^{-k}(K) \to [F_{2n}]$.

COROLLARY 3.3. *The closed orbits in* Q/M_g *are dense*

Proof of the corollary. Apply Propositions 2.2 and 3.2.

Before proving the proposition we make a few remarks. Thurston's theorems 5 and 10 in [14] include the first statement as a special case. The statements on the convergence of $g^k(K)$ and $g^{-k}(K)$ are also essentially there. The main ingredient in the proof given here is the unique ergodicity of the foliations. Marcus [11] proved generally that the flow of a one-dimensional unstable manifold of an Anosov diffeomorphism is uniquely ergodic. His proof would likely carry over to the present case of foliations with singularities. For completeness we give a self-contained proof.[*]

Proof of Proposition 3.2. We explicitly calculate $g = \tau_\alpha^n \tau_\beta^{-n}$. Let C be

---

[*]Note added in proof. A proof of unique ergodicity for general pseudo-Anosov foliations and the example of Proposition 3.2 have since appeared in [15].

the cylinder determined by a, C' determined by β. It is clear we may assume C and C' have height one and circumference c, the intersection number of a and β, as a change of scale will then prove the result for arbitrary heights. Both C and C' are parametrized by cylindrical coordinates (h, θ) and (h', θ') resp. with $0 \leq h$, $h' \leq 1$ and $0 \leq \theta$, $\theta' \leq c$. There are c vertical lines B_j in C whose union is the vertical critical trajectories of q. They bound c vertical strips whose union is C'. There are also c horizontal segments A_j whose union is the horizontal critical trajectories. We fix one boundary component of C to be $h = 0$ and one B_j to be $\theta = 0$. The coordinates (h', θ') are uniquely determined by requiring that $(h, \theta) = (1, 0)$ corresponds to $(h', \theta') = (0, 0)$. Now τ_a^n and τ_β^n are given by $(h, \theta) \to (h, \theta + chn)$ and $(h', \theta') \to (h', \theta' + chn)$. In each strip the change of coordinates is either $(h', \theta') = (\theta + k_1, k_2 - h)$ or $(h', \theta') = (k_1 - \theta, k_2 + h)$ for integers k_1, k_2. The first case occurs if the segment $h' = 0$ is on the left of the strip, the second if it is on the right. In the first case, the derivative is $\begin{bmatrix} 0 & -1 \\ 1 & 0 \end{bmatrix}$, in the second, $\begin{bmatrix} 0 & 1 \\ -1 & 0 \end{bmatrix}$. Therefore, the derivative of τ_β^{-n} with respect to (h, θ) is

$$\begin{bmatrix} 0 & -1 \\ 1 & 0 \end{bmatrix} \begin{bmatrix} 1 & -nc \\ 0 & 1 \end{bmatrix} \begin{bmatrix} 0 & 1 \\ 1 & 0 \end{bmatrix} = \begin{bmatrix} 1 & 0 \\ nc & 1 \end{bmatrix}$$ or its negative. Then g has

derivative $\begin{bmatrix} 1 + n^2 c^2 & nc \\ nc & 1 \end{bmatrix}$ or again, possibly its negative. In either case the eigenvalues are

$$\frac{2 + n^2 c^2 \pm nc \sqrt{n^2 c^2 + 4}}{2}$$

and the eigenvectors are

$$\begin{bmatrix} 1 \\ \pm \dfrac{\sqrt{c^2 n^2 + 4}}{2} - \dfrac{cn}{2} \end{bmatrix}.$$

This means that the quadratic differential $q_n = \left(1 - \left(\dfrac{\sqrt{c^2 n^2 + 4} - cn}{2}\right)\sqrt{-1}\right)^2 q$

has horizontal and vertical foliations F_{1n} and F_{2n} tangent to the eigen-vectors which means g is pseudo-Anosov with λ the eigenvalue greater than one, and $g(F_{2n}) = \lambda F_{2n}$, $g(F_{1n}) = \frac{1}{\lambda} F_{1n}$. It is easy to check that $q_n \to q$ proving the first two statements.

Let B be any of the vertical segments of unit length with endpoints on opposite components of C. For any $x = x_0 \in B$, let $x_1, \cdots, x_n$ be the first n return points in B on a trajectory through x passing from left to right. This is defined except at those finitely many points which meet a zero before returning n times. At a zero we follow the trajectory leaving the zero to the right of where it entered in order to define the re-turn map everywhere. The proposition will follow from the next lemma, which expresses the unique ergodicity of the return map.

LEMMA 3.4. *Let* $I \subset B$ *be any open interval of length* $|I|$. *Then*

$$\lim_{n\to\infty} \frac{1}{n} \sum_{i=1}^{n} \chi_I(x_i) = |I| \quad \text{uniformly for } x \in B.$$

Postponing the proof of the lemma we finish the proof of Proposition 3.2. Since the simple closed curves δ are dense in PR^S, it is enough to show $g^k(\delta) \to [F_{1n}]$ uniformly for $\delta \in K$. Suppose now ρ is any seg-ment with endpoints on the boundary of C' crossing C' once. Then ρ is composed of a leaf $\tilde{\rho}$ of F_{1n} crossing C' once followed by a vertical segment $\bar{\rho}$ on one component. There exists M such that for all ρ, $V_{F_{1n}}(\rho) = M|\bar{\rho}|$ where $|\ |$ is the Euclidean length. Now let δ_k be any sequence of simple closed curves represented by geodesics and ρ_1, ρ_2 any two segments as indicated above. If the slope of δ_k converges to μ the slope of F_{1n} as $k \to \infty$, then it is easy to see that

$$\lim_{k\to\infty} \frac{i(\delta_k, \rho_1)}{i(\delta_k, \rho_2)} = \lim_{k\to\infty} \frac{i(\delta_k, \bar{\rho}_1)}{i(\delta_k, \bar{\rho}_2)}.$$

However the uniform convergence given in the lemma shows that

$$\lim_{k\to\infty} \frac{i(\delta_k, \overline{\rho}_1)}{i(\delta_k, \overline{\rho}_2)} = \frac{|\overline{\rho}_1|}{|\overline{\rho}_2|} = \frac{V_{F_{1n}}(\rho_1)}{V_{F_{1n}}(\rho_2)}.$$ Since any simple closed curve is a

union of such ρ_i, this shows $\delta_k \to [F_{1n}]$ in PR^S. A similar argument

shows that if the slope of δ_k converges to $-1/\mu$, the slope of F_{2n},

then $\delta_k \to [F_{2n}]$.

Now suppose there is a sequence $\delta_i \in K$ containing segments δ_i'

such that the slope of δ_i' has limit $-1/\mu$ as $i \to \infty$. Some subsequence

of δ_i' converges to a critical leaf γ_1 of F_{2n}. Since δ_i has no trans-

verse self intersections and γ_1 is dense, all segments of δ_i have

slopes with limit $-1/\mu$ and $\delta_i \to F_{2n}$. This is impossible since K does

not contain F_{2n}, and therefore the slopes of δ_i are bounded away from

$-1/\mu$. Then if we let $\delta_k = g^k(\delta)$ the slope of $\delta_k \to \mu$ uniformly for $\delta \in K$

as $k \to \infty$. The argument in the previous paragraph shows $g^k(\delta) \to [F_{1n}]$

uniformly for $\delta \in K$. The second half of (iii) follows similarly.

Proof of Lemma 3.4. We remark first that the return map is similar to

what is known as an interval exchange map. (To define an interval ex-

change map the next return point must always be on the opposite side from

which it left.) It is an open question of which exchange maps are uniquely

ergodic, although there are counterexamples [9].

To prove the lemma it is enough to prove the limit exists uniformly for

a dense set of points on B. These are the points on a leaf γ_0 which

leaves a critical point. Since the critical points are fixed points of g,

$g(\gamma_0) = \gamma_0$. Let $\mu = \dfrac{\sqrt{c^2 n^2 + 4}}{2} - \dfrac{cn}{2}$ which is the slope of γ_0 with re-

spect to the coordinates of C. Notice $\mu^2 + nc\mu = 1$. For each positive

integer r, we subdivide the segments B_j and A_j r times inductively.

After r divisions there will be segments of length μ^r and μ^{r+1}. To

begin, divide B_j into a segment C_j^1 of length μ^2 with one endpoint on

$h = 1$ and nc segments D_j^1 of length μ. The first segment D_{j1}^1 has

an endpoint on $h = 0$. Suppose after r stages there are segments

$D_j^r \subset B_j$ of length μ^r and segments C_j^r of length μ^{r+1}. Divide each D_j^r

into one segment C_j^{r+1} of length μ^{r+2} with an endpoint coinciding with the upper endpoint of D_j^r and the complement into nc segments D_j^{r+1} of length $(\mu^r - \mu^{r+2})/nc = \mu^{r+1}$. At this point each previous C_j^r is re-labelled as some D_j^{r+1} completing the induction. Let n_r and m_r be the number of segments of length μ^r and μ^{r+1} contained in B_j and n_r^I and m_r^I the number contained in I.

Similarly we divide the segments A_i on $h = 0$. First we letter these segments so that the right endpoint of A_j is the lower endpoint of B_j. Some of the A_i are identified on the surface so they will be divided in two different ways. For each A_j subdivide as before into segments D_j^r and C_j^r, only here we require one endpoint on C_j^{r+1} to coincide with an endpoint of D_j^r on the left.

The subdivisions have the following property. If $\gamma \subset \tau_{\bar\beta}^n(\gamma_0)$ is a segment crossing C once with an endpoint on a segment D_j^r (resp. C_j^r) then $\tau_\alpha^n(\gamma) \subset \gamma_0$ intersects for each k, n segments D_k^{r+1} (resp. C_k^{r+1}) unless D_j^r (resp. C_j^r) is contained in C_{j1}^1 in which case $\tau_\alpha^n(\gamma)$ inter-sects $n+1$ segments in B_j and n in B_i, $i \neq j$. In either case exactly one of the intersected segments in B_j is contained in D_{j1}^1.

The situation is analogous for $\gamma \subset \gamma_0$ crossing C' once with an end-point on B_j. Then $\tau_{\bar\beta}^{-n}(\gamma)$ intersects each A_i n times unless γ inter-sects D_{j1}^1 in which case it intersects A_j $n+1$ times. In either case D_{j1}^1 is intersected exactly once.

For $\gamma \subset \tau_{\bar\beta}^n(\gamma_0)$, $\tau_\alpha^n(\gamma)$ does not have endpoints on the segments B_i. We take the largest segment in $\tau_\alpha^n(\gamma)$ that does. Similarly for $\gamma \subset \gamma_0$ we take the largest segment in $\tau_{\bar\beta}^{-n}(\gamma)$ with endpoints on δC. Thus for each k by iteration we find $\gamma_k \subset g^k(\gamma) \subset \gamma_0$. This is done to keep the inter-section count simple.

We now let $\gamma \subset \gamma_0$ be a segment crossing C' once. If we let $a_{jk} = i(\gamma_k, B_j)$, $b_{jk} = i(\gamma_k, D_{j1}^1)$, $\widetilde{a_{jk}} = i(\tau_{\bar\beta}^{-n}(\gamma_k), A_j)$ and $\widetilde{b_{jk}} = i(\tau_{\bar\beta}^{-n}(\gamma_k), \tilde{D}_{j1}^1)$ then

$$\widetilde{a_{jk}} = n \sum_{i=1}^{C} a_{ik} + b_{jk}, \quad \widetilde{b_{jk}} = a_{jk}, \quad a_{j,k+1} = n \sum_{i=1}^{C} \widetilde{a_{ik}} + \widetilde{b_{jk}}$$

and $b_{j,k+1} = a_{jk}$. These last four equations show that as $k \to \infty$, a_{ik} and b_{ik} are unbounded while $a_{ik} - a_{jk}$ and $b_{ik} - b_{jk}$ are bounded for any i, j. Because of this, in the ratios to follow we will be able to ignore the subscript i in a_{ik}, b_{ik}. For example it is easy to see

1.
$$\lim_{k \to \infty} \frac{a_k}{b_k} = 1/\mu \;.$$

For $r > 0$ we consider the intersections of γ_{k+r} with the segments of the $2r^{\text{th}}$ subdivision of each B_j. Then for each D_j^{2r} and C_j^{2r}, $i(D_j^{2r}, \gamma_{k+r}) = a_{ik}$ and $i(C_j^{2r}, \gamma_{k+r}) = b_{ik}$ for some i. Because $a_{ik} - a_{jk}$ is bounded and also I is contained in $m_{2r}^I + 2$ and $n_{2r}^I + 2$ segments of lengths μ^r and μ^{r+1} we have

2.
$$|i(\gamma_{k+r}, B_j) - n_{2r}a_k - m_{2r}b_k| < Mn_{2r}$$

and

3.
$$|i(\gamma_{k+r}, I) - n_{2r}^I a_k - m_{2r}^I b_k| < 2a_k + 2b_k + Mn_{2r}^I$$

where M is independent of k and r. We also have

4.
$$\lim_{r \to \infty} \frac{n_{2r}^I + \mu m_{2r}^I}{n_{2r} + \mu m_{2r}} = \lim_{r \to \infty} \frac{\mu^{2r} n_{2r}^I + \mu^{2r+1} m_{2r}^I}{\mu^{2r} n_{2r} + \mu^{2r+1} m_{2r}} = \frac{|I|}{|B_j|} = |I| \;.$$

Given $\varepsilon > 0$ we pick r large enough so that $\dfrac{1}{n_{2r}} < \varepsilon/4$ and

$$\left| |I| - \frac{n_{2r}^I + \mu m_{2r}^I}{n_{2r} + \mu m_{2r}} \right| < \varepsilon/2 \;.$$

Then for fixed r, 1, 2, and 3 show that

$$\left| \lim_{k\to\infty} \frac{i(\gamma_{k+r}, I)}{i(\gamma_{k+r}, B_j)} - \frac{n_{2r}^I + m_{2r}^I \mu}{n_{2r} + m_{2r}\mu} \right| < \varepsilon/2 \;.$$

Then

$$\left| \lim_{k\to\infty} \frac{i(\gamma_{k+r}, I)}{i(\gamma_{k+r}, B_j)} - |I| \right| < \varepsilon \;.$$

There exists K so that for all $k > K$

5.
$$\left| \frac{i(\gamma_{k+r}, I)}{i(\gamma_{k+r}, B_j)} - |I| \right| < 2\varepsilon \;.$$

It is clear that r and K can be taken to be independent of the particular γ. For any sufficiently long $\gamma \subset \gamma_0$ there is a k, and γ', γ'' segments each crossing C' once such that $\gamma'_{k+r} \subset \bar{\gamma} \subset \gamma''_{k+1+r}$. Therefore the estimate 5 holds if γ_{k+r} is replaced by any sufficiently long $\bar{\gamma}$. Finally, there exists N so that for any $x_0 \in B_j \cap \gamma_0$ and $n \geq N$, $x_0, x_1, \cdots, x_n$ lie on a sufficiently long $\bar{\gamma}$. This completes the proof of the lemma.

§4. *Proof of the theorem*

By results in Chapter 7 of [4] (see also [7]), it is enough to show that for arbitrary open sets W_1 and W_2 in Q_0 there exists $t > 0$, $q_1 \in W_1$, $q_2 \in W_2$, and $h \in M_g$ such that $h(T_t \psi_1) = \psi_2$, and similarly for $t < 0$. We may assume by Corollary 2.6 that W_1 and W_2 contain ϕ_1 and ϕ_2 with simple zeroes, that ϕ_1 and ϕ_2 have closed horizontal trajectories in the homotopy classes a_1 and a_2 and nondividing closed vertical trajectories in the classes β_1 and β_2. Find $g \in M_g$ so that $g(\beta_2) = \beta_1$. Mapping by g, we may assume the vertical trajectories of ϕ_2 are in the class $\beta = \beta_1$ as well. Then ϕ_1 and ϕ_2 determine projective foliations $a_1 = [F\phi_1]$ and $a_2 = [F\phi_2]$ and $-\phi_1$ and $-\phi_2$ determine the same class $\beta = [F]$. By Proposition 3.2, $\tau_{a_1}^n \tau_\beta^{-n}$ fixes two classes of foliations $[F_{1n}]$ and $[G_{1n}]$ and for particular representatives the quadratic differentials $\phi_{1n} = (F_{1n}, F_{2n}) \to \phi_1$. Similarly $\tau_{a_2}^n \tau_\beta^{-n}$ fixes $[F_{2n}]$ and $[G_{2n}]$ and $\phi_{2n} = (F_{2n}, G_{2n}) \to \phi_2$. Moreover, $[F_{1n}]$ and $[F_{2n}]$ are attractive fixed points for $\tau_{a_1}^n \tau_\beta^{-n}$ and $\tau_{a_2}^n \tau_\beta^{-n}$ resp. in the sense of Proposition 3.2. Since the map $p : Q \to PR^S$ is continuous, $[F_{1n}] \to [F_{\phi_1}]$, $[F_{2n}] \to [F_{\phi_2}]$ and both $[G_{1n}]$ and $[G_{2n}] \to [F]$ as $n \to \infty$.

By Proposition 2.5 there are neighborhoods V_1 of $[F_{\phi_1}]$, V_2 of $[F_{\phi_2}]$ and V_3 of $[F]$ in $\mathcal{F}$ such that if $[H_1] \in V_1$, $[H_3] \in V_3$ there exist $H_1 \in [H_1]$ and $H_3 \in [H_3]$ such that $(H_1, H_3) \subset W_1$. Similarly for $[H_2] \in V_2$ and $[H_3] \in V_3$ there exist H_2 and H_3 such that $(H_2, H_3) \in W_2$.

Pick n large enough so that $[F_{1n}] \in V_1$, $[F_{2n}] \in V_2$ and both $[G_{1n}]$ and $[G_{2n}] \in V_3$. Let $U_{1n} \subset V_1$, $U_{2n} \subset V_2$ and $U_{3n} \subset V_3$ be disjoint compact contractible neighborhoods in $\mathcal{F}$ of $[F_{1n}]$, $[F_{2n}]$ and both $[G_{1n}]$ and $[G_{2n}]$ resp. Since $[F_{1n}]$ and $[F_{2n}]$ are attractive fixed points of $\tau_{a_1}^n \tau_\beta^{-n}$ and $\tau_{a_2}^n \tau_\beta^{-n}$ resp., for positive k large enough, $(\tau_{a_2}^n \tau_\beta^{-n})^k (U_{1n}) \subset U_{2n}$ and $(\tau_{a_1}^n \tau_\beta^{-n})^k (U_{2n}) \subset U_{1n}$. Since $[G_{1n}]$ and $[G_{2n}]$ are repulsive and U_{3n} is a neighborhood of both, for large positive k $(\tau_{a_1}^n \tau_\beta^{-n})^{-k} (U_{3n}) \subset U_{3n}$ and $(\tau_{a_2}^n \tau_\beta^{-n})^{-k} (U_{3n}) \subset U_{3n}$.

Let $g_{nk} = (\tau_{a_1}^n \tau_\beta^{-n})^k (\tau_{a_2}^n \tau_\beta^{-n})^k$ and $h_{nk} = (\tau_{a_2}^n \tau_\beta^{-n})^k (\tau_{a_1}^n \tau_\beta^{-n})^k$. Then $g_{nk}(U_{1n}) \subset U_{1n}$, $g_{nk}^{-1}(U_{3n}) \subset U_{3n}$, $h_{nk}(U_{2n}) \subset U_{2n}$ and $h_{nk}^{-1}(U_{3n}) \subset U_{3n}$. By the Brouwer fixed point theorem g_{nk} has fixed points in U_{1n} and U_{3n} and h_{nk} has fixed points in U_{2n} and U_{3n}. Proposition 2.5 and 3.1 imply there is a quadratic differential q_{nk} in W_1 whose orbit $T_t(q_{nk})$ is fixed by g_{nk}. Similarly, there is a $\widetilde{q_{nk}} \in W_2$ whose orbit $T_t(\widetilde{q_{nk}})$ is fixed by h_{nk}. However, $f_{nk} = (\tau_{a_2}^n \tau_\beta^{-n})^k$ conjugates g_{nk} and h_{nk}; in fact, mapping the fixed points of one to the fixed points of the other. Therefore, $f_{nk}(T_t q_{nk}) = T_s \widetilde{q_{nk}}$. There are two possibilities. The first is if g_{nk} and h_{nk} have finite order and are conformal mappings preserving q_{nk} and $\widetilde{q_{nk}}$. Then the orbits are fixed pointwise and there exists t such that $f_{nk} T_t q_{nk} = \widetilde{q_{nk}}$. We may not be able to guarantee both positive and negative t as n and k vary in this case. However, since M_g acts properly discontinuously [6], [10] on T_g, there can only be finitely many n and k such that g_{nk} is conformal. This means that for all but finitely many n and k, g_{nk} and h_{nk} are hyperbolic and act as translates of the orbits. Here the numbers λ_{nk} of Proposition 3.2 are greater than one and in fact $\to \infty$ again by the discontinuity of M_g. For each integer m, $(h_{nk})^m \cdot f_{nk}(q_{nk})$ is on the orbit of $\widetilde{q_{nk}}$. Composing with $(h_{nk})^m$ has the effect of translating $f_{nk}(q_{nk})$ along the orbit. For each m then there is a t such that $T_t((h_{nk})^m \cdot f_{nk}(q_{nk})) = \widetilde{q_{nk}}$. By taking $|m|$ large enough we can insure t both positive and negative. The theorem is proved.

UNIVERSITY OF ILLINOIS
AT CHICAGO CIRCLE

REFERENCES

[1] L. V. Ahlfors, On Quasiconformal mappings, Journal D'Analyse Mathematique, 4(1954), 1-58.

[2] L. Bers, Quasiconformal mappings and Teichmüller's theorem, in Analytic Functions. (R. Nevanlinna et al., eds.), Princeton University Press, 1960.

[3] _________, An extremal problem for quasiconformal mappings, Acta Math. 141:1-2, 73-98 (1978).

[4] G. D. Birkhoff, Dynamical Systems, American Mathematical Society Colloquium Publications, 9(1927).

[5] A. Douady and J. Hubbard, On the density of Strebel Differentials. Inventiones Math. 30(1975), 175-179.

[6] C. Earle and J. Eells, A fibre bundle description of Teichmüller theory. Journal of Diff. Geometry 3(1969), 19-45.

[7] G. Hedlund, The dynamics of geodesic flow, Bulls. AMS, 45(1939), 241-260.

[8] J. Hubbard and H. Masur, Quadratic differentials and foliations, Acta Math. 142(1979), 221-274.

[9] H. Keynes and D. Newton, A Minimal non-uniquely ergodic interval exchange transformation. Math Z. 148(1976), 101-105.

[10] S. Kravetz, On the geometry of Teichmüller spaces and the structure of their modular groups, Ann. Acad. Sci. Fenn. 278(1959), 1-35.

[11] H. Marcus, Unique ergodicity of the horocycle flow: Variable negative curvature case. Israel Journal of Math. 21 nos. 2-3(1975), 133-144.

[12] H. Masur, The Jenkins Strebel differentials with one cylinder are
      dense, Comm. Math. Helv. 54(1979), 179-184.

[13] K. Strebel, On Quadratic differentials and extremal quasi-conformal
      mappings, Univ. of Minnesota. Lecture Notes. (1967).

[14] W. Thurston, On the geometry and dynamics of diffeomorphisms of
      surfaces, to appear.

[15] A. Fathi et al., Travaux de Thurston sur les surfaces, Asterisque
      66-67 Société mathematique de France, 1979.

AUTOMORPHISMEN EBENER DISKONTINUIERLICHER GRUPPEN

Gerhard Rosenberger

Einleitung

A. Sei $G = \langle s_1, \cdots, s_m, a_1, \cdots, a_p \, | \, s_1^{\gamma_1} = \cdots = s_m^{\gamma_m} = s_1 \cdots s_m [a_1, a_2] \cdots$

$[a_{p-1}, a_p] = 1 \rangle$, $\gamma_i \geq 2$ sowie $m \geq 3$ und $m - 2 - \sum_{i=1}^{m} \frac{1}{\gamma_i} > 0$ falls $p = 0$,

eine Fuchssche Gruppe mit kompaktem Fundamentalbereich. In [15] hat
Zieschang gezeigt, dass jeder Automorphismus von G induziert wird von
einem Automorphismus der freien Gruppe vom Rang $p + m$ (vgl. auch [17]).
Hier erweitern und ergänzen wir diesen Satz, indem wir unter anderem
zeigen: Jeder Automorphismus von G wird induziert von einem Automorphismus der freien Gruppe vom Rang r, wobei r der Rang von G ist.
Dies beinhaltet gleichzeitig einen neuen Beweis für den erwähnten Satz
von Zieschang.

B. Diese Arbeit verwendet die Terminologie und Bezeichnungsweise von
[8] und [16], wobei $\langle \cdots | \cdots \rangle$ die Gruppenbeschreibung durch Erzeugende
und Relationen bedeutet. Unter $\langle a_1, \cdots, a_m \rangle$ verstehen wir die von
$a_1, \cdots, a_m$ erzeugte Gruppe. Wie in [8] und [16] gewinnen wir oft aus
einem System $\{x_1, \cdots, x_n\}$ durch freie Übergänge (Nielsen-
Transformationen) ein neues und bezeichnen es mit denselben Symbolen.
Es bedeute: $[a, b]. = aba^{-1}b^{-1}$ den Kommutator von a und b. (α, β)
den grössten gemeinsamen Teiler von $\alpha, \beta \in \mathbb{N}$.

§1. *Vorbemerkungen*

Sei $G = H_1 \underset{A}{*} H_2$ freies Produkt der Gruppen H_1 und H_2 mit Amalgam $A = H_1 \cap H_2$. Ferner sei in G eine Ordnung und eine Länge L wie in [8] und [16] definiert. Die Ordnung genüge der Bedingung, dass vor einem Produkt von Restklassenvertretern $1_1 \cdots 1_m$ von A in den H_i nur endlich viele Produkte $1_1 \cdots 1_{m-1} 1$ liegen, wobei 1 ein Restklassenvertreter aus demselben Faktor H_i wie 1_m ist. Aus Satz 1 von [16] und Korollar 2 von [8] erhalten wir

SATZ 1.1. *Ist* $\{x_1, \cdots, x_n\} \subset G$ *ein endliches System, so ist* $\{x_1, \cdots, x_n\}$ *frei äquivalent zu einem System* $\{y_1, \cdots, y_n\}$, *für das einer der folgenden Fälle vorliegt*:

(i) *Für jedes* $w \in \langle y_1, \cdots, y_n \rangle$ *gibt es eine Darstellung* $w = \prod_{i=1}^{q} y_{\nu_i}^{\varepsilon_i}$, $\varepsilon_i = \pm 1$, $\varepsilon_i = \varepsilon_{i+1}$ *falls* $\nu_i = \nu_{i+1}$ *mit* $L(y_{\nu_i}) \leq L(w)$ *für* $i = 1, \cdots, q$.

(ii) *Es gibt ein Produkt* $a = \prod_{i=1}^{q} y_{\nu_i}^{\varepsilon_i}$, $a \neq 1$, $y_{\nu_i} \in A\,(i = 1, \cdots, q)$ *und in einem Faktor* H_j *ein Element* $x \notin A$ *mit* $xax^{-1} \in A$.

(iii) *Einige der* y_i *liegen in einer zu* H_1 *oder* H_2 *konjugierten Untergruppe von* G, *nicht alle liegen in* A, *und ein Produkt in ihnen ist zu einem von* 1 *verschiedenen Element aus* A *konjugiert.*

BEMERKUNGEN:

1) Der freie Übergang kann in endlich vielen Schritten so gewählt werden, dass $\{y_1, \cdots, y_n\}$ kürzer ist als $\{x_1, \cdots, x_n\}$ oder die Längen der Elemente von $\{x_1, \cdots, x_n\}$ erhalten bleiben.

2) Ist $\{x_1, \cdots, x_n\}$ ein Erzeugendensystem von G, so folgt aus Fall (i): $L(y_i) \leq 1$ für $i = 1, \cdots, n$.

Im weiteren benötigen wir noch die folgende, geringfügige Verallgemeinerung von Satz 1 aus [6]:

SATZ 1.2. *Sei* $H = \langle s_1, \cdots, s_m | s_1^{\gamma_1} = \cdots = s_m^{\gamma_m} = 1 \rangle$, $m \geq 2$, $\gamma_i \geq 2$. *Sei* $\{x_1, \cdots, x_n\} \subset H(n < m)$ *und* X *die von* $\{x_1, \cdots, x_n\}$ *erzeugte Untergruppe von* H. *Gilt* $y(s_1 \cdots s_m)^a y^{-1} \in X$ *für ein* $a \neq 0$, $y \in H$, *so tritt einer der folgenden Fälle ein:*

(a) *Es gibt einen freien Übergang von* $\{x_1, \cdots, x_n\}$ *zu einem System* $\{y_1, \cdots, y_n\}$ *mit* $y_1 = z\, s_i^{\delta_i} z^{-1}$ *für ein* $i(1 \leq i \leq m)$, $2 \leq \delta_i < \gamma_i, (\delta_i, \gamma_i) \geq 2$ *und* $z \in H$.

(b) *Es gibt einen freien Übergang von* $\{x_1, \cdots, x_n\}$ *zu einem System, in dem ein Element zu einer Potenz von* $s_1 \cdots s_m$ *konjugiert ist.*

(c) *Es ist* m *ungerade,* $n = m-1$, *alle* $\gamma_i = 2$. *Ferner gilt es einen freien Übergang von* $\{x_1, \cdots, x_n\}$ *zu* $\{s_1 s_2, \cdots, s_1 s_m\}$, *und es liegt* $s_1 \cdots s_m$ *nicht in* X, *aber die geraden Potenzen von* $s_1 \cdots s_m$.

Sei nun F freie Gruppe vom Rang $p \geq 1$ mit den freien Erzeugenden $a_1, \cdots, a_p$ und $P(a_1, \cdots, a_p) \,. = a_1^{a_1} \cdots a_n^{a_n} [a_{n+1}, a_{n+2}] \cdots [a_{p-1}, a_p]$, $0 \leq n \leq p$, $a_i \geq 2$, ein alternierendes Produkt in F.

SATZ 1.3 ([12]). *Sei* $\{x_1, \cdots, x_r\} \subset F$ $(r \geq 1)$ *und* X *die von* $x_1, \cdots, x_r$ *erzeugte Untergruppe von* F. *Gilt* $y\, P^a(a_1, \cdots, a_p) y^{-1} \in X$ *für ein* $a \neq 0$, $y \in F$, *so tritt einer der folgenden Fälle ein:*

(a) $\{x_1, \cdots, x_r\}$ *ist frei äquivalent zu einem System* $\{y_1, \cdots, y_r\}$ *mit* $y_1 = zP^\beta(a_1, \cdots, a_p)z^{-1}$, $\beta > 0$, $z \in F$.

(b) *Es ist* $r \geq p$, *und* $\{x_1, \cdots, x_r\}$ *ist frei äquivalent zu einem System* $\{y_1, \cdots, y_r\}$ *mit* $y_i = za_i^{\gamma_i} z^{-1}$, $1 \leq \gamma_i < a_i, \gamma_i | a_i\, (i = 1, \cdots, n), y_j = za_j z^{-1} (j = n+1, \cdots, p)$, $z \in F$.

KOROLLAR 1.4. *Sei* $p \geq 2$ *und* $\{z_1, \cdots, z_p, P^a(a_1, \cdots, a_p)\}$, $a \geq 2$, *ein Erzeugendensystem von* F. *Dann gibt es einen freien Übergang von* $\{z_1, \cdots, z_p, P^a(a_1, \cdots, a_p)\}$ *zu* $\{a_1, \cdots, a_p, P^a(a_1, \cdots, a_p)\}$, *bei dem* $P^a(a_1, \cdots, a_p)$ *nicht ersetzt zu werden braucht.*

Beweis. Ist $n = 0$, $p = 2$, so folgt die Behauptung analog wie bei Hilfssatz 5 von [16]. Sei nun $p \geq 3$ oder $n \geq 1$.

Ist $n = 0$, $p \geq 3$, so betrachten wir die Faktorisierung $F = H_1 \underset{A}{*} H_2$ mit $H_1 = \langle t, a_1, a_2; \rangle$, $H_2 = \langle a_3, \cdots, a_p; \rangle$ und $A = \langle t\, [a_1, a_2] \rangle =$

$<([a_3, a_4] \cdots [a_{p-1}, a_p])^{-1}>$. Ist $n \geq 1$, so betrachten wir die Faktorisierung $F = K_1 \underset{B}{*} K_2$ mit $K_1 = <t, a_1;>$, $K_2 = <a_2, \cdots, a_p;>$ und

$B = <ta_1^{a_1}> = <(a_2^{a_2} \cdots a_n^{a_n} [a_{n+1}, a_{n+2}] \cdots [a_{p-1}, a_p])^{-1}>$. Ist ein z_1

konjugiert zu einem $a_j^{\gamma_j}$, so ist $\gamma_j = \pm 1$ (dies folgt nach Einssetzen von $P^a(a_1, \cdots, a_p)$ mittels Satz 1 von [11]). Nun verkürzen wir $\{z_1, \cdots, z_p,$ $P^a(a_1, \cdots, a_p)\}$ und erhalten die Behauptung mittels Satz 1.1 und Satz 1.3.

BEMERKUNG. Ist $n = 0$ oder $a_i \leq 6$, $a_i \neq 5$ $(i=1, \cdots, n)$ falls $n \geq 1$, so gilt die Aussage von Korollar 1.4 auch für $a = 1$.

Im allgemeinen gilt diese Aussage aber für $a = 1$ nicht, wie folgendes Beispiel zeigt: $p = 2 = n$; $z_1 = a_1$, $z_2 = a_2^7$, $P(a_1, a_2) = a_1^5 \, a_2^5$.

KOROLLAR 1.5. *Sei* $K = <s, a_1, \cdots, a_p | s^\gamma = 1>$, $\gamma \geq 2$, $p \geq 2$. *Sei* $\{x_1, \cdots, x_r\} \subset K(r \leq p+1)$ *mit* $x_1 = P^a(a_1, \cdots, a_p)$, $a \geq 1$, *und* X *die von* $x_1, \cdots, x_r$ *erzeugte Untergruppe von* K. *Gilt* $y(sP^a(a_1, \cdots, a_p))^\beta y^{-1} \epsilon X$ *für ein* $\beta \neq 0$, $y \epsilon K$, *so tritt einer der folgenden Fälle ein:*

(a) $\{x_1, \cdots, x_r\}$ *ist frei äquivalent zu einem System* $\{y_1, \cdots, y_r\}$ *mit* $y_1 = x_1$ *und* $y_2 \epsilon <a_1, \cdots, a_p;>$.

(b) $\{x_1, \cdots, x_r\}$ *ist frei äquivalent zu einem System, in dem mindestens zwei Elemente in zu* $<s, P(a_1, \cdots, a_p)>$ *konjugierten Untergruppen von* K *liegen.*

(c) *Es ist* $r = p+1$, *und* $\{x_1, \cdots, x_r\}$ *ist frei äquivalent zu einem System* $\{y_1, \cdots, y_r\}$ *mit* $y_i = za_i^{\gamma_i} z^{-1}$, $1 \leq \gamma_i < a_i, \gamma_i | a_i (i=1, \cdots, n)$, $y_j = za_j z^{-1} (j = n+1, \cdots, p)$, $y_r = x_1$.

Beweis. Wir betrachten die Faktorisierung

$$K = H_1 \underset{A}{*} H_2$$

mit

$H_1 = <t, s | s^\gamma = 1>$, $H_2 = <a_1, \cdots, a_p>$ und $A = <ts> = <(P(a_1, \cdots, a_p))^{-1}>$.

Nun verkürzen wir $\{x_1, \cdots, x_r\}$. Da x_1 schon die Länge null hat, bleibt es unverändert erhalten. Die Behauptung folgt nun mittels Satz 1.1 und Satz 1.3. q.e.d.

BEMERKUNG. Fall (c) von Korollar 1.5 kann nur für $\gamma = 2$ eintreten.

§2. *Über Erzeugendensysteme ebener diskontinuierlicher Gruppen*

In diesen Paragraphen beschäftigen wir uns mit Gruppen

$$G = \langle s_1, \cdots, s_m, a_1, \cdots, a_p | s_1^{\gamma_1} = \cdots = s_m^{\gamma_m} = s_1 \cdots s_m P^\gamma(a_1, \cdots, a_p) = 1 \rangle,$$

wobei $\gamma_i \geq 2$, $\gamma \geq 1$, $P(a_1, \cdots, a_p) = a_1^{\alpha_1} \cdots a_n^{\alpha_n} [a_{n+1}, a_{n+2}] \cdots [a_{p-1}, a_p]$ ein alternierendes Produkt mit $a_j \geq 2$, $p \geq 0$, $0 \leq n \leq p$ und $m \geq 3$ für $p = 0$ sowie $p \geq 2$ und $\gamma = 1$ für $m \leq 1$. Diese Gruppen sind deshalb von Interesse, weil sie als Spezialfälle die ebenen diskontinuierlichen Gruppen mit kompaktem Fundamentalbereich und ohne Spiegelungen enthalten. Ist $p = 0$, so ist G stets ebene diskontinuierliche Gruppe. Die Lösung des Rangproblems wird gegeben durch

SATZ 2.1 ([6]). G *hat den Rang*

(a) p *falls* $m = 0$, $p \geq 2$;

(b) $m - 2$ *falls* $p = 0$, m *gerade und alle* $\gamma_i = 2$ *bis auf eines, das* *ungerade ist*;

(c) $p + m - 1$ *in allen unter* (a) *und* (b) *nicht erfassten Fällen.*

Der Fall (b) von Satz 2.1 ist unerwartet; hier unterscheidet sich der Rang von dem, den man aus geometrischen Grunden erwartet:

Ist G eine ebene diskontinuierliche Gruppe, die kompakten Fundamentalbereich besitzt und keine Spiegelungen enthält, so gibt es zu einem zusammenhängenden abgeschlossenen Fundamentalbereich F von G mindestens p (für $m = 0$) bzw. $p + m - 1$ (für $m \geq 1$) Paare $x, x^{-1} \epsilon G$, für welche $xF \cap F$ eindimensional ist (vgl. [6]). Satz 2.1 sagt also, dass der Rang einer ebenen diskontinuierlichen Gruppe nicht mit dem geometrischen Rang zusammen fallen muss.

In diesen Paragraphen fragen wir nach einer Kennzeichnung der Erzeugendensysteme $\{x_1, \cdots, x_{r(G)}\}$, wobei $r(G)$ der Rang von G ist. Der Fall $m \leq 1$ wird in [9] behandelt. Hier beschäftigen wir uns mit dem Fall $m \geq 2$.

SATZ 2.2. *Sei* $p \geq 2$ *und* $m \geq 2$. *Ist* $\{x_1, \cdots, x_{p+m-1}\}$ *ein Erzeugendensystem von* G, *so gibt es einen freien Übergang von* $\{x_1, \cdots, x_{p+m-1}\}$ *zu einem System* $\{s_{\nu_1}^{\beta_1}, \cdots, s_{\nu_{m-1}}^{\beta_{m-1}}, a_1, \cdots, a_p\}$ *mit* $\nu_i \in \{1, \cdots, m\}$, $\nu_1 < \cdots < \nu_{m-1}$ *(falls* $m > 2$) *und* $1 \leq \beta_i < \gamma_{\nu_i}, (\beta_i, \gamma_{\nu_i}) = 1$.

Beweis. G lässt sich schreiben als freies Produkt $G = H_1 \underset{A}{*} H_2$ mit Amalgam, wobei $H_1 = \langle s_1, \cdots, s_m | s_1^{\gamma_1} = \cdots = s_m^{\gamma_m} = 1 \rangle$, $H_2 = \langle a_1, \cdots, a_p; \rangle$ und $A = \langle a \rangle = \langle s_1 \cdots s_m \rangle = \langle P^{-\gamma}(a_1, \cdots, a_p) \rangle$.

Sei $\{x_1, \cdots, x_{p+m-1}\}$ ein Erzeugendensystem von G. Ist x_1 konjugiert zu einem $s_j^{\beta_j}$, so ist $(\beta_j, \gamma_j) = 1$ nach Satz 2.1. Ist x_1 konjugiert zu einem $a_j^{\beta_j}$, $1 \leq j \leq n$, mit $1 \leq \beta_j < a_j, \beta_j | a_j$, so ist $\beta_j = 1$ ebenfalls nach Satz 2.1. Ist $m = 2$ und $\gamma_1 = \gamma_2 = 2$, so ist x_1 nicht konjugiert zu einer Potenz von a, da sonst das freie Produkt

$$\langle s | s^2 = 1 \rangle * \langle a_1, \cdots, a_p | P^{\gamma}(a_1, \cdots, a_p) = 1 \rangle$$

einen Rang $\leq p$ hätte (vgl. Satz 2.1). Letzteres besagt, dass nach Hilfssatz 2 von [16] eine Situation (ii) von Satz 1.1 für kein zu $\{x_1, \cdots, x_{p+m-1}\}$ frei äquivalentes System eintreten kann. Nun verkürzen wir $\{x_1, \cdots, x_{p+m-1}\}$. Nach Satz 1.1 und den obigen Ausführungen können wir annehmen, dass einer der folgenden Fälle vorliegt:

(2.3): Es ist $L(x_j) \leq 1$ für $j = 1, \cdots, p+m-1$.

(2.4): Es liegen $p, p \geq 2$, der x_j in einer zu H_1 oder H_2 konjugierten Untergruppen von G, und ein Produkt in ihnen ist zu einem von 1 verschiedenen Element aus A konjugiert.

Liegt (2.3) vor, so liegen notwendig p der x_j in H_2 und $m-1$ der x_j in H_1. Da H_1 den Rang m hat, müssen bei der Darstellung mindestens eines s_i solche x_j verwandt werden, die in H_2 liegen. Somit muss sich eine von 1 verschiedene Potenz von $s_1 \cdots s_m$ als

Produkt von solchen x_j schreiben lassen, die in H_2 liegen; d. h. es liegt (2.4) vor. Wir können uns daher auf die Untersuchung von (2.4) beschränken. Es liege also (2.4) vor.

BEH (2.5). Es gibt einen freien Übergang von $\{x_1, \cdots, x_{p+m-1}\}$ zu einem system $\{y_1, \cdots, y_{m-1}, a_1, \cdots, a_p\}$.

Beweis von (2.5). Angenommen es gibt keinen freien Übergang von $\{x_1, \cdots, x_{p+m-1}\}$ zu einem System $\{y_1, \cdots, y_{m-1}, a_1, \cdots, a_p\}$. Nach Satz 1.2 und Satz 1.3 tritt dann einer der folgenden Fälle ein (es können nach Satz 2.1 nicht mehr als $m-1$ Elemente in einer zu H_1 konjugierten Untergruppe liegen):

(2.6): Es ist m ungerade, $y_1 = \cdots = y_m = 2$ und es gibt einen freien Übergang von $\{x_1, \cdots, x_{p+m-1}\}$ zu einem System $\{s_1 s_2, \cdots, s_1 s_m, z_1, \cdots, z_p\}$.

(2.7): Es gibt einen freien Übergang von $\{x_1, \cdots, x_{p+m-1}\}$ zu einem System, in dem ein Element zu einer Potenz von $s_1 \cdots s_m = a$ konjugiert ist.

Zu (2.6): Sei $H = <s_1 s_2, \cdots, s_1 s_m>$. Es ist $|H_1 : H| = 2$, und es lassen sich mittels $s_1 s_2, \cdots, s_1 s_m$ nur Konjugierte von Potenzen des Quadrates $(s_1 \cdots s_m)^2$ darstellen. Also ist $a = s_1 \cdots s_m$ ein Restklassenvertreter von H_1 nach H, und es ist $H_1 = H + aH$. Wir dürfen daher annehmen, dass $z_1, \cdots, z_p$ mit einem Element aus $H_1 - \{a, a^{-1}\}$ weder beginnen noch enden, und es zwischen ihnen keine verkürzenden freien Übergänge mehr gibt (Fall (iii) von Satz 1.1 ist wegen Satz 2.1 und nach Annahme in zu H_2 konjugierten Untergruppen nicht möglich). Dann können grössere Verkürzungen zwischen den $z_1, \cdots, z_p$ nicht durch $s_1 s_2, \cdots, s_1 s_m$ bewirkt werden. Wegen $a \in H_2$ muss daher $L(z_j) \leq 1$, $j = 1, \cdots, p$, sein. Ferner ist $H_2 = <a^2, z_1, \cdots, z_p>$. Nach Korollar 1.4 gibt es einen freien Übergang von $\{a^2, z_1, \cdots, z_p\}$ zu $\{a^2, a_1, \cdots, a_p\}$, der sich zu einem von $\{z_1, \cdots, z_p, s_1 s_2, \cdots, s_1 s_m\}$ zu $\{s_1 s_2, \cdots, s_1 s_m, a_1, \cdots, a_p\}$ erweitern lässt. Dies widerspricht aber unserer Annahme.

Zu (2.7): Sei ohne Einschränkung $x_1 = a^\delta = (s_1 \cdots s_m)^\delta$, $\delta \geq 1$. Es ist $m \geq 3$, denn für $m = 2$ erhalten wir mit Satz 1.1 und Satz 1.3 sofort einen freien Übergang von $\{x_1, \cdots, x_{p+1}\}$ zu einem System $\{a_1, \cdots, a_p, y_1\}$ (mit Hilfe von Satz 2.1 ergibt sich sofort, dass kein x_i, $2 \leq i \leq p+1$, in derselben zu H_2 konjugierten Untergruppe liegen kann wie x_1).

Sei nun $m \geq 3$. Das freie Produkt

$$\langle s_1, \cdots, s_m | s_i^{\gamma_i} = s_1 \cdots s_m = 1 \rangle * \langle a_1, \cdots, a_p | P^\gamma(a_1, \cdots, a_p) = 1 \rangle$$

hat einen Rang $\leq p + m - 2$. Nach Satz 2.1 folgt daher: m ist gerade und alle γ_i sind gleich 2 bis auf eines, das ungerade ist. Sei etwa ohne Einschränkung $\gamma_m = 2$. Wir betrachten die Faktorisierung $G = K_1 \underset{B}{*} K_2$ mit

$$K_1 = \langle s_1, \cdots, s_{m-1} | s_i^{\gamma_i} = 1 \rangle,$$

$$K_2 = \langle s_m, a_1, \cdots, a_p | s_m^2 = 1 \rangle$$

und

$$B = \langle b \rangle = \langle s_1 \cdots s_{m-1} \rangle = \langle s_m P^\gamma(a_1, \cdots, a_p))^{-1} \rangle.$$

Nun verkürzen wir $\{x_1, \cdots, x_{p+m-1}\}$ bzgl. dieser Faktorisierung.

Da $x_1 = (s_1 \cdots s_m)^\delta = (P^{\gamma\delta}(a_1, \cdots, a_p))^{-1}$ die Länge 1 hat, bleibt es unverändert erhalten (vgl. Satz 2.1). Fall (a) von Korollar (1.5) kann nicht eintreten, da kein x_i, $2 \leq i \leq p+m-1$, in derselben zu $H_2 = \langle a_1, \cdots, a_p \rangle$ konjugierten Untergruppe liegen kann wie x_1 (dies folgt mit Hilfe von Satz 2.1; es genügt hierbei, den Fall $p = 2$ zu untersuchen). Fall (b) von Korollar (1.5) kann nicht eintreten, da das freie Produkt

$$\langle s_1, \cdots, s_{m-1} | s_i^{\gamma_i} = s_1 \cdots s_{m-1} = 1 \rangle * \langle a_1, \cdots, a_p | P^\gamma(a_1, \cdots, a_p) = 1 \rangle$$

den Rang $p + m - 2$ hat. Mit derselben Begründung kann es nach Satz 1.2 nicht sein, dass einige der x_j $(2 \leq j \leq p+m-1)$ in einer zu K_1 konjugierten Untergruppe von G liegen und ein Produkt in ihnen zu einem von 1 verschiedenen Element von B konjugiert ist. Da andererseits wieder eine

Situation (iii) von Satz 1.1 eintreten muss, liegt also Fall (c) von Korollar 1.5 vor; d.h. es gibt einen freien Übergang von $\{x_1, \cdots, x_{p+m-1}\}$ zu einem System $\{x_1, z_2, \cdots, z_{m-1}, za_1z^{-1}, \cdots, za_pz^{-1}\}$, $z \in G$. Dies widerspricht aber unserer Annahme. Also gilt Behauptung (2.5).

Damit können wir $x_{m-1+i} = a_i$, $i = 1, \cdots, p$, annehmen. Wir betrachten wieder die alte Faktorisierung $G = H_1 \underset{A}{*} H_2$ und verkürzen $\{x_1, \cdots, x_{m-1}, a_1, \cdots, a_p\}$ bzgl. dieser Faktorisierung.

Tritt eine Situation (iii) von Satz 1.1 ein, an der einige der x_i, $1 \leq i \leq m-1$, beteiligt sind, so tritt—analog wie bei (2.5)—einer der folgenden Fälle ein:

(2.8): Es ist m ungerade, $\gamma_1 = \cdots = \gamma_m = 2$, und es gibt einen freien Übergang von $\{x_1, \cdots, x_{m-1}, a_1, \cdots, a_p\}$ zu einem system $\{zs_1s_2z^{-1}, \cdots, zs_1s_mz^{-1}, a_1, \cdots, a_p\}$, $z \in G$.

(2.9): Es ist m gerade, alle γ_i gleich zwei bis auf eines, das ungerade ist, und es gibt einen freien Übergang von $\{x_1, \cdots, x_{m-1}, a_1, \cdots, a_p\}$ zu einem System $\{y_1, \cdots, y_{m-2}, za^{\delta}z^{-1}, a_1, \cdots, a_p\}$, $\delta \geq 1$, $z \in G$.

Zu (2.8): Man überlegt sich leicht, dass wir $zs_1s_iz^{-1} \in H_1$, $i = 2, \cdots, m$, annehmen können. Es ist notwendig $H_1 = \langle a, zs_1s_2z^{-1}, \cdots, zs_1s_mz^{-1} \rangle$. Nun folgt die Behauptung mit Hilfe von Satz 1 aus [13].

Zu (2.9): Für $m = 2$ können wir $za^{\delta}z^{-1} \in H_1$ annehmen, und die Behauptung folgt mit Hilfe von Satz 2 aus [13]. Sei nun $m \geq 3$. Sei diesmal ohne Einschränkung $\gamma_m \geq 3$. Betrachten wir die Faktorisierung $K_1 \underset{B}{*} K_2$ mit

$$K_1 = \langle s_1, \cdots, s_{m-1} \,|\, s_i^2 = 1 \rangle,$$

$$K_2 = \langle s_m, a_1, \cdots, a_p \,|\, s_m^{\gamma_m} = 1 \rangle$$

und

$$B = \langle s_1 \cdots s_{m-1} \rangle = \langle (s_m P^{\gamma}(a_1, \cdots, a_p))^{-1} \rangle,$$

so erhalten wir nach Satz 1.1 und Satz 1.2 einen freien Übergang von $\{y_1, \cdots, y_{m-2}, za^\delta z^{-1}, a_1, \cdots, a_p\}$ zu einem System $\{xs_1 s_2 x^{-1}, \cdots,$ $xs_1 s_{m-1} x^{-1}, ya^\delta y^{-1}, a_1, \cdots, a_p\}$, $x, y \in G$. Wieder können wir entsprechend $xs_1 s_i x^{-1} \in H_1$, $i = 2, \cdots, m-1$, und $ya^\delta y^{-1} \in H_1$ annehmen. Es ist notwendig $H_1 = \langle xs_1 s_2 x^{-1}, \cdots, xs_1 s_{m-1} x^{-1}, ya^\delta y^{-1}, a \rangle$. Nun folgt die Behauptung mit Hilfe von Satz 2 aus [13].

Tritt keine Situation (iii) von Satz 1.1 ein, an der einige der x_i, $1 \leq i \leq m-1$, beteiligt sind, so dürfen wir annehmen, dass $x_1, \cdots, x_{m-1}$ mit einem Element aus H_2 weder beginnen noch enden und es zwischen ihnen keine verkürzenden freien Übergänge mehr gibt. Dann können grössere Verkürzungen zwischen den $x_1, \cdots, x_{m-1}$ nicht durch $a_1, \cdots, a_p$ bewirkt werden. Es muss daher $L(x_i) \leq 1$, $i = 1, \cdots, m-1$, sein.

Ferner ist $H_1 = \langle x_1, \cdots, x_{m-1}, a \rangle$. Nach Satz 1 von [13] gibt es einen freien Übergang von $\{x_1, \cdots, x_{m-1}, a\}$ zu einem System $\{s_{\nu_1}^{\beta_1}, \cdots,$ $s_{\nu_{m-1}}^{\beta_{m-1}}, a\}$, $\nu_i \in \{1, \cdots, m\}$, $\nu_1 < \cdots < \nu_{m-1}$ (falls $m > 2$) und $1 \leq \beta_i < \alpha_{\nu_i}$, $(\beta_i, \gamma_{\nu_i}) = 1$, der sich zu einem von $\{x_1, \cdots, x_{m-1}, a_1, \cdots, a_p\}$ zu $\{s_{\nu_1}^{\beta_1}, \cdots, s_{\nu_{m-1}}^{\beta_{m-1}}, a_1, \cdots, a_p\}$ erweitern lässt. q.e.d.

KOROLLAR 2.10. *Sei* $p \geq 2$ *und* $m \geq 2$. *Ist* $\{x_1, \cdots, x_{p+m-1}\}$ *ein Erzeugendensystem von* G *mit*

$$x_i = c_i s_{\nu_i}^{\beta_i} c_i^{-1}, \quad \nu_i \in \{1, \cdots, m\}, \quad i = 1, \cdots, m-1,$$

so gibt es einen freien Übergang von

$$\{x_1, \cdots, x_{p+m-1}\} \text{ zu } \{s_{\nu_1}^{\beta_1}, \cdots, s_{\nu_{m-1}}^{\beta_{m-1}}, a_1, \cdots, a_p\}.$$

Beweis. Wir betrachten die Faktorisierung $H_1 \underset{A}{*} H_2$ wie beim Beweis von Satz 2.2. Nun folgt die Behauptung entsprechend wie dort, wenn wir folgendes berücksichtigen: Ist einmal $L(x_j^\varepsilon c_i s_{\nu_i}^{\eta\beta_i} c_i^{-1}) < L(c_i s_{\nu_i}^{\beta_i} c_i^{-1})$,

$i \in \{1, \cdots, m-1\}$, $i \neq j$; $\varepsilon, \eta = \pm 1$, so ist auch $L(x_j^\varepsilon c_i s_{\nu_i}^{\eta\beta_i} c_i^{-1} x_j^{-\varepsilon}) <$

$L(c_i s_{\nu_i}^{\beta_i} c_i^{-1})$, oder es liegen x_j und $c_i s_{\nu_i}^{\beta_i} c_i^{-1}$ in einer zu H_1 oder H_2

konjugierten Untergruppe und $x_j^\varepsilon c_i s_{\nu_i}^{\eta\beta_i} c_i^{-1}$ ist zu einem von 1 ver-

chiedenen Element von A konjugiert. q.e.d.

BEMERKUNG. Für $p \leq 1$, $m \geq 2$ und Rang $(G) = p+m-1$ gilt eine
Aussage wie in Satz 2.2 im allgemeinen nicht (vgl. [8]). Es kann hier

Erzeugendensysteme $\{x_1, \cdots, x_{p+m-1}\}$ geben mit $x_1 = s_1^{\beta_1}$, $(\beta_1, a_1) > 1$

oder $x_1 = s_1^{\beta_1}, x_2 = g s_1^{\beta_2} g^{-1}$, $g \in G$, oder $x_1 = a_1^\beta$, $\beta \geq 2$. Indem wir
analog wie beim Beweis von Korollar 2.10 und Satz 2.2 vorgehen, erhalten
wir aber:

KOROLLAR 2.11.

1) *Sei* $p = 1$, $m \geq 2$. *Ist* $\{x_1, \cdots, x_m\}$ *ein Erzeugendensystem von* G
mit
$$x_i = c_i s_{\nu_i}^{\beta_i} c_i^{-1}, \quad \nu_i \in \{1, \cdots, m\}, \quad i = 1, \cdots, m-1,$$
so gibt es einen freien Übergang von

$$\{x_1, \cdots, x_m\} \quad zu \quad \{s_{\nu_1}^{\beta_1}, \cdots, s_{\nu_{m-1}}^{\beta_{m-1}}, a_1\}.$$

2) *Sei* $p = 0$, $m \geq 4$ *und Rank* $G = m-1$. *Ist* $\{x_1, \cdots, x_{m-1}\}$ *ein
Erzeugendensystem von* G *mit*

$$x_i = c_i s_{\nu_i}^{\beta_i} c_i^{-1}, \quad \nu_i \in \{1, \cdots, m\}, \quad i = 1, \cdots, m-1,$$

so gibt es einen freien Übergang von

$$\{x_1, \cdots, x_{m-1}\} \quad zu \quad \{s_{\nu_1}^{\beta_1}, \cdots, s_{\nu_{m-1}}^{\beta_{m-1}}\}.$$

Sowohl in 1) als auch in 2) gilt $(\beta_i, a_{\nu_i}) = 1$.

BEMERKUNGEN:

1) Weitere Resultate zur Kennzeichnung der Erzeugendensysteme $\{x_1, \cdots, x_{m-1}\}$ von G mit $p = 0$, $m \geq 6$ und alle $\gamma_i \geq 3$ werden in [5] gegeben.

2) Für $p = 0$, $m = 3$ gilt eine Aussage wie in Korollar 2.11 im allgemeinen nicht (vgl. [2] und [10]). Hier sind die Erzeugendenpaare durch Satz 4 von [10] gegeben.

SATZ 2.12. *Sei* $p = 0$, $m \geq 4$ *und Rang* $(G) = m-2$. *Dann gibt es genau eine Nielsen-Aquivalenz-Klasse minimaler Erzeugendensysteme.*
Konkreter: Sei $G = \langle s_1, \cdots, s_m | s_1^2 = \cdots = s_{m-1}^2 = s_m^{\gamma} = s_1 \cdots s_m = 1\rangle$, $m \geq 4$ *gerade,* $\gamma \geq 3$ *ungerade. Ist* $\{x_1, \cdots, s_{m-2}\}$ *ein Erzeugendensystem von* G, *so gibt es einen freien Übergang von* $\{x_1, \cdots, x_{m-2}\}$ *zu* $\{s_1 s_2, \cdots, s_1 s_{m-1}\}$.

Beweis. Für $m = 4$ folgt dies aus [1]. Sei nun $m \geq 6$. Wir schreiben G als freies Produkt $H_1 \underset{A}{*} H_2$ mit Amalgam, wobei $H_1 = \langle s_1, \cdots, s_{m-3} | s_i^2 = 1\rangle$,
$H_2 = \langle s_{m-2}, s_{m-1}, s_m | s_{m-2}^2 = s_{m-1}^2 = s_m^{\gamma} = 1\rangle$ und $A = \langle a\rangle = \langle s_1 \cdots s_{m-3}\rangle = \langle(s_{m-2} s_{m-1} s_m)^{-1}\rangle$. Nun erhalten wir mit Hilfe von Satz 1.1, Satz 1.2 und Satz 2.1 einen freien Übergang von $\{x_1, \cdots, x_{m-2}\}$ zu einem System $\{s_1 s_2, \cdots, s_1 s_{m-3}, y_1, y_2\}$. Sei $H = \langle s_1 s_2, \cdots, s_1 s_{m-3}\rangle$. Es ist $|H_1 : H| = 2$, und es lassen sich mittels $s_1 s_2, \cdots, s_1 s_{m-3}$ nur Konjugierte des Quadrates $(s_1 \cdots s_{m-3})^2$ darstellen. Also ist $a = s_1 \cdots s_{m-3}$ ein Restklassenvertreter von H_1 nach H, und es ist $H_1 = H + aH$. Nach Satz 1.2 und Satz 2.1 gibt es keine Situation (iii) von Satz 1.1, an der y_1, y_2 beteiligt sind. Wir dürfen annehmen, dass y_1, y_2 mit einem Faktor aus $H_1 - \{a, a^{-1}\}$ weder beginnen noch enden, und es zwischen ihnen keine verkürzenden freien Übergänge mehr gibt. Dann können

grössere Verkürzungen zwischen y_1 und y_2 nicht durch $s_1 s_2, \cdots, s_1 s_{m-3}$ bewirkt werden. Wegen $a \in H_2$ muss daher $L(y_i) \leq 1$, $i = 1, 2$, sein. Es ist sogar $L(y_i) = 1$ nach Satz 2.1 (Einssetzen von $s_1 \cdots s_{m-3}$). Weiter ist notwendig $H_2 = \langle a^2, y_1, y_2 \rangle$. Nach Satz 1 von [13] gibt es einen freien Übergang von $\{a^2, y_1, y_2\}$ zu $\{a^2, s_{m-1} s_{m-2}, s_{m-1} s_m\}$, der sich zu einem freien Übergang von $\{s_1 s_2, \cdots, s_1 s_{m-3}, y_1, y_2\}$ zu $\{s_1 s_2, \cdots, s_1 s_{m-3}, s_{m-1} s_{m-2}, s_{m-1} s_m\}$ erweitern lässt. Dieses System ist aber frei äquivalent zu $\{s_1 s_2, \cdots, s_1 s_{m-1}\}$. q.e.d.

§3. *Automorphismen ebener diskontinuierlicher Gruppen Zusammenfassung*

Sei G eine Gruppe wie in §2 gegeben, für die zusätzlich
$$\frac{1}{\gamma_1} + \frac{1}{\gamma_2} + \frac{1}{\gamma_3} < 1 \quad \text{falls} \quad p = 0 \quad \text{und} \quad m = 3.$$
Ist ϕ ein Automorphismus von G, so ist $\phi(s_i)$ für jedes $i \in \{1, \cdots, m\}$ als Element endlicher Ordnung konjugiert zu einer Potenz von einem s_j, $j \in \{1, \cdots, m\}$ (vgl. [17]).

SATZ 3.1. *Ist* $r(G)$ *der Rang von* G, *so wird jeder Automorphismus von* G *von einem Automorphismus der freien Gruppe vom Rang* $r(G)$ *induziert.*

Beweis.

1) Ist $m = 0$ und $p = 2$, so folgt die Aussage von 3.1 aus [14]. Ist $m = 0$ und $p \geq 3$, so folgt die Aussage von (3.1) mittels Satz 1.3 und Satz 3 von [9] (vgl. auch [4] und [16]); es gibt nur endlich viele Nielsen-äquivalenzklassen minimaler Erzeugendensysteme $\{x_1, \cdots, x_p\}$ und zu jedem System gibt es eine Präsentierung von G mit einer definierenden Relation. Ist $m = 1$ und $p \geq 2$, so folgt die Aussage von (3.1) aus Satz 2 von [9] (vgl. auch [7] für $p = 2$ und [11] für $p \geq 3$). Aus Theorem N 5 von [3] und Satz 2 von [9] erhalten wir darüberhinaus: Ist $m = 1$, $p \geq 2$ und ϕ ein Automorphismus von G, so ist $\phi(s_1)$ konjugiert zu s_1^ε, $\varepsilon = \pm 1$.

2) Sei nun $m \geq 2$, $p \geq 2$ und ϕ ein Automorphismus von G.

Ist $\phi(s_1) = c s_i^{\beta_i} c^{-1}$, $c \in G$, $i \in \{1, \cdots, m\}$, so ist $\gamma_i = \gamma_1$ und $\beta_i \equiv \pm 1 \pmod{\gamma_i}$. Denn: Wir können ohne Einschränkung $i = 1$ annehmen (Sei $\phi(s_j) = d s_k^{\beta_k} d^{-1}$; es gibt einen Automorphismus ψ von G mit $\psi(s_k) = d_1 s_j d_1^{-1}$; und es ist $\phi \circ \psi(s_k) = d_2 s_k^{\beta_k} d_2^{-1}$).

Sei also nun $i = 1$. Führen wir die Relationen $s_2 = \cdots = s_m = 1$ ein,

so induziert ϕ einen Automorphismus $\bar{\phi}$ der Gruppe $\langle s_1, a_1, \cdots, a_p | s_1^{\gamma_1} = s_1 P^\gamma(a_1, \cdots, a_p) = 1 \rangle$ mit $\bar{\phi}(s_1) = c' s_1^{\beta_1} c'^{-1}$. Nach den obigen Ausführungen in 1) ergibt sich $\beta_1 \equiv \pm 1 \pmod{\gamma_1}$. Induktiv folgt: Ist $\phi(s_k) = c_k s_j^{\beta_j} c_k^{-1}$, $c_k \in G$, $k, j \in \{1, \cdots, m\}$, so ist $\gamma_k = \gamma_j$ und $\beta_j \equiv \pm 1 \pmod{\gamma_j}$. Nun folgt für $m \geq 2$ und $p \geq 2$ die Aussage von (3.1) aus Satz 2.2 und Korollar 2.10.

3) Sei nun $m \geq 2$, $p = 1$ und ϕ ein Automorphismus von G.

Sind $m - 1$ der $\gamma_i \leq 4$, so folgt die Aussage von (3.1) unmittelbar aus Korollar 2.11. Seien nun wenigstens zwei der $\gamma_i \geq 5$.

a) Sei $m = 2$. Dann ist $\gamma_1, \gamma_2 \geq 5$. Analog wie in 2) können wir $\phi(s_1) = s_1^{\beta_1}$, $\phi(s_2) = c s_2^{\beta_2} c^{-1}$, $c \in G$, annehmen, um $\beta_i \equiv \pm 1 \pmod{\gamma_i}$ nachzuweisen.

Wir setzen $\phi(a_1) = b_1$; es ist $b_1^{\alpha_1} s_1^{\beta_1} = c s_2^{-\beta_2} c^{-1}$. Nun betrachten wir die Faktorisierung $G = H_1 \underset{A}{*} H_2$ mit $H_1 = \langle s_1, s_2 | s_1^{\gamma_1} = s_2^{\gamma_2} = 1 \rangle$,

$H_2 = \langle a_1 ; \rangle$ und $A = \langle s_1 s_2 \rangle = \langle a_1^{-\alpha_1} \rangle$ sowie das Erzeugendensystem $\{s_1^{\beta_1}, b_1\}$. Wegen $\gamma_1, \gamma_2 \geq 5$ und Korollar 2.11 kann keine Situation (ii) eintreten. Die Auswahl der Linksrestklassenvertreter der Gruppen H_i nach A sei so durchgeführt, dass $s_1^{\beta_1}$ selbst einer ist. Seien

$$c s_2^{-\beta_2} c^{-1} = t_1 \cdots t_{m_1} k_1 t_{m_1}^{-1} \cdots t_1^{-1}, \quad k_1 \in H_1, \quad \text{und} \quad b_1 = 1_1 \cdots 1_{m_2} k_2 r_{m_2} \cdots r_1$$

die (symmetrischen) Normalformen von $c s_2^{-\beta_2} c^{-1}$ und b_1 (vgl. [8] und [16]). Wir können ohne Einschränkung $t_1^{-1} \neq s_1^{\beta_1}$ annehmen falls $m_1 \geq 1$. Weiter können wir annehmen, dass für $m_2 \geq 1$ die Restklassenvertreter 1_1 und r_1 in verschiedenen Faktoren H_i liegen (Für H_2 ist das klar, und für H_1 erreichen wir das eventuell nach einer geeigneten Konjugation und Umbenennung).

Es gilt die Gleichung

$$t_1 \cdots t_{m_1} k_1 t_{m_1}^{-1} \cdots t_1^{-1} = 1_1 \cdots 1_{m_2} k_2 r_{m_2} \cdots r_1 \cdots 1_1 \cdots 1_{m_2} k_2 r_{m_2} \cdots r_1 s_1^{\beta_1} .$$

Angenommen $m_1 \geq 1$ und $m_2 \geq 1$. Zwischen r_1 und 1_1 ist keine Kürzung möglich. Es ist $r_1 \epsilon H_1$, da aus $r_1 \epsilon H_2$ notwendig $1_1 = t_1$, $t_1^{-1} = s_1^{\beta_1}$ folgt (nach Wahl der Restklassenvertreter). Es ist weiter $r_1 s_1^{\beta_1} \epsilon A$, da aus $r_1 s_1^{\beta_1} \epsilon H_1 \setminus A$ notwendig $1_1 \epsilon H_1$ folgt. Also ist $r_1 s_1^{\beta_1} \epsilon A$ und damit sogar $r_1 = s_1^{-\beta_1}$ nach Wahl der Restklassenvertreter. Es folgt notwendig $m_1 \geq m_2$ sowie $1_1 \cdots 1_{m_2} = t_1 \cdots t_{m_2}$ und $r_{m_2} \cdots r_2 = t_{m_2-1}^{-1} \cdots t_1^{-1}$, d. h.

$$t_{m_2} \cdots t_{m_1} k_1 t_{m_1}^{-1} \cdots t_{m_2}^{-1} = t_{m_2} k_2 t_{m_2-1}^{-1} \cdots t_1^{-1} s_1^{-\beta_1} t_1 \cdots t_{m_2} k_2 t_{m_2-1}^{-1} \cdots t_1^{-1} s_1^{-\beta_1} \cdots t_1 \cdots t_{m_2} k_2 .$$

Es ist $k_2 \epsilon A$, da aus $k_2 \epsilon H_i \setminus A$ notwendig $t_{m_2} \epsilon H_i$ folgt. Damit ergibt sich $t_{m_2} \epsilon H_1$ aus der Lösung des Wortproblems, d. h.

$$b_1 = t_1 \cdots t_{m_2-1} k t_{m_2-1}^{-1} \cdots t_1^{-1} s_1^{-\beta_1}$$

mit $k \epsilon H_1 \setminus A$. Das ist aber ein Widerspruch zu $G = \langle s_1^{\beta_1}, b_1 \rangle$, da sonst die Gruppe $\langle a_1 | a_1^{\alpha_1} = 1 \rangle$ trivial wäre. Also ist $m_1 = 0$ oder $m_2 = 0$. In beiden Fällen folgt aus Korollar 2.11, dass wir $\phi(s_1) = s_1^{\beta_1}$, $\phi(s_2) = s_2^{\beta_2}$ und $b_1 = \phi(a_1) = a_1$ annehmen können. Aus $1 = s_1 s_2 a_1^{\alpha_1} = \phi(s_1 s_2 a_1^{\alpha_1}) = s_1^{\beta_1} s_2^{\beta_2} a_1^{\alpha_1}$ ergibt sich $\beta_i \equiv \pm 1 \pmod{\gamma_i}$. Nun folgt für $p = 1$ und $m = 2$ die Aussage von (3.2) aus Korollar 2.11.

b) Sei $m \geq 3$. Analog wie in 2) können wir $\phi(s_i) = c_i s_i^{\beta_i} c_i^{-1}$, $c_i \epsilon G$, $i = 1, 2$, annehmen, um $\beta_i \equiv \pm 1 \pmod{\gamma_i}$ nachzuweisen. Führen wir die Relationen $s_3 = \cdots = s_m = 1$ ein, so induziert ϕ einen Automorphismus $\overline{\phi}$ der Gruppe

$$\langle s_1, s_2, a_1 \mid s_1^{\gamma_1} = s_2^{\gamma_2} = s_1 s_2 a_1^{\alpha_1} = 1 \rangle \quad \text{mit} \quad \overline{\phi}(s_i) = c_i' s_i^{\beta_i} c_i'^{-1}.$$

Nach a) ergibt sich $\beta_i \equiv \pm 1 \pmod{\gamma_i}$.

Induktiv folgt: Ist $\phi(s_k) = c_k s_j^{\beta_j} c_k^{-1}$, $c_k \in G$, $k, j \in \{1, \cdots, m\}$, so ist $\gamma_k = \gamma_j$ und $\beta_j \equiv \pm 1 \pmod{\gamma_j}$. Nun folgt für $p = 1$ und $m \geq 3$ die Aussage von (3.1) aus Korollar 2.11.

4) Sei nun $m \geq 3$, $p = 0$, $r(G) = m-1$ und ϕ ein Automorphismus von G. Für $m = 3$ folgt die Aussage von (3.1) leicht mittels [2] und Satz 4 von [10]. Wir erhalten insbesondere: Ist $m = 3$ und $\phi(s_i) = c s_j^{\beta_j} c^{-1}$, $c \in G$, $i, j \in \{1, 2, 3\}$, so ist $\gamma_i = \gamma_j$ und $\beta_j \equiv \pm 1 \pmod{\gamma_j}$. Sei nun $m \geq 4$. Sind $m-1$ der $\gamma_i \leq 4$, so folgt die Aussage von (3.1) unmittelbar aus Korollar 2.11. Seien nun wenigstens zwei der $\gamma_i \geq 5$; etwa $\gamma_1 \geq 5$ und $\gamma_2 \geq 5$. Analog wie in 2) können wir $\phi(s_i) = c_i s_i^{\beta_i} c_i^{-1}$, $c_i \in G$, $i = 1, 2, 3$, annehmen, um $\beta_i \equiv \pm 1 \pmod{\gamma_i}$ nachzuweisen. Führen wir die Relationen $s_4 = \cdots = s_m = 1$ ein, so induziert ϕ einen Automorphismus $\overline{\phi}$ der Gruppe

$$\langle s_1, s_2, s_3 \mid s_i^{\gamma_i} = s_1 s_2 s_3 = 1 \rangle \quad \text{mit} \quad \overline{\phi}(s_i) = c_i' s_i^{\beta_i} c_i'^{-1}.$$

Wegen $\gamma_1 \geq 5$ und $\gamma_2 \geq 5$ ist $\dfrac{1}{\gamma_1} + \dfrac{1}{\gamma_2} + \dfrac{1}{\gamma_3} < 1$. Nun ist $\beta_i \equiv \pm 1 \pmod{\gamma_i}$, $i = 1, 2, 3$ nach den obigen Bemerkungen. Induktiv folgt: Ist $\phi(s_k) = c_k s_j^{\beta_j} c_k^{-1}$, $c_k \in G$, $j \in \{1, \cdots, m\}$, so ist $\gamma_k = \gamma_j$ und $\beta_j \equiv \pm 1 \pmod{\gamma_j}$. Nun folgt für $p = 0$, $m \geq 4$ und $r(G) = m-1$ die Aussage von (3.1) aus Korollar 2.11.

5) Sei nun $m \geq 3$, $p = 0$ und $r(G) = m-2$. Dann folgt die Aussage von (3.1) unmittelbar aus Satz 2.12. q.e.d.

BEMERKUNG. Für $\dfrac{1}{\gamma_1} + \dfrac{1}{\gamma_2} + \dfrac{1}{\gamma_3} > 1$ falls $m = 3$ ist G endlich, und die Aussage von Satz 3.1 ist im allgemeinen nicht richtig. Ist $\dfrac{1}{\gamma_1} + \dfrac{1}{\gamma_2} + \dfrac{1}{\gamma_3} = 1$ falls $m = 3$, so ist die Aussage von Satz 3.1 richtig. In [2] und [10] benötigten wir aus beweistechnischen Gründen die Voraussetzung

$$\dfrac{1}{\gamma_1} + \dfrac{1}{\gamma_2} + \dfrac{1}{\gamma_3} < 1 \quad \text{falls} \quad m = 3.$$

ABTEILUNG MATHEMATIK DER UNIVERSITÄT DORTMUND
POSTFACH 50 05 00
4600 DORTMUND 50

LITERATUR

[1] R. N. Kalia, G. Rosenberger, Automorphisms of the Fuchsian groups of type (0; 2,2,2,q; 0). Comm. in Alg. (6) 11 (1978), 1115-1129.

[2] A. W. Knapp, Doubly generated Fuchsian groups. Mich. Math. J. 15 (1968), 289-304.

[3] W. Magnus, A. Karrass, D. Solitar, Combinatorial group theory. New York: Wiley 1966.

[4] N. Peczynski, Eine Kennzeichnung der Relationen der Fundamentalgruppe einer nicht-orientierbaren Fläche. Diplomarbeit, Bochum 1972.

[5] N. Peczynski, Über Erzeugendensysteme von Fuchsschen Gruppen. Dissertation, Bochum 1975.

[6] N. Peczynski, G. Rosenberger, H. Zieschang, Über Erzeugende ebener diskontinuierlicher Gruppen. Inventiones math. 29 (1975), 161-180.

[7] S. J. Pride, The isomorphism problem for one-relator groups with torsion ist solvable. Trans. Amer. Math. Soc. 227 (1977), 109-139.

[8] G. Rosenberger, Zum Rang- und Isomorphieproblem für freie Produkte mit Amalgam. Habilitationsschrift, Hamburg 1974.

[9] —————, Zum Isomorphieproblem für Gruppen mit einer definierenden Relation. Ill. J. Math. 20 (1976), 614-621

[10] —————, Von Untergruppen der Triangel-Gruppen. Ill. J. Math. 22 (1978), 404-413.

[11] —————, Über Gruppen mit einer definierenden Relation. Math. Z. 155 (1977), 71-77.

[12] —————, Alternierende Produkte in freien Gruppen. Pacific J. Math. 78 (1978), 243-250.

[13] G. Rosenberger, F. Tessun, Eine Bemerkung zu den Nielsen-Transformationen. Mh. Math. 83 (1977), 43-56.

[14] O. Schreier, Über die Gruppen $A^a B^b = 1$. Abh. Math. Sem. Univ. Hamb. 3 (1924), 167-169.

[15] H. Zieschang, Über Automorphismen ebener diskontinuierlicher Gruppen. Math. Ann. 166 (1966), 148-167.

[16] —————, Über die Nielsensche Kürzungsmethode in freien Produkten mit Amalgam. Inventiones math. 10 (1970), 4-37.

[17] H. Zieschang, E. Vogt, H.-D. Coldewey, Flächen und ebene diskontinuierliche Gruppen. Lecture Notes in Math. 122, Springer 1970.

REMARKS ON THE GEOMETRY OF THE
SIEGEL MODULAR GROUP

Robert J. Sibner

I. *Introduction*

1.1. Thinking of a torus T (with modulus τ) as the quotient of the universal cover C by the group of translations $\{L_1, L_\tau\}$ where $L_a z = z + a$, the various types of conjugate holomorphic involutions of T can be obtained as projections of involutions of C as follows: (i) If $\mathrm{Re}\,\tau = 0$, reflect C in the line $y = \tau/2$. The quotient surface is orientable and has two boundary components (i.e. a cylinder). (ii) Again for $\mathrm{Re}\,\tau = 0$, translate by $L_{1/2}$ and reflect in $y = \tau/2$. The quotient surface is a Klein bottle—compact and non-orientable. (iii) For $\mathrm{Re}\,\tau = \frac{1}{2}$, one can again translate by $L_{1/2}$ and reflect in $y = \tau/2$. Now one obtains a non-orientable surface with one boundary component (a Moebius band). Note that a Moebius band can also be obtained for $|\tau| = 1$ by reflecting in the line through the origin and the point τ.

Given the above information, Bers once observed that $|\tau| = 1$ and $\mathrm{Re}\,\tau = \frac{1}{2}$ are the defining conditions for the boundary of the standard fundamental domain of the elliptic modular group and that $\mathrm{Re}\,\tau = 0$ is an axis of symmetry.

This paper is an attempt to make precise this observation and to obtain a corresponding statement about the period matrices for surfaces of arbitrary genus.

457

1.2. We define an involution J of the Siegel generalized half plane and show that any period matrix of a surface conformally equivalent to its conjugate is equivalent (modulo the Siegel modular group) to its image under J. Moreover, the Siegel fundamental domain F is invariant under the action of J, as are the various types of boundary sets.

An extended modular group is formed by adjoining J to the Siegel modular group and a fundamental domain is obtained for this group. We show that *every surface which is conformally equivalent to its conjugate has a period matrix located on the boundary of this fundamental domain.*

A symmetric Riemann surface is a surface which admits a conjugate holomorphic involution and as such is conformally equivalent to its conjugate. More precise information on the location of symmetric surfaces has been obtained and will appear in [2].

II. *The modular group*

2.1. We denote by H_g the Siegel (generalized) half plane [3] consisting of $g \times g$ complex matrices $Z = X + iY$ with Y positive definite. The group of (holomorphic) automorphisms of H_g is the symplectic group $Sp(g, R)$, whose elements have the form

$$Z \rightarrow (AZ + B)(CZ + D)^{-1}$$

with real $g \times g$ matrices A, B, C, and D satisfying $MTM^t = T$ where

$$M = \begin{pmatrix} A & B \\ C & D \end{pmatrix} \quad \text{and} \quad T = \begin{pmatrix} O & I \\ -I & O \end{pmatrix}.$$

2.2. A fundamental domain D for a group G acting discontinuously on H_g is the closure of an open connected subset of H_g with the following two properties: (i) D contains a point in every G-equivalence class and (ii) no two points in the interior of D are equivalent by a non-trivial element of G.

As is well known, the Siegel modular group (of degree g) $\Gamma = Sp(g, Z)$ $\subset Sp(g, R)$, consisting of symplectic transformations with matrix coefficients A, B, C, and D having rational integral entries, acts discontinuously on H_g and has a fundamental domain F which we now describe [3].

Denoting by $\|W\|$ the absolute value of the determinant of a matrix W, and by $L_1[W], \cdots, L_q[W]$ the homogeneous linear functions of the elements of W which arise in the Minkowski theory of reduction of quadratic forms, the Siegel fundamental domain F is the set of all $Z \in H_g$ satisfying

(a) $\|CZ + D\| \geq 1$ for all modular transformations $Z \to (AZ + B)(CZ + D)^{-1}$

(b) $L_r[Y^{-1}] \geq 0$ $\qquad r = 1, \cdots, q$

(c) $-\dfrac{1}{2} \leq x_{k\ell} \leq \dfrac{1}{2}$ $\qquad 1 \leq k, \ell \leq g$.

The boundary of F (where equality is attained in one of the above inequalities) consists of a finite number of real algebraic manifolds. We group them into the boundary sets F_a, F_b, or F_c according as equality is attained in a, b, or c.

III. *The symmetry of the Siegel space*

3.1. The Siegel half plane H_g admits a conjugate holomorphic involution

$$J: Z \to -\bar{Z}$$

and our first observation is

THEOREM 1. (i) *The Siegel fundamental domain F is invariant under the action of J.*

(ii) *The interior of F, as well as the boundary sets F_a, F_b, and F_c are each invariant under the action of J.*

Proof. Writing $Z = (x_{k\ell}) + i(y_{k\ell})$, we see that $JZ = (-x_{k\ell}) + i(y_{k\ell})$ so that condition (c) is satisfied for JZ if it is satisfied for Z. Since condition

(b) depends only on $\mathrm{Im}\,Z$, it also is satisfied for JZ. It remains to show that if, for $Z \in H_g$, one has $\|CZ+D\| \geq 1$ for all modular transformations, then the same is true for $JZ = -\bar{Z}$. Note however, that if $Z \to (AZ+B)(CZ+D)^{-1}$ is symplectic then so is $Z \to (AZ-B)(-CZ+D)^{-1}$. But then

$$\|C(JZ)+D\| = \|-C\bar{Z}+D\| = \|-CZ+D\| \geq 1 .$$

Clearly, if $Z \in \mathrm{int}\,F$ then strict inequality holds for Z in (a), (b), and (c) and hence it holds for JZ. Moreover, equality at Z in any of the defining inequalities for F implies the corresponding equality for JZ so that a point on one of the boundary sets F_a, F_b, or F_c is taken into the boundary set of the same type.

3.2. We now adjoin the involution J to the modular group Γ and describe a fundamental domain for the extended group.

DEFINITION. The *extended modular group* $\Gamma^* = \{\Gamma, J\}$ is the group generated by the elements of the Siegel modular group Γ and the involution $J : Z \to -\bar{Z}$.

Quite generally, if H is a normal subgroup of finite index in G, H is discontinuous with region of discontinuity Ω, and Ω is G-invariant then G is also discontinuous in Ω. Although the general proof of this is essentially the same, we will prove only the case at hand; $H = \Gamma$ and $G = \Gamma^*$.

PROPOSITION. Γ^* *is discontinuous on* H_g.

Proof. Suppose Γ^* were not discontinuous. Then, since $\Gamma^* = \Gamma \cup \Gamma J$, there would exist points Z, $Z' \in H_g$ and transformations $\gamma_n \in \Gamma$ for which either $\gamma_n Z' \to Z$ or $\gamma_n JZ' \to Z$. The former is impossible since Γ is discontinuous on H_g. But setting, in the latter case, $Z'' = JZ' \in H_g$ we have $\gamma_n Z'' \to Z$ which is again impossible for the same reason.

THEOREM 2. $F^* = \{Z \in F \mid \mathrm{Re\ trace}\,Z \geq 0\}$ *is a fundamental domain of the extended modular group* Γ^*. *Moreover* $F = F^* \cup JF^*$ *where*
$JF^* = \{W \in F \mid W = JZ, Z \in F^*\} = \{W \in F \mid \mathrm{Re\ trace}\,W \leq 0\}.$

Proof. Every point of H_g is Γ-equivalent to a point $Z \in F$. Either $Z \in F^*$ or $\operatorname{Re}\operatorname{trace} Z < 0$ in which case (since $\operatorname{Re}(Z + JZ) = 0$) $JZ \in F^*$. This shows that F^* contains a point in every Γ^*-equivalence class. Suppose now that $Z_1, Z_2 \in \operatorname{int} F^* \subset \operatorname{int} F$ are Γ^*-equivalent. Then Z_1 and either Z_2 or JZ_2 are Γ-equivalent. But by Theorem 1, JZ_2 is also in $\operatorname{int} F$ so that (since F is a fundamental domain of Γ) neither case is possible. Thus F^* is a fundamental domain for Γ^* and two points of F^* can be Γ^*-equivalent only if they are on the boundary of F^*.

IV. *Surfaces and their conjugates*

4.1. We recall briefly (see [1]) the definition of the intersection number $C_1 \times C_2$ of two cycles on a Riemann surface S. Each cycle C induces a closed form β such that $\int_C \nu = \iint_S \beta \wedge \nu$ for all closed forms ν.

Then $C_1 \times C_2 = \iint_S \beta_1 \wedge \beta_2$ where β_1 and β_2 are the induced forms.

Intuitively, two smooth curves intersecting at a point have intersection number $+1$ (resp. -1) if their tangent vectors form a right-handed (resp. left-handed) coordinate system.

A homology basis $\{A_1, \cdots, A_g, B_1, \cdots, B_g\}$ for S is said to be canonical if: $A_j \times A_k = B_j \times B_k = 0$ and $A_j \times B_k = \delta_{jk}$. The associated normalized basis $\{\omega_i\}$ of abelian differentials (of the first kind) is obtained by requiring that $\int_{A_j} \omega_i = \delta_{ij}$ and the Riemann period matrix Z is then defined by

$$Z_{ij} = \int_{B_j} \omega_i$$

As is well known, every period matrix is an element of the Siegel space H_g.

4.2. To every Riemann surface S there corresponds a conjugate surface $\bar{S}$ obtained from S by using the same underlying topological space, but replacing the coordinate maps by their conjugates.

LEMMA 1. *If* $\{A_j, B_j\}$ *is a canonical homology basis for* S *then* $\{A_j, -B_j\}$ *is a canonical homology basis for* $\overline{S}$.

Proof. Since the orientation of $\overline{S}$ is opposite that of S, the intersection numbers of two cycles differ in sign. Using $-B_j$ in the canonical basis for $\overline{S}$ compensates for this. Alternatively, the same sign change is apparent from the integral representation of the intersection number since the volume elements on S and $\overline{S}$ differ in sign.

Observe now that if ω, represented locally as $f(z)dz$ on S, is an abelian differential (f is a holomorphic function of its argument) then $\overline{\omega}$ represented locally by $\overline{f(\overline{\zeta})}d\zeta$ is an abelian differential on $\overline{S}$.

LEMMA 2. $\int_C \omega = \overline{\int_{\overline{C}} \omega}$ *where* C *and* $\overline{C}$ *denote the same cycle on* S *and* $\overline{S}$ *respectively.*

Proof. Writing $\zeta = \overline{z}$ we have

$$\int_{\overline{C}} \overline{\omega} = \int_{\overline{C}} \overline{f(\overline{\zeta})}\,d\zeta = \int_C \overline{f(z)}\,d\overline{z} = \overline{\int_C f(z)\,dz} = \overline{\int_C \omega}\,.$$

4.3. Suppose now that the Riemann surfaces S_i have period matrices Z_i $(i=1,2)$. We recall [4, 5] that S_1 and S_2 are conformally equivalent if and only if Z_1 and Z_2 differ by an element of the Siegel modular group (i.e. are symplectically equivalent).

THEOREM 3. (i) *If* S *has a period matrix* Z *then* JZ *is a period matrix of the conjugate surface* $\overline{S}$.

(ii) S *is conformally equivalent to its conjugate* $\overline{S}$ *if and only if* $JZ = Z \pmod{\Gamma}$.

Proof. Suppose that the period matrix Z for S has been obtained from the canonical homology basis $\{A_j, B_j\}$ and the associated normalized basis for the abelian differentials $\{\omega_i\}$. By Lemma 1, $\{A_j, -B_j\}$ is a canonical homology basis on $\bar{S}$ and by Lemma 2, $\int_{\bar{A}_j} \bar{\omega}_i = \overline{\int_{A_j} \omega_i} = \delta_{ij}$,

so that $\bar{\omega}_i$ is the associated normalized basis for the abelian differentials on $\bar{S}$. But then the period matrix of $\bar{S}$ is given by

$$\int_{\bar{B}_j} \bar{\omega}_i = \overline{\int_{-B_j} \omega_i} = -\bar{Z}_{ij} = (JZ)_{ij}$$

This shows (i) and (ii) follows from (i) and the above remarks.

THEOREM 4. *If a Riemann surface* S *is conformally equivalent to its conjugate then it has a period matrix* Z *located on the boundary of the fundamental domain* F^* *of the extended modular group* Γ^*.

Proof. We may suppose that Z is contained in the fundamental domain F of the Siegel modular group. Since by Theorem 2 $F = F^* \cup JF^*$, either $Z \in F^*$ or $JZ \in F^*$. Now if S is conformally equivalent to $\bar{S}$ then, by Theorem 3, $JZ = Z \pmod{\Gamma}$. If $JZ = Z$ then $\mathrm{Re}\, Z = 0$ and Z is on the boundary of F^*, while if the equivalence is by a non-trivial element of Γ then, since both Z and JZ are (by Theorem 1) in F, they cannot both be in the interior and hence must both (again by Theorem 1) be on the boundary of F. Thus S has a period matrix located on the boundary of F^*.

BROOKLYN COLLEGE
CITY UNIVERSITY OF NEW YORK

REFERENCES

[1] I. Kra, Automorphic forms and Kleinian groups, W. A. Benjamin, Reading, Massachusetts (1972).

[2] R. J. Sibner, Symmetric surfaces in the Siegel half plane, Symposium on Fuchsian groups and modular forms, Pittsburgh, Pa. (1978) to appear.

[3] C. L. Siegel, Symplectic Geometry, Amer. J. Math. 65 (1943), 1-86.

[4] ________, Lectures on the analytic theory of quadratic forms, Inst. for Adv. Study and Princeton Univ. (1934-35)

[5] R. Torelli, Sulle varieta di Jacobi, Rendiconti d.R.A.d. Lincei 22 (1913).

This research was partially supported by NSF grant MCS78-03268.

ON THE ERGODIC THEORY AT INFINITY OF AN ARBITRARY
DISCRETE GROUP OF HYPERBOLIC MOTIONS

Dennis Sullivan

We will describe an ergodicity phenomenon for the action of an
arbitrary discrete group of hyperbolic isometries on the tangent spaces of
the sphere at ∞. One application (Section VII) is a maximal extension of
Mostow's rigidity theorem "rough isometry $\Longrightarrow$ isometry" from finite
volume hyperbolic manifolds to manifolds whose volume grows slower than
that of hyperbolic space. Another application (Section V) allows a complete description of the finitely generated Kleinian groups in one quasiconformal conjugacy class in terms of a nice complex manifold Teichmüller
space. Along the way to the main theorem we characterize ergodicity of
the action of Γ on geodesics in terms of the divergence of a series of
solid angles (Sections II, III). We also characterize the conservative part
of the action on the sphere at ∞ in terms of the horospherical limit set
(Section IV). (See [Su] for extensions to Hausdorff measure.)

To describe the situation in more detail recall that geometry in hyperbolic $(n{+}1)$-space tends as we approach ∞ to conformal geometry on the
n-sphere at ∞. Thus a discrete group Γ of hyperbolic isometries is
equivalent to a discrete group of conformal transformations of S^n. The
dynamics of these actions is quite different. In hyperbolic space Γ is
essentially permuting convex fundamental domains freely. While on the

sphere at ∞ one finds the cluster set of an orbit of Γ in hyperbolic space on which one knows Γ acts minimally (each point has a dense orbit) and an open complement where Γ acts discontinuously.

We seek the dynamical picture of Γ in the context of Lebesgue measure or ergodic theory. Thus we heretofore neglect sets on the n-sphere of measure zero. In that case any group action whatsoever breaks into a dissipative piece and a conservative piece. The dissipative piece is the disjoint union of measurable sets permuted by the group. The conservative (or recurrent) piece has the property that for any set A of positive measure $\gamma A \cap A$ has positive measure for infinitely many group elements. We will find this partition for discrete conformal groups and make use of it.

Now in terms of any measure of finite mass on S^n, say the solid angle measure as viewed from a point inside hyperbolic space, we can divide the points of S^n into those for which the sum over the group of area distortions is finite and those for which it is infinite. It follows as in Poincare's famous recurrence theorem that this is the partition into the dissipative and conservative parts. This statement is general and makes no use of our assumption that Γ is a discrete group of conformal transformations. These hypotheses imply more. Namely,

i) the area distortions are unbounded at almost all conservative points (Section IV),

ii) definite variation of the logarithm of area distortion takes place for appropriate individual group elements on small annuli of definite shape anywhere we look in the conservative part (Sections I, III).

The second property of area distortions is the key to the main result. Define a measurable conformal structure on S^n to be a measurable field of similarity structures on the tangent spaces. In Section VII we prove the following theorem. (The case $n = 2$ is treated completely in Section I.)

THEOREM. *Any measurable conformal structure on S^n which is kept a.e. invariant by a discrete group of conformal transformations of S^n must agree a.e. with the standard conformal structure outside the dissipative part of the action.*

The first property of area distortions above allows us to recognize the conservative part in geometric terms. For example, *the horospherical limit set,* by definition those points of S^n for which an orbit in hyperbolic space enters every horosphere (based at the point, see figure a) has full measure in the conservative part. Or the dissipative part is the union of the parts of the fundamental domains in hyperbolic space on the sphere at ∞ (Section IV). Thus if a fundamental domain has zero area at ∞ (figure b) the action is conservative, the measurable invariant conformal structure is unique and we have the needed step for the generalization of Mostow's theorem mentioned above (Section VII).

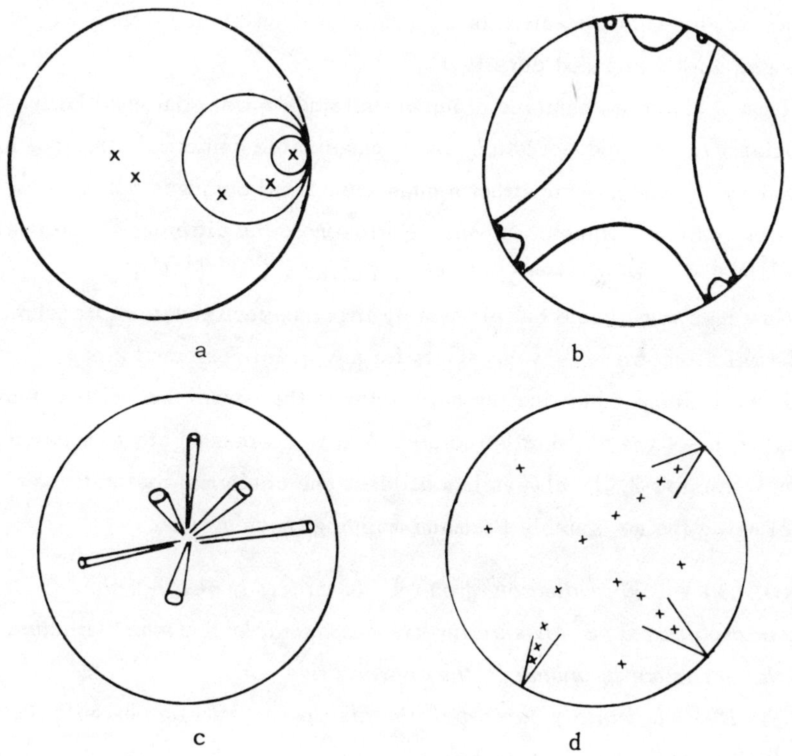

a

b

c

d

The argument of Section II uses random walks as in [G] and yields a new geometric characterization of the ergodicity of the geodesic flow in terms of infinite solid angle of the orbit in hyperbolic space (Section II). We recover E. Hopf's (1939) characterization in terms of conical approach (Section III) (figures c and d).

Finally, we turn the discussion to quasi-conformal homeomorphisms. First not only does hyperbolic geometry become conformal geometry at ∞ but quasi-(hyperbolic geometry) becomes quasi-conformal geometry on the sphere at ∞. (This phenomenon plus the ergodicity on tangent directions summarizes Mostow's theorem.)

Second, in dimension 2 quasi-conformal homeomorphisms of S^2 are parametrized [AB] by measurable conformal structures defined a.e. on S^2, and which are a bounded distance away from the standard structure. (Reminiscent of the parametrization of Lipschitz homeomorphisms of S^1 by measures with a bounded density.)

Thus if Γ is a countable group of uniformly quasi-conformal homeomorphisms of S^2 and ν is a bounded measurable conformal structure invariant by Γ one can construct a quasi-conformal conjugacy of Γ to a group of conformal transformations. Furthermore the different "conformal models" of Γ are parametrized by such invariant ν ([AB], [B]).

Now it is remarkable but elementary that one such invariant measurable conformal structure ν always exists for any quasi-conformal group Γ. (One merely forms fiberwise the barycenter of the convex hull of the transforms by Γ of the standard structure. And this works in any dimension.) So in dimension 2, Γ always has at least one conformal realization or model using the measurable Riemann mapping theorem [AB].

COROLLARY . i) *A discrete group* Γ *of uniformly quasi-conformal homeomorphisms of* S^2 *has an invariant measurable conformal structure and this structure is unique on the conservative part.*

ii) *If* Γ *is finitely generated, the dissipative part agrees with the topological domain of discontinuity.*

iii) *A general* Γ *always has conformal models and these are parametrized by varying the invariant measurable conformal structure on the dissipative part.*

iv) *In the finitely generated case, again, this parametrization is by a Teichmüller space associated to a Riemann surface of finite type (Section V).*

In Section VI we discuss further dynamical properties of finitely generated discrete groups of conformal transformations on S^2. The first part of Section I and Sections III, IV, VII also work for S^n $n \geq 1$ as we explain somewhat in Section VII. The proof of the main theorem for $n = 2$ is actually completed in the first section.

We record our debt of motivation to Ahlfors' papers "Remarks on the limit point set of a finitely generated Kleinian groups," Annals of Math. Studies 66, and especially "Some Remarks on Kleinian Groups" from the unpublished Tulane proceedings on Kleinian groups. In the latter paper Ahlfors establishes the topological limit set is conservative for finitely generated Kleinian groups (with a domain of discontinuity). In his well-known finiteness paper (1965) he establishes the (domain of discontinuity)/Γ is a Riemann surface of finite type.

For me these two theorems of Ahlfors are almost the axioms for a good theory of Kleinian groups. Together they impose a tight structure on the situation.

Finally, we dedicate this paper to Lucy Garnett who supplied the random walk idea for Theorem II, pointed out a non-obvious point about horocycle limit points, Section IV, and generally sustained the work in this paper.

Section I. *The variation of area distortion lemma and groups with finite solid angle,[1] $\sum_\Gamma 1/\lambda^2 < \infty$*

THEOREM I. *For groups Γ of finite solid angle[1] in $\mathbf{H}^3$ there is on the conservative part of the action of Γ on S^2 no measurable tangent line field invariant a.e. by Γ.*

---

[1] This finite solid angle condition can be dropped in two ways, see addendum to Section I or Section II.

Proof. Any conformal transformation γ of the sphere is the composition of a rotation followed by a hyperbolic transformation with antipodal fixed points. Thus the linear distortion d_γ of the spherical metric varies in an interval $[1/\lambda, \lambda]$ with $\lambda \geq 1$. The extreme values of the distortion are taken at antipodal points. The intermediate values are taken on concentric circles interpolating between these points. We need a uniform picture of the distortion for λ large.

Let r be the natural radial parameter for these circles with $r = 0$ corresponding to the value of distortion λ. For any given $k > 1$ and $\lambda^2 > k$ let D_γ be the unique continuous function on the sphere with values $1/\lambda r^2$ on the circles of radius r with $k/\lambda \leq r \leq 1/k$ and which is constant otherwise. Thus

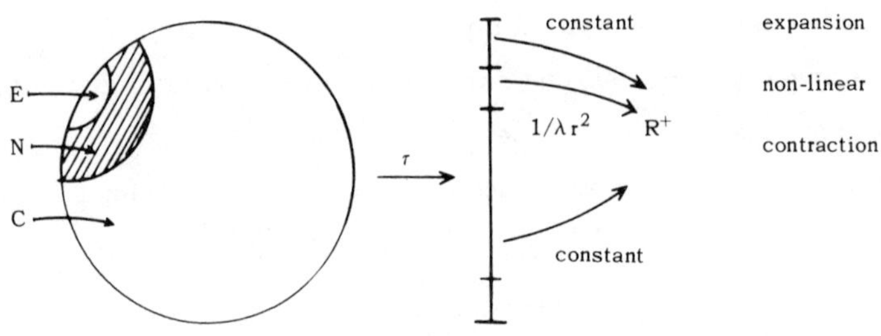

$$
D_\gamma = \begin{cases}
\lambda/k^2 & \text{on} & \{0 \leq r \leq k/\lambda\} & = E(\gamma) \\
1/\lambda r^2 & \text{on} & \{k/\lambda \leq r \leq 1/k\} & = N(\gamma) \\
k^2/\lambda & \text{on} & \{1/k \leq r \leq 1\} & = C(\gamma) .
\end{cases}
$$

LEMMA 1. *The ratio of the actual distortion* d_γ *to the approximation* D_γ *is bounded in terms of* k *for all conformal transformations* γ *with* $\lambda(\gamma) \geq k^2$. *Moreover on the nonlinear part* $N(\gamma)$ *this ratio is arbitrarily close to 1 for* k *sufficiently large.*

Proof. i) We compute for the affine transformation $x \to \lambda x$ on the line the distortion of the measure coming via stereographic projection from the uniform measure on the circle of diameter 1. This transposed measure is

$\dfrac{dx}{1+x^2}$. Thus the distortion of $x \to \lambda x$ is $d(x) = \dfrac{\lambda(1+x^2)}{1+(\lambda x)^2}$, which is monotone decreasing in x.

ii) For $0 \le x \le k \cdot 1/\lambda$, $d(x)$ lies in $\left[\dfrac{\lambda(1+k^2/\lambda^2)}{1+k^2}, \lambda \right] \subset \lambda[1/1+k^2, 1]$.

For $1/k \le x \le \infty$, $d(x)$ lies in $\left[1/\lambda, \dfrac{\lambda(1+1/k^2)}{1+\lambda^2/k^2} \right] \subset 1/\lambda[1, 1+k^2]$.

For $k/\lambda \le x \le 1/k$, $d(x) \cdot \lambda x^2$ lies in

$$\left[\frac{(\lambda x)^2}{1+(\lambda x)^2}, \frac{\lambda(1+1/k^2) \cdot \lambda x^2}{(\lambda x)^2} \right] \subset \left[\frac{1}{1+1/k^2}, 1+1/k^2 \right] .$$

iii) Now we reinterpret these inequalities replacing the variable x on the line by the variable r on the circle. The assertions for the parts $C(\gamma)$ and $E(\gamma)$ follow. For $N(\gamma)$ note that for k large stereographic projection is almost an isometry between $0 \le r \le 1/k$ and $0 \le x \le 1/k$.

COROLLARY 2. *Area* $E(\gamma) \le 1/\lambda^2 \cdot a_k$, *area* $\gamma C(\gamma) \le 1/\lambda^2 \cdot a_k$, *diameter* $N(\gamma) \le 2/k \cdot c_k$, *diameter* $\gamma N(\gamma) \le 2/k \cdot c_k$, *where* a_k *is a constant depending only on* k *and* c_k *is arbitrarily close to* 1 *for* k *large.*

Proof. The first three follow from Lemma 1 and the definitions. The last follows because the circles of radius r filling up $N(\gamma)$ map (by conformality) to circles of radius $r \cdot 1/\lambda r^2$ (times a factor near 1 for k large). Since $k/\lambda \le r \le 1/k$ in $N(\gamma)$ these radii lie in the interval $[k/\lambda, 1/k]$ (times c_k).

REMARK. If " $\sim$ " denotes an equality up to a factor (or discrepancy) which is near 1 (or negligible) when k is chosen large, then

$$\gamma(N(\gamma) \sim N(\gamma^{-1}),$$

$$\gamma E(\gamma) \sim C(\gamma^{-1}) \quad \text{and} \quad \gamma C(\gamma) \sim E(\gamma^{-1})$$

although we don't use these facts explicitly.

Now suppose R is a *recurrent* (conservative) set on S^2 for the discrete group Γ of conformal transformations. Thus $X \subset R$ of positive measure implies $X \cap \gamma X$ has positive measure for infinitely many γ. Suppose also that Γ has "finite solid angle": $\sum_\Gamma 1/\lambda^2 < \infty$. The *modulus* of a concentric annulus on the sphere is by definition the logarithm of the ratio of radii.

LEMMA 3. *Let $X \subset R$ of positive measure and two positive constants δ and Δ be given. Then there is a point* x, *an element* γ *and a concentric annulus* A *so that,*

    i) $x \in X$, $x \in A$, *and* $\gamma(x) \in X$.

    ii) A *and* $\gamma(A)$ *have diameter* $< \delta$.

    iii) A *and* $\gamma(A)$ *have modulus as close as we like to* Δ.

    iv) *The distortion of* γ *is constant on the concentric circles of* A *and the* log *of the distortion varies in* A *by an amount as close as we like to* 2Δ.

Proof. i) Fix k so large that $2/k \cdot c_k < \delta$ (see Lemma 1).

    ii) Remove finitely many elements from the group to make $\log \lambda/k^2 > \Delta$ for the rest (possible since Γ is discrete).

    iii) Remove finitely many elements so the sum of area $(E(\gamma) \cup \gamma C(\gamma))$ for the rest is less than area X (possible since $\sum_\Gamma 1/\lambda^2 < \infty$ using Lemma 1).[2] Remove this infinite union from X to obtain X' which still has positive measure.

    iv) Find a γ outside the finite sets above so that $\gamma^{-1}X' \cap X'$ has positive measure (using recurrence of the action). If $x \in \gamma^{-1}X' \cap X'$, by construction $x \notin E(\gamma)$ and $\gamma(x) \notin \gamma C(\gamma)$. Thus $x \in N(\gamma)$.[2]

    v) On $N(\gamma)$ the distortion d_γ varies between k^2/λ and λ/k^2 (Lemma 1), so $d_\gamma(x)$ is somewhere in this interval. On the log scale we

---

[2]See the addendum to Section I for an alternative to this step.

can fit an interval of length 2Δ about the value $\log d_\gamma(x)$, and stay in $\log d_\gamma N(\gamma)$, because $\log \lambda/k^2 > \Delta$ by ii) above. Let such an interval define the annulus A in question, so that $x \in A \subset N(\gamma)$ and $\log(d_\gamma)$ varies through $\sim 2\Delta$ in A, by construction.

vi) Since $A \subset N(\gamma)$, diameter A and diameter $\gamma A < \delta$ by i) and Corollary 2.

vii) Since $d_\gamma \sim 1/\lambda r^2$ on $N(\gamma)$ and thus on A, variation $\log d_\gamma \sim 2\Delta$ on A implies modulus $A \sim \Delta(\log r_2^2/r_1^2 = 2\Delta$ iff $\log r_2/r_1 = \Delta)$. Similarly γA has radii $dr_1 e^{2\Delta}$ and dr_2 (for some d) because we know d_γ on A. Thus the modulus of γA is

$$\log \frac{dr_1 e^{2\Delta}}{dr_2} \sim 2\Delta - \log r_1/r_2 \sim \Delta \, .$$

Now we pass from the sphere to the plane.

LEMMA 4. *If* e *is an absolutely continuous isomorphism of the plane (relative to Lebesgue measure) carrying* B *to* B$'$, *then a subset* $A \subset B$ *with a proportion* η *of area is carried to a subset* $e(A) = A' \subset B'$ *of proportion at least* η' *of area where* $\eta' = 1 - d(1-\eta)$ *and* d *is the maximum ratio of area distortion at various points of* B.

Proof. By an affine scaling we can assume area $B = 1$, the low value of area distortion is 1, and the high value is d. The worst case occurs when 1 occurs on all of A and d occurs on all of the complement of A. Then,

$$\eta' = \text{area } A'/\text{area } B' = \eta/\eta + d(1-\eta) \geq 1 - d(1-\eta) \, .$$

LEMMA 5. *Let* X *be a set in the plane of positive measure and let* η *and* Δ *be given positive numbers. Consider sectorial boxes of shape* Δ,

$$\{(r, \theta): r_0 \leq r \leq e^\Delta r_0, \; \theta_0 \leq \theta \leq \theta_0 + \Delta\} \, .$$

Then there is a $\delta > 0$ *and a subset* X$'$ *of* X *of positive measure so that each box of shape* Δ *and diameter* $< \delta$ *containing a point of* X$'$ *also contains at least the proportion* η *of* X.

Proof. The class of sectorial boxes of shape Δ are generated by similarity transformations from one of them. Thus Lebesgue's theorem concerning density points is true using these instead of round disks. (See E. Stein "Singular Integrals ..." pp. 11, 12.)

So for almost all x there is a largest positive δ_x such that the proportion of X in boxes containing x of diameter $\leq \delta_x$ is at least η. But then $x \to \delta_x$ is a positive measurable function which has to be greater than some $\delta > 0$ on a set $X' \subset X$ of positive measure. This proves the lemma.

Now consider a conformal transformation of the plane

$$\gamma : z \to \frac{az+b}{cz+d} \qquad ad - bc = 1, \qquad c \neq 0$$

and sectorial boxes B_Δ of shape $\Delta < \pi/2$ centered at $-d/c$.

LEMMA 6. *If* $A \subset B_\Delta$ *is any subset with the proportion* η *of area, then the variation on* A *of the real and imaginary parts of* $\log \gamma'z$ *is at least* $2\Delta(1 - e^{2\Delta}(1-\eta))$.

Proof. i) On a unit square the function $(x,y) \to x$ has variation at least η' on any subset whose proportion of area is at least η'.

ii) Introduce the variable $e^\xi = z + d/c$ so that the variation of

$$\log \gamma'z = \log \frac{1}{(cz+d)^2} = -2 \log(z+d/c) + \text{constant on } A \subset B_\Delta \text{ is just the}$$

variation of -2ξ on a corresponding subset A' of a square in the ξ-plane of side Δ.

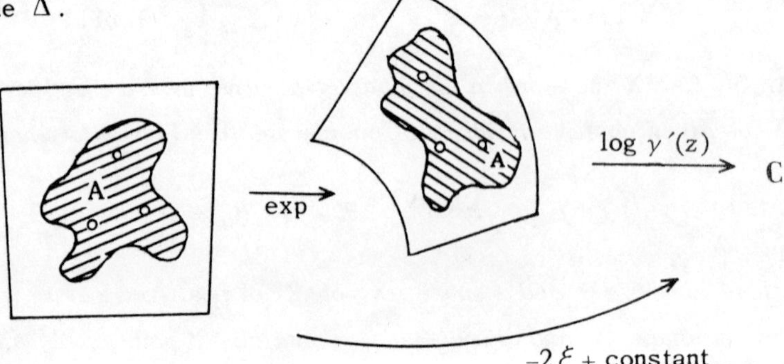

iii) The ratio of area distortion of exp at different points of the square is at most $e^{2\Delta}$. By Lemma 4 the proportion of $A' = \exp^{-1}(A)$ in the square is at least $\eta' = 1 - e^{2\Delta}(1-\eta)$.

iv) Applying i) the result follows.

Now we are ready to prove the nonexistence of invariant measurable line fields for groups of finite solid angle.

i)   Choose a small number, $\pi/2 > \Delta > 0$ and a set of positive measure X in the plane where the hypothetical invariant line field varies only in an interval of inclinations of length $\frac{1}{2}\Delta$.

ii)   Choose $0 < \eta < 1$ so that $1 - e^{2\Delta}(1 - \eta + e^{2\Delta}(1-\eta)) > 1/2$.

iii) Find $X' \subset X$ of positive measure satisfying a $\delta' > 0$ uniform density relative to X, η , and sectorial boxes of shape Δ (as in Lemma 5).

iv) Choose a point of density of X' and stereographically project the action of Γ on the plane to a sphere resting on this point.

v)   Let Y denote the intersection of X' with a ball B' about this point sufficiently small so that the distortion of stereographic projection on 2B is as close to 1 as we need for the following. Let $\delta = \min(\delta',$ radius $B')$.

vi) Relative to δ, Δ and Y (put Y on the sphere) find the element γ and the concentric annulus A satisfying Lemma 3 (and put A back on the plane). In A choose a sectorial box B of shape $\sim \Delta$ containing x and centered at the pole of γ (possible because we know the variation of $\log d_\gamma$ on A).

vii) Since the diameters of B' and $\gamma B'$ are less than δ' (even δ) they each contain the proportion η (at least) of X . ($x \in B \subset A$, $\gamma x \in \gamma B' \subset A$, and x and γx belong to X' .)

Since the ratio of area distortion of $B' \to \gamma B'$ is at most $e^{2\Delta}$ the proportion of y in B' so that $\gamma y \in X$ is at least $\eta' = 1 - e^{2\Delta}(1-\eta)$ by Lemma 4 again. Thus the proportion of $y \in B'$ so that $y \in X$ and $\gamma y \in X$ is at least $1 - [(1-\eta) + (1-\eta')] = \eta + \eta' - 1 = \eta''$.

viii) By Lemma 6 the variation of the argument of $\gamma'(z)$ on this subset $B \cap X \cap \gamma^{-1}X$ of B is at least $2\Delta(1 - e^{2\Delta}(1-\eta''))$. This variation is greater than $\sim \Delta$ by ii) which contradicts the definition of X i) and the invariance of the line field under γ. This proves the Theorem I for discrete groups of finite solid angle, $\sum_{\Gamma} 1/\lambda^2 < \infty$.

Addendum

After writing this paper I learned from Klaus Schmidt's notes "Cocycles of ergodic transformation groups" Warwick 1976 §4, that for any conservative group action, any set A of positive measure, and any $\varepsilon > 0$ there are infinitely many group elements γ so that $\{x \epsilon A \cap \gamma^{-1}A$ and the area distortion of γ at x is ε-close to $1\}$ has positive measure. For Kleinian groups such x belong to N_γ by the formula above for D_γ.

Section II. *Ergodicity of the geodesic flow and groups with infinite solid*
$$\text{angle, } \sum_{\Gamma} 1/\lambda^2 = \infty$$

THEOREM II. *If* $\sum_{\Gamma} 1/\lambda^2 = \infty$ *for a discrete group of conformal transformations on* S^2 *the action of* Γ *on* $S^2 \times S^2$ *is ergodic.* $(\phi(x,y) = \phi(\gamma x, \gamma y)$ *for all* $\gamma \epsilon \Gamma$ *and* ϕ *measurable implies* ϕ *is constant a.e.* $)$[1]

COROLLARY.[2] *For groups* Γ *of infinite solid angle in* $\mathbf{H}^3$ *there is on* S^2 *no measurable tangent line field invariant by* Γ *a.e.*

Proof. If there were such, for almost all pairs of points on S^2 we would have a measurable angle difference function (measured along connecting geodesic). By ergodicity one deduces this angle is constant a.e. The line field is seen to be the restriction of a continuous line field invariant by Γ. This is absurd.

Proof of Theorem. First we gather some facts from hyperbolic geometry. What we need follows from the complete symmetry of hyperbolic space

---

[1]This theorem holds for S^n using $\sum 1/\lambda^n$, $n \geq 1$, and the proof is the same.

[2]This proof is independent of Section I and is akin to Mostow's original discussion (see point iii), Section VII).

together with the positive lower bound on the convexity of spheres of arbitrarily large radius. Let $\theta(x,y)$ be the reciprocal of the area of the sphere passing through y with center x. So $\theta(x,y)$ is the (density of) solid angle of one point as viewed from the other.

LEMMA 1. $\underset{\Gamma}{\Sigma}\, 1/\lambda^2 = \infty$ *if and only if the total solid angle of an orbit,* $\underset{\Gamma}{\Sigma}\, \theta(x,\gamma y)$, *viewed from any point* x *not on the orbit is infinite.*

Proof. Consider a homothety γ in the upper half space model with fixed point at zero and a very small linear derivative $1/\lambda$. If x and y_0 are on the z axis above zero, then clearly $\theta(x,y_0)$ and $\theta(x,\gamma y_0)$ are in the approximate ratio $1/\lambda^2$. Thus in general the order of the term $\theta(x,\gamma y)$ for $\lambda(\gamma)$ large and x fixed is in a bounded ratio to $1/\lambda^2$. This proves the lemma.

EXTRA REMARK. If $g_x(y)$ denotes the Green's function of hyperbolic space with pole at x ($g_x(y)$ is positive, harmonic, tending to zero at ∞, and symmetric about x), there is the exact formula

$$g_x(y) = \int_\infty^y \theta(x,y')\,dy'$$

obtained by integrating in along a radius from ∞. Since $\theta(x,y)$ is exponentially decreasing in the distance (x,y), the integral is approximately the upper limit so $g_x(y) \sim \theta(x,y)$, at large distances.

Thus in case hyperbolic space mod Γ is a manifold V one sees the condition $\underset{\Gamma}{\Sigma}\, 1/\lambda^2 < \infty$ is equivalent to V has a finite Green's function $g_x(\bar{y}) = \underset{\gamma \in \Gamma}{\Sigma}\, g_x(\gamma y)$. Or in other words $\underset{\Gamma}{\Sigma}\, 1/\lambda^2 = \infty$ if and only if random motion on V is recurrent. This idea is the point of the ensuing discrete time proof.

Now to each point p in hyperbolic space associate the *Poisson measure* μ_p on the sphere S^2 at ∞. If $A \subset S^2$ then $\mu_p(A)$ is the *solid angle* of A viewed from p.

LEMMA 2. *The measure* μ_p *is the spherical average of the measure* μ_q *where q ranges over a sphere with center p.*

Proof. Each of the two measures on the sphere at ∞ is invariant by the full rotation group about p. Thus they are equal.

We can fill in bounded measurable functions ϕ on the sphere at ∞ to bounded harmonic functions on hyperbolic space. Namely define

$$h(p) = \int \phi \, d\mu_p \equiv \langle \phi, \mu_p \rangle, \quad \text{``Poisson formula''}.$$

LEMMA 3. ϕ *non-constant a.e. implies h non-constant.*

Proof. Take density points x and y in S^2 of sets where ϕ has values in disjoint intervals. For points p in hyperbolic space near x , μ_p sees mostly the values in the interval associated, so h(p) lies nearly in this interval. Similarly for points q near y , h(q) nearly lies in the disjoint interval.

Let P denote the (averaging) operator on functions and measures on hyperbolic space $f \mapsto Pf$ and $\mu \mapsto \mu P$, where $Pf(x)$ is the average of f over a ball of radius η centered at x , $B(x, \eta)$ and if δ_x denotes the dirac mass at x , $\delta_x P$ is the uniform measure on $B(x, \eta)$ of total mass 1. Note $\langle \mu P, f \rangle = \langle \mu, Pf \rangle$ when both make sense.

By Lemma 2 the functions h constructed by the Poisson formula are P-harmonic, namely $Ph = h$. Also P clearly commutes with isometries.

LEMMA 4. *The density at the point y of the measure* $\sum_{n=1}^{N} \delta_x P^n$ *is for N large at least a fixed constant times* $\theta(x, y)$.

Proof. It is clear that the sequence of measures $\delta_x P^n$ $n = 0, 1, 2, \cdots$ begins at x, spreads out symmetrically by steps of length at most η and converges to ∞ in the sense that almost all the mass is eventually outside any sphere centered at x. The last part follows since a ball of radius η centered on a sphere which is itself centered at x has a proportion of area definitely more than $1/2$ outside the sphere.

Thus all the mass of $\delta_x P^n$ as $n \to \infty$ passes through any spherical shell of thickness η centered at x. Since the density function of the measure $\sum_{n=1}^{\infty} \delta_x P^n$ is *decreasing*, only depends on the radial coordinate from x, and puts at least mass 1 in each shell of width η whose volume is proportional to $1/\theta(x,y)$ when y lies in the shell, the result follows. Q.E.D.

Now form the space T from the product of hyperbolic space with the sphere at ∞ by dividing by the diagonal action of Γ, $T = H \times S^2/\text{mod } \Gamma$. Provide T with an averaging operator $\tilde{P}$ on measures and functions which is defined using P in each H-level. Provide T with a natural smooth measure dm using the family of Poisson measures μ_p on the factors $(p \times S^2)$ and the natural measure on H to obtain a Γ invariant volume element dm' on $H \times S^2$. Let μ_p also denote the image measure resting on a sphere of T. Now we come to the key lemma.

LEMMA 5. *The density of the measure* $\sum_{n=1}^{N} \mu_p P^n$ *relative to* dm *converges to* $+\infty$ *at almost all points of* T *as* $N \to \infty$.

Proof. We compute the density of $\nu_N = \sum_{n=1}^{N} \mu_p (P \times \text{id})^n$ in $H \times S^2$ relative to dm and add these densities up along an orbit of Γ: $(y, s), (\gamma y, \gamma s), \cdots$.

Let $g_x(y)$ denote the density of the Green's measure $\delta_x + \delta_x P + \delta_x P^2 + \cdots$ in terms of the natural volume dh on hyperbolic space. In terms of the product measure $dh \times \mu_p$ on $H \times S$ (which is not Γ invariant) the measure $\lim_{N \to \infty} \nu_N$ has density $g_p(y)$ at each point (y, s) of the product

space. Rewriting in terms of dm' (which is Γ invariant) introduces the

Radon Nikodym factor $\dfrac{d\mu_p}{d\mu_y}(s)$ at each point (y, s). The desired density

is thus $\sum_{\Gamma} g_p(\gamma y) \dfrac{d\mu_p}{d\mu_{\gamma y}}(\gamma s)$.

Using γ^{-1} to transform $(p, \gamma y, \gamma s)$ into $(\gamma p^{-1}, y, s)$ and the symmetry

$g_x(y) = g_y(x)$ changes the sum to one of the form $\sum_{\Gamma} g_y(\gamma p) \dfrac{d\mu_{\gamma p}}{d\mu_y}(s)$. By

Lemmas 1 and 4, $\sum_{\Gamma} g_y(\gamma p) = \infty$ so Lemma 5 results from the following

lemma.

LEMMA 6. $\sum_{\gamma \in \Gamma} g_y(\gamma p) = \infty$ *implies that for almost all* s *on the sphere*

$$\sum_{\gamma \in \Gamma} g_y(\gamma p) \frac{d\mu_{\gamma p}}{d\mu_y}(s) = \infty.$$

Proof. i) Denote by B a small ball centered at p, and by ΓB the

disjoint union $\underset{\gamma \in \Gamma}{\cup}\, \gamma B$. Define a Γ invariant function $\pi_B(y)$ to be the

probability that a random walk, whose transition operator is P, starting

at y hits ΓB. Clearly $\pi_B(y) \le 1$ and $P\pi_B \le \pi_B$. If π_B is not identi-

cally 1, the Γ invariant function $\pi_B - P\pi_B$ is greater than $\epsilon > 0$ on

the Γ orbit of some smaller ball. Now consider the identity

$$\langle \delta_y + \delta_y P + \cdots \delta_y P^N, \pi_B - P\pi_B \rangle = \langle \delta_y, \pi_B \rangle - \langle \delta_y P^{N+1}, \pi_B \rangle.$$

The right-hand side is uniformly bounded wrt N. The left-hand side is at

least $\epsilon \sum_{\Gamma} g_y(\gamma p)$ as $N \to \infty$. We conclude that $\pi_B = P\pi_B$ or $\pi_B(y) = 1$.

Thus almost all paths starting at y hit ΓB when $\sum_{\gamma \in \Gamma} g_y(\gamma p) = \infty$.

ii) For fixed s_0 and x_0 the function of x, $\dfrac{d\mu_x}{d\mu_{x_0}}(s_0)$ is a

P-harmonic function (Lemma 2), whose boundary values are $+\infty$ at s_0

and zero at other points of the sphere. One sees by a standard limiting procedure [K] that $g_x(y) \frac{d\mu_y}{d\mu_x}(s_0)$ is the corresponding Green's density for the random walk conditioned so that the limit at ∞ is s_0. By Fubini's theorem and i) for almost all s_0, almost all paths starting at y and conditioned to end up at s_0 must also hit ΓB.

Now if for one of these s_0 $\sum_{\gamma \in \Gamma} g_y(\gamma p) \frac{d\mu_{\gamma p}}{d\mu_y}(s_0) < K < \infty$, the same inequality would hold for p in a small ball B. It would follow that the conditioned Green's measure would give finite measure to ΓB. This contradicts the fact that almost all conditioned paths hit ΓB starting from any point and thus hit it infinitely often by the Markov property. This completes the proof of Lemma 6.

Proof of Theorem II. Let $\phi(x,y)$ be a non-constant characteristic function on $S^2 \times S^2$ invariant by Γ and suppose $\sum_{\Gamma} 1/\lambda^2 = \infty$. We can suppose by Fubini (after interchanging x and y if necessary) that for a set of y of positive measure $\phi(x,y)$ is a non-constant a.e. function of x.

Fill in each $\phi(x,y)$ to a P-harmonic function on $\mathbf{H} \times y$ by the Poisson formula above. We obtain a function $h(x,y)$ on T by invariance of ϕ, harmonic on each level (Lemma 2) which implies $\tilde{P}h = h$, and finally $h(x,y)$ is non-constant on a set of levels of positive measure (Lemma 3).

Break $h - 1/2$ into positive and negative parts h_+ and h_- and add the inequality $\tilde{P}|h-1/2| \geq |\tilde{P}(h-1/2)| = |h-1/2|$ to the equality $\tilde{P}(h-1/2) = h-1/2$ to obtain $Ph_+ \geq h_+$. Using Lemma 5 and the cancellation argument of part i) Lemma 6 deduce $Ph_+ = h_+$ a.e. The latter implies the subregions of leaves (levels) where h_+ is zero don't communicate via posers of P with their complements. This is absurd and the theorem is proved.

NOTE. This method of proving measurable functions harmonic along the leaves of a foliation are constant was borrowed from Lucy Garnett's thesis

482   DENNIS SULLIVAN

[G] which contains a very simple proof of the ergodicity of the geodesic
flow in the finite volume case.

Section III. *Conical approach and E. Hopf's theorem*[1]

Say that Γ has *conical approach* at a point s on the sphere at
infinity in hyperbolic space if an orbit of Γ has infinitely many points in
a circular cone with vertex at s. E. Hopf proved in 1939 that Γ acts
ergodically on the (sphere $\times$ sphere) if and only if Γ has conical approach
at almost all points of the sphere. We study conical approach and derive
E. Hopf's theorem (cf. [BM]).

LEMMA 1. *If Γ has conical approach to a density point of a Γ-invariant
set on the sphere, this set has full measure.*

Proof. We integrate the characteristic function χ_A of the set A against
the Poisson measure of p inside hyperbolic space. Choose a sequence
of group elements γ_i so that $\gamma_i p$ has conical approach to a density point
of A.

$$\text{area } A = \mu_p A = \int \chi_A d\mu_p = \int \gamma_i \chi_A d\mu_p = \int \chi_A d\mu_{\gamma_i p} \, .$$

The easy classical estimate shows the right hand side approaches 1
since a is a density point of A.

COROLLARY. *If Γ has conical approach to a set of positive measure
on the sphere then Γ has conical approach to a set of full measure on
the sphere.*

Proof. Apply Lemma 1 to the set where Γ has conical approach.

LEMMA 2. *If Γ only has conical approach to a set of measure zero, the
action of Γ on $S^2 \times S^2$ is dissipative (i.e. it has a fundamental domain
a.e.).*

---

[1]This section works on S^n, $n \geq 1$, without change.

Proof. To each point q of the orbit of p under Γ associate A_q, those pairs of points on the sphere so that the connecting geodesic is closer to q than to any other point of the orbit. For almost all pairs there is a q because given a pair we merely swell up a geodesic tubular nghd of the connecting geodesic until it first meets the orbit. There is a first time because of the no conical approach assumption. The point q is unique after we throw out from the pairs countably many lower dimensional sub-manifolds. The A_q provide the desired partition of almost all the pairs into sets permuted freely by Γ.

COROLLARY. *If Γ has infinite solid angle then Γ has conical approach at a set of full measure.*[2]

Proof. Infinite solid angle implies the action on pairs is ergodic (Theorem II) which would be contrary to the conclusion of Lemma 2 if Γ did not have full conical approach.

LEMMA 3. *If Γ has conical approach at a set of positive measure on the sphere then Γ has infinite solid angle.*

Proof. The best angle of approach is a positive measurable function so is at least α for a set of positive measure which by Lemma 1 can be assumed to be the entire sphere.

α determines the size of a ball B appropriate for what follows. If the total solid angle of the orbit of B were finite, we could cast out finitely many balls so the remaining solid angle would be arbitrarily small. But the α conical approach implies the balls near infinity block infinity from view.

COROLLARY (E. Hopf). *A group Γ has conical approach either at a set of measure zero or at a set of full measure. In the first case the action on pairs of points on the sphere is dissipative. In the second the action on pairs is ergodic.*

---

[2]Proven by Tsuji (1944) for the case of Fuchsian groups using complex function theory.

HISTORICAL REMARK:

Theorem II (for hyperbolic 2-space) has an interesting history in function theory and the theory of Fuchsian groups. There were the implications

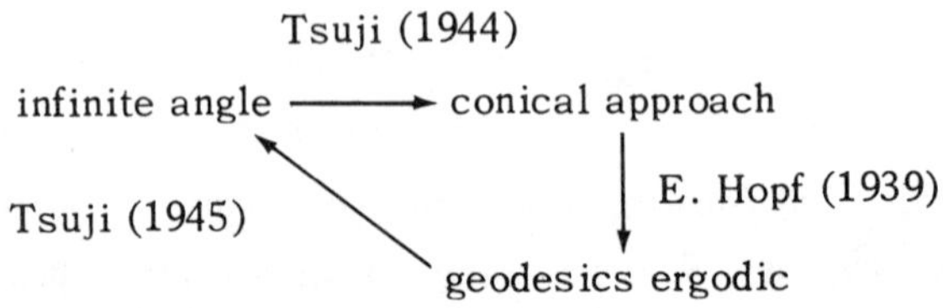

However, the implication

$$\text{infinite angle} \longrightarrow \text{geodesics ergodic}$$

was called *Tsuji's problem*, Shimada (1960). Tsuji apparently only knew the earlier Hopf theorem

$$\text{finite area} \longrightarrow \text{geodesics ergodic}$$
$$\text{E. Hopf (1937)}$$

which he reproved using potential theory rather than Birkhoff's ergodic theorem. Hopf's stronger (1939) theorem became more accessible after his Gibbs Lecture on the topic, AMS Bull 1971. P. J. Nicholls connected up Hopf (1939) and Tsuji (1944), corrected by Yujobo (1949), and Tsuji (1951), in a 1976 paper (see $[N_1]$ and p. 531 of Tsuji's book).

REMARK. The action of Γ on pairs is orbit equivalent to the geodesic flow on H/Γ. E. Hopf used the latter model, the associated asymptotic foliations, and his extension of Birkhoff's individual ergodic theorem to the case of an infinite invariant measure for the proof of the 1939 theorem.

For Fuchsian groups the condition infinite angle is just the divergence of the Blaschke product associated to an orbit. For the 1944 result Tsuji used complex function theory in an essential way, in particular the Riemann mapping theorem and the theorem of F. and M. Riesz, (Stockholm 1925).

Recall that our path around the triangle was different,

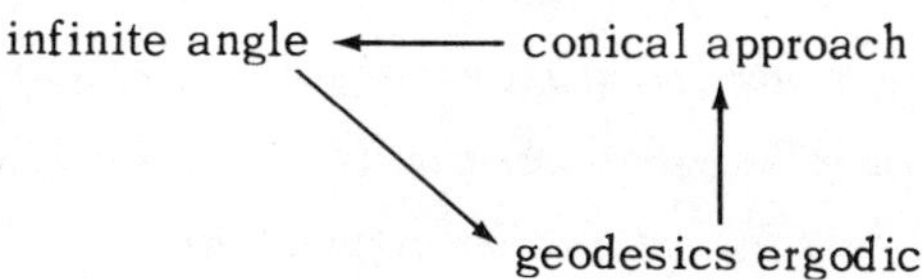

Only the slating arrow using random walks was not obvious.

Perhaps it is worth mentioning a conceptual anatomy of our proof. One rearrangement is

i)   There is a general theorem that a discrete group Γ of symmetries of the state space E underlying a symmetric Markov process acts ergodically on pairs of points of the boundary of the process if and only if the induced random walk on E/Γ is not transient [G].

ii)   This abstract theorem applies to random motion on an infinite regular covering of any Riemannian manifold complete for random motion.

iii) For a complete constant negatively curved manifold the boundary of the natural random process can be identified because a random path hits a definite point on the sphere at ∞ with probability 1. Also transience can be identified with finite solid angle.

Section IV. *Horospherical approach and recurrence*

Say that Γ has *horospherical approach* to a point s on the sphere at ∞ if the Γ orbit of a point in hyperbolic space enters every horosphere based at s. We call such points of the sphere the "horospherical limit set of Γ." It is geometrically clear the horospherical limit set is independent of the choice of reference orbit. These points were studied by Hedlund and later Eberlein [H], [E].

If s is not a horospherical limit point we can try to swell up horospheres at s until one first hits a point q of a fixed reference orbit. If there is a unique first hit we define this q to be the "closest orbit point to s."

THEOREM III. *For any group Γ and any choice of reference orbit in hyperbolic space the horospherical limit set union the points of the sphere which have a closest orbit point form a set of full measure. Furthermore this dichotomy is precisely the partition of the action of Γ on the sphere into its conservative part and its dissipative part.*

Proof. Assuming the first part for the moment, let us prove the second. To each orbit point q, associate the points A_q of the sphere to which it is closest. The A_q are freely permuted by Γ, so their union is a dissipative set.

On the other hand the horospherical limit set is easily seen to be characterized as those points s at which the derivatives of elements of Γ evaluated at s are unbounded. It follows there is no dissipative part in the horospherical limit set because along almost all dissipative orbits the sum of Jacobians is finite (the union of areas is finite).

Let us return to the first part. Throw out countably many lower dimensional submanifolds of the sphere which are bases of horospheres containing more than one point of the reference orbit. Now a point ℓ on the sphere not in the horospherical limit set fails to have a unique closest orbit point only if there is a critical horosphere $h(\ell)$ so that every larger horosphere based at ℓ contains infinitely many orbit points. We call such points "Garnett points."[1] The proof of the theorem is completed by the following lemma.

LEMMA 1. *The Garnett points have measure zero.*

Proof. Let s be a point of density in a set G of Garnett points of positive measure where the radius of the critical horosphere is a continuous function. Now there are infinitely many orbit points x_i outside the critical horosphere converging to s but entering every larger horosphere. For

---

[1]Lucy Garnett pointed out that these points have to be considered. Later John Garnett independently found a simple proof they have measure zero. In $[N_2]$ this possibility is overlooked (line 5, p. 310) invalidating the proof of Theorem 1, i) $\implies$ ii), there.

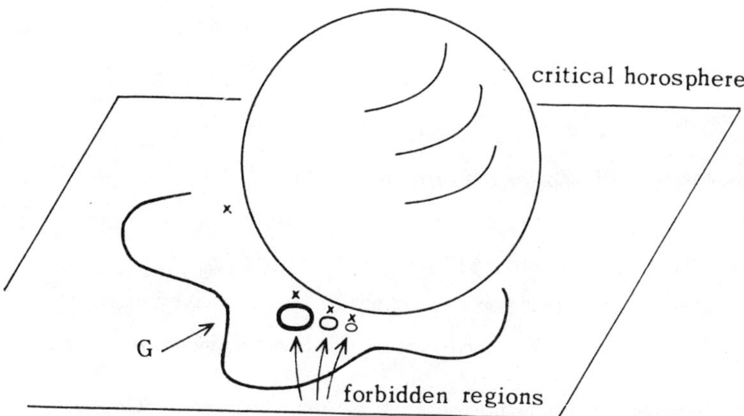

each x_i there is an associated forbidden region on the sphere at ∞ which does not contain Garnett points of G. For a point of G is the base of a horosphere whose radius is approximately that of h(s) by continuity and *which contains no orbit points.*

One calculates easily the forbidden region associated to x_i is a definite proportion of the area of the smallest disk centered at s containing the vertical projection of x_i. But then this picture contradicts the fact that s is a Lebesgue density point of G. Q.E.D.

Associated to the reference orbit is the partition of hyperbolic space into convex fundamental domains. To each orbit point q one assigns all the points D_q closer to it than to any other orbit point. The above discussion has extended this notion of closest point to the sphere at ∞. One has the

COROLLARY. *The dissipative part of the action of the discrete group* Γ *on the sphere is just the union of the "fundamental domains intersect the sphere at* ∞.*" (s $\epsilon\, D_q \cap S$ if there is a geodesic in D_q hitting s.)*

COROLLARY. *The action of* Γ *on the sphere is conservative if the area of one fundamental domain intersect the sphere at* ∞ *is zero.*

REMARK. The condition of the last corollary may be reformulated in terms of the growth of volume of the quotient manifold $V = H/\Gamma$. Let $V(r)$

denote the volume of the points of V within a distance r of some base point p. Let H(r) denote the volume of the ball of radius r in hyperbolic space.

THEOREM IV. *The following are equivalent*

  i)   *The fundamental domain has zero area at* ∞.

  ii) *The ratio* $V(r)/H(r) \to 0$ *as* $r \to \infty$.

  iii) *The action of* Γ *on the sphere at* ∞ *is conservative.*

  iv) *The horospherical limit set of* Γ *has full measure on the sphere.*

Proof. Only i) $\Rightarrow$ ii) is not obvious or already proven. Consider the solid angle $\omega(r)$ of the sphere of radius r centered at q within the fundamental domain D_q. By convexity of D_q $\omega(r)$ is decreasing. By i) the limit at $r = \infty$ is zero. If $a(r)$ denotes the area of the sphere of radius r we have

$$V(r)/H(r) = \frac{\displaystyle\int_0^r a(r)\,\omega(r)\,dr}{\displaystyle\int_0^r a(r)\,dr}\,.$$

Since $\omega(r) < \varepsilon$ for $r > r_0$ we have

$$V(r)/H(r) \leq \frac{\displaystyle\int_0^{r_0} a(r)\,\omega(r)\,dr}{\displaystyle\int_0^r a(r)\,dr} + \varepsilon\,\frac{\displaystyle\int_{r_0}^r a(r)\,dr}{\displaystyle\int_0^r a(r)\,dr}\,.$$

The first term is zero in the limit of $r \to \infty$ and the second is at most ε.

NOTE. In the 1939 treatise E. Hopf asked whether condition ii) or others like it might imply the ergodicity of the geodesic flow. (Equivalently the action of Γ on $S \times S$). In the fifties a Fuchsian group Γ was constructed[2] (after a long search) whose Riemann surface carried a Green's function but no bounded harmonic function. Thus the action of this Γ on $S^1 \times S^1$ is not ergodic but the action of Γ on S^1 is ergodic. In particular condition ii) is satisfied but there is no sectorial approach. By Theorem IV there is horocycle approach almost everywhere.

Section V. *Ahlfors-Bers quasi-conformal deformations of discrete subgroups of* $\mathrm{PS}\ell(2, \mathbf{C})$

If Γ is a discrete group of isometries of hyperbolic space $\mathbf{H}^3$, one knows how to deform Γ in a nice way using a measurable line field invariant by Γ and a positive bounded measurable function invariant by Γ. By Ahlfors-Bers [AB] there is a quasi-conformal homeomorphism ϕ of the sphere whose conformal distortion lies a.e. in the direction of the line field with strength given by the function.

One may view the measurable data consisting of the line field and function as a new measurable conformal structure invariant by Γ (which is a bounded distance away from the smooth conformal structure). The homeomorphism ϕ converts the measurable structure back to the smooth conformal structure on S^2.

Bers [B] conjugates the conformal transformations of Γ by ϕ to obtain new conformal transformations, $\{\gamma\} \to \{\phi\gamma\phi^{-1}\}$, and a new discrete group Γ'. If Γ and Γ' are also conjugate by a conformal transformation it follows the measurable data we started with vanishes a.e. on the topological limit set (because fixed points are carried to fixed points by any conjugacy, so any topological conjugacy between Γ and Γ' is determined uniquely on the topological limit set). $\Gamma' = \phi\Gamma\phi^{-1}$ is called a quasi-conformal deformation of Γ.

From Section I we find there can only exist invariant line fields on the dissipative part of the action of Γ on S^2.

---

[2]See Ahlfors-Sario, p. 256-257.

Now the dissipative part of the action splits into two pieces: the domain of discontinuity of Γ (if any) whose quotient by Γ is called the Riemann surface associated to Γ, and the dissipative part (if any) of the action on the topological limit set.

Any dissipative piece in the limit set will lead to an infinite dimensional space of inequivalent quasi-conformal deformations by the uniqueness of the conjugacy remark above. The domain of discontinuity leads to a Teichmüller space of deformations which can be nicely described if the Riemann surface of Γ is obtained from a compact by removing at most finitely many points, [B] and [M].

For *finitely generated* groups Γ one has Ahlfor's result [A] that the associated Riemann surface is obtained from a compact surface by removing at most finitely many points. One also knows the embeddings of the abstract group into $PS\ell(2, C)$ is a finite dimensional space.

Thus the infinite dimensional linear space is not present for finitely generated groups and the Teichmüller space in question is a nice complex manifold about which much is known, [B].

THEOREM V. *For a finitely generated discrete subgroup Γ of $PS\ell(2, C)$ the classes of quasi-conformal deformations consists of a Teichmüller space associated to the uniformized Riemann surface of finite type. For example, if Γ has a dense orbit on S^2, then Γ is quasi-conformally rigid.*

Proof. One only needs to read [B] and [M] and forget their hypothesis that the deformation data be zero on the limit set. For it must be by the uniqueness theorem and the above deformation remark.

Section VI. *Dynamical properties of finitely generated Kleinian groups*

Let us collect here the information about finitely generated groups following from the foregoing general discussion. The main point of the finite generation of Γ for us is the deformation remark of Theorem V which is mostly due to Ahlfors.

THEOREM VI. *For a finitely generated discrete group of isometries of* H^3 *, the horospherical limit set has full measure in the topological limit set.*

Proof. The deformation remark of Section V shows the topological limit set contains no dissipative piece.[1] Thus by Theorem III, Section IV the horospherical set has full measure in the topological limit set.

We record the proof as a

COROLLARY. *The action of a finitely generated* Γ *on the topological limit set is conservative.*[1]

COROLLARY. *There is no measurable field of regular tangent n-crosses defined on a positive area subset of the limit set and invariant by* Γ.

Proof. We have proved this result for tangent line fields (= tangent 2 – cross) in Section I. The same proof works for n-gons.

REMARK. A formal ergodic consequence of the previous corollary is that the "angular ratio set" is the entire circle. Namely given θ on the circle, a positive number ε, and a subset A of the limit set of positive measure there is an element $\gamma \in \Gamma$ so the subset of $A \cap \gamma^{-1}A$ where the angular part of the derivative of γ is within ε of θ has positive measure. In other words one sees every twist (θ) up to any approximation (ε) anywhere one looks (A).

COROLLARY. *There is no absolutely continuous measure on the limit set invariant by* Γ.

Proof. If ρ were the density function of such a measure, work in a subset where ρ is approximately constant using the variation of distortion

---

[1]Ahlfors (1967) gave a "Cauchy transform" proof of this fact when there was a non-trivial domain of discontinuity. See "Remarks on Kleinian Groups" Tulane Conference, 1967.

produced by Lemma 3 of Section I to derive a contradiction, as in the last part of Section I.

NOTE. The density function ρ is not assumed to be locally integrable. For it is elementary that for an arbitrary group Γ there is no invariant probability measure. The corollary means for finitely generated groups there is never a σ-finite invariant measure in the smooth measure class. This fact means that the measurable equivalence relation induced by Γ on the limit set is type III in the sense of Murray and Von Neumann. One can hope to classify the equivalence relation up to Lebesgue isomorphism by a further study of the derivatives (going beyond Lemma 3 of Section I).[2] For example when are there non-trivial elements in the "ratio set", defined as in the previous remark using the area distortion of the derivative rather than the angular part?[2]

Recall that we can pretend to understand the ergodic theory of groups where $\sum_{\Gamma} 1/\lambda^2 = \infty$. If then $\sum_{\Gamma} 1/\lambda^2 < \infty$, what is the significance of the of the function $\sum_{\Gamma} 1/\lambda^s$ for other values of s? For $s = 1$ one has the

COROLLARY. *If the topological limit set has positive area then* $\sum_{\Gamma} 1/\lambda = \infty$.

Proof. The area of the isometric disk of γ union the image of its complement has the order $1/\lambda$ using Lemma 1 of Section I. Using recurrence as in the final argument of Section I after casting out finitely many elements to make $\sum 1/\lambda$ small we arrive at a contradiction. Namely, $\sum 1/\lambda < \infty \Rightarrow$ the action of any Γ on S^2 is totally dissipative.

---

[2] Actually the addendum to Section I and Lemma 3 implies the ratio set for the area distortion is all the positive reals. Thus any Kleinian group has type III_1 on its conservative part. This theorem will be explained in more detail in a future publication.

Section VII. *Rigidity in the sense of Mostow for infinite volume discrete groups in* $O(n, 1)$

Say that a complete hyperbolic n-manifold $V = H^n/\Gamma$ is Mostow-rigid if any pseudo-isometry between V and another hyperbolic manifold V' is homotopic to an isometry.

A pseudo-isometry is by definition a continuous map $V \to V'$ inducing an isomorphism between the fundamental groups and keeping the distances between sufficiently farlying pairs of points in a bounded ratio.

Mostow showed [Mo] that

i)   a pseudo-isometry of hyperbolic n-space is onto and has a continuous extension to the sphere at ∞ which is a quasi-conformal homeomorphism;

ii)   a quasi-conformal homeomorphism of S^{n-1} ($n \geq 3$) whose derivative is a.e. a similarity is a conformal transformation;

iii) for discrete groups Γ whose fundamental domains have finite volume there is only one a.e. measurable conformal structure on the sphere invariant by Γ.

MOSTOW'S THEOREM. *Complete hyperbolic* n-manifolds of finite volume are rigid in the above sense, $n \geq 3$.

Proof. Mostow lifts the pseudo isometry to hyperbolic space, extends to the boundary by i) to find a quasi-conformal homeomorphism which must be conformal by iii) then ii), and obtains an equivariant isometry of hyperbolic space with these boundary values.

Celebrated corollaries are: compact hyperbolic n-manifolds M even complete finite volume hyperbolic n-manifolds are determined up to isometry ($n \geq 3$) by their fundamental groups.

Proof. The fundamental group determines the pseudo-isometry type in these cases. This is clear for the compact case and follows from an analysis of the cusps in the finite volume case (Margulis, Prasad).

Using Theorem VI below to extend point iii) we obtain by the same proof an extension of Mostow's theorem to the infinite volume case.

THEOREM VI. *If* V *is a complete hyperbolic* n-*manifold* $(n \geq 3)$ *and satisfies any of the equivalent conditions of Theorem IV (for example if the volume of the part of* V *out to distance* r *grows slower than the ball of radius* r *in hyperbolic space), then* V *is rigid in the sense of Mostow.*

COROLLARY. *If* V^3 *is defined by any discrete finitely generated group of isometries of* $\mathbf{H}^3$ *with a dense orbit on* S^2, *then* V^3 *is rigid in the sense of Mostow.*

Proof of Corollary. Section VI reduces the corollary to Theorem VI.

Proof of Theorem VI. Same as above proof using Theorem VII.

THEOREM VII. *Let* Γ *be a discrete group of conformal transformations of* S^n, $n \geq 2$ *and let* ν *be a measurable conformal structure (on the tangent spaces), which is a.e. invariant by* Γ. *Then* ν *agrees a.e. with the standard conformal structure on the conservative part of action of* Γ *on* S^n.

Proof. By comparing ν with the standard structure we find a field of ellipsoids, each defined up to similarity. We can work on an invariant set of positive measure where the smallest axes form a proper subspace of constant dimension. We find then an invariant tangent k-plane field invariant by Γ, $k < n$.

We ruled out this possibility for $k = 1$, $n = 2$ in Section I by an argument based on the variation of distortion (Lemma 3). We will deduce the general case from this same argument. Also note that Sections I, II, III go word for word the same for general n, using $\sum_{\Gamma} 1/\lambda^n$ instead of $\sum_{\Gamma} 1/\lambda^2$.

Now working in the $\mathbf{R}^n$-model choose a set of positive measure X where the k-plane field $\{P\}$ is within ε of being parallel. Now consider any transformation γ keeping $\{P\}$ invariant. We can factor γ into a composition

$$\gamma = (\text{inversion in circle}) \cdot (\text{Euclidean isometry}) = I_\gamma R_\gamma .$$

Then $\{P'\} = R_\gamma\{P\}$ is another almost parallel field (on $X' = R_\gamma(X)$) which is carried by the inversion I_γ to $\{P\}$, $I_\gamma\{P'\} = \{P\}$.

Now the tangent maps to an inversion are just the reflections in the tangent planes to spheres concentric about the center of inversion. Near to the center of inversion these tangent planes turn rapidly and it is difficult for the inversion to carry an almost parallel field on X' to an almost parallel field on X (if X' has many points sufficiently close to the center which are also carried to X).

Now if $\gamma \epsilon \Gamma$ we can determine where X' is relative to the center of inversion of I_γ by knowing the distortion of γ on X' (since R_γ is an isometry, and in polar coordinates based at the center the distortion of the inversion $(r, \theta) \to (1/\lambda r, \theta)$ is $1/\lambda r^2$). What we will know from conservativity and the argument of Section I is that for any $Y \subset X$ of positive measure there is a $\gamma \epsilon \Gamma$ so that $\gamma^{-1}Y \cap Y$ has a point y, $\log(\text{planar distortion } \gamma)$ varies in an interval of length $\sim \Delta$ over an $\sim$ concentric annular shell A containing y (being constant on $\sim$ concentric spheres), A and γA have diameter $\leq \delta'$, and "$\sim$", Δ, and δ' are at our disposal.

Writing $A' = R_\gamma A$ we see from these values of distortion that A' is a nearly concentric annular shell about the center of inversion of I_γ of diameter δ'. Since the shape of A is determined by Δ and "$\sim$" we can use such sets to describe (η, δ) uniform density points of X as in Lemma 3, Section I.

We choose η so close to 1 that a contradiction results from the following chain of considerations. Y is chosen, using Lemma 3, Section I for (X, η, δ). Then γ is chosen as above for Δ, δ', Y with $\delta' < \delta$. A will be almost filled with points of X which map to X by γ (as in the argument of Section I). So the same holds for I_γ, namely A' is almost filled with points of X' which map to X (note R_γ alters nothing being an isometry).

The above description of the tangent map forces a contradiction, I_γ cannot carry the ε-almost parallel filled on X' to the ε-almost parallel field on X.

INSTITUT DES HAUTES ETUDES SCIENTIFIQUES
35, ROUTE DE CHARTRES
91440 BURES-SUR-YVETTE, (FRANCE)

REFERENCES

[A]   Ahlfors, L. "Finitely Generated Kleinian Groups." American Journal of Mathematics, volume 86 (1964), pp. 413-429.

[AB] Ahlfors, L. and Bers, L. "Riemann's Mapping Theorem for Variable Metrics." Annals of Math. vol. 72 (1960), pp. 385-404.

[BM] Beardon, A. and Maskit, B. "Limit points of Kleinian groups and finite sided fundamental polyhedra." Acta Math. 132 (1974), pp. 1-12.

[B]   Bers, L. "Spaces of Kleinian Groups." Several Complex Variables I Maryland (1970). Springer vol. 155 (1970).

[E]   Eberlein, P. "Geodesic Flows on negatively curved Manifolds." II, Trans. Amer. Math. Soc., vol. 178 (1973).

[G]   Garnett, Lucy. "Measures and Functions Harmonic along the Leaves of a Foliation and the Ergodic Theorem." I.H.E.S. preprint June, 1980.

[H]   Hedlund, H. "Transitive Horocycles for Fuchsian groups." Duke Journal of Math. 2 (1936), pp. 530-542.

[K]   Kemeny, J., Knapp, A., Snell, L. "Countable Markov Chains." Van Nostrand, (1966).

[M]   Maskit, B. "Self Mappings of Kleinian Groups." Amer. J. Math. XCIII (1971), pp. 840-856.

[Mo] Mostow, D. "Rigidity of locally symmetric spaces." Princeton University Press (1972).

[N_1] Nicholls, P. J. "Transitivity properties of Fuchsian groups." Can. J. Math. 28 (1976), pp. 805-814.

[N_2] __________. "Transitive horocycles for Fuchsian groups." Duke Math. J. 42 (1975), pp. 307-312.

[S]   Stein, E. M. "Singular Integrals and Differentiability Properties of Functions." Princeton University Press (1970).

[Su] Sullivan, Dennis. "The density at infinity of a discrete group of hyperbolic motions." I.H.E.S. publications, vol. 50.

[T]   Tsuji, M. "Potential theory in modern function theory." (Maruzen, Tokyo, 1959).

ON INFINITE NIELSEN KERNELS

Judith C. Wason[*]

§1. *Introduction*

Let S_{-1} be the Nielsen kernel of a Riemann surface $S = S_0$ of finite
type (g, n, m), $6g - 6 + 2n + 3m > 0$, $m > 0$. S_{-1} is then also a Riemann
surface of type (g, n, m), and, in general, we may consider S_{k-1}, the
Riemann surface of type (g, n, m) which is the Nielsen kernel of S_k,
$k = 0, -1, -2, \cdots$. In [1] Bers described the relationship of S_{-1} and S,
and suggested a study of $S_{-\infty} = S_{-1} \cap S_{-2} \cap S_{-3} \cdots$, the infinite Nielsen
kernel of S.

We may consider two metrics on S_k; the usual Poincaré metric, and
the intrinsic metric, which is the restriction to S_k of the Poincaré metric
on the Schottky double S_k^d of S_k. Recall that the intrinsic metric on S_k
agrees with the restriction to S_k of the Poincaré metric of S_{k+1}, and
that a boundary curve of S_k will be the geodesic homotopic to the corresponding boundary curve of S_{k+1}. We will discuss the lengths of two
types of curves on the S_k's.

§2. *Lengths of the boundary curves*

As in Theorem 3 of [1], the proof of the following theorem is based on
an application of Grötzsch's inequality.

---

[*]This research was conducted while the author was a visitor at Harvard
University.

THEOREM 1. *Let $C_{k,i}$ be a boundary curve of S_k. Let $L_{k,i}$ be its length in the intrinsic metric on S_k (or Poincaré metric of the Nielsen extension S_{k+1}). Then $\lim\limits_{k \to -\infty} L_{k,i} = \infty$, $i = 1, \cdots, m$.*

Proof. At each stage of taking the kernel we have in effect removed the funnels $\{A_{k-1,i}\}$ from the Riemann Surface S_k, where $A_{k-1,i}$ has module $M_{k-1,i}$. Hence $S - S_{-\infty} = F_1 \cup \cdots \cup F_m$, where $F_i = \bigcup\limits_{k=-1}^{-\infty} A_{k,i}$. The interiors of the $\{A_{k,j}\}$ are all disjoint, and at any stage, $A_{k-1,i}$ separates the boundary continua of $A_{-1,i} \cup A_{-2,i} \cdots \cup A_{k-1,i}$. If M_i is the module of F_i, by Grötzsch's inequality,

$$M_i \geq M_{-1,i} + M_{-2,i} + \cdots$$

$$= \pi^2 (L_{-1,i}^{-1} + L_{-2,i}^{-1} + \cdots) \text{ by Lemma 1 of [1]}.$$

In particular, $M_i \geq \pi^2 (L_{-1,i}^{-1} + L_{-2,i}^{-1} + \cdots L_{-n,i}^{-1})$. Now, by [1], $L_{k,i} > L_{k+1,i}$, and so $M_i \geq \pi^2(n)(L_{-n,i}^{-1})$. Suppose the $L_{k,i}$ remained bounded, $L_{k,i} \leq P$. Then $M_i \geq \pi^2(n)(1/P)$ becomes infinite as n becomes large. But this would imply that F_i was, in fact, a punctured disk to begin with, a contradiction. Hence $\{L_{k,i}\}$ must have a subsequence which is unbounded; since $L_{k,i} > L_{k+1,i}$, the entire sequence must go to infinity as $k \to -\infty$, while the M_i's approach zero.

Note that we actually can make a somewhat stronger statement regarding the divergence. If $\{L_{k,i}\}$ has a subsequence $\{\tilde{L}_{-n,i}\}$ such that $\tilde{L}_{-n,i} \leq n$ for $n \geq 1$, then M_i would be larger than the "sum" of the harmonic series, and F_i would again be a punctured disk.

§3. *Distances between the boundary curves*

We now assume that $m \geq 2$. Represent S as U/G where G is a finitely generated torsion-free Fuchsian group of the second kind. Let G have a convex fundamental polygon T with the property that the intersections of two sides with a given free side are also the intersections of R

and the isometric circles of the transformations pairing the two sides. (If necessary, use a conjugate of G to insure that these isometric circles are well defined.) Furthermore, with at most two exceptions, sides ending in free sides should intersect only other sides ending in free sides [2].

Let γ_i be a curve contained in T with endpoints on two adjacent free sides C_i and C_{i+1} for $i = 1, \cdots, m-1$. Let p_i be the projection to S^d of $\gamma_i \cup \bar{\gamma}_i$. Label the corresponding funnels of S, $F_1, \cdots, F_m$. Then the axes of the transformations pairing free sides are representatives of part of a canonical homotopy basis for S. Since the corresponding boundaries and hence geodesics of the kernels are homotopic, we in fact may at any stage obtain, in the same order, representatives of part of a canonical homotopy basis. Consequently, there is a sequence of fundamental polygons T_k satisfying the requirements of T, which preserves the relative order, and hence adjacency, of corresponding free sides [2], [3].

Let $p_{k,i}$ be the unique shortest geodesic in the homotopy class of $p_i \cap S_k$ and its mirror image in S_k^d.

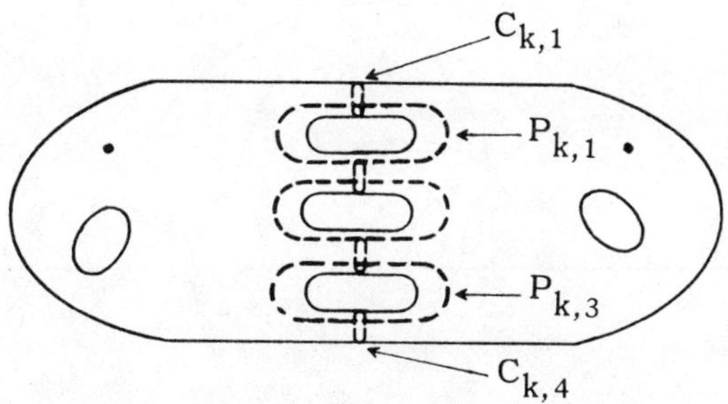

Fig. 1. S_k^d has type $(5, 2, 0)$.

THEOREM 2. *Let S_k have at least two boundary curves. Then there are curves* $p_{k,i}$, $i = 1, \cdots, m-1$ *on S_k^d of length $P_{k,i}$, connecting pairs of boundary curves of S_k such that* $\lim_{k \to -\infty} P_{k,i} = 0$, $i = 1, \cdots, m-1$.

Proof. The geodesics $C_{k-1,i}$ of S_k will lift in T_k to axes of hyperbolic elements generating the holes of S_k. Since T_k is convex, a particular $P_{k-1,i} \cap S_k$ lifts to a circular arc $\overset{\frown}{R_1 R_2}$ crossing two non-intersecting axes, at the points R_1 and R_2. We will show that for a fixed, but arbitrary i, the length $|\overset{\frown}{R_1 R_2}|$ of $\overset{\frown}{R_1 R_2}$ goes to zero. Since the Poincaré metric of S_k agrees with the intrinsic metric of S_{k-1}, the intrinsic distance between the corresponding boundary curves of S_{k-1}, $\frac{1}{2} P_{k-1,i}$, will be equal to $|\overset{\frown}{R_1 R_2}|$ and hence also go to zero.

When it is clear that we are considering only a particular stage, the subscript k will be omitted.

Let R_0 be the midpoint of $\overset{\frown}{R_1 R_2}$. Draw non-Euclidean straight lines from R_0 to Q_1, Q_2, Q_3, and Q_4, the intersection points of the boundary of the fundamental polygon R and the axes connected by $\overset{\frown}{R_1 R_2}$ (figure 2). If the boundary of R coincides with $\overset{\frown}{R_1 R_2}$, draw n.E. lines from R_0 to the other intersection points, Q_1 and Q_4 (figure 3). If necessary, use the appropriate combination of constructions.

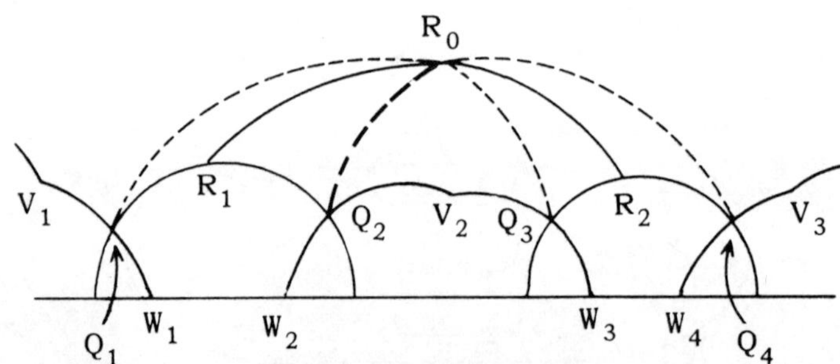

Fig. 2

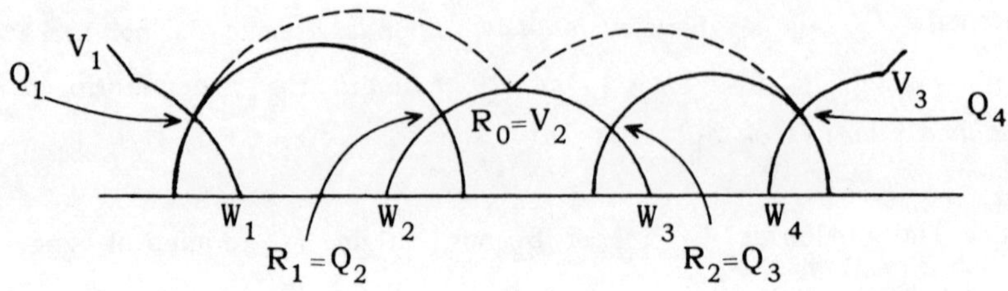

Fig. 3

We next look more closely at the situation involving the n.E. triangle $R_0Q_1Q_2$ (figures 4 and 5). The other is analogous.

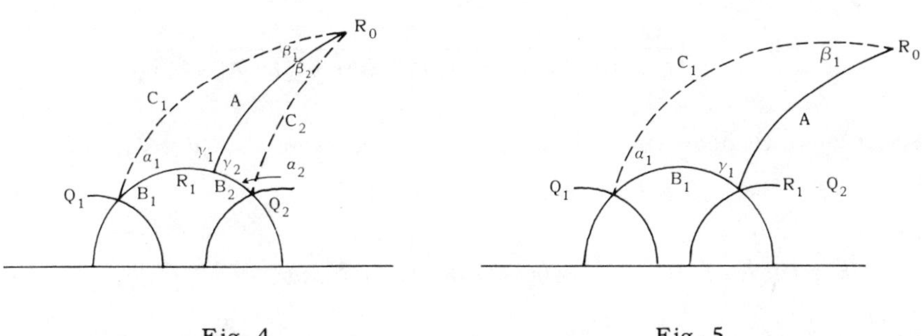

Fig. 4   Fig. 5

The triangle $R_0Q_1Q_2$ degenerates if and only if all three points are collinear. But then R_0 must lie on the axis $A(\gamma)$ of γ, and $0 = |\overparen{R_0R_1}|$ $= \frac{1}{2} |\overparen{R_1R_2}|$ and the two axes intersect. Similarly, at least one triangle $R_0Q_jR_1$ is non-degenerate.

The length of the portion of axes $\overparen{Q_1Q_2}$ represents the translation lengths of the transformations generated by the geodesic around a hole of S_k. Hence as $k \to -\infty$, $|\overparen{Q_1Q_2}|$ becomes arbitrarily large, and so either $|\overparen{Q_1R_1}|$ or $|\overparen{R_1Q_2}|$ must also become large. Consequently, we may form a sequence $\{Q_{k,j}R_{k,1}\}$ in which every k is represented, and which satisfies $|Q_{k,j}R_{k,1}| \to \infty$. This sequence consists of two subsequences, each with infinite limit, $\{Q_{k,1}R_{k,1}\}$ and $\{Q_{k,2}R_{k,1}\}$.

LEMMA 1. *If* $\lim_{k\to-\infty} |\overparen{Q_{k,j}R_{k,1}}| = \infty$, *then* $a_{k,j}$, *the angle* $R_{k,0}Q_{k,j}R_{k,1}$, *has limit zero.*

Proof. Let a be the angle $R_0Q_jR_1$, and $A = |\overparen{R_0R_1}|$. Let β be the angle $Q_jR_0R_1$, and $B = |\overparen{Q_jR_1}|$. Finally, let γ be the angle $R_0R_1Q_j$, and let $C = |\overparen{R_0Q_j}|$. The hyperbolic sine law implies $\dfrac{\sin a}{\sin \gamma} = \dfrac{\sinh A}{\sinh C} < 1$ since $\overparen{R_0R_1}$ is part of the unique shortest geodesic $\overparen{R_1R_2}$. Since the sum of the angles of a non-degenerate hyperbolic triangle is strictly less than π, $\sin a < \sin \gamma$. In fact, a must be less than $\pi/2$ for if $a = \pi - \varepsilon$, $0 < \varepsilon \le \pi/2$, then $\gamma < \varepsilon \le \pi/2$, and $\dfrac{\sin a}{\sin \gamma} > \dfrac{\sin(\pi-\varepsilon)}{\sin \varepsilon} = 1$, a contradiction. By a formula in [4], p. 85,

$$|\cosh B| = \left|\frac{\cos a \cos \gamma + \cos \beta}{\sin a \sin \gamma}\right| \le \frac{2}{|\sin a \sin \gamma|} < \frac{2}{|\sin^2 a|}\ .$$

Since $\cosh B$ becomes arbitrarily large as $k \to -\infty$, $\sin^2 a \to 0$. Since $a \le \pi/2$, $\displaystyle\lim_{k \to -\infty} a_{k,j} = 0$.

A computation using tangent vectors then shows:

LEMMA 2. *Let* $\{C_{n,1}, C_{n,2}\}$ *be a sequence of pairs of hyperbolic straight lines which intersect at points* (u_n, v_n) *with angle of intersection* a_n, *where* $0 \le a_n \le \pi/2$. *Then* $a_n \to 0$ *if and only if* $\{C_{n,1}, C_{n,2}\} = S_1 \cup S_2$ *where* S_1 *is a (possibly empty) subsequence for which* $v_n \to 0$, *and* S_2 *is a (possibly empty) subsequence for which* $C_{n,1} \to C_{n,2}$.

Note that if $\overparen{R_{k,0}Q_{k,j}}$ and $\overparen{Q_{k,1}Q_{k,2}}$ are members of a subsequence with coinciding limit, then R_0 must be arbitrarily close to $A(\gamma_k)$, and so $|\overparen{R_1 R_2}|$ must go to zero. Hence we assume that $v_{k,j} \to 0$. (Recall that by the choice of subsequences $\{\overparen{Q_{k,1}R_{k,1}}\}$ and $\{\overparen{Q_{k,2}R_{k,1}}\}$, every remaining k will be represented in the respective subsequences $\{v_{k,1}\}$ or $\{v_{k,2}\}$, both with limit zero.)

Consider the arrangement of the isometric circles $I(\gamma)$ and $I(\gamma^{-1})$ and the axis $A(\gamma)$ of the hyperbolic transformation $\gamma(z) = \dfrac{az+b}{cz+d}$. Computations show that the distance D between either endpoint of the axis of γ and the nearer of the endpoints of the isometric circles under the axis of γ satisfies $D \ge \dfrac{1}{|c_k|} - \dfrac{\varepsilon}{2|c_k|}$, $\varepsilon > 0$, where $\varepsilon \to 0$ as the length of the geodesic determined by γ goes to infinity. Hence, if any subsequence of $\{|c_k|\}$ remains bounded, $|c_n| \le M$, then, for $|k| > K$, $D_k \ge \dfrac{1}{|c_k|} - \dfrac{\varepsilon}{2|c_k|} > 0$, $\varepsilon > 0$, and $D_k \ge \dfrac{1}{M}\left(1 - \dfrac{\varepsilon}{2}\right) > 0$ for $|k| > K$. By Lemma 2, as $k \to -\infty$, $v_{k,j}$ (ordinate of the intersection point $Q_{k,j}$ of $A(\gamma)$ and $\widehat{R_0 Q_j}$) must go to zero. Since $D_k \geq \frac{1}{M}\left(1-\frac{\epsilon}{2}\right)$, $Q_{k,j}$ cannot collapse to

$W_{k,j}$. Hence for $|k| > K$, the arc $\widehat{W_{k,j} Q_{k,j} V_{k,j}}$ (a side of the fundamental polygon) must be contained in $I(\gamma)(I(\gamma^{-1}))$. But then $\widehat{W_{k,h} Q_{k,h} V_{k,h}}$, $h \neq j$, must be an arc outside $I(\gamma_k^{-1})(I(\gamma_k))$, which is tangent to and outside a circle of radius $\frac{1}{M}$ centered at least $\frac{1}{M}\left(1-\frac{\epsilon}{2}\right)$ units away from

$W_{k,i}$. Since γ is conformal, and $\widehat{R_{k,0} Q_{k,j}}$, $\widehat{Q_{k,j} R_{k,1}}$ are distinct, the image of $a_{k,j}$, $a_{k,j}^1$ with vertex $Q_{k,h}$ also must go to zero, and hence $v_{k,h} \to 0$. Then $\widehat{Q_{k,1} Q_{k,2}}$ must approach the x axis, with small Euclidean length. This clearly cannot be done via a sequence of convex fundamental polygons. Hence $\{|c_n|\}$ must be unbounded.

We now must consider a sequence of transformations $\{\gamma_k\}$ where the radii of $I(\gamma)$ and $I(\gamma^{-1})$ become arbitrarily small. Suppose $\{\widehat{W_{k,j} Q_{k,j} V_{k,j}}\}$ has a subsequence which remains inside or on the isometric circle $I(\gamma_k^{\pm 1})$. Since $\frac{1}{|c_k|} \to 0$, $\{\widehat{W_{k,j} Q_{k,j} V_{k,j}}\}$ must collapse to a point, and $V_{k,j}$ and hence $V_{k,h}$ must be arbitrarily close to points on $\{A(\gamma) \cap R\}$. If $\{\widehat{W_{k,j} Q_{k,j} V_{k,j}}\}$ remains outside (or on) $I(\gamma_k^{\pm 1})$, then $\{\widehat{W_{k,h} Q_{k,h} V_{k,h}}\}$ must be inside (resp. on) $I(\gamma_k^{\pm 1})$ and the same argument holds. The corresponding situation exists for the neighboring γ (by the remark following Lemma 2). A computation then shows that $|\widehat{R_1 R_2}|$ must go to zero.

WELLESLEY COLLEGE
WELLESLEY, MASS. 02181

REFERENCES

[1] Bers, L., Nielsen Extensions of Riemann Surfaces, *Ann. Acad. Sci. Fenn.*, 2 (1976), 17-22.

[2] Keen, L., Canonical polygons for finitely generated Fuchsian Groups, *Acta Math.*, 115 (1966), 1-16.

[3] ________, Intrinsic Moduli on Riemann Surfaces, *Ann. of Math.*, 84 (1966), 404-416.

[4] Meschkowski, H., *Noneuclidean Geometry*, Academic Press, New York, 1964.

HYPERBOLIC 3-MANIFOLDS WHICH SHARE A FUNDAMENTAL POLYHEDRON

Norbert J. Wielenberg[*]

1. *Introduction*

Let G be a discrete, torsion-free subgroup of $PSL(2, C)$. Then the quotient under the action of G of the region of discontinuity of G in the closure of hyperbolic 3-space is a 3-manifold, possibly with boundary. The boundary is $\Omega(G)/G$ where $\Omega(G)$ is the ordinary set of G in $C \cup \{\infty\}$. Marden [2] has discussed group constructions which topologically consist of gluing two boundary components of the same or of different 3-manifolds. This is of particular interest if each component of each lift of a boundary component is a euclidean disc. Geometrically the constructions then consist of gluing totally geodesic surfaces which are incompressible and boundary-parallel in the 3-manifolds. Algebraically, the group constructions are free products with amalgamation or HNN extensions. See [8] for link groups constructed in this way and Thurston [6] for more general examples.

As corollaries of the theorems of Gromov and Jørgensen, the following results are given in [6]. The set of volumes of hyperbolic 3-manifolds is well ordered. The volume is a finite-to-one function of hyperbolic manifolds with finite volume.

———————
[*] Research supported in part by the National Science Foundation.

We will discuss some groups which are constructed by HNN extensions and which justify the following.

THEOREM. *For each integer* N *, there is a polyhedron of finite volume in hyperbolic 3-space which is a fundamental polyhedron for at least* N *different 3-manifolds.*

In our examples, the corresponding groups are subgroups of finite index in the Picard group $PSL(2, Z(i))$.

2. *Hyperbolic isometries*

As a model for the simply-connected hyperbolic 3-space we will use $H^3 = \{z + tj : j = (0, 0, 1) \text{ and } t > 0\}$. The Riemannian metric is $t^{-2}(dx^2 + dy^2 + dt^2)$ and the volume element is $dV = t^{-3}(dx\,dy\,dt)$. The operation of inversion in the sphere of radius 1 with center at the origin can be written as $z + tj \to (z + tj)^*$ where

$$(z + tj)^* = \frac{z + tj}{|z + tj|^2} .$$

This is an isometry of H^3.

For $c \in C$, we can define an orthogonal transformation T_c of R^2 by $T_c(z) = -(\bar{c}/c)\bar{z}$. An orthogonal transformation of R^2 extends in an obvious way to an orthogonal transformation of R^3 which is an isometry of H^3, i.e., leave the t-coordinate unchanged. We remark that if $A(z) = \frac{az + b}{cz + d}$ with $ad - bc = 1$, then an elementary calculation shows that

$$A(z) = \frac{a}{c} + \left(\frac{1}{|c|}\right)^2 T_c\left(z + \frac{d}{c}\right)^* .$$

We then extend A to act on H^3 by

$$(1) \qquad A(z + tj) = \frac{a}{c} + \left(\frac{1}{|c|}\right)^2 T_c\left(z + tj + \frac{d}{c}\right)^* .$$

This is an isometry of H^3 and the group of all orientation preserving

isometries of H^3 can be identified with $PSL(2, C)$, where A corre-
sponds to $\pm \begin{pmatrix} a & b \\ c & d \end{pmatrix}$ and A^{-1} to $\pm \begin{pmatrix} d & -b \\ -c & a \end{pmatrix}$.

The circle where $|cz + d| = 1$ is called the isometric circle of A. So
we have isometric spheres $|z + tj + \frac{d}{c}| = \frac{1}{|c|}$ and $|z + tj - \frac{a}{c}| = \frac{1}{|c|}$ for A

and A^{-1}. When c is real, note that $T_c(z) = -\overline{z}$, that is T_c is reflection in the plane $x = 0$. Recall that a parabolic, elliptic, or hyperbolic
transformation can be conjugated into $PSL(2, R)$. We obtain the action of
A on H^3 as discussed by Riley in [4]. It is possible to read off from
the matrix of A how the points of the two isometric spheres are paired by
A and A^{-1}.

If $c \neq 0$, then by (1) A consists of the composition of inversion in
the isometric sphere of A, reflection in a plane through $-\frac{d}{c}$ and perpendicular to the line between $-\frac{d}{c}$ and $\frac{a}{c}$, a rotation leaving $-\frac{d}{c}$ fixed
(possibly the identity), and translation of the isometric sphere of A to
the isometric sphere of A^{-1}. The reflection and rotation can be combined
into a reflection in a plane through $-\frac{d}{c}$ and perpendicular to the complex
plane. Note that if A is parabolic, the isometric spheres are tangent at
the fixed point. If A is hyperbolic, they are disjoint. If A is elliptic,
they intersect in a circle. If A is loxodromic, the isometric spheres may
be disjoint, tangent, or intersecting.

3. *Some link groups*

If the volume of H^3/G is finite, then H^3/G is homeomorphic to $N-\ell$
where N is a closed 3-manifold and $\ell = C_1 \cup \cdots \cup C_n$ where each C_i is
homeomorphic to S^1. (See [7] or Thurston [6].) The fundamental group of
$S^3 - \ell$ is called a link group. We refer to Riley [5] for a discussion of
representations of link groups in $PSL(2, C)$. In particular, an anti-
isomorphism between a discrete finite-volume subgroup of $PSL(2, C)$ and
a link group implies that the link complement is homeomorphic to the
corresponding quotient space.

By a cusp torus we mean a set homeomorphic to $\{z : 0 < |z| < 1\} \times S^1$. Cusp tori occur in H^3/G as deleted neighborhoods of the components of a link [2] and correspond to conjugacy classes of rank two parabolic subgroups of G. The number of cusp tori in H^3/G is equal to the number of cycles of parabolic fixed points on the boundary of a Poincaré region or Ford domain for a finite-volume group. We refer to Maskit [3] for a discussion of fundamental polyhedra for discrete isometry groups of H^3.

Let $M(G) = (H^3 \cup \Omega(G))/G$. Recall that a surface S in a 3-manifold M is incompressible if the inclusion $S \to M$ induces an injection of fundamental groups, $\pi_1(S) \to \pi_1(M)$.

It follows from Fine [1] that the Picard group is generated by

$$ t = \begin{pmatrix} 1 & 1 \\ 0 & 1 \end{pmatrix} \qquad u = \begin{pmatrix} 1 & i \\ 0 & 1 \end{pmatrix} $$

$$ v = \begin{pmatrix} 1 & 0 \\ -1 & 1 \end{pmatrix} \qquad w = \begin{pmatrix} 1 & 0 \\ i & 1 \end{pmatrix}. $$

The group G_1 generated by $t_4 = t^4$, $u_2 = u^2$ and v is a free product $G_1 = \langle t_4, u_2, v; t_4 u_2 = u_2 t_4 \rangle$. $\partial M(G_1)$ has two components, each a 3-punctured sphere. These components are covered by euclidean discs stabilized by fuchsian subgroups of G_1, e.g.,

$$ F_1 = \langle v, u_2^{-1} v u_2; \rangle \text{ which preserves } |z - (-i)| = 1 $$

$$ F_2 = \langle v t_4, u_2^{-1}(v t_4) u_2; \rangle \text{ which preserves } |z - (-2 - i)| = 1. $$

(See Figure 1 which shows the isometric circles of v and v^{-1} and the generators of F_1 and F_2 and lines paired by t_4 and u_2. Imagine hemispheres over the circles and half-planes perpendicular to C along the lines.)

Let Σ_i be the hemisphere over the disk Δ_i preserved by F_i. Then Σ_i/F_i is incompressible, totally geodesic, and parallel to Δ_i/F_i. We

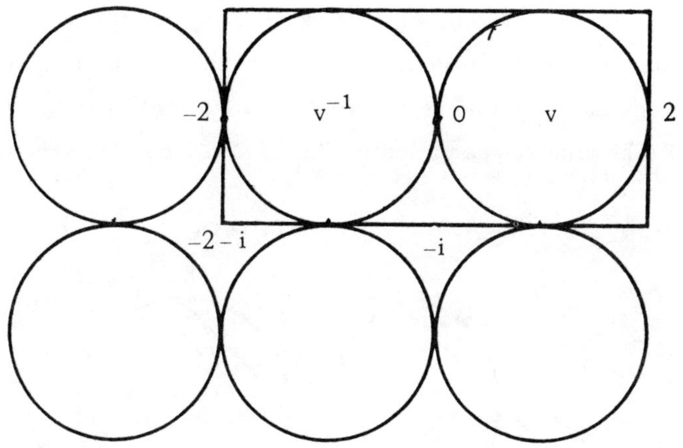

Fig. 1

can glue Σ_1/F_1 to Σ_2/F_2 by adding a generator s such that $sF_2s^{-1} = F_1$ and s takes the outside of Σ_2 to the inside of Σ_1. By Seifert-Van Kampen the result G is an HNN extension of G_1, $G = <G_1, s; sF_2s^{-1} = F_1>$.

It was shown in [8] that if s is parabolic, then $H^3/G = S^3 - B$ where B is the Borromean rings. The surface Σ_i/F_i occurs in $S^3 - B$ as a twice-punctured open disc which spans a component of the link (Figure 2(a)). The group presentation reduces by Tietze transformations to

$$<u_2, v, s; [u_2, v^{-1}svs^{-1}] = [s, u_2^{-1}vu_2v^{-1}] = 1> .$$

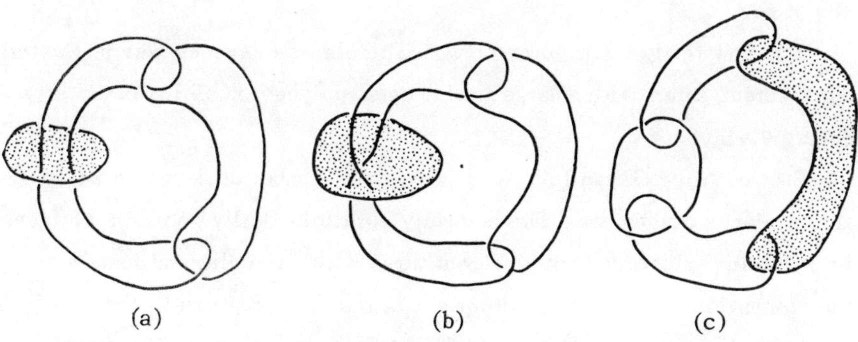

(a)   (b)   (c)

Fig. 2

A loxodromic s , which can be thought of as a parabolic transformation followed by a $180°$ rotation, gives a half-integer twist of a spanning
disc. The result is the link in Figure 2(b). (See also Thurston [6], Section 6.8.) The group presentation in this case reduces to

$$< u_2, v, s;\ [sv^{-1}, vu_2^{-1}v^{-1}u_2] = [u_2, svs^{-1}u_2^{-1}v] = 1 > .$$

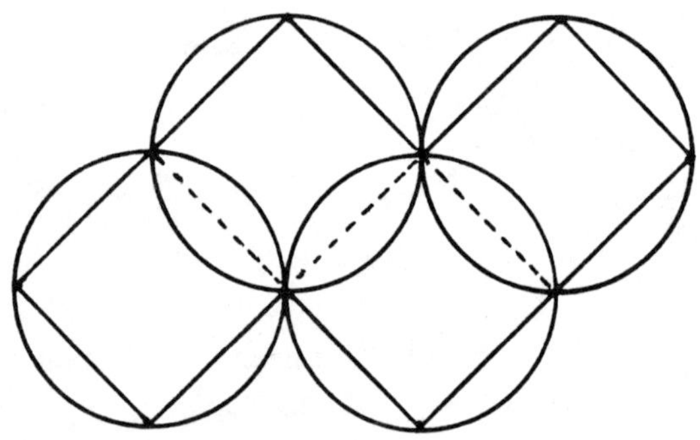

Fig. 3

The fundamental polyhedron for both groups is shown in Figure 3. It
lies above the hemispheres indicated by circles and between the portions
of vertical planes indicated by lines. A third group also shares this polyhedron. It is discussed in detail in [8]. The link complement is shown in
Figure 2(c) and can be obtained from 2(b) by a twist along a different spanning disc.

It is now fairly clear how to proceed to obtain examples which illustrate
our theorem. Start with a large fundamental polyhedron as indicated in
Figure 4(a).

The resulting 3-manifold M has an even number of 3-punctured spheres
as boundary components. The boundary-parallel, totally geodesic surfaces
in M can be glued in many different ways with parabolic and loxodromic
transformations. The only requirement is that the additional generators
conjugate the fuchsian groups which stabilize the corresponding hemispheres
and interchange inside and outside of the hemispheres.

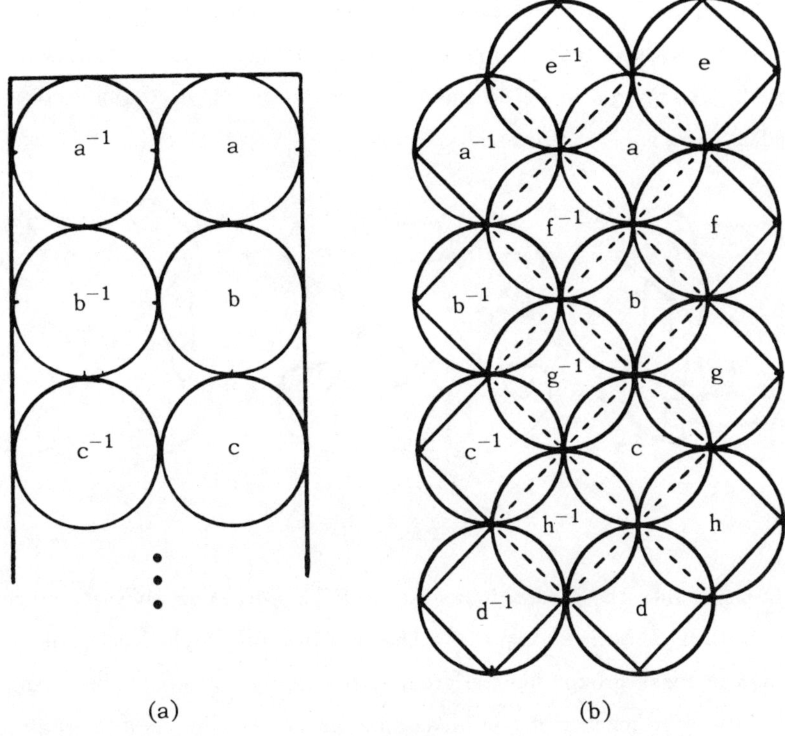

Fig. 4

If we use all parabolic generators to get a cyclic n-fold covering of
the Borromean rings in the obvious way, we obtain a link complement like
the one in Figure 5. The point at the center indicates a component of the
link which can be thought of as a straight line perpendicular to the plane

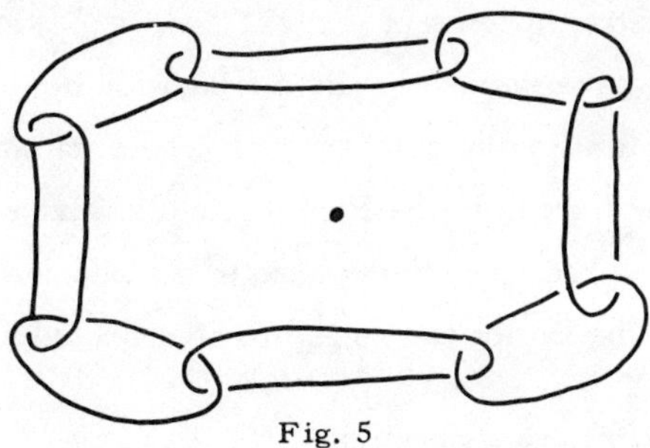

Fig. 5

of the paper and passing through ∞. If one of the parabolic generators is replaced by a loxodromic, the result is a half-integer twist along a twice-punctured spanning disk. The link complements in Figure 6 can be obtained, for example.

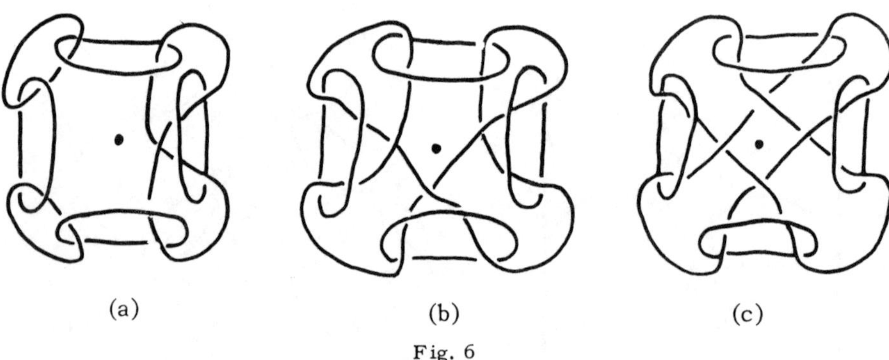

(a)   (b)   (c)

Fig. 6

In particular, the groups indicated by Figure 4(b) can give link complements with 6, 7, 8, or 9 cusp tori. The fundamental polyhedron is the same for each of these groups but the groups are clearly not isomorphic. An arbitrarily large number of non-isomorphic groups can be seen to share a fundamental polyhedron in this way.

The following considerations show that these groups are subgroups of the Picard group. Form the composition of inversion in a hemisphere of radius 1 and reflection in a vertical plane parallel to $x = 0$ through the center of the hemisphere. Following this with translation two units to the right gives a parabolic, translation two units up gives a loxodromic. The centers of the isometric spheres of v, v^{-1}, w, w^{-1} are at 1, -1, i, $-i$ and the spheres all have radius 1. By conjugation by translations, we can move these centers to any point $m + ni$ where m and n are integers.

The index of each group in the Picard group is given by the ratio of the volume of its polyhedron to the volume of a fundamental polyhedron for the Picard group. The Borromean rings are of index 24.

UNIVERSITY OF WISCONSIN-PARKSIDE
KENOSHA, WISCONSIN 53141

BIBLIOGRAPHY

[1] B. Fine, The structure of $PSL(2, R)$; R, the ring of integers in a euclidean quadratic imaginary number field, *Discontinuous Groups and Riemann Surfaces*, ed. L. Greenberg, Ann. of Math. Studies 79, 1974.

[2] A. Marden, Geometrically finite kleinian groups and their deformation spaces, *Discrete Groups and Automorphic Functions*, ed. W. J. Harvey, Academic Press, 1977.

[3] B. Maskit, On Poincaré's theorem for fundamental polygons, Adv. in Math. 7(1971), 219-230.

[4] R. Riley, A quadratic parabolic group, Math. Proc. Cambridge Phil. Soc. 77(1975), 281-288.

[5] _________, Discrete parabolic representations of link groups, Mathematika 22(1975), 141-150.

[6] W. P. Thurston, The Geometry and Topology of 3-Manifolds, lectures at Princeton University, 1977-78.

[7] N. Wielenberg, Discrete Moebius groups: Fundamental polyhedra and convergence, Amer. Jour. of Math. 99(1977), 861-877.

[8] _________, The structure of certain subgroups of the Picard group, Math. Proc. Cambridge Phil. Soc., to appear.

THE LENGTH SPECTRUM AS MODULI FOR
COMPACT RIEMANN SURFACES

Scott Wolpert[*]

The purpose of this note is to announce results concerning a moduli problem. A complete description will appear elsewhere, [1]. Denote the Teichmüller space for compact Riemann surfaces of genus g, $g \geq 2$ as T_g. A point $\{S\}$ of T_g is the equivalence class of a pair, a Riemann surface S and the homotopy class of a homeomorphism from a base surface S_0 to S. Let $\overline{M}_g$ be the extended Teichmüller modular group. A point of $T_g/\overline{M}_g$ is the conformal and anticonformal equivalence class of a Riemann surface. Let $\mathcal{F}(S)$ be the set of free homotopy classes of closed curves on S. The marking homeomorphism of $\{S\} \in T_g$ is used to identify $\pi_1(S_0)$ (resp. $\mathcal{F} = \mathcal{F}(S_0)$) and $\pi_1(S)$ (resp. $\mathcal{F}(S)$). Each element $[\gamma]$ of $\mathcal{F}(S)$ contains a unique Poincaré geodesic; denote by $[\gamma](S)$ the length of this geodesic. Let Γ be the group of deck transformations for the covering of S by the upper half plane. It was noted by Fricke Klein that the information $\{([\gamma], [\gamma](S)) \mid [\gamma] \in \mathcal{F}\}$ determines Γ modulo conjugation by $PGL(2; R)$ and S modulo $\overline{M}_g$. Later it was asked if the set (with multiplicities) of numbers $\{[\gamma](S) \mid [\gamma] \in \mathcal{F}\}$ determines Γ modulo conjugation. We emphasize in the former case one is provided with considerable

---

[*]Partially supported by National Science Foundation Grant #MCS75-07403-A03.

topological data and in the latter case no topological information is given. The set $\{[\gamma](S) \mid [\gamma] \in \mathcal{F}\}$ will be called the length spectrum of S and denoted $\text{Lsp}(S)$. We have obtained the following result.

THEOREM 1. *A real analytic subvariety* V_g *of* T_g *is defined. Let* $\{R\}$, $\{S\} \in T_g$ *be such that* $\text{Lsp}(R) = \text{Lsp}(S)$. *Then either* $\{R\} \equiv \{S\}$ *mod* $\overline{M}_g$ *or* $\{R\}$, $\{S\} \in V_g$.

The argument involves the theory of amalgamation of Fuchsian groups. Criterion are developed to identify $[\gamma_1], \cdots, [\gamma_p] \in \mathcal{F}$ as specific classes from knowledge (only) of the graphs of the functions $[\gamma_1](\hat{S}), \cdots, [\gamma_p](\hat{S})$, $\hat{S} \in T_g$. Our argument involves the following results.

LEMMA 2. *Let* $f : R \to S$ *be a* K *quasiconformal map and* γ *a closed curve on* R. *Denote by* ℓ *(resp.* ℓ'*) the length of the Poincaré geodesic freely homotopic to* γ *(resp.* $f(\gamma)$*). Then*

$$\ell/K \le \ell' \le \ell K .$$

A deformation is a continuous map of $[0, 1)$ into T_g.

LEMMA 3. *Let* $[\gamma_1], \cdots, [\gamma_p] \in \mathcal{F}$ *be primitive. Then* $[\gamma_1], \cdots, [\gamma_p]$ *are simple and disjoint if and only if there exists a deformation* $\hat{S}_t$ *with* $[\gamma_j](\hat{S}_t) \to 0$, $1 \le j \le p$.

An element $[\gamma] \in \mathcal{F}(S)$ corresponds to a conjugacy class in Γ. Let $|\text{tr}\,[\gamma]|(S)$ be the common absolute trace of the elements of this conjugacy class. Given A, $B \in SL(2; R)$ the identity

$$\text{tr}\,A\,\text{tr}\,B = \text{tr}\,AB + \text{tr}\,A^{-1}B$$

holds. As a partial converse we have the following.

LEMMA 4. *Let* $[\gamma_0]$, $[\gamma_1]$, $[\gamma_2]$ *and* $[\gamma_3]$ *distinct in* $\mathcal{F}$ *exist such that* $[\gamma_1]$, $[\gamma_2]$, $[\gamma_3]$ *are disjoint and simple with*

$$|\text{tr}\,[\gamma_1]|(\hat{S})|\text{tr}\,[\gamma_2]|(\hat{S}) = \epsilon|\text{tr}\,[\gamma_3]|(\hat{S}) + |\text{tr}\,[\gamma_0]|(\hat{S})$$

for all $\hat{S} \in T_g$ *and* ε *fixed,* $\varepsilon = \pm 1$. *The geodesics* $[y_1]$, $[y_2]$ *and* $[y_3]$ *bound a triply connected subdomain and* $\varepsilon = -1$. *Furthermore* $a, \beta \in \pi_1(\hat{S})$ *exist such that* $[y_1] = [a]$, $[y_2] = [\beta]$, $[y_3] = [a\beta]$ *and* $[y_0] = [a^{-1}\beta]$.

LEMMA 5. *Let* $[a]$ *and* $[y]$ *be simple geodesics intersecting* p *times. A constant* $c > 0$ *is determined. Then* $-p(\log [y](\hat{S})) \leq [a](\hat{S}) + c$ *for all* $\hat{S} \in T_g$. *A deformation* $\hat{S}_t$ *is determined such that* $[y](\hat{S}_t) \to 0$ *and* $[a](\hat{S}_t) \leq -p(\log [y](\hat{S}_t)) + c$.

A characterization of the first eigenvalue of the (Poincaré) Laplace Beltrami operator is also required. A recent result of M. F. Vignéras establishes that V_g is not empty for certain choices of the genus g, [2].

REFERENCES

[1] S. Wolpert, The length spectrum as moduli for compact Riemann surfaces, Annals of Math., 109(1979), 323-351.

[2] M. F. Vignéras, Exemples de sous-groupes discrets non conjugués de PSL(2, $\hat{R}$) qui ont même fonction zêta de Selberg, C. R. Acad. Sc. Paris, t. 287, 1978.

Library of Congress Cataloging in Publication Data

Riemann surfaces and related topics: proceedings of
the 1978 Stony Brook conference.

1. Riemann surfaces—Congresses.   I. Kra, Irwin.
II. Maskit, Bernard.   III. Title.
QA333.C593   1978   515′.223   79-27823
ISBN 0-691-08264-2